AF820193

NEUROMETHODS ☐ 4

Receptor Binding

NEUROMETHODS

Program Editors: Alan A. Boulton and Glen B. Baker

Series I: Neurochemistry

1. **General Neurochemical Techniques**
   Edited by *Alan A. Boulton* and *Glen B. Baker,* 1985
2. **Amines and Their Metabolites**
   Edited by *Alan A. Boulton, Glen B. Baker,* and *Judith M. Baker,* 1985
3. **Amino Acids**
   Edited by *Alan A. Boulton, Glen B. Baker,* and *James D. Wood,* 1985
4. **Receptor Binding**
   Edited by *Alan A. Boulton, Glen B. Baker,* and *Pavel D. Hrdina,* 1986
5. **Enzymes**
   Edited by *Alan A. Boulton, Glen B. Baker,* and *Peter H. Yu,* 1986

NEUROMETHODS

Series I: Neurochemistry

Program Editors: Alan A. Boulton and Glen B. Baker

NEUROMETHODS ☐ 4
Receptor Binding

Edited by

Alan A. Boulton, Glen B. Baker,
and Pavel D. Hrdina

Humana Press • Clifton, New Jersey

Library of Congress Cataloging in Publication Data
Main entry under title:
Receptor binding.

(Neuromethods. Series I, Neurochemistry; 4)
Includes bibliographies and index.
1. Neurotransmitter receptors. 2. Neural receptors.
I. Boulton, A. A. (Alan A.) II. Baker, Glen B.,
1947– . III. Hrdina, Pavel D. IV. Series.
QP364.7.R427 1986 599'.0188 86–7142
ISBN 0–89603–078–4

© 1986 The Humana Press Inc.
Crescent Manor
PO Box 2148
Clifton, NJ 07015

All rights reserved

No part of this book may be reproduced, stored in a retrieval system, or transmitted in any form or by any means, electronic, mechanical, photocopying, microfilming, recording, or otherwise without written permission from the Publisher.

Printed in the United States of America

Foreword

Techniques in the neurosciences are evolving rapidly. There are currently very few volumes dedicated to the methodology employed by neuroscientists, and those that are available often seem either out of date or limited in scope. This series is about the methods most widely used by modern-day neuroscientists and is written by their colleagues who are practicing experts.

Volume 1 will be useful to all neuroscientists since it concerns those procedures used routinely across the widest range of subdisciplines. Collecting these general techniques together in a single volume strikes us not only as a service, but will no doubt prove of exceptional utilitarian value as well. Volumes 2 and 3 describe all current procedures for the analyses of amines and their metabolites and of amino acids, respectively. These collections will clearly be of value to all neuroscientists working in or contemplating research in these fields. Similar reasons exist for Volume 4 on receptor binding techniques since experimental details are provided for many types of ligand-receptor binding, including chapters on general principles, drug discovery and development, and a most useful appendix on computer programs for Scatchard, nonlinear, and competitive displacement analyses. Volume 5 provides procedures for the assessment of enzymes involved in biogenic amine synthesis and catabolism.

Volumes in the NEUROMETHODS series will be useful to neurochemists, -pharmacologists, -physiologists, -anatomists, psychopharmacologists, psychiatrists, neurologists, and chemists (organic, analytical, pharmaceutical, medicinal); in fact, everyone involved in the neurosciences, both basic and clinical.

Preface

The purpose of this volume, which is part of a series on "Neuromethods," is to critically review the contribution of receptor binding studies to the fields of neuropharmacology and neurochemistry. Emphasis is being placed on the methodological aspects of these studies and on the problems in integrating the binding data with the function of receptors in physiological or pathological processes.

Current trends in receptor research aim at (i) correlation of receptor binding with the electrophysiological, biochemical, or behavioral responses, (ii) adaptive changes in the state of receptors reflected in altered binding, and (iii) the search for endogenous ligands for binding sites characterized by drugs. All this is reflected in individual chapters of this book. They are not proceedings of a symposium or a workshop. The contributors were selected on the basis of their work in a particular field. The chapters share a common organizational scheme in order to provide the reader with a familiar pattern of information when searching through various topics. The introductory chapter is intended as a brief review of general principles in receptor binding assays for uninitiated readers. It is complemented by the last chapter on the role of receptor binding in drug discovery and development. Examples of computer programs for estimation of binding parameters are given in an appendix to chapter 10. Reasonable attempts were made to minimize unnecessary repetition and overlap. However, some overlap was unavoidable, particularly where assays for various neurotransmitters are described in detail.

This volume reviews the current state of knowledge of receptor (recognition) sites in the brain that mediate the action of putative neurotransmitters or modulators such as GABA, excitatory amino acids, dopamine, noradrenaline, serotonin, acetylcholine, purines, and opioid and nonopioid peptides, and of some psychotropic drugs such as antidepressants, amphetamine, phencyclidine, and benzodiazepines. It will be of interest particularly to neuropharmacologists, neurophysiologists, neurobiologists, neurochemists, neurologists, and biological psychiatrists, but also to researchers or physicians in other fields of biology and medicine interested in receptor-mediated events.

Pavel D. Hrdina

Contents

CHAPTER 3
ADRENERGIC RECEPTORS
John Marvin May and Kenneth P. Minneman

CHAPTER 4
SEROTONIN RECEPTORS
Stephen J. Peroutka

CHAPTER 5
TRYPTAMINE AND PHENYLETHYLAMINE RECOGNITION SITES IN BRAIN
Kenneth J. Kellar and Caren S. Cascio

CHAPTER 6
MUSCARINIC CHOLINERGIC RECEPTORS
Andrew P. Braun and Prakash V. Sulakhe

CHAPTER 7
EXCITATORY AMINO ACID RECEPTORS
John W. Ferkany and Joseph T. Coyle

CHAPTER 8
GABA RECEPTOR BINDING
Kenneth George Lloyd

CHAPTER 9
RECEPTORS FOR NONOPIOID NEUROPEPTIDES
Michel Goedert

CHAPTER 10
MULTIPLE OPIOID RECEPTORS IN THE CENTRAL NERVOUS SYSTEM
Paul L. Wood

CHAPTER 12
THE BENZODIAZEPINE RECEPTOR
Ian L. Martin

CHAPTER 13
BINDING SITES FOR ANTIDEPRESSANTS
Pavel D. Hrdina

CHAPTER 15
RECEPTOR BINDING IN DRUG DISCOVERY AND DEVELOPMENT
Michael Williams and Paul L. Wood

Contributors

ANDREW P. BRAUN • *Department of Physiology, College of Medicine, University of Saskatchewan, Saskatoon, Saskatchewan, Canada*

CAREN S. CASCIO • *Department of Pharmacology, Georgetown University School of Medicine and Dentistry, Washington, DC, USA*

JOSEPH T. COYLE • *Departments of Neurosciences, Pharmacology, and Experimental Therapeutics, Johns Hopkins University, School of Medicine, Baltimore, Maryland, USA*

JOHN W. FERKANY • *Nova Pharmaceutical Corporation, Baltimore, Maryland, USA*

MICHEL GOEDERT • *MRC Neurochemical Pharmacology Unit, Medical Research Council Centre Medical School, Cambridge, UK*

PAVEL D. HRDINA • *Department of Pharmacology, School of Medicine, University of Ottawa, Ottawa, Ontario, Canada*

KENNETH J. KELLAR • *Department of Pharmacology, Georgetown University School of Medicine and Dentistry, Washington, DC, USA*

KENNETH GEORGE LLOYD • *Laboratoire d'Etudes et de Recherches, Synthelabo, Bagneaux, France*

IAN L. MARTIN • *MRC Neurochemical Pharmacology Unit, Medical Research Council Centre Medical School, Cambridge, UK*

KENNETH MARVIN MAY • *Department of Pharmacology, Emory University School of Medicine, Atlanta, Georgia, USA*

KENNETH P. MINNEMAN • *Department of Pharmacology, Emory University School of Medicine, Atlanta, Georgia, USA*

RAM K. MISHRA • *Departments of Psychiatry and Neurosciences, McMasters University, Hamilton, Ontario, Canada*

REMI QUIRION • *Department of Psychiatry, McGill University and Douglas Hospital Research Centre, Verdun, Quebec, Canada*

PRAKASH V. SULAKHE • *Department of Physiology, College of Medicine, University of Saskatchewan, Saskatoon, Saskatchewan, Canada*

MICHAEL WILLIAMS • *Pharmaceuticals Division, Neurosciences Research, CIBA-GEIGY Corporation, Summit, New Jersey, USA*

PAUL L. WOOD • *Pharmaceuticals Division, Neurosciences Research, CIBA-GEIGY Corporation, Summit, New Jersey, USA*

Chapter 1
General Principles of Receptor Binding

Pavel D. Hrdina

1. Introduction

The increasing availability over the past 20 yr of biologically active radiolabeled compounds of high-specific activity has enabled investigators to label and quantitatively assay the receptor (recognition) sites that are thought to mediate the agonistic or antagonistic effects of putative neurotransmitters and many psychotropic drugs in the brain. Such direct studies of ligand–receptor interaction have resulted in a significant increase in our understanding of the processes at the molecular level involved in the action of various neurotransmitters, peptides, and drugs at their respective receptive sites in target tissues.

These studies have also led to the discovery of high-affinity binding sites for psychoactive drugs, such as benzodiazepines (Squires and Braestrup, 1977; Möhler and Okada, 1977), opiates (Pert and Snyder, 1973), neuroleptics (Creese et al., 1975; Seeman et al., 1975), and antidepressants (Raisman et al., 1979), opening new approaches to investigation of the mode of action of these drugs, as well as of mechanisms underlying disorders for which they are used. The fundamental principles and methodology of neurotransmitter receptor binding were first reviewed in an excellent monograph by Yamamura et al. (1978). Since then, a wealth of neuropharmacological data has accumulated from neurotransmitter receptor binding studies. They have been the subject of a recent volume edited by Iversen et al. (1983).

The amount of information in this fast-growing field of research continues to increase exponentially, and with new ligands being introduced, new sites of potential biological significance in the mode of action of drugs or in pathophysiological states (e.g., imipramine sites) are being characterized. The interpretation of binding data is becoming more sophisticated, but also more vulnerable to challenge and criticism (Laduron, 1983). Serious attempts are being made to relate the results of binding studies to changes in physiological function and behavior.

The purpose of this volume, which is part of a series on "neuromethods," is to critically review the contribution of receptor binding studies to the field of neuropharmacology and neurochemistry. Emphasis is being placed on the methodological aspect of these studies and on the problems in integrating the binding data with the function of receptors in physiological or pathological processes.

1.1. Receptor as a Supramolecular Complex

Before the advent of receptor binding studies, receptor activity was studied and characterized by measurement of physiological responses such as muscle contraction or secretion of endogenous substances. The introduction of radioligands initially resulted in an explosion of new "receptors" and subclasses of "receptors." Many binding studies simply equated the newly discovered binding sites to receptors without considering whether or not the binding resulted in a physiological or biochemical response. An example of this was the introduction of several subclasses of opiate and dopaminergic receptors. Recently, however, more and more studies are focusing on integrating the binding data with the measurement of physiological or biochemical responses triggered or modulated by the specific binding of the ligand to membrane sites. A new concept of the receptor as a supramolecular complex is emerging (Costa, 1982; Snyder, 1983). Transfer or initiation of a biological signal by a neurotransmitter or drug molecule is now being envisaged as a process involving binding to recognition sites (detection) and coupling to a transducing (and amplifying) mechanism resulting in a biological response (Fig. 1). Binding of a ligand to a receptor recognition site is thus the first step in ligand–receptor interaction. Alterations in ionic permeability and cyclic nucleotide formation can be considered to be major transducing mechanisms. A nucleotide binding protein (G-protein), first described for β-adrenergic receptors (Pfeuffer, 1977), could represent a coupling mechanism that may be similar in several receptor complexes. Methylation of phospholipids has recently been shown to be an important event in the receptor-mediated transmission, and perhaps in amplification of the biological signal across membranes (Hirata and Axelrod, 1980).

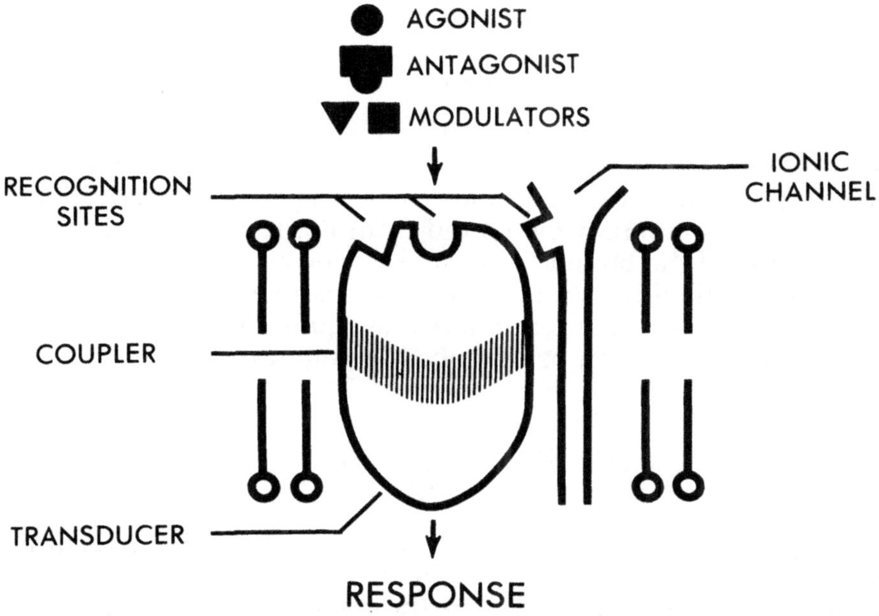

Fig. 1. A schematic model of a receptor complex and the transmission of a signal.

Although the integration of binding studies with measurement of biological responses has increased our understanding of neurotransmission in some systems (e.g., β-adrenergic, serotonergic), the exact sequence of events and the functioning of the various components of the receptor complex are yet to be elucidated.

In addition to primary transmitter mechanisms responsible for the direct transfer of the biological signal across the postsynaptic membranes, one can now postulate the existence of regulatory sites (Fig. 1) that would modulate the transfer of the biological signal by altering any of the steps involved in the direct transfer of signal. Many drugs (e.g., antidepressants, anxiolytics) and some endogenous substances (e.g., peptides, purines) act by modifying neurotransmission or its components (opening of ionic channels, activation of enzymes). The concept of dynamic regulation of neurotransmission after repeated exposure to drugs (development of supersensitivity, desensitization) derived from physiological experiments has been confirmed by results of binding studies that in many cases revealed a parallel up- or down-regulation of corresponding receptor density.

2. Ligand–Receptor Interaction

2.1. The Direct Binding Assay

The binding assays employed in most direct studies of ligand–receptor interaction involve incubation of a suitable tissue receptor preparation with an isotopically labeled ligand (agonist or antagonist), separation of the bound from the free ligand, and determination of the bound radioactivity. The assumption made in most receptor binding studies is that the ligand–receptor binding is a reversible bimolecular reaction that at equilibrium obeys the law of mass action and can be described as follows:

$$[L] + [R] \underset{k_2}{\overset{k_1}{\rightleftharpoons}} [LR] \tag{1}$$

where $[L]$ is the concentration of the free ligand, $[R]$ the concentration of the free receptor, $[LR]$ is the concentration of the ligand–receptor complex, and k_1 and k_2 are the association and dissociation rate constants, respectively.

The equilibrium dissociation constant (K_d) for the ligand–receptor interaction, which is used as a measure of affinity of the ligand for the receptor, is defined by the law of mass action:

$$K_d = \frac{[L]\,[R]}{[LR]} = \frac{k_2}{k_1} \tag{2}$$

2.1.1. The Ligand

With the increasing use of receptor binding assays, a vast number of radioligands has become commercially available. They are either receptor agonists or antagonists, or drugs and other substances used to characterize receptor (recognition) sites. The choice of the radioligand will, of course, depend on the aim of the study. There are, however, several basic requirements that should be met in any radioreceptor assay. These include: purity and stability of the radioligand, biological activity, sufficiently high specific activity, and specificity for the given site. Purity and stability of the radioligand are essential for correct interpretation of binding data. Impurities or decomposition products may result in binding to tissue components other than specific sites in question and thus make the interpretation of results difficult. Stability may be a problem with oxygen-sensitive compounds (such as some biogenic amines). Oxidation can be prevented by addition of antioxidants, but some of these additives (e.g., ascorbic acid) have been shown to alter the binding characteristics of some ligands. Biological activity is an important requirement, particularly when analogs of biogenic compounds or antagonist drugs are being used as ligands.

High-specific activity of the ligand is important to allow a reliable measurement of radioactivity at low concentrations, since in most binding assays high-affinity binding sites become saturated in the nanomolar concentration range and only a small fraction of the added radioligand is bound. The specific activity of stock solutions should be checked before each assay to avoid changes from decomposition or evaporation of solvent. The most commonly used isotope to label ligands is tritium. Tritium-labeled compounds attain sufficiently high specific activity (20–50 Ci/mmol), retain biological activity, are stable for several months, and have a relatively long radiochemical half-life (>12 yr). When very high specific activity is required, ^{131}I can be used to label the ligand. The advantage of ^{131}I-labeled compounds is that, because of very high specific activity (>2000 Ci/mmol), they can be used when the size of the biological sample is limited (e.g., cells in culture). The disadvantages include a relatively short half-life, the possibility of radiation-induced damage to tissue structure and, obviously, possible biological hazard to personnel.

Ideally, the radioligand should be selective and specific for one type of binding sites (receptors). In practice, however, this criterion is becoming difficult to meet and questionable to maintain since many ligands, including neurotransmitter substances or their antagonists (e.g., ^{3}H-spiroperidol), have been shown to bind to more than one set of recognition sites. Varying experimental conditions (e.g., employing different displacers) can be used to characterize the binding of these ligands to particular sets of recognition sites.

2.1.2. The Tissue Preparation and Incubation Conditions

The ideal preparation for binding studies would be a homogenous population of cells with preserved biological function so that the binding to specific receptor sites could be compared with the biological response. In practice, this is seldom the case. The vast majority of neuroreceptor binding studies have been carried out with tissue homogenates or membrane preparations. Some preparations enriched in membrane material maintain the biochemical responsiveness (e.g., β-receptor-mediated response of adenylate cyclase), which can thus be correlated with binding characteristics. Recently, an increasing number of binding studies is being carried out in intact cells in culture or in tissue slices, with the obvious advantage of maintaining the structure and metabolic processes closer to the "in vivo" situation.

Solubilization of tissue by various detergents can be particularly useful in studies of the physicochemical properties of receptors. Many solubilized receptors show binding characteristics similar to those seen with membrane preparations. However, disruption of membrane protein conformation by solubilizing agents can result in distortion of binding characteristics, and these solubilized preparations may thus no longer reflect the properties of the receptor in its physiological *"in situ"* milieu.

Ideally, the receptor-containing tissue preparations should be incubated with the radioligand using the temperature, pH, and medium resembling as much as possible the *"in situ"* situation. In practice, optimum experimental conditions are often arrived at by trial and error. Optimum conditions are usually considered to be those that allow the highest degree of specific binding. Whether this truly reflects the situation "in vivo" is not known. Various buffers such as Tris-HCl, phosphate, and Hepes are used as media in binding assays, and the pH is usually maintained at 7.4–7.6. The ionic composition of the incubation media depends on the requirements of particular binding assays. Sodium ion appears to be essential for binding to many receptor sites and differentially regulates binding of agonists and antagonists (Snyder, 1983).

The concentration of radioligand should be kept relatively low (0.1–20.0 n*M*), since the nonspecific component of total binding increases with higher ligand concentration. Radioligands with sufficiently high specific activity will allow reliable measurement of radioactivity even at low concentrations. Larger incubation volumes (600–1000 μL) and low concentration of membrane protein are useful in assays with relatively high levels of "nonspecific" binding.

Measurement of ligand bound to tissue requires separation of the bound and free ligand. This can be achieved by filtration, centrifugation, or equilibrium dialysis. The most widely used and fastest method has been filtration. After incubating the preparation with the radioligand, an aliquot of the incubation mixture is rapidly filtered under negative pressure through a filter (usually of 2.4 or 2.5 cm diameter to fit the opening of the manifold units). The aliquot is often diluted into a larger volume of buffer to allow homogenous filtering. Filters are usually washed three times with buffer (to remove any free or trapped ligand), and dried before counting the particulate radioactivity. The type of filters used for separation of bound and free ligand varies with the type of assay.

Whatman glass fiber filters GF/B, GF/C, or GF/F are often used in assays employing brain tissue preparations.

The concentration of bound ligand is usually expressed as per milligram of protein. The protein content of the preparation is most often determined by Lowry's method (Lowry et al., 1951). When Tris-HCl is being used as incubation medium, it could be advantageous to use Peterson's (1977) modification of this protein assay method to avoid possible interference of Tris with protein measurement. Binding assays should be performed at tissue concentrations that show linearity between binding and protein content. Most binding assays use a protein concentration of 0.5–3.0 mg/mL. However, in the case of highly lipid-soluble ligands, it is desirable to use lower protein concentrations (about 100 μg/mL), since the free ligand may be depleted by the tissue; this may then yield distorted K_d and B_{max} values and nonlinear Scatchard plots (Hartley and Seeman, 1978).

Although most methodological principles outlined in section 2.2 below apply in general to all radioligand binding studies, every receptor system examined has some particular, individualized experimental condition(s) under which the measurement of receptor-ligand binding appears to be optimal. These will be described in more detail in individual chapters of this book dealing with particular receptor systems.

2.2. Criteria for Specific Binding of Biological Consequence

Since, to some extent, "everything binds to everything," the binding of the radiolabeled ligand to the biological preparation must meet some basic criteria in order to be recognized as a biologically meaningful binding to a specific receptor (recognition) site. The most important of these criteria are: (1) saturability and reversibility, and (2) physiological and pharmacological specificity.

2.2.1. Saturability and Reversibility

The basic assumption in a direct binding assay is that tissues contain a finite number of specific receptor (recognition) sites that become saturated when the preparation is exposed to increasing concentrations of the radioligand. The binding of the ligand can be envisaged as the sum of at least two processes—"specific" binding to saturable sites and "nonspecific" binding that is nonsaturable and includes low-affinity binding to tissue components other than "specific" sites, trapping by tissue pellet, and absorption to the material (e.g., filters). It is assumed that in a physiologically meaningful range of ligand concentrations most of the binding occurs to specific receptor (recognition) sites (Fig. 1). The binding to these specific sites is usually saturable and reversible. Addition of an excess amount of nonradioactive molecules of the same (or related) compound will, by competition, displace a portion of bound radioactive molecules and result in a decrease in bound radioactivity.

The saturation experiment is the cornerstone of studies employing binding assays since it is the basis for estimation of binding parameters such as the equilibrium dissociation constant (K_d) and binding site density (B_{max}). In practice, the saturation experiment is performed in a set of pairs of tubes in which a fixed amount of tissue preparation is incubated with increasing concentrations of the radioactive ligand in the presence (nonspecific binding) or absence (total binding) of an excess of nonradioactive drug (displacer). The specific binding at each concentration is then calculated as the difference between the total and nonspecific binding (Fig. 2).

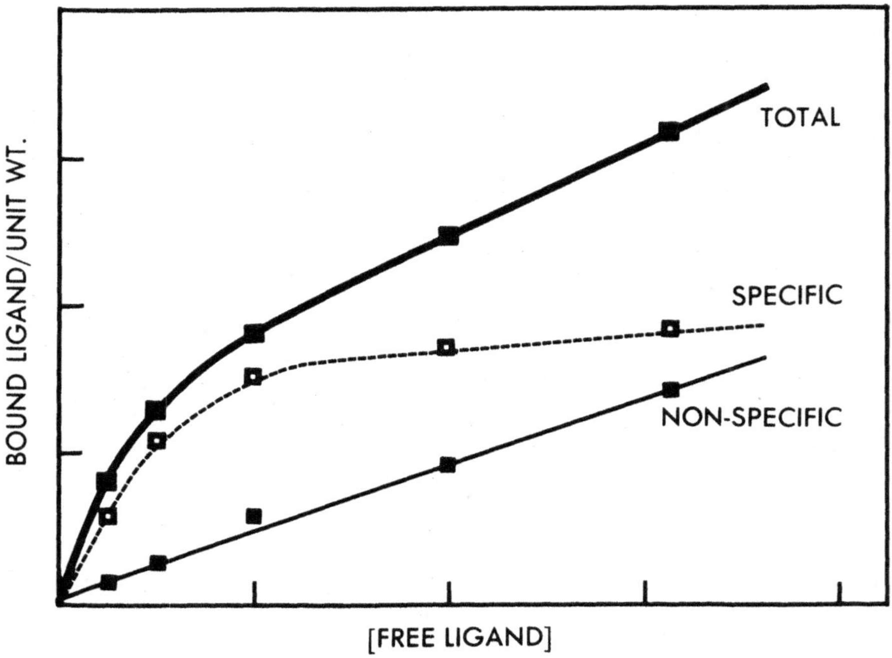

Fig. 2. Typical curves generated from a direct binding assay. The concentration of bound radioactive ligand (usually in fmol/mg protein) is plotted against the increasing concentrations of the free radioactive ligand. Specific binding is calculated as the difference between total binding and nonspecific binding (in the presence of displacer) (*see* text for details).

2.2.2. Physiological and Pharmacological Specificity

Physiological or pharmacological specificity is an essential requirement for a ligand or receptor binding characterized by binding studies. The ligand should bind selectively to a receptor site in concentrations similar to those that elicit physiological or pharmacological responses. The high affinity of the ligand for the specific binding site is no longer considered a necessary prerequisite to prove agonist binding to a neurotransmitter receptor since recognition sites involved in neurotransmission show generally low affinity for neurotransmitters. On the other hand, receptor antagonist and endogenous hormones bind with high affinity to their respective recognition sites (Laduron, 1984).

Stereospecificity of binding is an important indication of selectivity, particularly when correlated with physiological activity "in vivo." Examples of this are (+) and (−)butaclamol and dopamine receptors. Similarly, correlation between the affinity of the drug for its binding site "in vitro" and its pharmacological potency "in vivo" is an important criterion of receptor specificity. Displacement profiles of binding should also correspond to the rank order of potencies of displacing drugs "in vivo."

3. Kinetics of Ligand–Receptor Interaction

The analyses of kinetics of ligand–receptor interactions are based on the law of mass action and Michaelis-Menten analysis of enzyme–substrate interaction.

The discussion of the theory of ligand–receptor interaction and mathematical methods used to derive various equations for estimation of kinetic parameters is beyond the scope of this chapter; this has been a subject of several recent reviews (Bennet, 1978; Williams and Lefkowitz, 1978; Braestrup and Nielsen, 1980). This introductory chapter will therefore focus on the application of these principles to the estimation of binding parameters commonly used in radioligand binding studies.

3.1. Scatchard Plot

In most binding studies, the equilibrium binding affinity constant (K_d) and the maximum number of binding sites (B_{max}) are estimated from the Scatchard plot of saturation isotherm obtained by using the Scatchard equation (Scatchard, 1949) for analysis of binding data:

$$\frac{B}{F} = \frac{B_{\text{max}} - B}{K_d} \tag{3}$$

By measuring the specifically bound ligand (B) and knowing the concentration of the free ligand (F) in the incubation medium at equilibrium, one can plot the ratio of bound and free ligand (B/F) against the amount of bound ligand (B) and fit a best line by linear regression (Fig. 3). The equilibrium binding constant (K_d) is then estimated from the plot as a negative reciprocal of the slope ($-1/\beta$) and the maximum number of binding sites (B_{max}) is given by intercept of the line on the abscissa (x-intercept).

If the radioligand binds to a single set of non-interacting sites, the Scatchard plot yields a straight line. However, it is not uncommon that plotting the saturation binding data results in a nonlinear or curved Scatchard plot. This can be indicative of either binding of the ligand to two or more sites with different affinity or of cooperative interaction between the sites. If this cooperativity is positive (binding to one site facilitated by binding to the other site), the Scatchard plot is curvilinear upward. In contrast, a Scatchard

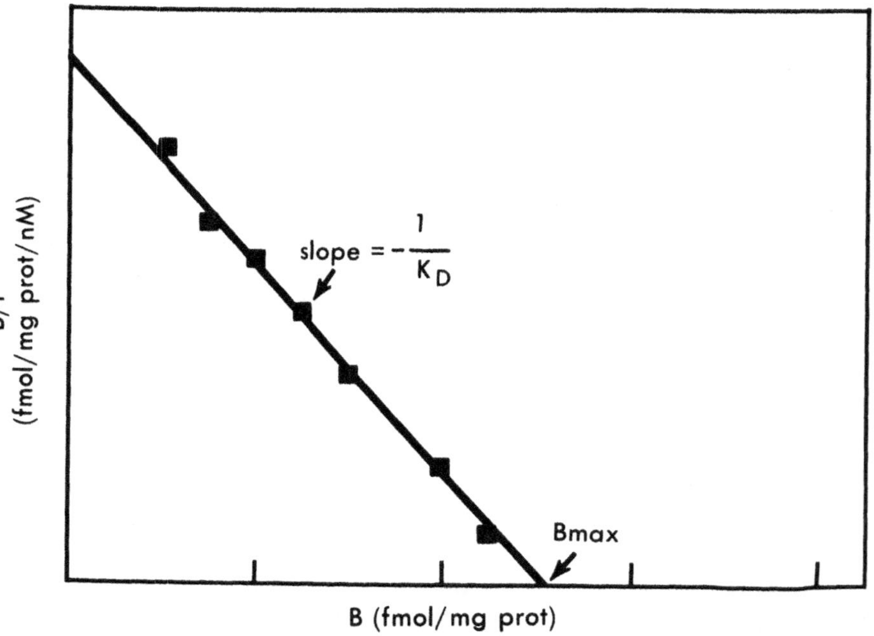

Fig. 3. A typical Scatchard plot of binding data. The straight line suggests the presence of one set of non-interacting specific binding sites for the ligand (*see* text for details).

plot curvilinear downward is indicative of a possible negative cooperativity. A nonlinear Scatchard plot in case of binding to two sets of independent, non-interacting sites can be resolved into two components by either a simple graphical analysis described by Rosenthal (1967) or by specially written computer programs now widely available (e.g., Munson and Rodbard, 1980). Examples of computer programs for Scatchard plots, IC_{50} determinations, and Hill coefficients are given in Chapter 10.

3.2. Hill Plot

The presence or absence of cooperativity can also be determined from Hill (1909) analysis of binding data. The Hill plot (Fig. 4) can be obtained from saturation binding data by plotting the log of radioligand concentration (log[L]) against the ratio log [$B/(B_{max}-B)$], where [L] is the concentration of the free ligand, B is the amount of ligand bound, and B_{max} is the maximum number of binding sites estimated by Scatchard analysis. The slope of the line fitted by linear regression is denoted as the Hill coefficient or Hill number (n_H). The degree of interaction can be quantitated by

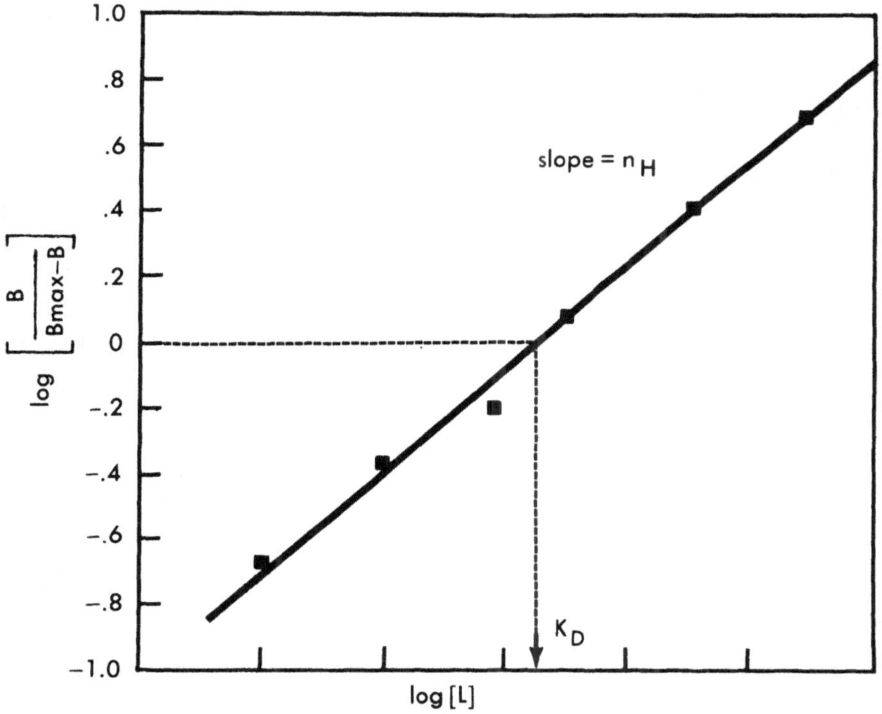

Fig. 4. Hill plot of binding data (*see* text for details).

deviation of the Hill coefficient (n_H) from unity. The Hill coefficient is greater than 1.0 in case of positive cooperativity and less than 1.0 in case of negative cooperativity. In the case of lack of cooperativity, the Hill coefficient is close to unity.

The equilibrium binding constant (K_d) can be estimated from the Hill plot as the abscissa value for log $[B/B_{max} - B)] = 0$, and should be similar to the K_d determined by Scatchard analysis. Hill plots can also be useful in analyzing the displacement profile of drugs for a particular binding site (*see* section 4).

3.3. Determination of Kinetic Constants

The equilibrium binding constant (K_d) can also be determined from experiments in which the rate constants for association (k_1) and dissociation (k_2) of ligand–receptor binding are estimated, since K_d is equal to the ratio k_2/k_1 (*see* Eq. 2). The K_d derived from such experiments should, within the experimental error, be similar to the K_d obtained from the saturation experiments (Scatchard plot).

In practice, it is easier to first make an estimate of the dissociation rate constant. This can be done by incubating the tissue preparation with a fixed concentration of the radioligand until steady-state is reached (when no more binding occurs) and then either dilute the ligand with an excess of buffer (100-fold) or add an excess (1000-fold) of nonradioactive ligand. An example of the latter method is illustrated in Fig. 2 of Chapter 13. Specific binding is measured at various time intervals after dilution (or addition of excess displacer) and the half-life ($t½$) of loss of specifically bound radioligand is estimated from a plot of log $[B]$ against time (Fig. 5). The dissociation rate constant (k_2) is then calculated as $0.693/t½$. Alternatively, the slope of log $[B]$ against time is multiplied by -2.303 to yield k_2 (Bennet, 1978). The dissociation rate constant can also be estimated from the negative slope of a plot of ln $[B_t/B_{eq}]$ against time.

The association rate constant (k_1) is difficult to estimate from binding studies in which usually only less than 10% of the ligand is bound at equilibrium, because it is subject to large error. According to the method described by Bennet (1978), k_1 can be calculated from the amount of the bound ligand at various time intervals (B_t) until equilibrium (B_{eq}) from the plot of ln ($B_{eq} - B_t$) vs time (Fig. 6). The slope of this line (k_{obs}) relates to both the association (k_1) and dissociation (k_2) rate constants and free ligand concentration $[L]$ as follows:

$$k_1 = \frac{k_{obs} - k_2}{[L]} \tag{4}$$

This method can be used to estimate the association rate constant when the amount of free ligand bound is less than 10% of initial free ligand.

4. Pharmacological Characterization of Binding Sites

One of the important criteria for recognizing a binding site as a receptor is pharmacological specificity. The binding of the ligand should be displaced by nonradioactive functional congeners—agonists of the same class, or antagonists, with a rank order of potencies shown to mimic or inhibit the pharmacological response to the ligand. In practice, the IC_{50}s of displacing drugs are calculated from ligand displacement curves where increasing amounts of unlabeled drugs are added to the incubation mixture and the

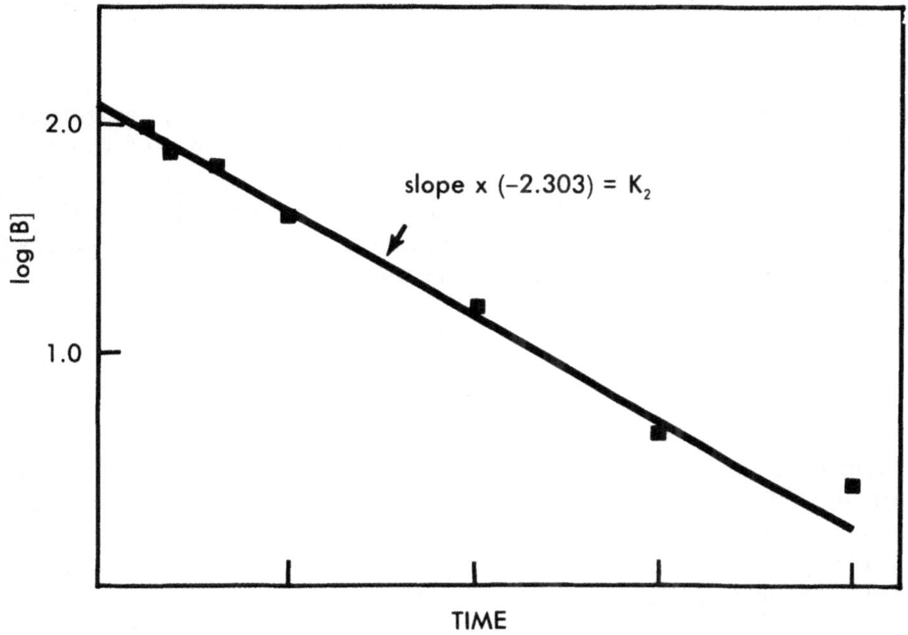

Fig. 5. Semilog plot of specific binding (log [B]) vs time after addition of excess displacer or dilution, from which the dissociation constant (K_2) can be calculated (*see* text for details).

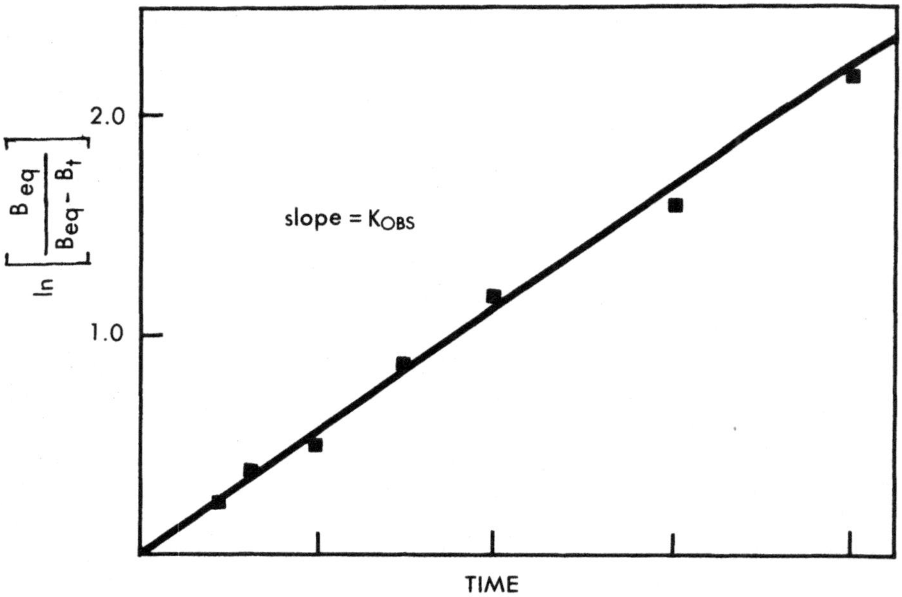

Fig. 6. Semilog (ln) plot of increase in specific binding vs time. Slope of the line represents the observed rate of association (K_{obs}), which is then used to calculate the association constant (k_1) (*see* text for details).

amount of specific binding is measured at each concentration point. The binding data are usually plotted as percent of specific binding remaining against log of displacing drug concentration. Examples of such displacement curves are illustrated in Chapter 13, Fig. 4. Binding affinities of displacing nonradioactive ligands (K_i) can be calculated from the relationship:

$$k_i = \frac{IC_{50}}{1 + [L]/K_d} \qquad (5)$$

where IC_{50} is the concentration of the unlabeled ligand required to inhibit 50% of the radioactive ligand binding, $[L]$ is the concentration of the free radioligand used in the assay, and K_d is the equilibrium binding constant of the radioligand as determined from saturation experiment (Cheng and Prusoff, 1973).

Displacement curves can also be analyzed by Hill plots. In this case, the specific binding of the radioligand in the absence of any displacing drug is taken as B_{max} and the binding remaining at various concentrations of displacing drug is expressed as percentage of B_{max} (% B_{max}). The Hill plot can then be made by plotting log

[% B_{max}/(100% − %B_{max})] against log of displacer concentration (Bennet, 1978). The Hill coefficient, calculated from the slope of the line, will be unity if the displacement is of a competitive type at a single set of binding sites. Hill coefficients below unity (and shallow displacement curves) usually indicate either negative cooperativity or multiple binding sites involved in the interaction of two ligands. The two cases, however, cannot be distinguished by Hill plot analysis of binding data. Displaceable binding may also occur to nonspecific sites (Laduron, 1983). It is important to consider this in interpreting biphasic displacement curves.

It is not unusual that centrally acting drugs or neurotransmitters bind to more than one class of recognition sites, depending on the brain region, and that their relative affinity may also change from region to region. Use of drugs with known selectivity for a particular receptor type can help in characterizing the binding site pharmacologically.

5. Distribution and Localization of Receptors and Correlation With Biological Response

One of the criteria for receptor site characterization is that the relative distribution of binding sites in various brain regions should roughly correlate with the density of synaptic junctions utilizing the transmitter in question. Numerous studies have been carried out correlating the density of receptor binding sites with various markers of density of innervation (e.g., neurotransmitter level, synthesis, turnover, reuptake). They will be discussed in individual chapters of this volume dealing with particular types of receptor. Although in most cases a reasonable correlation has been found between the levels of innervation and receptor binding, there are cases of discrepancy in which a high concentration of receptor sites has been found in a region known to contain very little respective neurotransmitters (e.g., β-adrenergic sites in the striatum). This, however, does not exclude the possibility that other, as yet unrecognized, endogenous substances act at these sites.

Localization of receptor sites has been studied primarily by using autoradiography and lesion techniques. Autoradiography, for instance, was instrumental in localizing opiate receptors in brain areas associated with pain perception. The recently introduced technique of position emission tomography (PET) has allowed in vivo labeling and visualizing of dopamine receptors in human brain. The lesion techniques (employing mechanical lesions or selective neurotoxins) have helped to localize the binding of ^{3}H-imipramine and ^{3}H-desipramine to presynaptic uptake sites on serotonergic and noradrenergic neurons, respectively (Hrdina et al., 1981; 1982).

Correlation of binding to receptor or recognition sites with receptor-mediated biological responses in the central nervous system has proved to be the most difficult part of receptor research. The reasons for this are several. First, the binding studies are performed "in vitro" or, at best, "*ex vivo,*" whereas the biological responses are usually measured "in vivo" or "*in situ.*" Second, whereas the binding data are derived from a tissue or membrane preparation, i.e., in isolation from the physiological milieu, biological responses, particularly in the brain, are likely to reflect a complex interaction of several systems in the target region. Third, the density of receptors or recognition sites need not necessarily correlate with the level of responsiveness of these sites to agonists or antagonists. Finally, modulatory influences that are in operation "in vivo" are absent in the binding assay. In spite of these difficulties, successful correlations have been found between binding and some biochemical functional parameters, for example, β-receptor-stimulated activity of adenylate cyclase and the density of β-receptors labeled by ^{3}H-dihydroalprenolol, inhibition by drugs of ^{3}H-imipramine binding, and the neuronal reuptake of serotonin (Langer et al., 1980). Attempts are also being made to correlate receptor occupation with electrophysiological events or behavioral responses elicited by stimulation of a particular receptor system. A detailed account of these types of studies is presented in individual chapters of this book.

6. Regulation of Receptor Sites

Receptor binding assays have become an important tool in studying the regulation of receptor activity. The most important factor regulating the density of postsynaptic receptors is the synaptic concentration of the neurotransmitter itself. In general, there is an inverse relationship between the levels of neurotransmitter and number of receptor sites. A decrease in the amount of active neurotransmitter (depletion, selective destruction of terminals) is usually accompanied by functional supersensitivity and an increase in number of postsynaptic receptor sites. Conversely, increased synaptic levels of neurotransmitter usually lead to a reduction of receptor density and development of subsensitivity. Some presynaptic sites, e.g., uptake sites for noradrenaline labeled by

^{3}H-desipramine, are also regulated by the synaptic level of neurotransmitter, but here the relationship between the transmitter concentration and number of uptake sites is a direct one.

Studies have shown that dynamic, time-related changes in receptor density occur after chronic treatment with some psychotropic drugs. For instance, repeated administration of tricyclic antidepressants results in down-regulation of β-adrenergic receptors in brain. Altered density of receptors has now been found in various pathophysiological states, for example a decrease in insulin receptors in hyperinsulinemia and a reduced number of ^{3}-H-imipramine binding sites in platelets of depressed patients (*see* Chapter 13, section 4.2). In most cases, changes only occur in binding site density, not in affinity. This is understandable if one considers that membrane proteins are in a state of continuous turnover, and thus any change in the rate of this turnover could be reflected in altered receptor density.

7. Conclusion and Trends

There are obvious advantages in employing the direct radioligand binding assay in studies of drug–receptor interaction. The labeled ligands enable us to "visualize" the drug–receptor complex and to characterize this interaction in kinetic terms. The eventual isolation of the ligand–receptor complex can help in better understanding the chemical nature of the receptor. Binding assays are also becoming important tools in studying the changes of receptors caused by adaptive or pathophysiological processes and in devising new types of compounds with selective action on specific sets of receptors. Suitable radioligands may thus become biological markers of the state of receptors. In addition, radioreceptor assays can be useful in measuring the effective drug concentration in extracellular fluids and in relating them to the dose and clinical effect. For the successful and reliable outcome of any binding assay, it is important to meet the following essential requirements: (a) a pure, stable, and relatively selective radioligand of sufficiently high specific activity, (b) suitable tissue preparation, (c) optimum incubation conditions, (d) the quickest possible separation of bound from free ligand, and (e) proper analysis of binding data by means of Scatchard and Hill plots and determination of kinetic constants. Examples of radioligands currently used in binding studies to label neurotransmitter receptors and drug recognition sites in the CNS are given in Table 1.

TABLE 1
Neurotransmitter and Drug Receptor (Recognition) Sites in the CNS[a]

Transmitter/drug	[^{3}H]-Endogenous ligand	Receptor activation	Subtype	Labeled by: Agonist	Labeled by: Antagonist
GABA	^{3}H-GABA	Opens Cl^- channels	A	^{3}H-muscimol	
			B	^{3}H-baclofen	
			Convuls.		^{35}S-TBPS
Glycine	^{3}H-glycine	Opens Cl^- channels			^{3}H-strychnine
Glutamate/aspartate	^{3}H-glutamic acid	Opens cation channels	Kainate		^{3}H-kainate
	^{3}H-*N*-acetyl aspartyl glutamate		Quisqualate		^{3}H-AMPA
			NMDA	^{3}H-glutamate	
	^{3}H-aspartic acid				
Acetylcholine	^{3}H-acetylcholine iodide	Closes K^+ channels	M	^{3}H-methoxotremorine	^{3}H-QNB
			N	^{3}H-nicotine	^{125}I-bungarotoxin
Dopamine	^{3}H-dopamine	Stimulates AC	D_1	^{3}H-ADTN	^{3}H-thioxanthines
			D_2	^{3}H-propylnorapomorphine	^{3}H-butyrophenones
Noradrenaline			α_1		^{3}H-prazosin
			α_2	^{3}H-aminoclonidine	^{3}H-yohimbine
		Stimulates AC	β	^{3}H-isoproterenol	^{3}H-DHA

Serotonin	^{3}H-5-HT	Stim. AC (?)	S_1	^{3}H-5-HT; ^{3}H-8-OH-DPAT	
			S_2		^{3}H-ketanserin; ^{3}H-spiperone (frontal cortex)
Histamine	^{3}H-histamine		H_1		^{3}H-pyrilamine
			H_2		^{3}H-tiotidine
Opiate			μ	^{3}H-dihydromorphine	^{3}H-naloxone
	^{3}H-enkephalin		δ	^{3}H-DADLE	
			κ	^{3}H-ethylketocyclazocine; ^{3}H-dynorphin	
			σ	^{3}H-SKF 10,047	
Benzodiazepines	?		BZ_1	^{3}H-diazepam	
			BZ_2	^{3}H-methyl β-carboline carboxylate	
Adenosine	^{3}H-adenosine	Suppresses AC	A_1	^{3}H-chloroadenosine	
		Stimulates AC	A_2	^{3}H-ethylcarboxamide adenosine	
Imipramine	?	Inhibition of 5-HT uptake		^{3}H-imipramine	
Desipramine	?	Inhibition of NA uptake		^{3}H-desipramine	

[a] *Abbreviations:* AC, adenylate cyclase; ADTN, 6,7-dihydro-2-aminotetralin; AMPA, R,S-amino-3-hydroxy-5-methylisoxazole-4-propionic acid; DADLE, [D-Ala^2, D-Leu^5]enkephalin; DHA, dihydroalprenolol; β-OH-DPAT, 8-hydroxy-2-(di-*n*-propylamino)tetralin; GABA, γ-aminobutyric acid; 5-HT, 5-hydroxytryptamine; NMDA, *N*-methyl-D-aspartic acid; QNB, quinuclinidyl benzilate; TBPS, *t*-butylbicyclophosphorothionate.

On the other hand, one has to keep in mind the inherent disadvantages and limitations of receptor binding studies to avoid "overinterpretation" of binding data. If binding assays are relatively easy to perform, the interpretation of binding data becomes increasingly difficult, particularly when one is attempting to correlate them with functional consequences triggered by ligand–receptor interaction. Specific binding need not always result in activation of receptors and many ligands can bind to several receptor (recognition) sites. Only critical interpretation of binding data will advance our understanding of receptor function in neurotransmission and drug action.

Current trends in receptor research, as documented by the chapters in this volume, aim at correlation of receptor binding with the evoked electrophysiological, biochemical, or behavioral responses, at the adaptive changes in the state of receptors reflected in altered binding, and at the search for endogenous ligands for binding sites characterized by drugs. This volume attempts to review the current state of knowledge of receptor sites in the brain that mediate the action of putative neurotransmitters or modulators such as GABA, excitatory amino acids, dopamine, noradrenaline, serotonin, acetylcholine, purines, opioid and non-opioid peptides, and some psychotropic drugs such as antidepressants, amphetamine, phencyclidine, and benzodiazepines.

Acknowledgments

This work was supported by the Ontario Mental Health Foundation grant #871 and Medical Research Council of Canada grant MA 6775. I thank Denise Mainville for the preparation of the manuscript.

References

Bennet J. P. (1978) Methods in Binding Studies, in *Neurotransmitter Receptor Binding.* (Yamamura H. I., Enna S. J., and Kuhar M. J., eds.) pp. 57–90, Raven Press, New York.

Braestrup C. and Nielsen M. (1980) Multiple benzodiazepine receptors. *Trends Neurosci.* **3,** 1–303.

Costa E. (1982) Perspectives in the Molecular Mechanisms and Antidepressant Action, in *Typical and Atypical Antidepressants: Molecular Mechanisms* (Costa E. and Racagni G., eds.), pp. 21–26, Raven Press, New York.

Cheng Y. C. and Prusoff W. H. (1973) Relationship between the inhibition constant (K_i) and the concentration of inhibitor which causes 50 percent inhibition (IC_{50}) of an enzymatic reaction. *Biochem. Pharmacol.* **22,** 3099–3108.

Creese I., Burt D. R., and Snyder S. H. (1975) Dopamine receptor binding: Differentiation of agonist and antagonist states with ^{3}H-dopamine and ^{3}H-haloperidol. *Life Sci.* **17,** 993–1002.

Hartley E. J. and Seeman P. (1978) The effect of varying ^{3}H-spiperone concentration on its binding parameters. *Life Sci.* **23,** 513–518.

Hill A. V. (1909) The mode of action of nicotine and curare determined by the form of the contraction curve and the method of temperature coefficients. *J. Physiol.* (Lond.) **39,** 361–373.

Hirata F. and Axelrod J. (1980) Phospholipid methylation and biological signal transmission. *Science* **209,** 1082–1090.

Hrdina P. D., Elson-Hartman K., Roberts D. C. S., and Pappas B. A. (1981) High affinity [^{3}H]desipramine binding in rat cerebral cortex decreases after selective lesion of noradrenergic neurons with 6-hydroxydopamine. *Eur. J. Pharmacol.* **73,** 375–376.

Hrdina P. D., Pappas B. A., Bialik R. J., and Ryan C. L. (1982) Regulation of [^{3}H]imipramine binding sites in rat brain regions: Effect of neonatal 5,7-dihydroxytryptamine treatment. *Eur. J. Pharmacol.* **83,** 343–344.

Iversen L. L., Iversen S. D., and Snyder S. H. (eds.) (1983) *Biochemical Studies on CNS Receptors.* Plenum Press, New York.

Laduron P. (1983) More binding, more fancy. *Trends Pharmacol. Sci.* **4,** 333–335.

Laduron P. M. (1984) Criteria for receptor sites in binding studies. *Biochem. Pharmacol.* **33,** 833–839.

Langer S. Z., Moret C., Raisman R., Dubocovich M. L., and Briley M. (1980) High affinity [^{3}H]imipramine binding in rat hypothalamus: Association with uptake of serotonin but not of norepinephrine. *Science* **210,** 1133–1135.

Lowry O. H., Rosebrough N. J., Farr A. G., and Randall R. J. (1951) Protein measurement with the Folin phenol reagent. *J. Biol. Chem.* **193,** 265–275.

Möhler H. and Okada T. (1977) Benzodiazepine receptors: Demonstration in the central nervous system. *Science* **198,** 849–851.

Munson P. J. and Rodbard D. (1980) Ligand: A versatile computerized approach for characterization of ligand-binding systems. *Anal. Biochem.* **107,** 220–239.

Pert C. and Snyder S. H. (1973) Opiate receptor: Demonstration in nervous tissue. *Science* **179,** 1011–1014.

Peterson G. L. (1977) A simplification of the protein binding assay method of Lowry et al. which is more generally applicable. *Anal. Biochem.* **83,** 346–356.

Pfeuffer T. (1977) GTP-binding proteins in membranes and the control of adenylate cyclase activity. *J. Biol. Chem.* **252,** 7224–7234.

Raisman R., Briley M., and Langer S. Z. (1979) Specific tricyclic antidepressant binding sites in rat brain. *Nature* (Lond.) **281,** 148–150.

Rosenthal E. H. (1967) A graphic method for the determination and presentation of binding parameters in a complex system. *Anal. Biochem.* **20,** 525–532.

Scatchard G. (1949) The attraction of proteins for small molecules and ions. *Ann. NY Acad. Sci.* **51,** 660–672.

Seeman P., Chan-Wong M., Tadesco J., and Wong K. (1975) Brain receptors for antipsychotic drugs and dopamine: Direct binding assays. *Proc. Natl. Acad. Sci. USA* **72,** 4376–4380.

Snyder S. H. (1983) Molecular Aspects of Neurotransmitter Receptors: An Overview, in *Biochemical Studies on CNS Receptors* (Iversen L. L., Iversen S. D., and Snyder S. H., eds.), pp. 1–12, Plenum Press, New York.

Squires R. F. and Braestrup C. (1977) Benzodiazepine receptors in rat brain. *Nature* (Lond.) **266,** 732–734.

Williams L. T. and Lefkowitz R. J. (1978) *Receptor Binding Studies in Adrenergic Pharmacology.* Raven Press, New York.

Yamamura H. I., Enna S. J., and Kuhar M. J. (eds.) (1978) *Neurotransmitter Receptor Binding.* Raven Press, New York.

Chapter 2
Central Nervous System Dopamine Receptors

Ram K. Mishra

1. Introduction

Dopaminergic neuronal systems constitute the principal catecholamine systems of the striatum of the brain (Carlsson, 1959) and the inner retina (Haggendal and Malmfors, 1965). In both of these regions dopamine (DA) appears to function mainly as an inhibitory neurotransmitter (Conner, 1970; Woodruff et al., 1976), although an excitatory role of DA has also been proposed for discrete brain regions (Cools and Van Rossum, 1976).

Alterations in the dopaminergic mechanisms of the central nervous system (CNS) underlie several disorders, such as schizophrenia, Parkinson's disease, and Huntington's chorea. In Parkinson's disease, the primary neuropathology is thought to be a degeneration of the DA-containing nigrostriatal system (*see* review by Barbeau, 1976; Price et al., 1978). Some of the symptoms, including akinesia, rigidity, and tremor, can be significantly alleviated by treatment with L-DOPA or other agents such as proline-leucine-glycinamide (PLG) (Mishra et al., 1983) and the direct DA receptor agonists, bromocryptine and lergotrile (Rinne et al., 1977). It has been hypothesized that schizophrenic symptoms result from overactivity of the DA systems in discrete regions of the forebrain (Crow, 1980). Moreover, the DA receptor antagonists (neuroleptics) are thought to hasten remission of acute psychotic illnesses and prevent later exacerbations of chronic psychotic symptoms by blocking DA receptors. The antipsychotics (neuroleptics) have also been used to treat the dyskinetic symptoms of Huntington's chorea and Gilles de la Tourette's syndrome, leading to the speculation that the behavioral symptoms of these disorders are a result of an overactivity of the DA system in the basal ganglia.

Catecholamines, as well as the majority of peptide hormones, act on their target cells by binding rapidly and reversibly to a set of specific proteins on the surface of the target cells called "receptors." The combination of the hormone with the receptor produces a signal that crosses the plasma membrane and activates intracellular processes. In the 1970s, with the introduction of radioreceptor assays, receptors for various hormones and catecholamines were identified and established. A few examples included: insulin (Freychet et al., 1971), growth hormone (Lesniak et al., 1973), β-adrenergic receptors (Lefkowitz et al., 1974), α-adrenergic receptors (Williams and Lefkowitz, 1976), DA receptors (Seeman et al., 1975; Creese et al., 1975), and opiate receptors (Goldstein et al., 1971; Terenius, 1973; Pert and Snyder, 1973). This new technology has been applied not only to establish and identify the receptors, but also to characterize and detect normal and abnormal states of receptors in health and disease (*see* reviews by Brown and Aurbach, 1982; Seeman, 1980; Gullick and Lindstrom, 1983). The fact that antipsychotic and antiparkinsonian agents block and activate DA receptors, respectively, and minimize symptoms of these disorders, has been a major impetus in the advancement of research devoted to the biochemical and pharmacological characterization of DA receptors. In the last few years, various investigators have shown the feasibility of directly identifying the dopaminergic receptors by using radioactive agonist and antagonist binding techniques. These receptors or binding sites have been designated and classified differently (D-0, D-1, D-2, D-3, D-4 . . . D_2 low, D_2 high, and so on), depending upon the individual preference of the investigators and the approaches they used [Costall and Naylor, 1981; Seeman, 1982 (commentary); Creese and Sibley, 1982 (comments on the Seeman's commentary)]. According to biochemical (adenylate cyclase-linked) and ligand binding studies, the receptors were mainly classified as D-1 cyclase-linked and D-2 noncyclase linked (Kebabian and Calne, 1979; Mishra and Cleghorn, 1980), although other sites (D3, D4) also received some notable attention. However, these classifications were recently disputed by a number of investigators (Laduron, 1980; Clements-Jewery and Robson, 1981; Costall and Naylor, 1981; Freedman et al., 1981; Offermeier and Van Rooyen, 1982). In addition to the D-1 (adenylate cyclase-linked) receptor, a D-2 receptor in the anterior pituitary has been suggested to exist in D-2 low and D-2 high states (the one that is induced by agonists, but not by antagonists, and modulated by guanine nucleotides) (Sibley et al., 1982). On the other hand, DeLean et al. (1982) have suggested that guanine nucleotides convert the D-2 agonist high-affinity to the agonist low-affinity state and the antagonist low-affinity to the antagonist high-affinity state. This finding is in contradiction to that of Sibley et al. (1982). The reason for this discrepancy is not known.

Inhibition by sulpiride of DA-stimulated adenylate cyclase has been reported when the tissue of castrated or immature rats is used. This suggests that there could be only one receptor, which behaves differently under different hormonal conditions (Gnegy et al., 1982). Sulpiride was a key factor when the D-1 and D-2 classification was introduced (Kebabian and Calne, 1979). In addition to D-1 (DA-stimulated adenylate cyclase), a DA-inhibited adenylate cyclase was also reported, initially in human pituitary adenomas (De Camilli et al., 1979). The pharmacological characteristics of the DA receptor causing inhibition of the adenylate cyclase are different from that receptor causing stimulation of this enzyme. In intact pituitary intermediate lobe cells, DA agonists inhibited the isoprenaline-stimulated adenylate cyclase with K_i values in the low nanomolar range (Munemura et al., 1980; Meunier and Labrie, 1982). In a cell-free homogenate, however, 3 μM apomorphine was necessary to elicit the same response (Cote et al., 1981), perhaps implying the loss of a cofactor. In intact cells, this inhibition could be reversed by very low (n*M*) concentrations of the neuroleptics. These receptors were classified as D-2 receptors negatively coupled to adenylate cyclase. They were also found in the caudate nucleus (Stoof and Kebabian, 1981); their physiological significance remains unclear, however.

In addition to membrane homogenates, Sibley et al. (1983) have also demonstrated the presence of high-affinity receptor binding sites on intact bovine anterior pituitary cells and the absence of high-affinity agonist receptor binding states. In these studies, Sibley et al. (1983) also suggested that, "D-2 and D-4 receptors correspond exactly with demonstrated high- and low-affinity agonist-receptor binding state of a single D-2 receptor and thus do not represent two discrete receptor subtypes." In addition, they further argued that "discrimination of dopamine receptor subtypes based solely on agonist binding affinities in brain membrane preparations (Seeman, 1980, 1982; Titeler et al., 1979) could be premature and should be interpreted with caution." These types of arguments will continue to exist until the DA receptor protein molecule is purified and biochemically characterized. In any event, the classification of DA-receptor as D-1 and D-2 described by Stoof and Kebabian (1984) is generally accepted.

In the present paper, some of the studies carried out on DA receptors, on DA-stimulated adenylate cyclase, and on the effect of peptides on DA receptors is summarized. In addition, the technique of solubilization and purification of DA receptors is also briefly described.

2. Dopamine Receptors

2.1. Membrane Binding Assay

In most of the studies, [^{3}H]-spiroperidol has been used as the ligand of choice, although [^{3}H]-haloperidol has also been utilized in some of our previous studies as well as those of other investigators. In general, the [^{3}H]-spiroperidol binding is carried out as follows:

Tissue is homogenized (Polytron setting 6, 20 s) in 50 vol (μL/mg wt wt) of ice-cold 50 m*M* Tris-HCl buffer (pH 7.7) and centrifuged at 50,000 *g* for 10 min. This procedure is repeated once. The tissue pellet is rehomogenized in 100 vol of ice-cold buffer containing 50 m*M* Tris-HCl (pH 7.6), 10 μ*M* pargyline, 0.1% ascorbic acid, 120 m*M* NaCl, 5 m*M* KCl, 2 m*M* $CaCl_2$, and 1 m*M* $mgCl_2$. This tissue homogenate is preincubated at 37°C for 5 min, then returned to ice. The assay mixture contains 0.5 mL of tissue homogenate, 50 μL of [^{3}H]-spiroperidol (varying concentrations), and 50 μL of either 0.1% ascorbic acid or drug solution. The reaction is started by the addition of the [^{3}H]-haloperidol or [^{3}H]-spiroperidol, followed by rapid vortex mixing and placement in the 37°C waterbath. After a 10-min incubation, the assay solution is filtered rapidly through Whatman GF/B filters, followed by three rinses with ice-cold 50 m*M* Tris-HCl buffer (pH 7.7). Wet filters are placed directly into scintillation vials with 5 mL of Phase Combining System (PCS) scintillation cocktail (Amersham, Chicago). Filters are allowed to stand in the scintillation cocktail for at least 5 h to allow the aqueous fraction to dissolve in the PCS. Samples are counted with a series of standard samples, using the External Standard Ratio mode in a Beckman LS-230 liquid scintillation counter. Data are converted from cpm to dpm by means of a quench correction program in a Hewlett-Packard calculator. Specific binding is defined as the difference in binding between incubations in the absence and in the presence of 1 μ*M* D-butclamol. The binding data are then analyzed by the computer using a ligand binding program to obtain the equilibrium dissociation constant (K_d) and maximum binding sites (B_{max}).

2.2. Dopamine-Stimulated Adenylate Cyclase Assay

Fresh striata, immediately after removal from the brain, are gently homogenized by hand in all-glass homogenizers at 0–4°C in medium containing 2 m*M* Tris-maleate buffer (pH 7.4), plus 0.8 m*M* EGTA at a dilution of 1:50 (wet wt/vol). Aliquots of 50 μL each of homogenate are then incubated in a shaking waterbath at 30°C for 2.5 min in a final volume of 200 μL incubation medium containing 80 m*M* Tris-maleate buffer (pH 7.4), 10 m*M* theophylline, 2 m*M* $MgSO_4$, 0.5 m*M* ATP, and appropriate hormones or other test substances. The reaction is terminated by placing assay tubes in a boiling waterbath for 2.5 min. Particulate matter is removed by low-speed centrifugation and aliquots of the supernatant fluids are assayed for cyclic AMP by a binding assay previously described by Gilman (1970). All values are corrected for the small amount of cyclic AMP present after incubation without ATP. Each experiment involves at least duplicate (usually triplicate) adenylate cyclase incubations for every condition studied, with replicate determinations of cyclic AMP content in each case.

For determination of cyclic AMP levels, striatal slices are pooled from 4–6 rats, suspended in Krebs-Ringer-biocarbonate buffer, and preincubated at 37°C for 60 min under 95% O_2 and 5% CO_2. The buffer is changed 3–4 times during the preincubation period. At the end of the preincubation, the slices are washed with fresh buffer and resuspended, test substances are added, and the tubes are further incubated for 5 min. All incubations are then terminated by addition of 50% acetic acid. The contents are homogenized and centrifuged. The cyclic AMP contents are then determined by Gilman's (1970) method.

3. Effects of Peptides on Dopamine Receptors

The number of neurotransmitter candidates in the CNS is rapidly increasing and their various physiological functions are only beginning to unfold. In our laboratory we have studied the interaction of L-prolyl-L-leucyl-glycinamide (PLG) and cholecystokinin (CCK) with dopaminergic systems in the striatum and mesolimbic regions. Some of the results obtained along these lines of investigation are described briefly below.

3.1. Effect of PLG on Dopamine Receptors

L-Prolyl-L-leucyl-glycinamide (PLG) is one of the shortest peptides discovered to date. Its role as a melanocyte-stimulating hor-

Table 1
Effect of Apomorphine on Locomotor Activity of Rats Treated With PLG and Haloperidol[a]

Treatment group[b]	Counts/h. mean ± SEM
Saline	768 ± 40
PLG	710 ± 36
Haloperidol	1011 ± 97
PLG + haloperidol	798 ± 54

[a]The locomotor activity was measured after all groups of animals were challenged with apomorphine (adapted with the permission of the publisher from Mishra, 1983, copyright © 1983, Pergamon Press).

[b]Apomorphine: 1 mg/kg ip
PLG: 10 mg/kg ip
Haloperidol: 3mg/kg ip.

mone release inhibitor factor remains questionable (Kastin et al., 1976). PLG, nevertheless, has proven to be a useful peptide in many paradigms (*see* review, Mishra et al., 1983). The effects of PLG on the locomotor activity in rats and on DA receptor (D-2) binding following chronic administration of neuroleptic drugs are shown in Tables 1 and 2. On the basis of these results and our earlier studies (Chiu et al., 1981, 1983), a possible mechanism of action of PLG is proposed in Fig. 1. Furthermore, these results indicate that PLG and its stable active analogs warrant clinical trials to establish their therapeutic efficiency in the treatment of neuroleptic drug-induced tardive dyskinesia, Parkinson's disease, and other related extrapyramidal disorders involving alterations of dopaminergic mechanisms.

3.2. Cholecystokinin and Dopamine Receptors

Cholecystokinin (CCK) is found in very high concentrations in the brain, with its highest concentration found in the striatum (Innis et al., 1977; Keely and Dodd, 1981). High concentrations of CCK binding sites are also found in the striatum and limbic regions where they are located primarily on small neurons (Hays et al., 1981; Mishra, 1983). Chronic administration of neuroleptic drugs (haloperidol) to rats results in an increase in CCK binding along with the spiroperidol binding sites, without change in the equilibri-

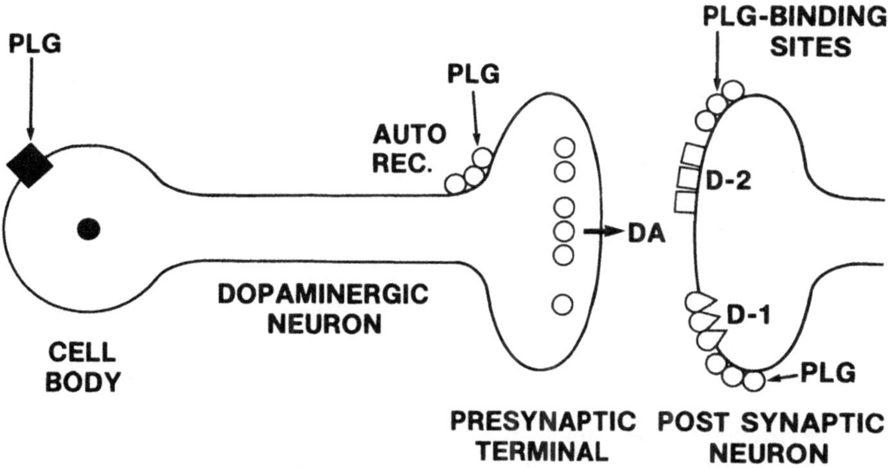

Fig. 1. PLG can interact with D-1 or D-2 receptors through its own specific binding sites at the postsynaptic neuron, or PLG can interact with the dopamine autoreceptors at the presynaptic neurons, causing release of dopamine; thus the dopamine can down-regulate receptors. Alternatively, PLG can interact at the cell body level causing increased synthesis of dopamine (adapted with the permission of the publisher from Mishra, 1983, copyright © 1983, Pergamon Press).

um dissociation constant (K_d). This increase was specific to the mesolimbic regions and was not observed in the striatum (Mishra, 1983). Most of this evidence is therefore compatable with the notion that CCK may selectively interact with the dopaminergic mechanisms in the mesolimbic regions of the brain. These results are significant in the clinical treatment of schizophrenia. Based on the results obtained in our laboratory, a working model for the interaction of CCK and dopaminergic system is proposed in Fig. 2.

3.3. Dopamine Receptor Blocking Agents and Substance P

Substance P occurs in very high concentrations in the substantia nigra and the globus pallidus in humans and other mammalian species (Gale et al., 1978). This putative neurotransmitter or neuromodulator also interacts with dopaminergic receptor mechanisms in the extrapyramidal regions. Chronic administration of neuroleptic drugs selectively decreases substance P content in the substantia nigra (*see* Table 3). There are no changes in other brain regions. This suggests a regional specificity for dopaminergic-substance P-ergic interactions.

Table 2
Blockade of Haloperidol (HAL)-Induced Increase in Specific ^{3}H-Spiroperidol Binding by PLG

Treatment group	N	B_{max}, fmol/mg protein	K_d, nM
Saline	6	215 ± 21	0.22 ± 0.07
HAL	6	512 ± 32[a]	0.24 ± 0.09
PLG	6	304 ± 16	0.28 ± 0.05
PLG and HAL	6	330 ± 18	0.21 ± 0.02

[a]($p > 0.01$) The various groups of animals received various drugs for 21 d. The dose of haloperidol was 3 mg/kg/d and PLG was administered 10 mg/kg/d. The animals were sacrificed 5 d after the last injection and ^{3}H-spiroperidol binding was carried out on striata (adapted with the permission of the publisher from Mishra, 1983, copyright © 1983, Pergamon Press).

Table 3
Effects of Chronic Administration of Haloperidol on Substance P Content of Various Regions of Rat Brain[a]

		Substance P content, ng/mg protein	
Brain part	Region	Control	Drug-treated
Basal ganglia	Caudate	3.5 ± 0.2	3.1 ± 0.09
	Globus pallidus	4.6 ± 0.9	4.1 ± 0.4
	Subthalamic nucleus	3.2 ± 0.1	3.5 ± 0.2
	Substantia nigra	21.0 ± 3.0	9.0 ± 0.8[b]
Limbic system	Olfactory tubercle	4.8 ± 0.6	5.2 ± 1.0
	Septum	6.8 ± 0.9	6.2 ± 0.6
	Nucleus accumbens	4.2 ± 0.2	3.8 ± 0.5
	Amygdala	6.2 ± 0.8	5.8 ± 0.5
Hypothalamus	Medial lobe	7.6 ± 2.0	6.8 ± 0.9

[a]Animals were treated with haloperidol (3 mg/kg ip) for 3 wk and were tested for apomorphine induced behavioral supersensitivity. After establishment of supersensitivity, the animals were sacrificed and the substance P content in various regions was determined.

[b]$p < 0.01$.

4. Studies of Dopamine Receptor Subsensitivity and Supersensitivity

The dopamine-responsive cyclic AMP-generating system in the CNS is affected by alterations of the normal neuronal input. We observed that following 6-hydroxydopamine (6-OHDA) lesions of the substantia nigra (*see* Mishra et al., 1980), the DA content of the caudate nucleus was decreased to 50% of the control side within 24 h, and to less than 3% 1 wk after surgery (Fig. 3) in two distinct denervation supersensitivity syndromes. The timecourse of the increase in cyclic AMP levels in striatal slices following 6-OHDA lesions was similar to the increase in the number of DA receptors (Figs. 4 and 5). The small but significant decrease in haloperidol binding sites 4 d postoperation (Fig. 6) and lack of change in cyclic AMP levels at this timepoint suggested that a minor but functionally significant population of haloperidol binding sites was localized on presynaptic terminals that were not associated with adenylate cyclase. On the other hand, the presynaptic functional pool of cyclic AMP may be small compared to the total pool of caudate cyclic AMP. These observations suggest that, following denerva-

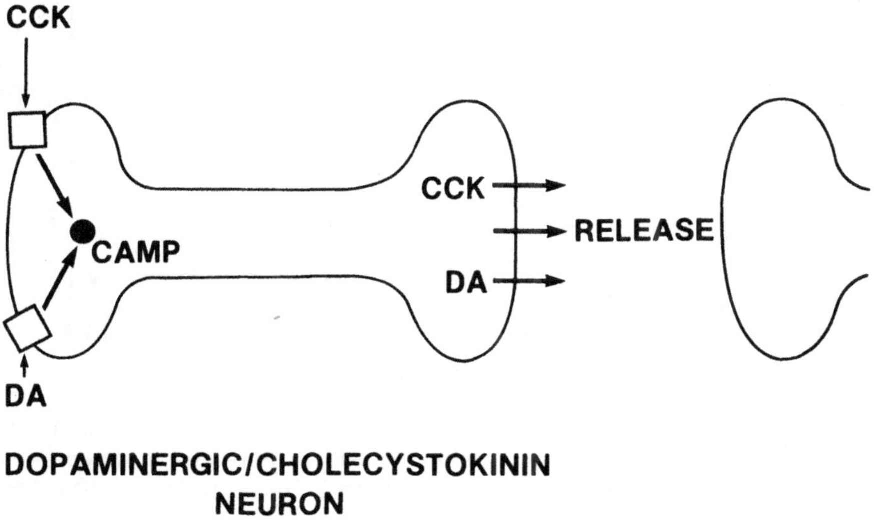

Fig. 2. The common localization of dopamine and CCK in the mesolimbic regions suggests that two such cotransmitters could modulate each other's release by interaction at the neuron. Such dual modulation might be highly advantageous in the case in which a neurotransmitter has the purpose of creating both short- and long-term postsynaptic changes (adapted with the permission of the publisher, from Mishra, 1983, copyright © 1983, Pergamon Press).

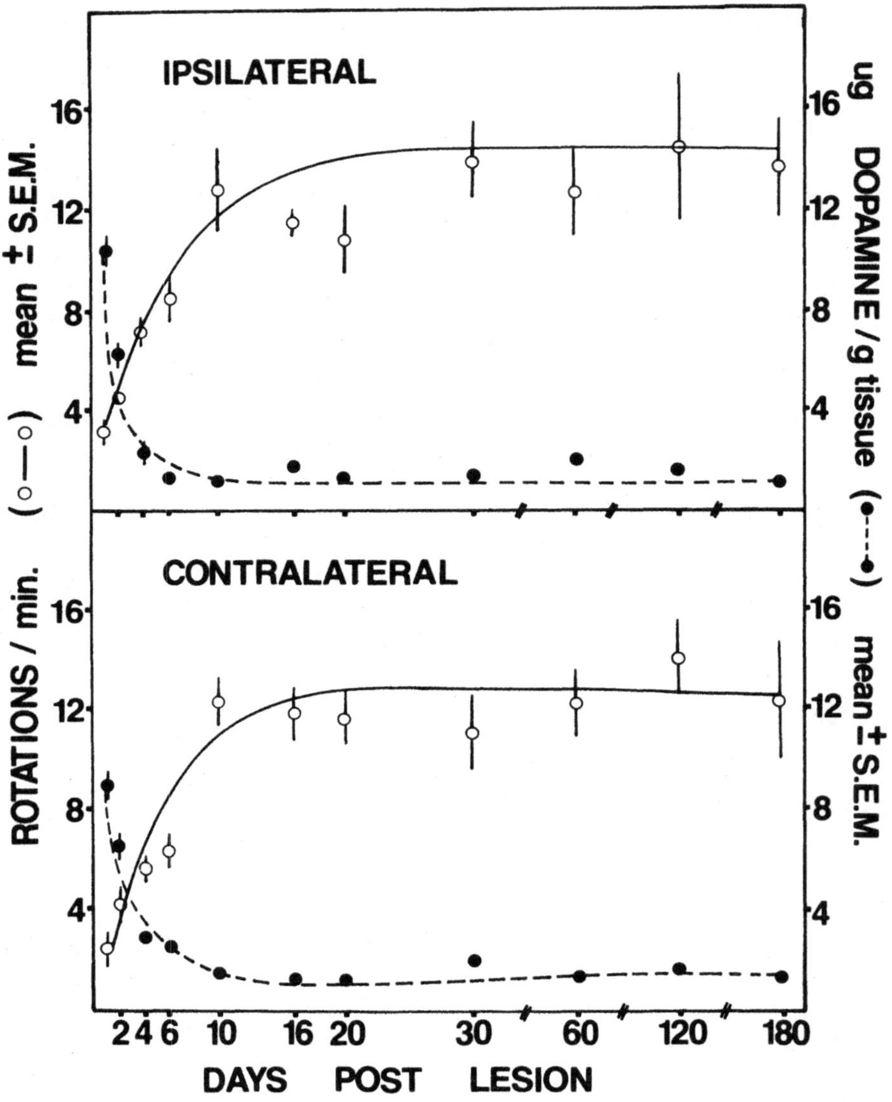

Fig. 3. Time-course of the development of behavioral supersensitivity and dopamine loss in two distinct caudate denervation syndromes. Each data point is an average of six separate experiments ± SEM. The rotational behavior was observed after subcutaneous injection of 2 mg/kg apomorphine (adapted with the permission of the publisher from Mishra et al., 1980, copyright © 1980, Elsevier/North Holland Biomedical Press).

Table 4
^{3}H-Spiroperidol Binding After Exposure of Rat Striatal Slices to Dopamine[a]

	Binding sites (B_{max}), fmol/mg protein	K_d, nM
Control	298	0.12
Dopamine-treated	164	0.14

[a]H-spiroperidol binding characteristics were measured in membranes prepared from brain slices incubated for 30 min with 100 μM dopamine. Specific binding was determined at various concentrations of radioactive ligand and the maximum binding densities and dissociation constants for ^{3}H-spiroperidol were calculated from Scatchard analysis.

tion, a new steady-state level of DA receptors is achieved and that this increased level of receptors causes an increase in postsynaptic sensitivity of the adenlyate cyclase system following temporal development of supersensitivity with time. Similarly, agonist-induced subsensitivity of DA receptors is also observed (Mishra et al., 1978); these studies were carried out by in vivo administration of drugs followed by in vitro biochemical analysis. In order to find out whether the desensitization brought about by chronic exposure to the DA agonists might in fact be mediated either totally or in part by changes induced in the receptors themselves, we have studied this problem in vitro.

The rat striatal slices were incubated with or without DA or apomorphine and the washed slices were used for the ^{3}H-spiroperidol binding. The effect of DA was tested in DA-preincubated washed slices by measuring cyclic AMP levels. Figure 7 shows that slices preincubated with DA agonist showed consistently decreasing formation of cyclic AMP at all DA concentrations tested. Similarly, the number of binding sites was also decreased. The affinity of [^{3}H]-spiroperidol binding was unaffected (Table 4).

It is not known whether: (a) the "desensitization" occurs following preexposure of tissue to DA receptor antagonists; (b) there is a change in affinity of receptor for agonist (the present experiments were carried out using antagonist binding); (c) the system is specific for DA; (d) GTP and ions (K^+) prevent the receptor from becoming desensitized; (e) whether GTP and ions (K^+) could resensitize the system; or (f) the solubilized receptor

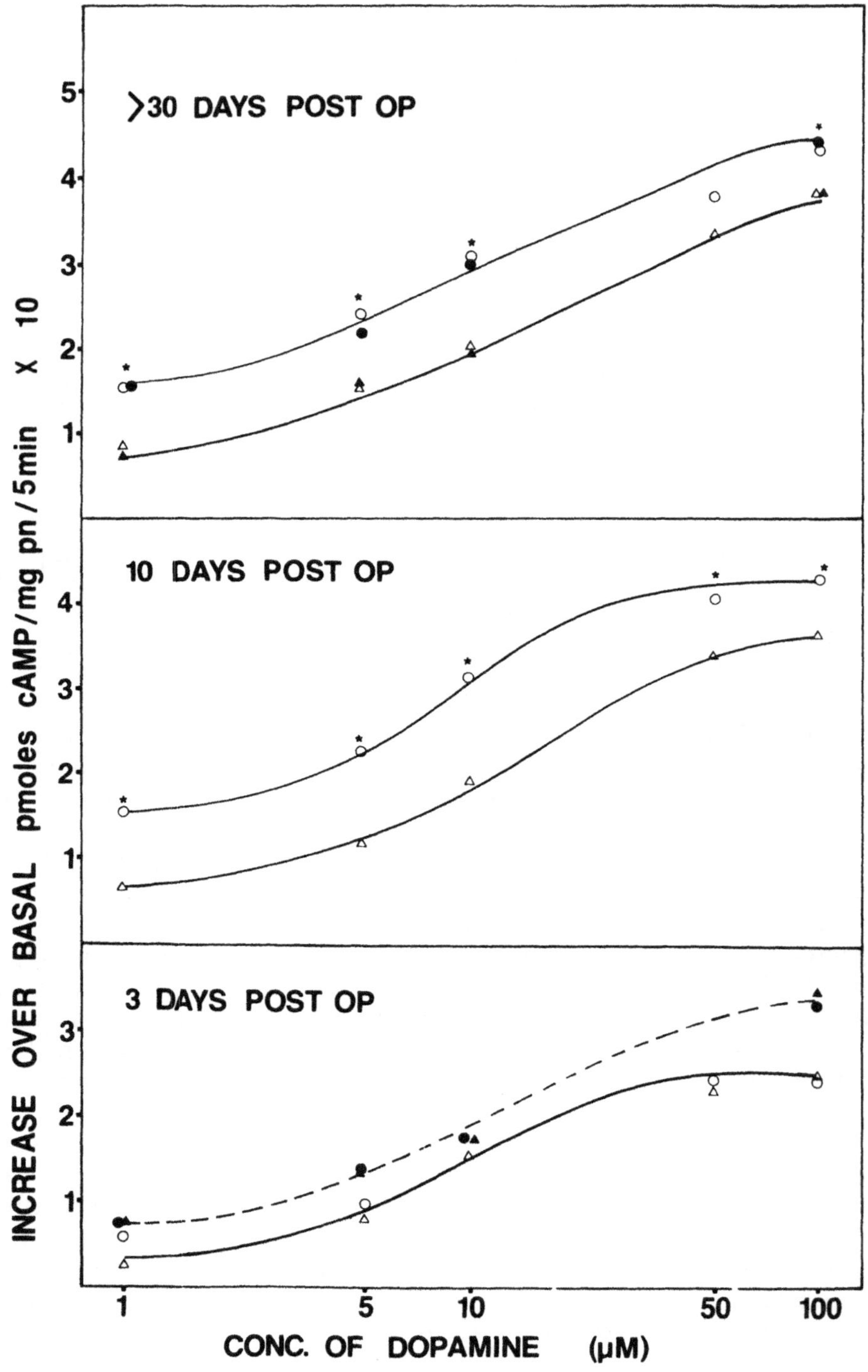

>30 DAYS POST OP
10 DAYS POST OP
3 DAYS POST OP
INCREASE OVER BASAL pmoles cAMP/mg pn/5min X 10
CONC. OF DOPAMINE (µM)

and adenylate cyclase show different responses after desensitization.

5. Effect of Guanine Nucleotides on Dopamine Receptor Binding

Several lines of evidence suggest that guanine nucleotides, in particular GTP, are involved in the coupling of the DA receptor with DA-sensitive adenylate cyclase. In rat striatal tissue preparations, guanine nucleotides regulate the sensitivity of adenylate cyclase to DA stimulation. Addition of GTP to the incubation medium restores the hormonal sensitivity of striatal adenylate cyclase, apparently uncoupled from its receptor by treatment with Russell's viper venom (Kebabian, 1978). Furthermore, results from receptor binding studies indicate that in the presence of GTP, the affinity of the DA receptor for agonists is decreased (Mishra and Cleghorn, 1980), and that, as for β-adrenergic receptors, this effect of GTP is possibly agonist-specific.

We have examined the effects of various nucleotides, in particular Gpp(NH)p (the nucleotide phosphohydrolase-resistant GTP analog), on DA agonist and antagonist binding properties and on DA-sensitive adenylate cyclase in rat striatal tissue preparation. Figures 8 and 9 show the inhibition of [^{3}H]-agonist binding by various nucleotides. The guanine nucleotides Gpp(NH)p, GTP, and GDP are far more potent inhibitors of [^{3}H]-apomorphine and [^{3}H]-ADTN binding than are GMP, ATP, and App(NH)p. At 1 μM

Fig. 4. Accumulation of cyclic AMP levels at various concentrations of dopamine in caudate slices after 6-hydroxydopamine injections into substantia nigra. ○, lesioned side, contralateral turners; ●, lesioned side, ipsilateral turners; △, nonlesioned side, contralateral turners; ▲, nonlesioned side, ipsilateral turners. Each point is an average of six separate experiments. The standard deviations at each point varied less than 5% and this represents a significant difference between the lesioned and the nonlesioned animals, according to a one-tailed *t*-test, with p ranging between 0.005 and 0.05. The data at 10 d postoperation for ipsilateral animals are not shown. The basal cyclic AMP levels ($\bar{X}$ ±SEM) in pmol cAMP/mg protein/5 min are: ○, 10.1 ± 2.38 (3 d); 9.3 ± 1.26 (10 d); 9.5 ± 1.04 (>30 d). ●, 10.5 ± 1.71 (3 d); 10.0 ± 1.08 (>30 d). △, 11.6 ± 0.88 (3 d); 10.5 ± 1.15 (10 d); 10.0 ± 1.82 (>30 d). ▲, 10.7 ± 2.07 (3 d); 10.4 ± 1.72 (>30 d) (adapted with the permission of the publisher, from Mishra et al., 1980, copyright © Elsevier/North Holland Biomedical Press).

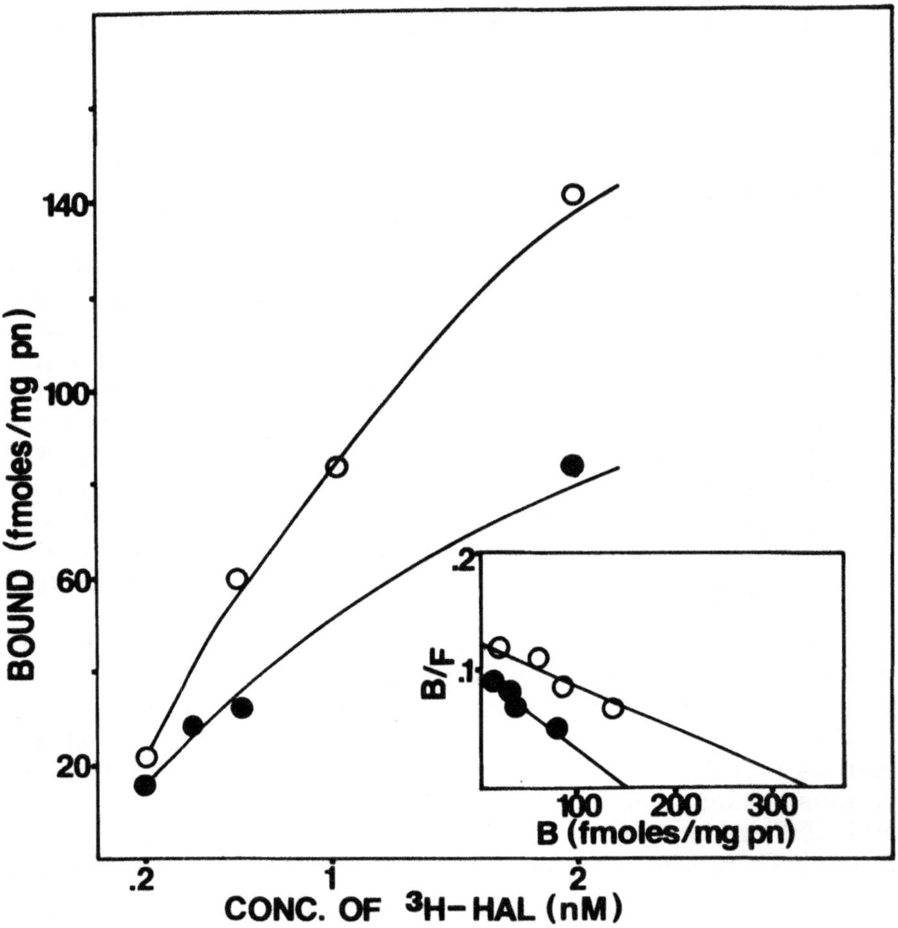

Fig. 5. Receptor binding of ^{3}H-haloperidol at 21 d postoperation in ipsilateral turners, Scatchard analysis was performed on binding data. ○, lesioned side (B_{max}, 343 fmol/mg protein; K_d 2.8 n*M*). ●, nonlesioned side (B_{max}, 151 fmol/mg protein, K_d 1.65 n*M*) (adapted with the permission of the publisher, from Mishra et al., 1980).

concentrations, Gpp(NH)p markedly increases the K_d values of both [^{3}H]-agonists, without having any effect on the number of binding sites (*see* Table 5). This decrease in receptor affinity is seen at relatively low concentrations of Gpp(NH)p and GDP (1.0 μ*M*), concentrations at which other nucleotides do not have an effect (Table 5). The same nucleotides, however, seem to be relatively ineffective inhibitors of the binding of the antagonists [^{3}H]-haloperidol and ^{3}H-spiroperidol. Gpp(NH)p exhibited an inhibitory effect on DA-sensitive adenylate cyclase at all con-

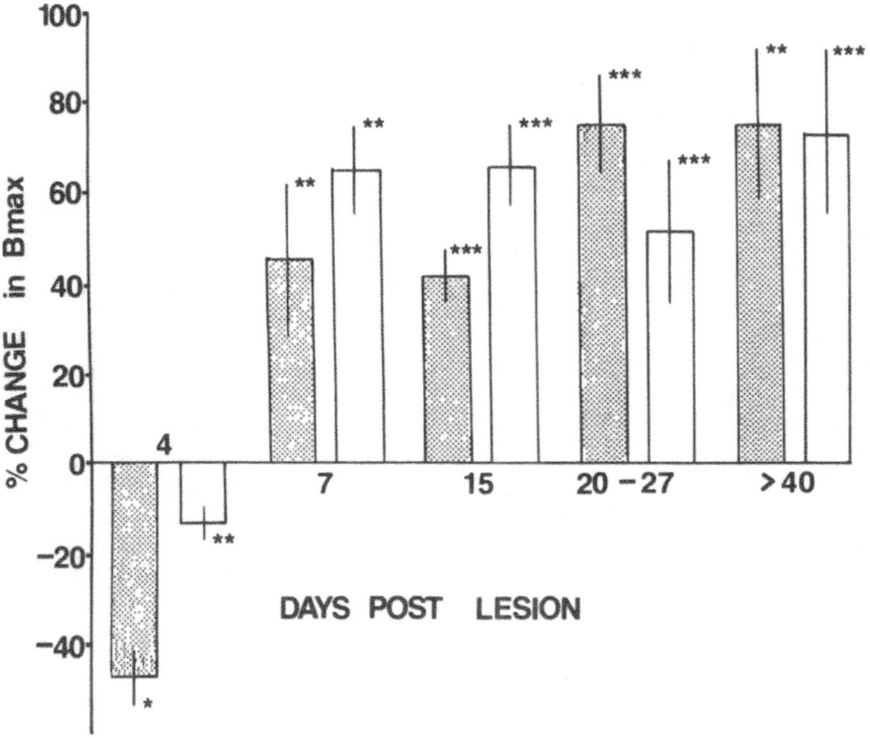

Fig. 6. Time-course of the development of dopamine receptor supersensitivity in caudate following 6-hydroxydopamine lesions of substantia nigra. Caudates from unilaterally lesioned animals were assayed either singly or pooled from two animals, for ^{3}H-haloperidol binding. The data were subjected to Scatchard analysis to determine B_{max} and K_d. The percentage change in B_{max} is:

$$100 \times \frac{B_{max}\text{ (lesioned side)} - B_{max}\text{ (nonlesioned side)}}{B_{max}\text{ (nonlesioned side)}}$$

Shaded bars represent data from contralaterally turning animals; unshaded bars from ipsilaterally turning points. Each point represents the mean of 5–8 separate experiments. The mean B_{max} ±SEM for nonlesioned caudates from contralaterally turning animals is 175.5 ± 20 fmol/mg protein and from ipsilaterally turning animals is 190.5 ± 16.7 fmol/mg protein. *, $p < 0.025$; **, $p < 0.005$ (adapted with the permission of the publisher, from Mishra et al., 1980, copyright © 1980, Elsevier/North Holland Biomedical Press).

Table 5
Effects of Various Nucleotides on K_d and B_{max} of ^{3}H-Apomorphine Binding in Rat Caudate[a]

	K_d, nM[b]	B_{max}, fmol/mg protein[b]
Control	1.2 ± .11	233 ± 25.4
GMP (1 μM)	1.52 ± .25	232 ± 23.9
GDP (1 μM)	2.35 ± .21	209 ± 36.2
Gpp(NH)p (1 μM)	2.35 ± .10	219 ± 30.25
Gpp(NH)p (10 μM)	6.85 ± .74	211 ± 19.15
ATP (1 μM)	1.09 ± .08	277 ± 15.8
App(NH)p (1 μM)	1.04 ± .18	188 ± 42.84
Cpp(NH)p (1 μM)	1.31 ± .34	224 ± 29.36
UMP PNP (1 μM)	1.18 ± .15	209 ± 21.55

[a]Data are expressed as the mean ±SEM of four separate experiments.
[b]K_d (dissociation constant) and B_{max} (maximum number of binding sites) were determined by Scatchard analysis of equilibrium binding data.

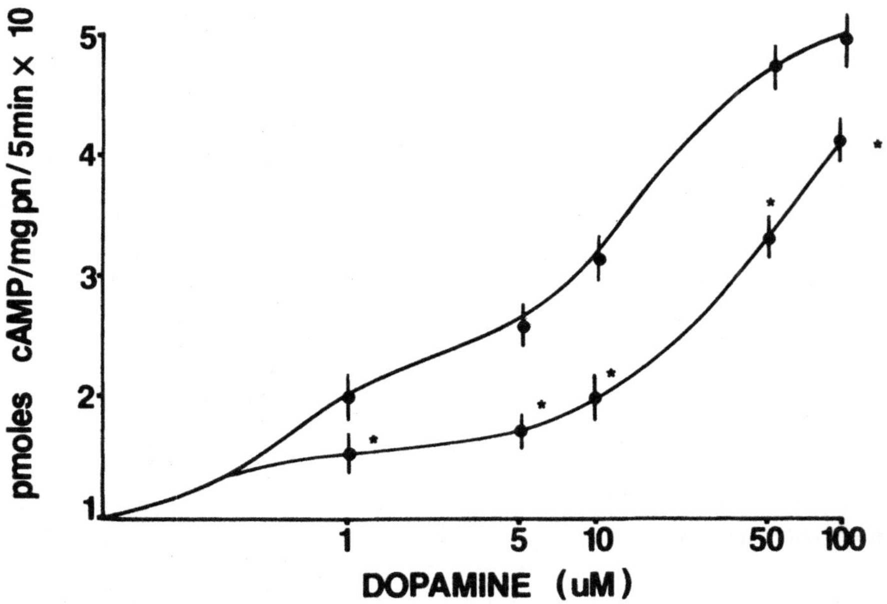

Fig. 7. Cyclic AMP accumulation in dopamine-pretreated (lower line) and control striatal slices (upper line). Slices were incubated for 30 min in the presence or absence of 100 μM dopamine, washed five times, and preincubated in the presence of buffer (basal) and 50 μM dopamine for an additional 5 min, after which cyclic AMP accumulation was measured. Each value is an average of four determinations ±SD *($p < 0.01$).

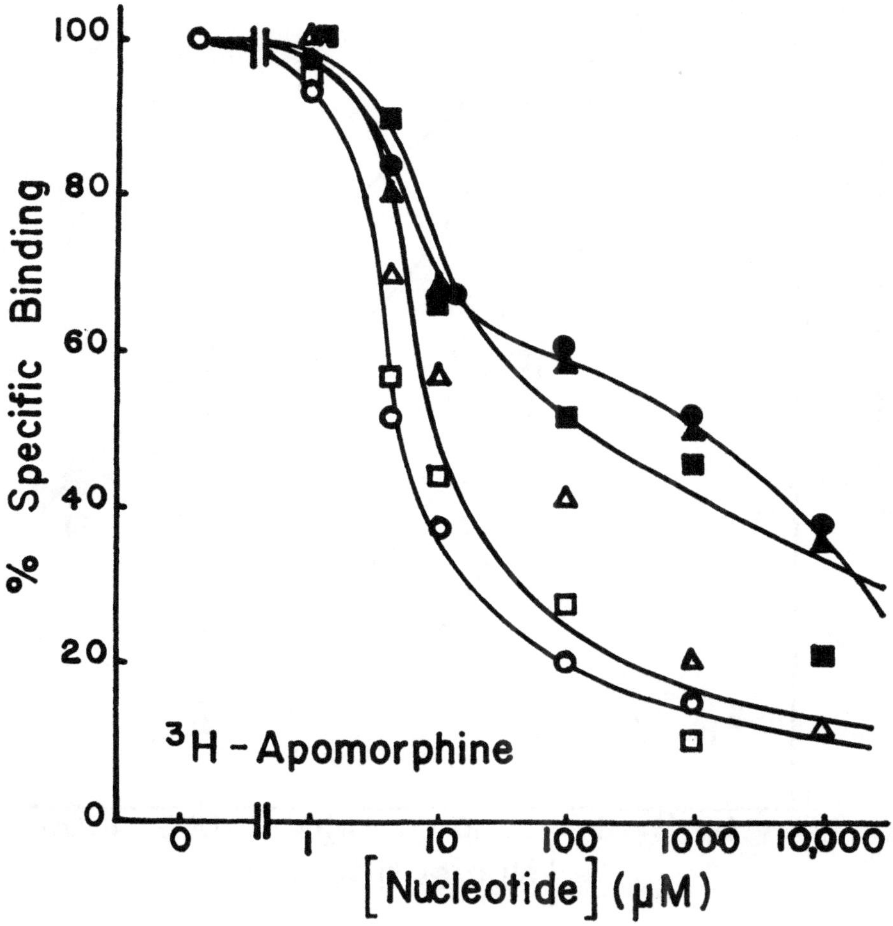

Fig. 8. Effect of increasing concentrations of GTP (□), GDP (△), Gpp(NH)p (○), GMP (▲), ATP (●), and App(NH)p (■) on the specific binding of ^{3}H-apomorphine to rat striatal membranes. Values are expressed as a percentage of the specific binding obtained in the absence of nucleotide and are from three replicate experiments (adapted with the permission of the publisher, from Mishra and Cleghorn, 1980).

centrations tested (Fig. 10). This inhibition was also observed in the absence of DA (Fig. 10), although the higher concentrations (10 and 100 μ*M*) caused a stimulation of basal adenylate cyclase after approximately 5 min of incubation at 30°C. An analysis of variance and covariance, including repeated measurements of pooled data from two separate experiments, indicated that the observed differences between Figs. 10 and 11 are highly significant ($p < 0.005$).

The results of the adenylate cyclase assays indicate that the

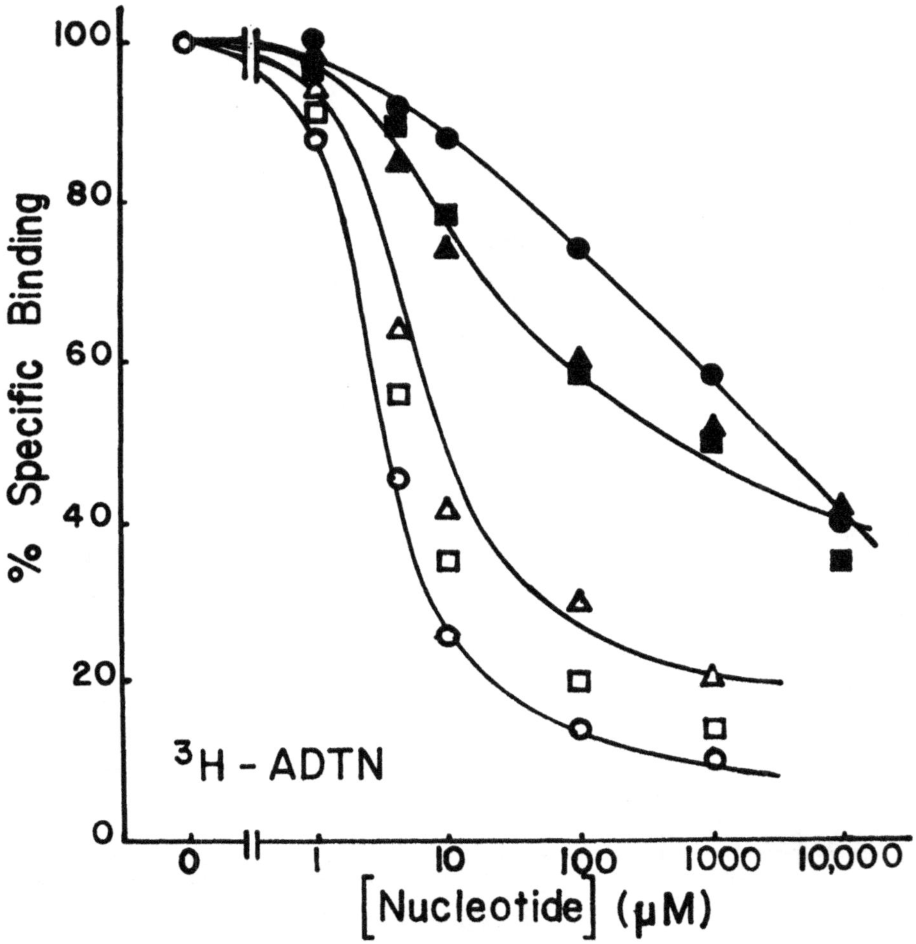

Fig. 9. Effect of increasing concentrations of GTP (□), GDP (△), Gpp(NH)p (○), GMP (▲), ATP (●), and App(NH)p (■) on the specific binding of ^{3}H-ADTN to rat striatal membranes. Values are expressed as a percentage of the specific binding obtained in the absence of nucleotide and are from three replicate experiments (adapted with the permission of the publisher, from Mishra and Cleghorn, 1980).

mechanism of action of guanine nucleotides on this system is highly complex. At first glance, the inhibition of DA-sensitive adenylate cyclase could be attributed to the inhibitory effect on DA receptor binding. However, there are several observations that tend to obscure such a mechanism:

1. The Gpp(NH)p was inhibitory, at the lower concentration and at the early timepoints for the higher concentrations, on the basal enzyme activity. This

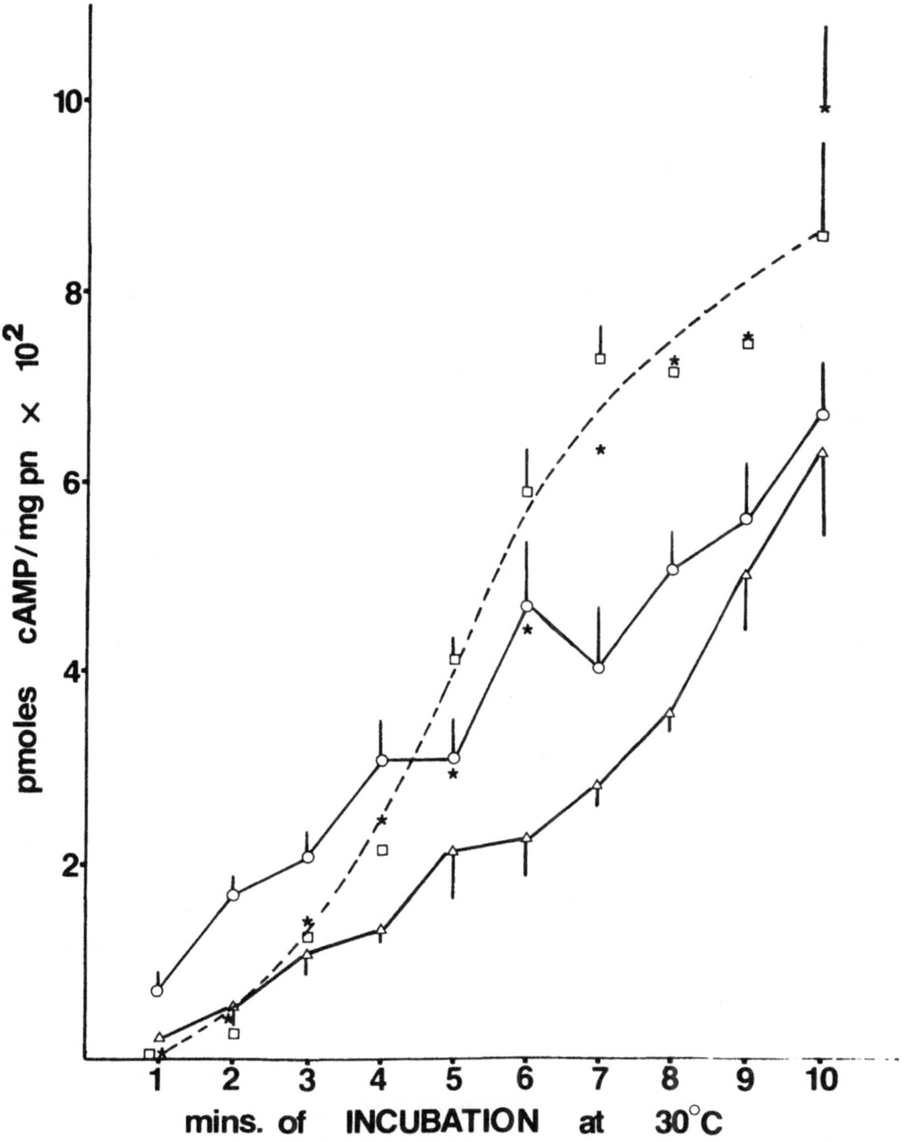

Fig. 10. Time-course of the effects of 1 μ*M* (△), 10 μ*M* (*), and 100 μ*M* (□) Gpp(NH)p on adenylate cyclase activity in striatal slices in the presence of 50 μ*M* dopamine. The stimulatory effect of 50 μ*M* dopamine alone is also shown (0). Each data point is an average of six determinations ±SEM. An analysis of variance with repeated measures was performed on a HP 2000 series 2 computer using the BMDP 1977 Biomedical package.

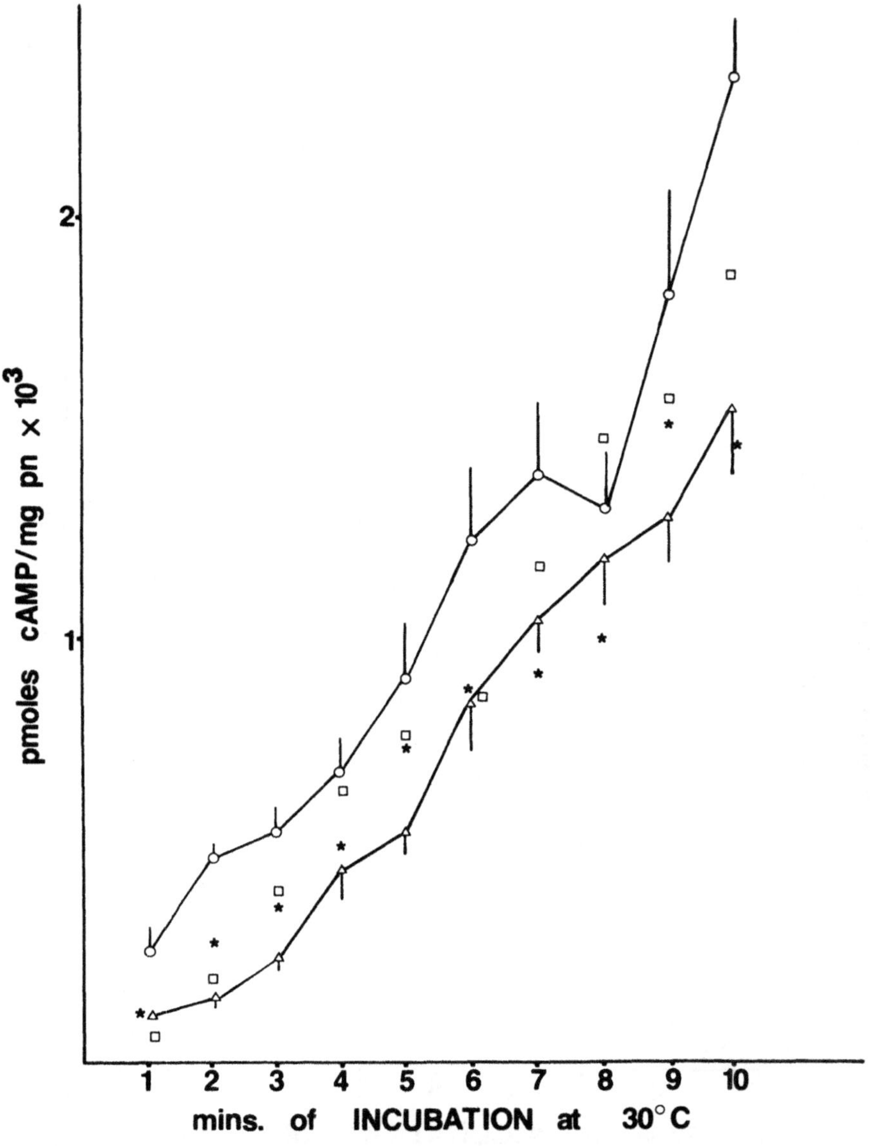

Fig. 11. Effects of guanine nucleotide on adenylate cyclase activity in striatal slices over time; (0) basal activity, (△) 1 μM Gpp(NH)p, (*) 10 μM Gpp(NH)p, (□) 100 μM Gpp(NH)p. Adenylate cyclase activity was determined as described in the text. Each data point is an average of six determinations ±SEM.

suggests that the responsiveness of the enzyme *alone* may be responsible for the observed results.

2. The Gpp(NH)p was inhibitory at *all* concentrations and times (up to 10 min) on the DA-sensitive adenylate cyclase. Thus, although the inhibition of the basal activity may account for some of this inhibition, some other mechanism is also introducing an inhibitory effect.
3. The lack of dose-responsiveness in the inhibition of DA-sensitive adenylate cyclase by Gpp(NH)p suggests that the mechanism whereby it interacts with the cyclase is different from its interaction with the receptor, which does respond in a dose-dependent manner. However, this may simply reflect a combination of a stimulation of the enzyme coupled with an inhibition of the receptor binding, thus leading to the intermediary effect of the higher concentrations of the Gpp(NH)p on the DA-sensitive adenylate cyclase.

Interpretation of these results is further obscured by the nature of Gpp(NH)p and its apparently irreversible activation of adenylate cyclase (Levitzky and Helmreich, 1979). In addition, there are differences between the responses to GTP and Gpp(NH)p, with the latter appearing to elicit responses more readily and the former requiring the presence of hormone (Londos and Rodbell, 1975) or the prior treatment of tissue with cholera toxin (Johnson and Bourne, 1977).

It is not clear from any studies published to date whether the guanine nucleotide effect on agonist binding serves a functional purpose or whether it is simply a side effect (artifact?) of guanine nucleotide interaction with adenylate cyclase. The fact that vastly different systems, which respond to guanine nucleotides at the adenylate cyclase level and agonists in different ways (i.e., inhibitory or stimulatory), all respond at the receptor level in a similar manner would seem to suggest a dichotomy of effects (*see* Table 6). Thus, several hypotheses can be formed:

1. Guanine nucleotides act on receptors and adenylate cyclases through two different mechanisms (and possibly two different components) that are linked but independent.
2. The effect of guanine nucleotides on receptors is a consequence of the effect on adenylate cyclase (i.e., by virtue of the close association between receptors

Table 6
Effect of Guanine Nucleotides on Various Receptor–Adenylate Cyclase Systems

Receptor	Tissue	Effect of guanine nuc. on receptor binding	Effect of guanine nuc. on adenylate cyclase	Effect of receptor on adenylate cyclase	Reference
Glucagon	Rat liver	↓	↑	↑	Rodbell et al., 1971a,b
Angiotensin II	Bovine adrenal cortex	↓	↑	↑ (?)	Glossmann et al., 1974
Opiate	Neuroblastoma × glioma cells	↓	↑	↓	Blume, 1978a
Opiate	Rat brain	↓	↓	none	Blume, 1978b Carenzi et al., 1978
α-Adrenoreceptor	Human platelets	↓	ND[a]	↓ PGE_1	Tsai and Lefkowitz, 1979
β-Adrenoreceptor	S-49 lymphoma cells	↓	↑	↓	Sternweiss and Gilman, 1979
Dopamine	Rat brain	↓	↓	↑	Clement-Cormier et al., 1978; Mishra et al., 1980

[a]ND = no data.

and adenylate cyclase and through some allosteric change in the receptor that causes more rapid dissociation of the agonist). This model would necessitate a separate receptor for antagonists.

3. The effect of the guanine nucleotides on adenylate cyclase is a consequence of the effect on receptors. This model would not explain, however, the effect of guanine nucleotides on adenylate cyclase in the absence of agonist.
4. Guanine nucleotides exert their effects on both receptors and cyclase through dynamic changes in the various components, as would be seen during the coupling process.

Lefkowitz's group (Lefkowitz and Williams, 1978; Wessels et al., 1979) favors the last hypothesis in their "dynamic receptor affinity" model. However, the DA receptor data presented here would not fit this model well because (a) DA does not appear to require GTP for activation (although this may be a result of compensation by ATP, which is at a concentration of 0.5 m*M* in the reaction mix); and (b) Gpp(NH)p appears to have a mixed inhibitory/stimulatory effect on the rat caudate adenylate cyclase, suggesting a more complex mechanism. Clearly, more detailed experiments are required to clarify some of the issues surrounding the GTP coupling component. It is hoped that studies of solubilized, purified components will provide some insight into this complex system.

Further, studies on the G-protein are needed in the light of recently published clinical results (Memo et al., 1983) showing changes in G-protein in schizophrenia. These authors have demonstrated that coupling of D-1 recognition sites with adenylate cyclase is more efficient in the brain of schizophrenic patients, perhaps because of an elevated affinity of the G-protein for the GTP.

6. Solubilization of Dopamine Receptors

Although there is a general consensus regarding the existence of D-1 (DA-stimulated adenylate cyclase coupled) and D-2 (non-cyclase coupled) receptors, there are still questions regarding the presence of two distinct receptor proteins in the membrane (Nielsen et al., 1984). In addition, the proposed D-3 and D-4 sites (Seeman, 1980, 1982) have been seriously questioned (Creese, 1982; Sibley et al., 1983). Physical separation of the various sites would clarify the issue of DA receptor subtypes. Recently, at-

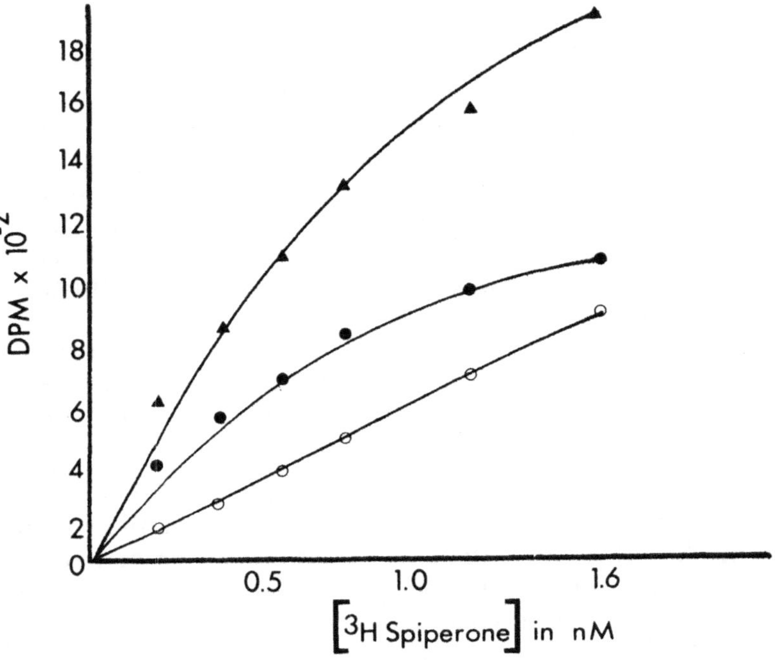
DPM x 10^-2
0
2
4
6
8
10
12
14
16
18
0.5
1.0
1.6
[3H Spiperone] in nM

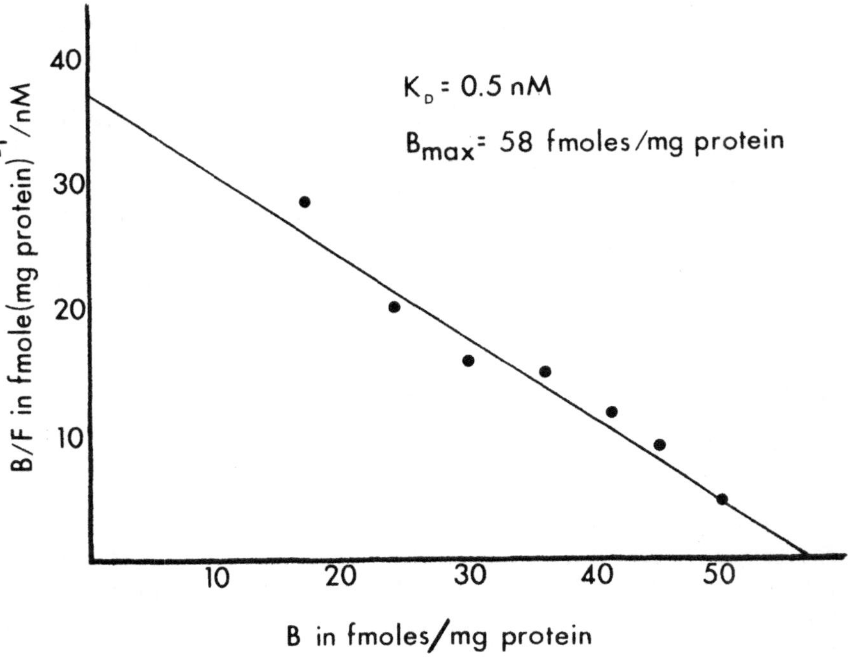
B/F in fmole(mg protein)^-1 /nM
10
20
30
40
K_D = 0.5 nM
B_max = 58 fmoles/mg protein
10
20
30
40
50
B in fmoles/mg protein

tempts have been made to solubilize DA receptors from DA-rich regions of the CNS for further purification.

In our laboratory, the DA D-2 receptor was solubilized using several solubilizing agents (cholic acid–ammonium sulfate, digitonin, lysolecithin, and a combination of cholic acid and sodium chloride) (*see* Ramwani and Mishra, 1983, for details). Table 7 shows the comparative solubilization of bovine D-2 receptor by different detergents or a combination of detergents. The cholic acid–NaCl combination yielded more than 30% recovery of the specific binding. The binding of solubilized D-2 receptor was saturable, stereospecific, and of high affinity. The representative results are shown in Fig. 12. This soluble preparation of D-2 receptors was subjected to further purification by haloperidol–sepharose affinity chromatography (Ramwani and Mishra, 1982, 1984).

The solubilized preparation was mixed with haloperidol–sepharose coupled gel (5 mg protein/mL of gel) in a flask and shaken for 1 h at room temperature. The slurry was then packed in a column (0.5 cm × 7 cm). The flow rate of the column was adjusted to 15–20 mL/h and bound receptors were eluted with several different agents (0.5*M* NaCl, 1 μ*M* spiroperidol, 10 μ*M* DA, 10 μ*M* haloperidol). Spiroperidol (1 μ*M*) or DA (10 μ*M*) resulted in efficient elution of bound receptors from the affinity gel. The receptor preparations eluted from the columns were assayed for [^{3}H]-spiroperidol binding as described by Ramwani and Mishra (1983). The purification achieved from this technique ranged between 2000- and 5000-fold. Further experiments are underway for detailed biochemical characterization.

7. Conclusion

The data and techniques presented in this manuscript indicate the usefulness of DA receptor binding studies in investigation of the detailed functioning of the dopaminergic mechanisms in the CNS. The data clearly support the existence of two different subtypes of DA receptors; whether the two different subtypes are two

Fig. 12. Top: The saturation binding curves from the solubilized bovine striatal dopamine receptors. ▲–▲ total binding; ●–● specific binding; 0–0 nonspecific binding. Bottom: Scatchard analysis of ^{3}H-spiroperidol binding in solubilized bovine receptors (adapted with the permission of the publishers from Ramwani and Mishra, 1983, copyright © 1983, Pergamon Press).

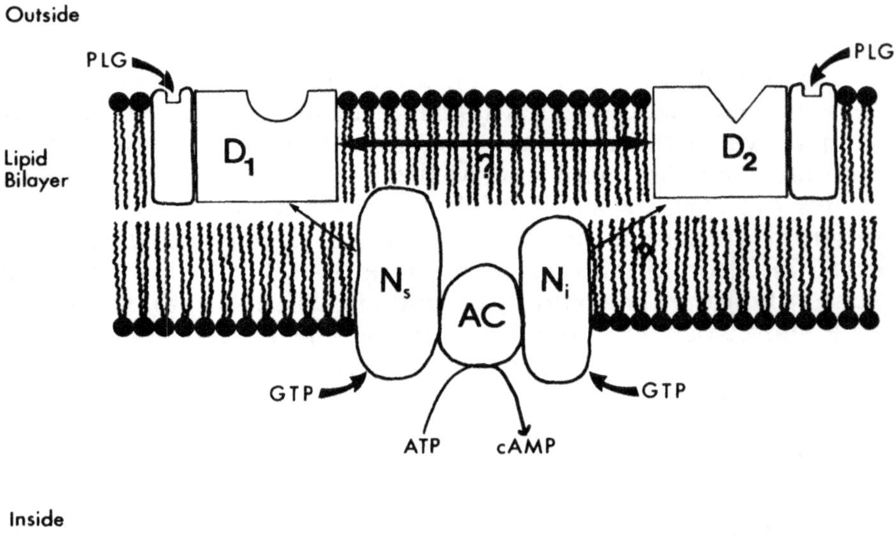

Fig. 13. Schematic representation of the D-1 and D-2 dopamine receptors on the striatal postsynaptic membrane. The dopamine stimulated-adenylate cyclase activity (D-1 receptor coupled) is exclusively located on postsynaptic neurons and is modulated by guanine nucleotide (adenylate cyclase) stimulatory protein (N_s). Other peptides such as L-prolyl-L-leucyl-glycinamide (PLG). The compound R-SK & F 38393 acts as an agonist and SCH 23390 acts as an antagonist. The D-2 dopamine receptor is probably influenced by the guanine nucleotide inhibitory protein (Ni) and PLG. *Abbreviations:* GTP, guanosine triphosphate; ATP, adenosine triphosphate; cAMP, cyclic adenosine 3',5' cyclic monophosphate; Ns, guanine nucleotide stimulatory subunit; Ni, guanine nucleotide inhibitory subunit; D-1, dopamine 1 receptor; D-2, dopamine 2 receptor; PLG, L-prolyl-L-leucyl-glycinamide.

distinct membrane proteins remains to be investigated. Further elucidation of the biochemical nature of the various components of the DA receptor complex (*see* Fig. 13), the production of monoclonal antibodies specific for DA receptors, and identification of the DA receptor gene(s) remain exciting and interesting topics for future research.

Acknowledgments

The work reported in this manuscript was supported by The Medical Research Council of Canada, The Ontario Mental Health

Table 7
Comparative Solubility of Dopamine (D_2) Receptors of Bovine Striatum by Various Detergents[a]

Detergent	% Specific binding	% Binding sites recovered
0.25% Cholic acid ± 1*M* NaCl	85	35
0.5% Cholic acid + 49% ammonium sulfate	65	15
1% Digitonin	30	8
1:5 Digitonin: cholic acid	29	12
0.25% Cholic acid	25	8

[a]Values shown are the means of triplicate determination for 2–4 separate experiments. ±SEM varied less than 10% (adapted with the permission of the publisher from Ramwani and Mishra, 1983, Copyright © 1983, Pergamon Press).

Foundation, The Parkinson Foundation of Canada, Abstainer's Life Insurance, and National Institute of Health (NIH) USA. I wish to thank Ms. Marjorie Cullen for valuable advice and constructive criticism, and Sara DeSilvio for skillfully typing the manuscript.

References

Barbeau A. (1976) Parkinson's Disease: Etiological Considerations in Basal Ganglia (Yarhr M. D., ed.) pp. 281–289 Raven New York.

Blume A. (1978a) Interaction of ligands with the opiate receptors of brain membranes: regulation by ions and nucleotides. *Proc. Natl. Acad. Sci. USA* **75,** 1713–1717.

Blume A. (1978b) Opitate binding to membrane preparations of neuroblastoma × glioma hybrid cells NG108–15: effects of ions and nucleotides. *Life Sci.* **22,** 1843–1852.

Brown E. M. and Aurbach G. D. (1982) Receptors and Second Messengers in Cell Function and Clinical Disorders, in *Contemporary Metabolism.* Vol. 2, (Freinkel N., ed.), pp. 247–299, Plenum, New York.

Carenzi A., Frigeni V., and Della Bella D. (1978) Synaptic Localization of Opiate Receptors in Rat Striatum, in *Advances in Biochemical Psychopharmacology.* Vol. 18 (Costa E. and Trabucchi T., eds.) pp. 265–270, Raven, New York.

Carlsson A. (1959) The occurrence, distribution and physiological role of catecholamines in the nervous system. *Pharmacol. Rev.* **11,** 490–493.

Chiu S., Paulose C. S., and Mishra R. K. (1981) Neuroleptic drug-induced dopamine receptor supersensitivity: antagonism by prolyl-leucyl-glycinamide. *Science* **214,** 1261–1262.

Chiu S., Wong Y. W., Ferris J. A., Johnson R. L., and Mishra R. K. (1983) Binding studies of L-propyl-leucyl-glycinamide (PLG), a novel anti-parkinsonian agent, in normal human brain. *Pharmacol. Res. Commun.* **15,** 41–51.

Clement-Cormier Y., Rudolph F. B., and Robinson G. A. (1978) Dopamine-sensitive adenylate cyclase from the rat caudate nucleus: regulation by guanyl nucleotides and the interaction of magnesium and magnesium ATP. *J. Neurochem.* **30,** 1163–1172.

Clements-Jewery S. and Robson P. A. (1981) Are striatal (^{3}H)-spiperone radioreceptor assays indicative of multiple dopamine receptor interactions? *Neuropharmacology* **20,** 1295–1298.

Conner J. D. (1970) Caudate nucleus neurons: correlation of the effects of substantia nigra stimulation with iontophoretic dopamine. *J. Physiol.* (Lond.) **208,** 691–703.

Cools A. R. and Van Rossum J. M. (1976) Excitation-mediating and inhibition-mediating dopamine receptors: A new concept towards a better understanding of electrophysiological, biochemical, pharmacological, functional and clinical data. *Psychopharmacology* **45,** 243–254.

Costall B. and Naylor R. J. (1981) The hypothesis of different dopamine receptor mechanisms. *Life Sci.* **28,** 215–224.

Cote T. E., Grewe C. W., and Kebabian J. W. (1981) Stimulation of the D-2 dopamine receptor in the intermediate lobe of the rat pituitary gland decreases the responsiveness of the beta-adrenoceptor: biochemical mechanism. *Endocrinology* **108,** 420–426.

Creese I. (1982) Dopamine receptors explained. *Trends Neurosci.* **5,** 40–43.

Creese I. and Sibley D. R. (1982) Comments on the commentary by Dr. Seeman. *Biochem. Pharmacol.* **31,** 2568–2569.

Creese I., Burt D. R., and Snyder S. H. (1975) Dopamine receptor binding: differentiation of agonist and antagonist states with ^{3}H-dopamine and ^{3}H-haloperidol. *Life Sci.* **17,** 933–1001.

Crow T. J. (1980) Molecular pathology of schizophrenia: more than one disease process? *Br. Med. J.* **280,** 66–68.

De Camilli P., Macconi D., and Spado A. (1979) Dopamine inhibits adenylate cyclase in human prolactin-secreting pituitary adenomas. *Nature* (Lond.) **278,** 252–254.

DeLean A., Kilpatric B. F., and Caron M. C. (1982) Guanine nucleotides regulate both dopaminergic agonist and antagonist binding in porcine anterior pituitary. *Endorcinology* **110,** 1064–1066.

Freedman S. B., Poat J. A., and Woodruff G. N. (1981) (^{3}H)-Sulpride, a ligand for neuroleptic receptors. *Neuropharmacology* **20,** 1323–1326.

Freychet P., Roth J., and Neville D. M. (1971) Insulin receptors in the liver: specific binding of (^{125}I)-insulin to the plasma membrane and its relation to insulin bioactivity. *Proc. Natl. Acad. Sci. USA* **68,** 1833–1838.

Gale J. S., Bird E. D., Spokes E. G., Iversen L. L., and Jessel T. M. (1978) Human brain substance P: distribution in controls and Huntington's chorea. *J. Neurochem.* **30,1 633–634.**

Gilman A. G. (1970) A protein binding assay for adenosine 3' : 5'-cyclic monophosphate. *Proc. Natl. Acad. Sci. USA* **67,** 305–312.

Glossmann H., Baukal A., and Catt K. (1974) Angiotensin II receptors in bovine adrenal cortex. Modification of angiotensin II binding by guanyl nucleotides. *J. Biol. Chem.* **249,** 664–666.

Gnegy M., Treisman G., and Bernabei A. (1982) Chronic sulpiride treatment produces supersensitivity of striatal adenylate cyclase to dopamine in sexually immature or adult castrated rats. *J. Pharmacol. Exp. Ther.* **224,** 627–633.

Goldstein A., Lowney O., and Pal B. K. (1971) Stereospecific and nonspecific interactions of the morphine congener levorphanol in subcellular fractions of mouse brain. *Proc. Natl. Acad. Sci.* USA **68,** 1742–1747.

Gullick W. and Lindstrom J. (1983) Comparison of the subunit structure of acetylcholine receptors from muscle and electric organ of *Electrophorus electricus. Biochemistry* **22,** 3801–3807.

Haggendal J. and Malmfors T. (1965) Identification and cellular localization of the catecholamines in the retina and choroid of the rabbit. *Acta Physiol. Scand.* **64,** 58–66.

Hays S. E., Meyer D. K., and Paul S. M. (1981) Localization of CCK receptors on neuronal elements in rat caudate nucleus. *Brain Res.* **219,** 208–213.

Innis R. B., Correa I. M., Uhl G. R., Schneider B., and Snyder S. H. (1977) CCK-octapeptide-like immunoreactivity: histochemical localization in the rat brain. *Proc. Natl. Acad. Sci. USA* **76,** 521–525.

Johnson G. and Bourne H. (1977) Influence of cholera toxin on the regulation of adenylate cyclase by GTP. *Biochem. Biophys. Res. Commun.* **78,** 792–798.

Kastin A. J., Sandman C. A., Miller L. H., and Schally A. V. (1976) Some questions related to melanocyte-stimulating hormone. *Mayo Clin. Proc.* **51,** 632–636.

Kebabian J. (1978) Dopamine-Sensitive Adenylyl Cyclase: A Receptor Mechanism for Dopamine. *Advances in Biochem. Psychopharmacol.* Vol. **19,** (Roberts P. J., Woodruff G. N., Iverson L. L., eds.) pp. 131–154, Raven, New York.

Kebabian J. W. and Calne D. B. (1979) Multiple receptors for dopamine. *Nature* (Lond.) **277,** 93–97.

Keely J. S. and Dodd J. (1981) CCK and Gastrin as Transmitters in the Mammalian Central Nervous System. *Adv. Biochem. Psychopharmacol.* (Martin J., Reichlim S., Bick K. L., eds.) **28,** 133–144. Raven, New York.

Laduron P. (1980) Dopamine receptor: From an in vivo concept towards a molecular characterization. *Trends Pharmacol. Sci.* **2,** 471–474.

Lefkowitz R. and Williams L. (1978) Molecular mechanisms of activation and desensitization of adenylate cyclase-coupled β-adrenergic receptors. *Advances in Cyclic Nucleotide Research.* Vol. 9, (George W. J. and Ignarro L. J., eds.) pp. 1–17, Raven, New York.

Lefkowitz R., Mukkerjee C., Coverstone M., and Caron M. (1974) Stereospecific (^{3}H)-(−)-alprenolol binding sites, beta-adrenergic receptors and adenylate cyclase. *Biochem. Biophys. Res. Commun.* **60,** 703–709.

Lesniak M., Roth J., Gordon P. and Gavin J. (1973) Human growth hormone radioreceptor assay using cultured human lymphocytes. *Nature* (Lond.) **241,** 20–23.

Levitzki A. and Helmreich E. (1979) Hormone-receptor-adenylate cyclase interactions. *FEBS Lett.* **101,** 213–219.

Londos C. and Rodbell M. (1975) Multiple inhibitory and activating effects of nucleotides and magnesium on adrenal adenylate cyclase. *J. Biol. Chem.* **250,** 3459–3465.

Memo M., Kleinman J., and Hanbauer I. (1983) Coupling of dopamine D-1 recognition sites with adenylate cyclase in nuclei accumbens and caudatus of schizophrenics. *Science* **221,** 1304–1307.

Meunier H. and Labrie F. (1982) The dopamine receptor in the intermediate lobe of the rat pituitary gland is negatively coupled to adenylate cyclase. *Life Sci.* **11,** 963–968.

Mishra R. K. (1983) Modulation of CNS dopamine receptors by peptides. *Prog. Neuro-psychopharmacol. Biol. Psychiat.* **6,** 437–441.

Mishra R. K., Wong Y.-W., Varmuza S., and Tuff L. (1978) Chemical lesion and drug induced supersensitivity and subsensitivity of caudate dopamine receptors. *Life Sci.* **23,** 443–446.

Mishra R. K. and Cleghorn J. M. (1980) Dopamine Receptors in the Central Nervous System: Basic and Clinical Neuroendocrine Studies, in *Recent Advances in Canadian Neuro-psychopharmacology* (Grof P. and Saxena B., eds.) pp. 50–65, S. Karger Press, Basle.

Mishra R. K., Marshal A. M., and Varmuza S. L. (1980) Supersensitivity in rat caudate nucleus: effects of 6-hydroxydopamine on the time course of dopamine receptor and cyclic AMP changes. *Brain Res.* **200,** 47–57.

Mishra R. K., Chiu S., Chiu P., and Mishra C. P. (1983) Pharmacology of prolyl-Leucyl-glycinamide. *Methods and Findings in Exptl. and Clin. Pharmacol.* **5,** 203–233.

Munemura M., Eskay R. L., and Kebabian J. W. (1980) Release of alpha-melanocyte-stimulating hormone from dispersed cells of the intermediate lobe of the rat pituitary gland: involvement of catecholamines and adenosine 3'-5'-monophosphate. *Endocrinology* **106,** 1795–1803.

Nielsen M., Klimek V., and Hyttel J. (1984) Distinct target size of dopamine D-1 and D-2 receptors in rat striatum. *Life Sci.* **35,** 325–332.

Offermeier J. and Van Rooyen J. M. (1982) Is it possible to integrate dopamine receptor terminology? *Trends Pharmacol. Sci.* **3,** 326–328.

Pert C. and Snyder S. H. (1973) Opiate receptor: demonstration in nervous tissue. *Science* **179,** 1011–1014.

Price K. S., Farley I. J., and Hornykiewicz O. (1978) Neurochemistry of Parkinson's Disease: Relation Between Striatal and Limbic Dopamine. *Advances in Biochem. Psychopharmacol.* Vol. 19 (Robert P. J., Woodruff G. N., Iversen L. L., eds.) pp. 293–300, Raven, New York.

Ramwani J. and Mishra R. K. (1982) Purification of dopamine (D2) receptor by affinity chromatography. *Fed. Proc.* **41,** 1325.

Ramwani J. and Mishra R. K. (1983) A simple and efficient method for solubilization of the dopamine receptor. *Prog. Neuro-psychopharmacol. Biol. Psychiat.* **7,** 769–772.

Ramwani J. and Mishra R. (1984) Purification and characterization of dopamine D-2 receptor (submitted for publication).

Rinne U. KL., Sonnienen V., and Mortila R. (1977) Brain Dopamine Metabolism and Relief of Parkinsonism, in *Parkinson's Disease—Concepts and Prospects.* (Lakke J. P. W., Krof J., and Wessling H., eds.) pp. 73–83, Excerpta Medica, New York.

Rodbell M., Birnbaumer L., Pohl S., and Krans H. (1971a) The glucagon-sensitive adenyl cyclase system in plasma membranes of rat liver. An obligatory role of guanyl nucleotides in glucagon action. *J. Biol. Chem.* **246,** 1877–1882.

Rodbell M., Krans H., Pohl S., and Birnbaumer L. (1971b) The glucagon-sensitive adenyl cyclase system in plasma membranes of rat liver. IV. Effects of guanylnucleotides on binding of ^{125}I-glucagon *J. Biol. Chem.* **246,** 1872–1876.

Seeman P., Chau-Wong M., and Lee T. (1975) Brain receptors for antipsychotic drugs and dopamine: Direct binding assays. *Proc. Natl. Acad. Sci. USA* **72,** 4376–4380.

Seeman P. (1980) Brain dopamine receptors. *Pharmacol. Rev.* **32,** 229–313.

Seeman P. (1982) Nomenclature of central and peripheral dopaminergic sites and receptors. *Biochem. Pharmacol.* **31,** 2563–2568.

Sibley D. R., DeLean A., and Creese I. (1982) Anterior pituitary dopamine receptors: Demonstration of interconvertible high and low affinity states of the D-2 dopamine receptor. *J. Biol. Chem.* **257,** 6351–6361.

Sibley D., Mahan L. C., and Creese L. (1983) Dopamine receptor binding on intact cells: Absence of a high-affinity agonist-receptor binding state. *Mol. Pharmacol.* **23,** 295–302.

Sternweiss P. and Gilman A. (1979) Reconstitution of catecholamine-sensitive adenylate cyclase. Reconstitution of the uncoupled variant of the S40 lymphoma cell. *J. Biol. Chem.* **254,** 3333–3340.

Stoof J. C. and Kebabian J. W. (1981) Opposing roles for D-1 and D-2 dopamine receptors in efflux of cyclic AMP for rat neostriatum. *Nature* (Lond.) **294,** 366–368.

Stoof J. C. and Kebabian J. W. (1984) Two dopamine receptors: Biochemistry, physiology and pharmacology. *Life Sci.* **35,** 2281–2296.

Terenius L. (1973) Characteristics of the "receptor" for narcotic analgesics in synaptic plasma membrane fraction from rat brain. *Acta Pharmacol. Toxicol.* **33,** 377–384.

Titeler M., List S., and Seeman P. (1979) High affinity dopamine receptors (D_3) in rat brain. *Commun. Psychopharmacol.* **3,** 411–420.

Tsai B. and Lefkowitz R. (1979) Agonist-specific effects of guanine nucleotides on alpha-adrenergic receptors in human platelets. *Mol. Pharmacol.* **16,** 61–68.

Wessels M., Mullikin D., and Lefkowitz R. (1979) Selective alteration in high affinity agonist binding: A mechanism of β-adrenergic receptor desensitization. *Mol. Pharmacol.* **16,** 10–20.

Williams L. and Lefkowitz R. J. (1976) Alpha-adrenergic receptor identification by (^{3}H)-dihydroergocrytine binding. *Science* **192,** 791–793.

Woodruff G. N., McCarthy P. S., and Walker R. J. (1976) Studies on the pharmacology of neurones in the nucleus accumbens of the rat. *Brain Res.* **115,** 233–242.

Chapter 3
Adrenergic Receptors

John Marvin May and Kenneth P. Minneman

1. Introduction

Epinephrine (adrenaline, A) and norepinephrine (noradrenaline, NA) are naturally occurring catecholamines that are important in many physiological processes. Adrenaline is found in high concentrations in the adrenal medulla, where it is released into the blood to function as a hormone. Noradrenaline is the major neurotransmitter of postganglionic sympathetic nerves and is particularly important in control of cardiovascular function. Both catecholamines also control metabolic processes in many glands and muscles. In addition, both A and NA are found in specific neurons of the central nervous system, where they function as neurotransmitters and play important roles in a variety of integrative functions, such as control of autonomic outflow, mood, and appetite.

Adrenaline and NA cause their effects by binding reversibly to receptors located on the external surface of target cells. Since A (and in some species, NA) is found in high concentrations in the adrenal medulla, these receptors are called adrenergic receptors. The effects of A and NA have been studied for many years and an extensive literature exists concerning the receptors through which these compounds cause their effects. In recent years, the methodology for studying adrenergic receptors has become increasingly sophisticated, and has led to an increased understanding of the actions of these compounds and other sympathomimetic drugs, as well as the mechanisms underlying these actions.

2. Classification

Early investigations on the actions of sympathomimetic drugs demonstrated that some drugs could produce either excitatory or inhibitory responses, depending on the tissue under study. Dale (1906) found that after administration of certain ergot alkaloids, the response to intravenously injected A changed from an increase in vascular resistance to a decrease in vascular resistance. Furthermore, not all of the responses to sympathetic nerve stimulation were the same as those produced by injected A. These differences led Cannon and Rosenblueth (1937) to propose that there were two sympathetic mediator substances, an excitatory compound that they called sympathin E, and an inhibitory substance called sympathin I.

Clarification of the paradoxical actions of certain sympathomimetic drugs on different systems did not occur until later, when Ahlquist (1948) demonstrated that there were two types of receptors, rather than two mediators, that were responsible for the different responses. Ahlquist noted that there were differences in the orders of potency of sympathomimetic amines in producing different responses. The order of potency of these drugs in activating smooth muscle contraction and inhibiting intestinal motility was A > NA > α-methylnoradrenaline > α-methyladrenaline > isoproterenol. The same drugs inhibited smooth muscle contraction and increased cardiac contractility with a potency order of isoproterenol > A > α-methyladrenaline > α-methylnoradrenaline > NA. Ahlquist (1948) proposed that the receptors responsible for the former responses be called α-adrenergic receptors, and the receptors responsible for the latter responses be called β-adrenergic receptors.

Support for Ahlquist's proposal came with the introduction of the non-equilibrium adrenergic blocking drugs of the β-haloalkylamine type (Nickerson, 1949), such as dibenamine and phenoxybenzamine. These drugs blocked the responses mediated by the receptors that Ahlquist had called α-adrenergic receptors, but did not block the responses mediated by the receptors he had called β-adrenergic receptors. Conclusive evidence for the existence of at least two distinct types of adrenergic receptor was obtained when the dichloro derivative of isoproterenol was introduced by Slater and Powell (1957). Moran and Perkins (1958) showed that dichloroisoproterenol was a competitive antagonist of responses that Ahlquist had proposed to be mediated by β-adrenergic receptors, and suggested that such drugs be called β-adrenergic receptor blocking drugs.

2.1. β-Adrenergic Receptor Subtypes

In the 1960s it became clear that the properties of β-adrenergic receptors were not identical in all tissues. Evidence for the existence of more than one kind of β-adrenergic receptor came with the observation that α-methyl derivatives of methoxamine blocked some, but not all, of the responses attributed to stimulation of β-adrenergic receptors (Burns et al., 1964; Levy, 1966). Careful characterization of the responses of different tissues to β-adrenergic receptor stimulation was performed by Lands et al. (1967), who demonstrated a difference in the sensitivity of β-adrenergic receptors in different tissues to NA. β-Adrenergic receptors in some tissues were equally sensitive to both A and NA, whereas in other tissues A was much more potent than NA. Lands et al. (1967) proposed that the β-adrenergic receptors stimulated equipotently by A and NA be classified as β_1-adrenergic receptors, and the β-adrenergic receptors at which A was much more potent than NA be classified as β_2-adrenergic receptors. The development of drugs that were selective agonists or antagonists at β_1- or β_2-adrenergic receptors (Brittain et al., 1968; Dunlop and Shanks, 1968) provided more conclusive evidence for the existence of two distinct receptor subtypes. As the technology for direct labeling of β-adrenergic receptors using radioligand binding assays advanced, the distribution of β-adrenergic receptor subtypes in various tissues was determined using radioligand binding techniques (Barnett et al., 1978; Minneman et al., 1979c). It is now clear that the two different subtypes of β-adrenergic receptor can coexist in a single tissue (Barnett et al., 1978; Minneman et al., 1979b) and on a single cell (Homburger et al., 1981), although the physiological implications of this coexistence is not yet clear (*see* section 8).

2.2. α-Adrenergic Receptor Subtypes

Further evidence on the subclassification of adrenergic receptors was obtained during the early 1970s when it was observed that some α-adrenergic receptors were more sensitive than others to irreversible blockade by phenoxybenzamine (Cubeddu et al., 1974). The α-adrenergic receptors less sensitive to phenoxybenzamine inhibition were those that influenced the release of neurotransmitters from presynaptic nerve terminals. Langer (1974) suggested that postsynaptic α-adrenergic receptors causing smooth muscle contraction might have different pharmacological characteristics than presynaptic α-adrenergic receptors affecting neurotransmitter release. He proposed that these receptors be classified as α_1- and α_2-adrenergic receptors, respectively. Although the concept of two distinct subtypes of α-adrenergic receptor gained experimental support over the next few years, it soon became clear that α_2-adrenergic receptors could also exist on tissue elements other than presynaptic nerve terminals. Evidence accumulated that these receptors could also exist postsynaptically, modulating processes such as renin release and contraction of smooth muscle (Berthelsen and Pettinger, 1977; Timmermans et al., 1979; Langer et al., 1980; Docherty and McGrath, 1980; Drew, 1980). For this reason, pharmacological criteria for the classification of α-adrenergic receptors based on the potencies of drugs in stimulating or inhibiting these receptors was developed. The α-adrenergic receptor-selective agonist drugs methoxamine and phenylephrine are more potent at α_1-adrenergic receptors, whereas α-methylnoradrenaline and clonidine are more potent at α_2-adrenergic receptors (Starke, 1981).

The development of selective antagonists of α-adrenergic receptor subtypes provided more evidence for distinct receptor classes. Prazosin (Cambridge et al., 1977) was shown to be very selective in antagonizing α_1-adrenergic receptors, and yohimbine was shown to be relatively selective in antagonizing the α_2-adrenergic receptor subtype (Starke et al., 1975). Other antagonists selective for α_1-(indoramin, ARC 239) or α_2-(piperoxan) adrenergic receptors have also been developed.

Given our current understanding of the multiple subtypes of adrenergic receptors, their coexistence in different tissues, and their heterogeneous distribution on different tissue elements, the confusion of the early adrenergic pharmacologists over the effects produced by sympathomimetic drugs is understandable. By using differences in the rank order of potencies of agonists in producing responses and selective antagonism of responses mediated by a particular subtype, four distinct types of adrenergic receptors have been defined. It is becoming increasingly clear that each of these receptors has a different molecular structure and can exist independently of the other subtypes, or can coexist with any or all of the other subtypes, depending on the particular tissue or cell under study.

3. Radioligands for Direct Measurements

Until recently, all data concerning the binding properties of adrenergic receptors have been derived from indirect physiological measurements of the properties of a receptor-mediated response in isolated tissues. The method developed by Arunlakshana and Schild (1959) provided a tool for determining the binding constants of competitive antagonists at receptors in isolated tissue preparations. A method developed by Furchgott (1966) allowed the determination of binding constants for agonists at receptors in isolated tissue preparations, although this method requires the availability of suitable irreversible receptor-blocking drugs that are not available for all receptors. These measurements do not allow quantitation of receptor density in tissues. Labeling receptors directly with radioactive compounds with a high specificity and specific activity has added a powerful tool for investigating the recognition properties and tissue density of these molecules, as well as the initial events involved in signal transduction from the cell-surface receptor into the cell.

3.1. β-Adrenergic Receptors

The β-adrenergic receptor was one of the first receptors to be successfully labeled using radioligand binding assays. Early attempts using ^{3}H-labeled catecholamines were generally unsuccessful, since receptor binding could not always be separated from binding to metabolic enzymes and nonspecific binding of metabolic degradation products of catecholamines (Cuatrecasas et al., 1974; Wolfe et al., 1974). Other investigators (Levitzki et al., 1975; Nahorski, 1976) successfully used the antagonist radioligand ^{3}H-propranolol to label β-adrenergic receptors. Propranolol is a very lipophilic molecule, however, and nonreceptor (nonspecific) binding of this compound is fairly high, making it difficult to work with. The development of the antagonist radioligand ^{3}H-dihydroalprenolol by Lefkowitz et al. (1974) provided a useful tool for quantification of β-adrenergic receptors in a number of tissues. The introduction of radioiodinated compounds such as ^{125}I-iodohydroxybenzylpindolol (Aurbach et al., 1976), which have a much higher specific activity (2200 Ci/mmol) than tritiated drugs (10–80 Ci/mmol), permits quantitation of receptors in smaller amounts of tissue. A number of radioligands have been developed for labeling β-adrenergic receptors, and many of those in current use are listed in Table 1.

3.2. α-Adrenergic Receptors

Table 1 also lists compounds that are commonly used to label α-adrenergic receptors with radioligand binding techniques. The ergot derivative ^{3}H-dihydroergocryptine (Williams et al., 1976), an antagonist drug that may possess some intrinsic sympathomimetic activity (Innes, 1962), was the first compound to be used successfully to directly label α-adrenergic receptors. Other antagonists, such as ^{3}H-prazosin (Greenglass and Bremner, 1979) and ^{3}H-yohimbine (Tharp et al., 1981; Rouot et al., 1982), as well as several agonist radioligands, have also been used to label α-adrenergic receptors successfully in various tissues (Table 1).

Table 1
Radioligands Commonly Used To Label Adrenergic Receptors

Radioligand	Subtype selectivity	Agonist/ antagonist	References
β-*Adrenergic*			
^{3}H-dihydroalprenolol	None	Antagonist	Lefkowitz et al., 1974
^{125}I-hydroxybenzylpindolol	None	Antagonist	Aurbach et al., 1976
^{125}I-cyanopindolol	None	Antagonist	Engel et al., 1981
^{125}I-pindolol	None	Antagonist	Barovsky and Brooker, 1980
^{3}H-hydroxybenzylisoproterenol	None	Agonist	Lefkowitz and Williams, 1977
α-*Adrenergic*			
^{3}H-clonidine	α_2	Agonist	Greenberg et al., 1976
^{3}H-*para*-aminoclonidine	α_2	Agonist	Rouot and Snyder, 1979

^{3}H-epinephrine	Not clear	Agonist	U'Prichard and Snyder, 1977
^{3}H-norepinephrine	Not clear	Agonist	U'Prichard and Snyder, 1977
^{3}H-dihydroergocryptine	None	Antagonist	Williams et al., 1976
^{3}H-WB 4101	Controversial	Antagonist	Greenberg et al., 1976
^{3}H-yohimbine	α_2	Antagonist	Tharp et al., 1981 Rouot et al., 1982
^{3}H-rauwolscine	α_2	Antagonist	Perry and U'Prichard, 1981 Cheung et al., 1982
^{3}H-prazosin	α_1	Antagonist	Barns et al., 1979 Hornung et al., 1979
^{125}I-BE 2254 (^{125}I-HEAT)	α_1	Antagonist	Engel and Hoyer, 1981 Glossman et al., 1981

Many of the radioligands used to label α-adrenergic receptors have different affinities for the α_1- and α_2-adrenergic receptor subtypes, and are used to selectively label only a single subtype. ^{3}H-Prazosin and ^{125}I-BE 2254 are antagonist radioligands specific for α_1-adrenergic receptors, whereas ^{3}H-yohimbine and its most selective isomer, ^{3}H-rauwolscine, are antagonist radioligands that selectively label α_2-adrenergic receptors. ^{3}H-Dihydroergocryptine labels both subtypes, and can also label dopamine and serotonin receptors in brain (Davis et al., 1977), confounding results obtained with this compound. ^{3}H-Clonidine and ^{3}H-*para*-aminoclonidine are agonist radioligands that bind preferentially to α_2-adrenergic receptors and are commonly used to label this receptor population. However, the complexity of agonist binding to α_2-adrenergic receptors, and the consequent uncertainty as to which particular form(s) of the receptor is being detected in filtration binding assays, makes data obtained with these compounds difficult to interpret quantitatively.

In contrast to radioligands for α-adrenergic receptors, the compounds used to measure β-adrenergic receptors generally do not distinguish between the β_1- and β_2-adrenergic receptor subtypes. Thus, differentiation of β-adrenergic receptor subtypes generally involves labeling the entire receptor population with a nonselective antagonist radioligand and using subtype-selective antagonists to displace the radioligand from its binding sites. Such data are then analyzed by computer-assisted curve fitting techniques to determine the ratio of β_1- and β_2-adrenergic receptor subtypes in a particular tissue (Minneman et al., 1979b,c; Hancock et al., 1979).

4. Kinetic and Equilibrium Binding Properties

One assumption of early receptor theory was that the binding of drugs to receptors is a simple noncooperative bimolecular reaction governed by the principles of mass action. Direct measurements of the binding of radioligands to receptors, or of the competition for the radioligand binding site by other drugs, have permitted a more detailed examination of the nature of the receptor–drug interaction and the effects of physiologically important ions and cofactors on the binding event.

4.1. Binding Properties of β-Adrenergic Receptors

The binding of antagonists to β-adrenergic receptors generally appears to be a bimolecular reaction obeying the law of mass action. Saturation isotherms of the binding of radioligands to β-adrenergic receptors analyzed by the method of Scatchard (1949) are usually linear, with Hill coefficients approximately equal to 1.0, suggesting a single class of non-interacting binding sites (Maguire et al., 1976; Minneman et al., 1980). Equilibrium dissociation constants (K_d) derived from dividing the backward rate constant k_{-1} (a first-order reaction) by the forward rate constant k_{+1} (a second-order process) agree well with the K_d obtained directly from equilibrium saturation analysis of antagonist binding. Antagonist binding has not generally been found to be affected by changes in the ionic composition of the incubation buffer, although Wolfe and Harden (1981) reported small effects of guanine nucleotides on the equilibrium binding constants of antagonists for β-adrenergic receptors in several tissues.

In contrast to the binding of antagonists, agonist interactions with β-adrenergic receptors are markedly dependent upon experimental conditions. Competition curves for the displacement of radiolabeled antagonists from β-adrenergic receptors by agonists are often shallow, with apparent Hill coefficients substantially less than one. Guanine nucleotides (GTP and nonhydrolyzable analogs) change the shape of agonist inhibition curves, resulting in Hill coefficients much closer to one, and also decrease the potencies of agonists in competing for the receptor site, often by 10–100-fold (Lefkowitz et al., 1976; Maguire et al., 1976).

Magnesium ions have an effect on agonist binding opposite from that of guanine nucleotides. The presence of magnesium increases the affinity of the β-adrenergic receptor for agonists (Maguire et al., 1976). In the absence of GTP, Mg^{2+} increases the apparent negative cooperativity of agonist binding, reflected in a decrease in the apparent Hill coefficient (Bird and Maguire, 1978; Williams et al., 1978). Some other divalent cations, such as Mn^{2+} and Ca^{2+}, will also produce this effect to a lesser extent, but other cations will not.

These phenomena were all discovered by examining the inhibition of the binding of antagonist radioligands by unlabeled agonist drugs. The use of the agonist radioligand ^{3}H-hydroxybenzylisoproterenol (^{3}H-HBI) has extended and confirmed these observations. The nonhydrolyzable GTP analog guanylimidodiphosphate facilitates the dissociation of ^{3}H-HBI from the receptor (Lefkowitz and Williams, 1977), although the presence of magnesium increases specific binding (Williams, et al., 1978). Scatchard analysis of the binding ^{3}H-HBI to rat lung tissue results in curvilinear plots that become linear if the assays are performed in the presence of sodium, guanine nucleotides, or magnesium, with either a decrease (sodium and guanine nucleotides) or increase (magnesium) in the apparent affinity of the receptor for the drug (Heidenreich et al., 1980).

These data are consistent with a ternary complex model in which agonist binding induces the formation of an agonist–receptor–effector (or regulatory component) complex (DeLean et al., 1980). Thermodynamic studies (Weiland et al., 1979) support this. Antagonist binding is entropy driven, whereas agonist binding proceeds with a change in the enthalpy of the system at the expense of a decrease in entropy. These results might be expected if agonists induce ordering of membrane components and antagonists do not. The effector molecule involved in β-adrenergic receptor-mediated signal transduction will be discussed in more detail in section 5.

4.2. Binding Properties of α-Adrenergic Receptors

The binding of drugs to α_2-adrenergic receptors is qualitatively similar to the binding of drugs to β-adrenergic receptors, although there are some differences. Guanine nucleotides decrease the affinity of agonists for α_2-adrenergic receptors (Tsai and Lefkowitz, 1978a,b; U'Prichard and Snyder, 1978), as does the presence of sodium (Limbird et al., 1982). Magnesium ions also increase the affinity of agonists in a manner similar to that seen with β-adrenergic receptors. However, effects of cations and guanine nucleotides on the binding of antagonists to α_2-adrenergic receptors are also seen. Sodium increases the binding (both the affinity and apparent number of sites) of ^{3}H-yohimbine (Limbird et al., 1982), whereas magnesium decreases binding (Rouot et al., 1982). As discussed above, under some circumstances similar effects of guanine nucleotides can be observed on the binding of antagonists to β-adrenergic receptors (Wolfe and Harden, 1981), although the effects are observed less commonly than with α_2-adrenergic receptors.

The similarity between the binding properties of β and α_2-adrenergic receptors probably involves their similar signal transduction mechanisms. Both types of receptors can produce changes in the activity of adenylate cyclase (EC 4.6.1.1), with β-adrenergic receptors activating, and α_2-adrenergic receptors inhibiting, the enzyme. Both receptors do this by interacting with guanine nucleotide regulatory proteins in the cell membrane, and it is the interaction of the receptors with this protein that confers guanine nucleotide sensitivity on agonist binding. The interaction of these receptors with the cyclic AMP generating system is discussed below.

Effects of guanine nucleotides and monovalent cations on the binding properties of α_1-adrenergic receptors are controversial. Goodhardt et al. (1982) have shown that guanine nucleotides produce a small but significant reduction in the affinity of agonists for the α_1-adrenergic receptors, measured by competition for ^{3}H-prazosin binding sites. This effect has also been observed by Snaveley and Insel (1982) in rat renal cortex. Other investigators, however, have not seen this effect (Hoffman et al., 1980).

5. Signal Transduction

Activation of adrenergic receptors by agonists results in specific alterations in cell function, such as a change in membrane potential, contraction, relaxation, secretion, and so on, depending on the cell type. These changes are often a secondary result of changes in the concentration of specific compounds inside the cell, referred to as second messengers. It is by changing the metabolism of these second messengers that receptor activation initiates the signal that ultimately produces a physiological response. Since not all physiological responses lend themselves to easy measurement, the changes in second messenger production in response to receptor activation can often provide a biochemical correlate to receptor occupation that can be easily measured. This is particularly true in brain tissue, in which responses to receptor activation often consist of increases or decreases in the firing rate of individual cells, effects that do not lend themselves to macroscopic measurement in vitro. Quantitating changes in second messenger production is often a valuable tool for assessing receptor responsiveness.

Perhaps the most thoroughly studied second messenger system is the receptor-activated adenylate cyclase system. Sutherland and Rall (1960) discovered that A increased the intracellular concentration of adenosine 3′5′-monophosphate (cyclic AMP) by activating the enzyme adenylate cyclase. Walsh et al. (1968) discovered a protein kinase enzyme that would increase its ability to phosphorylate proteins in the presence of cyclic AMP. The ubiquitous distribution of the cyclic AMP-dependent protein kinase led Kuo and Greengard (1969) to hypothesize that the ultimate result of increased cyclic AMP accumulation inside a cell was an increase in protein phosphorylation by activation of this protein kinase, a mechanism that is now generally accepted.

The mechanism by which A and other sympathomimetic drugs increase adenylate cyclase activity has been studied extensively. Occupation of β-adrenergic receptors causes an activation of adenylate cyclase in a rather complex manner. Guanosine 5′-triphosphate (GTP) is an essential cofactor for receptor-mediated activation of adenylate cyclase (Rodbell et al., 1971) and it is clear that GTP does not directly affect the receptor or the adenylate cyclase enzyme, but acts on a third membrane protein that is also essential to this response. Pfeuffer and Helmreich (1975) and Pfeuffer (1977) partially purified a guanine nucleotide binding protein distinct from the adenylate cyclase enzyme, which was later shown to be essential for receptor-mediated activation of adenylate cyclase activity. Limbird and Lefkowitz (1978) showed that occupation of the β-adrenergic receptor by the agonist radioligand ^{3}H-hydroxybenzylisoproterenol resulted in an increase in the apparent Stokes radius of digitonin-solubilized receptors, compared to receptors labeled with the antagonist radioligand ^{3}H-dihydroalprenolol. They interpreted this data to suggest that the agonist-occupied receptor interacts with another membrane component to form a ternary complex of agonist, receptor, and this other component. These investigators later showed that this membrane component was probably the guanine nucleotide regulatory protein (Limbird et al., 1980). The interaction of the guanine nucleotide regulatory protein with the agonist-occupied receptor can be destabilized with GTP, and it is likely that this destabilization of the agonist–receptor–nucleotide regulatory protein ternary complex is responsible for the effects of guanine nucleotides on agonist affinity for β-adrenergic receptors discussed above (Haga et al., 1977). Thus activation of adenylate cyclase by receptor occupation requires three distinct membrane proteins—the receptor itself, the guanine nucleotide regulatory protein, and the adenylate cyclase (Ross and Gilman, 1980).

Over the last several years, a detailed description of the molecular events involved in receptor-mediated increases in cyclic AMP production has emerged. Binding of agonists to β-adrenergic receptors results in the formation of a complex between the receptor and a guanine nucleotide binding protein (G/F). The nucleotide binding protein has guanosine 5′-diphosphate (GDP) tightly bound to it, and the interaction with the receptor allows the exchange of this GDP for GTP. In the presence of GTP, the nucleotide binding protein dissociates from the receptor and then dissociates into two distinct subunits, α and β. The α subunit of the nucleotide binding protein then binds to adenylate cyclase and activates the enzyme (Sternweis et al., 1981). This activation is terminated by the nucleotide binding protein hydrolyzing GTP to GDP, which promotes reassociation of the two subunits of the nucleotide binding protein. At this point, the cycle can be reinitiated by further receptor occupation. It has been known for many years that cholera toxin causes an irreversible activation of the adenylate cyclase enzyme. This is now known to be caused by covalent transfer of an ADP-ribose moiety to the nucleotide binding protein that prevents the hydrolysis of GTP to GDP, resulting in irreversible activation of adenylate cyclase activity (Cassel and Selinger, 1977).

The apparent affinity of the β-adrenergic receptor for agonists, as determined in radioligand binding assays, depends on the molecular interaction between the receptor and the nucleotide regulatory protein. The complexity of binding observed with agonists is probably caused by the receptor existing in at least two different states, either alone or in complex with the nucleotide regulatory protein. The complexing of the receptor with the nucleotide regulatory protein increases the apparent affinity of the receptor for agonists by stabilizing a high-affinity ternary complex of agonist–receptor–nucleotide binding protein. Addition of exogenous GTP converts all the receptors to the low-affinity state that is not complexed with the nucleotide binding protein, thus abolishing the complex binding characteristics. The mechanism by which magnesium increases the apparent affinity of agonists for β-adrenergic receptors is not well understood.

Both β_1- and β_2-adrenergic receptors activate adenylate cyclase activity in various tissues (Minneman et al., 1981), and it is likely that this is the major molecular mechanism by which activation of β-adrenergic receptors initiates signal transduction in most cells. There is some evidence that β-adrenergic receptor activation increases membrane phospholipid metabolism, particularly by increasing the methylation of various types of phospholipid and altering their orientation in the cell membrane (Hirata and Axelrod, 1980). Thus, there may be other biochemical events associated with β-adrenergic receptor occupation that may contribute to initiation of a signal in the cell, although this area is still under investigation.

α_2-Adrenergic receptors have been shown to inhibit adenylate cyclase activity in many tissues, particularly if the enzyme is previously activated by activation of another receptor type. This inhibition has been shown to involve an inhibitory guanine nucleotide binding protein that has many similarities to the stimulatory guanine nucleotide-regulatory protein involved in activation of adenylate cyclase by β-adrenergic receptor occupation (Katada et al., 1984a). GTP is also an essential cofactor for inhibition of adenylate cyclase activity by α_2-adrenergic receptor activation, and it is likely that the molecular mechanisms involved in receptor-mediated inhibition and stimulation of adenylate cyclase are similar. Smith and Limbird (1981) have shown that agonist occupancy of α_2-adrenergic receptors causes an alteration in the sedimentation characteristics of the solubilized receptor consistent with a stabilized ternary complex, similar to that observed with β-adrenergic receptors. It seems likely therefore that α_2-adrenergic receptor-mediated inhibition of adenylate cyclase activity is mediated by an inhibitory nucleotide-binding protein. Gilman and coworkers (Katada et al., 1984b) hypothesize that stimulation of α_2-adrenergic receptors, or other receptors that decrease adenylate cyclase activity, results in the dissociation of the two subunits of an inhibitory nucleotide-binding protein. The β subunit released from the inhibitory nucleotide-binding protein appears to be identical to the β subunit of the stimulatory nucleotide binding protein, and combines with the free α subunit from the stimulatory protein to prevent (decrease) the activation of adenylate cyclase. It appears that the number of inhibitory nucleotide-binding protein molecules is much greater than the number of stimulatory nucleotide-binding protein molecules, making this type of stoichiometric inhibition possible in reconstituted cell preparations.

Signal transduction by α_1-adrenergic receptors apparently does not involve the adenylate cyclase system. Activation of α_1-adrenergic receptors is known to increase calcium influx in many cells through the opening of receptor-operated calcium channels. It is also known that activation of α_1-adrenergic receptors increases the metabolism of a minor phospholipid in cell membranes, phosphatidylinositol (Michell, 1975; Fain and Garcia-Sainz, 1980). Increases in the metabolism of phosphatidylinositol associated with activation of many different types of membrane receptors is thought to play a role in signal transduction, and may be involved in the initiation of calcium influx by these receptors. Increases in phosphatidylinositol hydrolysis releases diacylglycerol into the intracellular space, and this compound is a potent activator of a particular calcium-dependent protein kinase, protein kinase C (Nishizuka, 1984). In addition, phosphatidylinositol usually contains arachidonic acid as one of its acyl chains, and metabolism of phosphatidylinositol is also associated with increased availability of arachidonic acid. Arachidonic acid is considered to be rate-limiting for the synthesis of several groups of important biological compounds, such as prostaglandins, thromboxanes, and leukotrienes (Berridge, 1981). Finally, there is increasing evidence that the first step in receptor-mediated increases in phosphatidylinositol metabolism is breakdown of triphosphoinositides to release 1,4,5-inositol triphosphate (Berridge, 1983), a compound that has been shown to mobilize intracellular calcium stores in low concentrations (Streb et al., 1983). Thus activation of α_1-adrenergic receptors and the subsequent activation of phosphatidylinositol metabolism may initiate several intracellular or extracellular signals that might be responsible for the observed cellular responses to receptor stimulation.

6. Relationship Between Receptor Occupancy and Tissue Response

The relationship between the fractional occupancy of a receptor population by an agonist and the observed response of the tissue has been studied for many years. This is an important characteristic of the drug–receptor interaction, since the fractional occupancy of the receptor population required to produce a maximal response will be an important factor in determining the potency of an agonist in activating a receptor-mediated response in that tissue. If an agonist is equally potent in binding to receptors in two different tissues, but a much smaller fraction of receptors needs to be occupied to produce a response in one tissue, then the agonist will be more potent in causing a response in that tissue. Therefore the potency of an agonist in causing a tissue response may be quite different in different tissues, even if the agonist is acting on identical receptor populations. It is important to understand how the fractional occupancy of the receptor is related to the response of the tissue, and how this parameter can vary between different systems.

The theory of drug–receptor interactions has usually assumed that the binding of drugs to receptors is a simple bimolecular event kinetically similar to surface adsorption phenomena. In this case, the binding of a drug to its receptor is dependent only on the concentrations of the drug and the receptor. When A. J. Clark (1937) first formalized receptor theory, he assumed that the response observed upon addition of a drug to a tissue was a linear function of receptor occupancy. One receptor would be occupied by one agonist molecule and a maximum response would be reached only when all of the receptors were occupied. Under these assumptions, the concentration of drug necessary to produce a 50% maximal response would occupy approximately 50% of the total receptor population, and the dose–response curves for receptor occupancy and tissue response would be almost superimposable (Fig. 1). In this case, the dose of drug causing a half-maximal response (EC_{50}) would approximately correspond to the equilibrium dissociation constant of the drug binding to its receptor (K_d).

Noting that some drugs produced a greater maximum re-

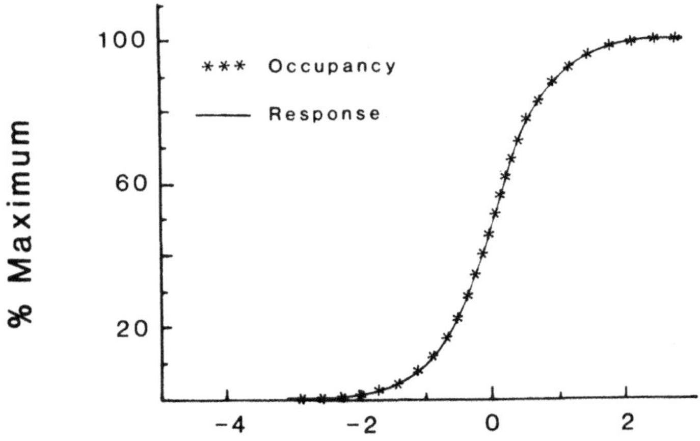

log [Agonist]/K_D

Fig. 1. Relationship between receptor occupancy (***) and observed tissue response (—), assuming a linear relationship between the two parameters as assumed by Clark (1937). When the concentration of agonist is equal to its K_d value [log (agonist)/K_d = 0], 50% of the receptors are occupied and the observed tissue response is about 50% of maximum. At an agonist concentration 100 times greater than the K_d, 99% of the receptors are occupied and the tissue response is nearly maximum.

sponse than did others that appeared to activate the same population of receptors, Ariens (1954) introduced the concept of intrinsic activity. The ability of a drug to produce a response was separated into two parameters: the potency in binding to the receptor (affinity), and the ability to produce a response once bound to the receptor (intrinsic activity). Antagonists cannot activate the receptor, and have an intrinsic activity of zero, whereas full agonists can activate the receptor completely and have intrinsic activities of one. Drugs that produce a lesser maximum response than full agonists fall somewhere in between, and their intrinsic activities are determined as a fraction of the maximum response produced by a full agonist (Fig. 2). Although Ariens' concept of intrinsic activity allowed for some undefined property of a drug in producing a maximal tissue response, he still assumed response to be a linear function of receptor occupancy.

Stephenson (1956) suggested that tissue response was not necessarily a linear function of receptor occupancy and that a maximum tissue response might be achieved when only a small fraction of the receptor pool was occupied by an agonist. He proposed that different drugs might have different abilities to

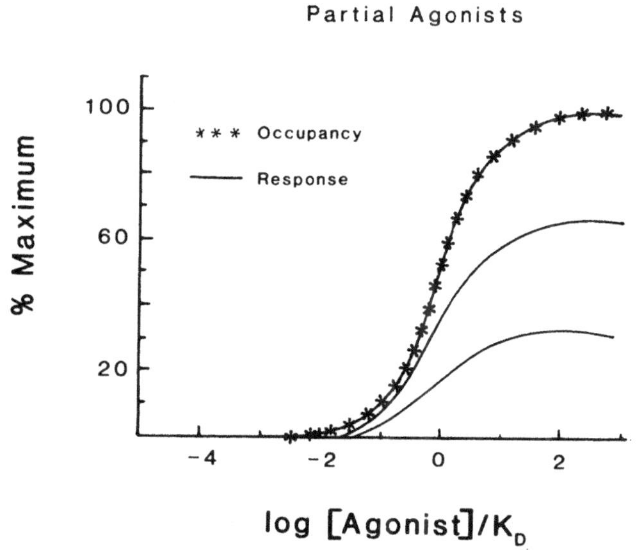

Fig. 2. Relationship between receptor occupancy (***) and tissue response (—) for three drugs with the same affinity in binding to the receptor, but with different intrinsic activities as proposed by Ariens (1954). The receptor occupancy curve is the same for all three drugs since they all have the same affinity for the receptor binding site. The top curve represents a drug with an intrinsic activity of 1, the middle curve represents a drug with an intrinsic activity of 0.66, and the bottom curve represents a drug with an intrinsic activity of 0.33. This model also assumes a linear relationship between receptor occupancy and response.

initiate a response, and might occupy different proportions of the receptor pool to produce the same response. He suggested that the ability of a drug to produce a response be called efficacy. Drugs of high efficacy would need to occupy a smaller fraction of the receptor population than drugs of lower efficacy to produce comparable responses. He proposed that drugs of low efficacy that could produce a response, but that would also antagonize the actions of high-efficacy drugs, be called partial agonists.

Modifications of Stephenson's theories (Furchgott, 1966) have taken into account the differences between tissue receptor populations. The efficacy of a drug at a receptor in a particular tissue would be related to the ability of the drug to stimulate the receptor and also to the proportion of spare receptors in that tissue. This proportion varies from tissue to tissue. If there is a large receptor reserve in a tissue, a partial agonist that could not fully activate each receptor might still be able to provoke a maximal response. In addition, the potency of a full agonist in causing a response would

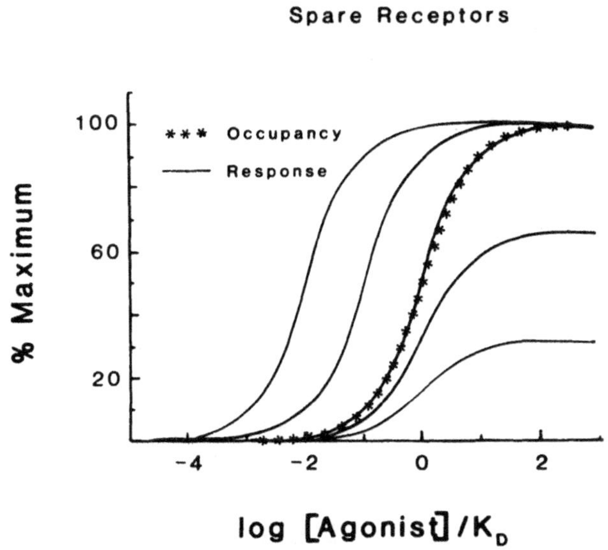

Fig. 3. The relationship between receptor occupancy (***) and tissue response (—) for agonists with different efficacies and identical affinities in a tissue containing spare receptors based on Stephenson (1956). An agonist with a very high efficacy, pictured in the first curve on the left, can stimulate 50% of the maximal response with occupation of only around 1% of the total receptor pool in this tissue. An agonist with a lower efficacy must occupy around 10% of the total receptor pool to cause a similar response (second curve). If the efficacy of the agonist is just sufficient to cause a maximal response in the tissue, then the dose–response curves for occupancy and response will be superimposed (third curve). Agonists with even lower efficacies will not be able to provoke a maximal response at any concentration, even with maximal receptor occupancy (fourth and fifth curve). This figure illustrates that in the presence of spare receptors the concentration of agonist necessary to cause a half-maximal response is not necessarily a measure of the equilibrium binding constant of the agonist at the receptor.

vary between different tissues with different proportions of spare receptors, and a partial agonist could appear to have a different intrinsic activities in such tissues. Stephenson's concepts changed the notion of a direct relationship between receptor occupancy and tissue response and demonstrated that the dose of a drug causing a half-maximal tissue response could not be assumed to be an estimate of the affinity of the drug in binding to the receptor (Fig. 3) (Furchgott, 1972; Ruffolo, 1982).

Direct support for Stephenson's proposal came when Nickerson (1956) used irreversible receptor antagonists to reduce receptor density. He demonstrated that irreversible receptor blockade often resulted in a rightward shift of the dose–response curve for a receptor-mediated response, with no reduction in maximum response. Receptor inactivation produced a further rightward shift as well in the dose–response curve until all of the spare receptors were inactivated. Any further inactivation of the receptor population produced a decrease in maximal response, with no change in the EC_{50} of the agonist in causing the response. Furchgott (1966) developed a method for determining agonist binding constants using this technique, since when all of the "spare" receptors have been inactivated by an irreversible antagonist, the EC_{50} should approximately represent the equilibrium dissociation constant for receptor activation (K_a) by the agonist.

A number of investigators have used irreversible receptor blockade with β-haloalkylamine drugs, such as dibenamine and phenoxybenzamine, to demonstrate that there are a large proportion of spare α-adrenergic receptors in a variety of smooth muscle preparations (Ariens et al., 1960; Furchgott, 1966; Sheys and Green, 1972; Besse and Furchgott, 1976). It is remarkable that less than 1% of the total receptor pool must apparently be occupied by an agonist to produce a maximal response in some tissues. It has been more difficult to provide direct evidence for the existence of spare β-adrenergic receptors in isolated tissues caused by the lack, until recently, of irreversible β-adrenergic receptor blocking drugs that could be used to reduce receptor density. Using functional antagonism to effectively but not actually reduce receptor density, evidence for spare β-adrenergic receptors has been obtained in a variety of tissues (Van den Brink, 1973; O'Donnell and Wanstall, 1979; Emmerson and Mackay, 1981).

With the development of radioligand binding techniques for measuring receptor occupation and receptor density directly, more substantial evidence for the existence of spare adrenergic receptors has been obtained in a variety of tissues. The concentrations of agonists necessary to half-saturate both α- and β-adrenergic receptors are often much higher than the concentrations needed to elicit half-maximal responses in isolated tissues, consistent with the existence of a large pool of spare receptors in these tissues (Lefkowitz et al., 1974; Sporn et al., 1976; Williams et al., 1976; Kaumann, 1982; Minneman et al., 1983).

A few studies have examined adrenergic receptor density directly following receptor inactivation with irreversible antagonist drugs. Venter (1979) showed that inactivation of β-adrenergic receptors in cat heart with a highly reactive derivative of propranolol decreased the potency of the β-adrenergic receptor agonist isoproterenol in causing an inotropic response, and correspondingly decreased the density of β-adrenergic receptor binding sites. This study suggested that greater than 99% of the β-adrenergic receptors in cat papillary muscle were "spare," that is they did not contribute to the maximal response of the tissue to agonist. Terasaki et al. (1979) obtained similar data studying β-adrenergic receptor-mediated cyclic AMP accumulation in cultured glioma cells. Irreversible receptor blockade shifted the dose–response curve for agonist stimulation to the right without any depression of the maximum.

Direct evidence for the existence of spare α_1-adrenergic receptors in rat vas deferens has been obtained by Minneman and Abel (1984a,b). Various α_1-adrenergic receptor agonist drugs were shown to be more potent in causing contraction of the isolated vas deferens than in competing for α_1-adrenergic receptor binding sites in membrane preparations of vas deferens (Minneman et al., 1983). Inactivation of these receptors with the irreversible alkylating drug phenoxybenzamine decreased the potencies of agonists in causing contractile responses. When all of the spare receptors had been inactivated, the potencies of agonists in causing contraction of isolated vas deferens were similar to their potencies in competing for the receptor binding sites (Minneman and Abel, 1984a). Therefore, the difference in agonist potencies observed before receptor inactivation must be caused by the existence of spare α_1-adrenergic receptors in rat vas deferens. Comparison of the decrease in the density of functional receptors after phenoxybenzamine treatment, calculated by the indirect method of Furchgott (1966) from contractile data, with the observed decrease in the density of α_1-adrenergic receptor binding sites in vas deferens showed that the decrease in functional receptors was much greater than the decrease in total receptor density (Minneman and Abel, 1984b), suggesting that the receptors mediating contraction represent only a fraction of the total α_1-adrenergic receptor pool in rat vas deferens. The selective inactivation of functional receptors may be caused by diffusional barriers in the intact tissue and the geometrical arrangement of the muscle layers in this tissue. These studies have presented direct evidence that the density of α_1-adrenergic receptors in rat vas deferens is greater than is necessary for a maximal contractile response, and that this greater density of receptors results in a greater potency of agonists in causing a functional response, supporting the ideas originally put forth by Stephenson (1956).

Thus, there is growing evidence that the number of receptors in a tissue may be greater than necessary to produce a maximal response of the tissue. Drugs with a high efficacy may be non-linearly coupled to the production of a response, whereas weak partial agonists must occupy all of the receptor population in order to produce a comparable response.

7. Regulation of Receptor Density and Responsiveness

The homeostatic mechanisms by which a tissue regulates its response to sympathomimetic drugs are not well understood. Experimental manipulations that alter the normal stimulation to a tissue change the tissue response to agonists. Treatments of animals that result in a decreased adrenergic input to a tissue result in an increased tissue sensitivity to agonists (supersensitivity), whereas treatments that increase normal adrenergic input cause a decreased tissue sensitivity (subsensitivity) (Trendelenburg, 1963). The exact mechanisms underlying these phenomena are not understood, although changes in receptor density have been suggested to contribute to the altered tissue responsiveness.

7.1. Alterations in β-Adrenergic Receptors

Intraventricular injections of 6-hydroxydopamine to destroy catecholaminergic nerves cause an increase in cyclic AMP production in response to β-adrenergic receptor stimulation in rat cerebral cortex (Palmer, 1972). Sporn et al. (1977) found that the increased cyclic AMP production was associated with an increase in the density of β-adrenergic receptors in rat cortex, an effect that was later shown to be restricted only to the β_1-adrenergic receptor subtype (Minneman et al., 1979a). Treatment of rats with reserpine to deplete catecholamines also increases β-adrenergic receptor density and responsiveness in rat cerebral cortex (Nahorski, 1977; Dolphin et al., 1979). Similar treatment with guanethidine has been reported to increase β-adrenergic receptor density in rat heart muscle (Glaubiger et al., 1978). These and other studies suggest that removal of the normal adrenergic neuronal input can increase both the density of β-adrenergic receptors and their ability to stimulate cyclic AMP production both centrally and peripherally.

Increasing the concentration of NA in adrenergic synapses produces effects opposite to those described above. Chronic treatment of animals with desmethylimipramine to block neuronal uptake of NA, or with a monoamine oxidase inhibitor to block catecholamine degradation, reduces β-adrenergic receptor density and responsiveness in rat cerebral cortex (Vetulani and Sulser,

1975; Wolfe, et al., 1978). The effects of desmethylimipramine are also restricted to the β_1-adrenergic receptor population in cerebral cortex (Minneman et al., 1979a). Electrophysiological studies by Olpe and Schellenberg (1980) and Yeh and Woodward (1983) have demonstrated a decreased sensitivity of central neurons to iontophoretically applied NA following chronic treatment with desmethylimipramine. Recent studies suggest that the effects of this drug may be dependent on the presence of an intact serotonergic input to the cortex (Brunello et al., 1982; Janowsky et al., 1982).

An increase in the availability of NA at adrenergic synapses only results in changes in receptor density if the transmitter is allowed to interact with the receptors. Wolfe et al. (1978) demonstrated that the effects of desmethylimipramine were prevented by chronic blockade of β-adrenergic receptors by concurrent administration of propranolol. Chronic blockade of the β-adrenergic receptor by propranolol causes an increase in receptor density in brain (Wolfe et al., 1978) and heart (Glaubiger and Lefkowitz, 1977), although some authors have observed no effect of chronic propranolol treatment on β-adrenergic receptor density in heart (Baker and Potter, 1980).

The metabolic activity of cells also has apparent effects on receptor expression on the cell surface. Adrenalectomy increases β-adrenergic receptor density and receptor-mediated cyclic AMP accumulation in rat liver, an effect that can be prevented by exogenous administration of glucocorticoids (Wolfe et al., 1976). Increases in thyroid hormone level can also increase β-adrenergic receptor density and responsiveness (Williams et al., 1977; Tse et al., 1980).

These experiments demonstrate that β-adrenergic receptor density can be increased or decreased depending upon the amount of transmitter available to interact with the receptor, or the metabolic state of the cells. The studies mentioned above utilized chronic (i.e., several days) treatment schedules to produce effects on receptor density. In cell cultures, exposure to agonists for short time periods can apparently uncouple the β-adrenergic receptors from adenylate cyclase, resulting in a receptor that when occupied by agonist does not cause a response. Exposure to agonist for longer periods of time, i.e., several hours, does reduce receptor density (Su et al., 1979, 1980). Using a hydrophilic radioligand, Hertel et al. (1983) have shown that treatment of C-6 glioma cells with isoproterenol induced a rapid apparent loss of part of the receptor pool that was not observed when the usual hydrophobic radioligands were used to measure receptor density. This data suggests that initial agonist occupation of β-adrenergic receptors results in a redistribution of part of the receptor pool so that it is not accessible to hydrophilic compounds, indicating a possible sequestration of these receptors into intracellular vesicles. However, longer-term stimulation of receptors causes further adaptive responses, resulting in a greatly decreased receptor density on the cell surface (for a review of desensitization, *see* Harden, 1983).

Overall, it is clear that alterations in adrenergic input to a tissue often alter the density and responsiveness of β-adrenergic receptors in that tissue. It is not clear, however, that these two parameters change in a parallel fashion. There are several reports in both cultured cells (Su et al., 1979) and isolated tissues (Hawthorn and Broadley, 1982) in which receptor-mediated responsiveness changes without a corresponding change in receptor density. It seems clear, therefore, that alterations in receptor density are only one of several adaptive mechanisms by which cells respond to long-term alterations in synaptic or hormonal input.

7.2. Alterations in α-Adrenergic Receptors

α-Adrenergic receptor density is also regulated by changes in normal transmitter–receptor interaction. U'Prichard et al. (1979) showed that the number of binding sites for a variety of α-adrenergic receptor radioligands was increased in the brain following 6-hydroxydopamine treatment. Several authors have found increases in α_1-adrenergic receptor density in rat cerebral cortex following 6-hydroxydopamine treatments (Reader and Briere, 1982). These increases are small and have not been observed in some studies (Skolnick et al., 1978). Morris et al. (1981) showed a decrease in α_2-adrenergic receptor density and an increase in α_1-adrenergic receptor density in rat cerebral cortex following frontal lobotomy, suggesting that some of the α_2-adrenergic receptors were localized presynaptically in this tissue.

The effect of chronic treatment with antidepressant drugs on α_1- and α_2-adrenergic receptor density in rat brain is the subject of considerable debate (Sugrue, 1983). This treatment has been reported to alter α_1- and α_2-adrenergic receptor density in various reports (Sugrue, 1982). Chronic treatment with antidepressant drugs has been shown to increase α_1-adrenergic receptor responsiveness both on individual neurons and in the whole animal acoustic startle response (Menkes and Aghajanian, 1981), but to have no effect on in vivo sensitivity of brain α_2-adrenergic receptors (Sugrue, 1982; Schoffelmeer and Mulder, 1982). Asakura et al. (1982) report an increase in α_2-adrenergic receptor density following treatment with some, but not all, clinically efficacious antidepressants. In addition, Menkes et al. (1983) report that in thalamic membranes there was no change in the density of α_1-adrenergic receptors or their affinity for ^{3}H-prazosin, but that the potency of phenylephrine in competing for these binding sites was increased following a 3-wk treatment with antidepressant drugs. This report suggests a supersensitivity that does not correlate with an increased receptor density.

Using cultured smooth muscle cells from rabbit aorta, Wikberg et al. (1983) have shown that α_1-adrenergic receptor density decreases following prolonged (26 h) exposure to NA. The phenomenon appears to be a desensitization phenomena similar to that observed with β-adrenergic receptors. There are similarities between the changes in α- and β-adrenergic receptors resulting from altered transmitter–receptor interactions, but how similar the changes are and the significance of simultaneous alterations in one or more types of receptor population remain unclear.

8. Functional Implications of Multiple Receptor Subtypes

It has become increasingly clear that multiple subtypes of a receptor can coexist in the same tissue and, in some cases, on the same cell (Homberger et al., 1981). The physiological significance of this coexistence is not clear, since the functions of the different receptor subtypes in a given tissue are not yet delineated. Both α_1- and α_2-adrenergic receptors are found in some smooth muscles, as are β_1- and β_2-adrenergic receptors, and in many cases they may affect the same physiological function, such as muscle contraction. Several possible explanations for the existence of these multiple receptor subtypes have been considered.

Lands et al. (1967) differentiated β_1- and β_2-adrenergic receptors primarily on their differing responsiveness to NA. Since NA is the major neurotransmitter in postganglionic sympathetic nerves, it has been suggested that the β_1-adrenergic receptor subtype, which is most responsive to NA, receives a neuronal innervation and that the β_2-adrenergic receptor subtype, which is less sensitive to NA, may be normally stimulated by A released from the adrenal (Carlsson and Hedberg, 1977; Ariens and Simonis, 1976). Investigations addressing this hypothesis have yielded conflicting results. Viveros et al. (1968) found that sympathetic nerve stimulation caused vasodilation in the dog gracilis muscle in the presence of both α-adrenergic and muscarinic cholinergic receptor-blocking drugs. However, Russell and Moran (1980) found no evidence for such vasodilation in this preparation. Rosell and Belfrage (1975) showed that the decrease in vascular resistance in dog adipose tissue caused by NA was not affected by sympathetic denervation or blockade of neuronal NA uptake, suggesting that β-adrenergic receptors of the vasculature (typically thought to be β_2) were not sympathetically innervated. Hyman et al. (1981) found that in the feline pulmonary vascular bed, β_2-adrenergic receptors responded to sympathetic nerve stimulation. Whether this reflects species differences or differences between vascular beds is not clear.

Although the evidence is still equivocal, the theory that β_1-adrenergic receptors represent the receptors for neuronally released NA and β_2-adrenergic receptors represent the receptors for circulating A is gaining support. Since the β_2-adrenergic receptor is less responsive to NA, if these receptors are innervated does this mean that the neurons release much more transmitter at β_2-receptors than β_1-receptors? It is interesting to note that there are a substantial number of β_2-adrenergic receptors in brain (Minneman et al., 1979c), and that A released from the adrenal can not cross the blood–brain barrier. Therefore, β_2-adrenergic receptors in the central nervous system would need to be stimulated by neuronally released A or very high concentrations of NA. Hokfelt et al. (1974) have described A-containing neurons in the brain, although these are few in number and not associated in any obvious manner with β_2-adrenergic receptor density in the brain.

A similar distinction between α_1 and α_2-adrenergic receptors is also developing. Although originally classified as presynaptic (Langer, 1974), α_2-adrenergic receptors have also been shown to exist on smooth muscle and other tissues in which α_1-adrenergic receptors also exist (Timmermans et al., 1979; Drew, 1980; Langer et al., 1980; Docherty and McGrath, 1980). Langer et al. (1980) and Wilffert et al. (1982) have presented data that α_2-adrenergic receptors are extrasynaptic; that is, not located near the myoneuronal varicosities. Circulating A and/or corticosteroids appear to be required for the proper function of α_2-adrenergic receptors, as adrenalectomized animals have attenuated responses to α_2-adrenergic receptor stimulation (van Meel et al., 1983). These attenuated responses were restored by administration of A and aldosterone. Baker and Drew (1981) found little effect of adrenal demedullation on the responsiveness of α_2-adrenergic receptors, strongly suggesting that the adrenal cortex is involved in the regulation. Wilffert et al. (1982) presented evidence that all four subtypes of adrenergic receptors (α_1, α_2, β_1, β_2) are stimulated by intravenously administered A, but that only α_1- and β_1-adrenergic receptors were stimulated by neuronally released NA, consistent with an extrasynaptic localization for both α_2- and β_2-adrenergic receptors in the cardiovascular system.

As with β_2-adrenergic receptors, α_2-adrenergic receptors may be normally stimulated more by A than by NA, although unlike β_2-adrenergic receptors, there is little difference in the potency of the two compounds in stimulating α_2-adrenergic receptors. The evidence is controversial, but these hyotheses provide a basis for further experimentation. It is important to know which endogenous compounds stimulate the various adrenergic receptor subtypes in order to make appropriate pharmacological interventions in disease states, elucidate the mechanism of information transfer and integration on a cellular level, and determine what substance regulates the density of a particular receptor type.

It is probably more important to clarify the interactions between different receptor types that mediate opposing functional responses. For example, in some tissues α_1-adrenergic receptors mediate contraction and β_2-adrenergic receptors mediate relaxation of the same smooth muscle. In such systems, there appears to be a natural functional antagonism that maintains the appropriate balance of receptor-mediated actions. How the tissue responds to opposing receptor-mediated stimuli and how the response of the tissue affects the receptor properties are fundamental questions that have yet to be addressed. The mechanism by which β-adrenergic receptor-mediated responses inhibit or alter α-adrenergic receptor-mediated responses (or vice versa) will lead to an increased understanding of signal transduction processes involved in transforming receptor stimulation into an observable tissue response.

9. Summary and Future Directions

The data available concerning adrenergic receptors is very extensive and this chapter attempts to summarize some of the salient points in the development of concepts that are currently accepted in this field. The realization that there are multiple subtypes of receptor for a single neurotransmitter or hormone has encouraged the development of selective drugs, and indeed, the development of selective drugs has furthered the delineation of different receptor subtypes. New techniques have allowed an exploration of fundamental questions regarding the signal transduction process. Investigators can now isolate and purify the receptor molecules that were only theoretical concepts 15 yr ago.

As much progress as we have made in the study of adrenergic receptors, there are still many unanswered questions. A more detailed, mechanistic understanding of how a sympathomimetic drug produces a cellular response is still needed. Many smooth muscles respond to α_1-adrenergic receptor stimulation with a measurable contraction, but between the binding of an agonist to the receptor and the contraction of the smooth muscle cell are many unknowns. As discussed above, the physiological significance of the multiple types of adrenergic receptor and their potentiative or antagonistic interactions are still not understood. Further studies on the relationship between receptor occupancy and the subsequent tissue response may provide information about the factors controlling the potencies of agonists in causing responses in the whole organ, obviously an important factor in understanding drug action.

Purification and biochemical characterization of adrenergic receptors should provide useful data concerning the initiation of responses. The cloning of the DNA code for receptors and elucidation of posttranslational modifications of adrenergic receptors will help investigators understand how receptor density and responsiveness are regulated. It is currently not clear how receptor synthesis by cells is controlled.

The importance of adrenergic receptors in physiological control processes makes it important to understand what their properties are, how they initiate tissue responses, and how these responses are regulated. Much progress has been made in this area, and much progress is still to come.

References

Ahlquist R. P. (1948) A study of the adrenotropic receptors. *Am. J. Physiol.* **153,** 586–600.

Ariens E. J. (1954) Affinity and intrinsic activity in the theory of competitive inhibition. Part I. Problems and theory. *Arch. Int. Pharmacodyn.* **99,** 32–49.

Ariens E. J., van Rossum J. M., and Koopman P. C. (1960) Receptor reserve and threshold phenomena. I. Theory and experiments with autonomic drugs tested on isolated organs. *Arch. Int. Pharmacodyn.* **127,** 459–478.

Ariens E. J. and Simonis A. M. (1976) Receptors and Receptor Mechanisms, in *Beta-Adrenoceptor Blocking Agents* (Saxena P. R. and Forsyth R. P., eds.), pp. 3–27, North Holland, Amsterdam.

Arunlakshana O. and Schild H. O. (1959) Some quantitative uses of drug antagonists. *Brit. J. Pharmacol.* **14,** 48–58.

Asakura M., Tsukamoto T., and Haseqawa K. (1982) Modulation of brain α_2-and β-adrenergic receptor sensitivity following long-term treatment with antidepressants. *Brain Res.* **235,** 192–197.

Aurbach G. D., Fedak S. A., Woodard C. J., Palmer J. S., Hauser D., and Troxler F. (1976) Beta-adrenergic receptor: Stereospecific interaction of iodinated beta-blocking agent with a high affinity site. *Science* **186,** 1223–1224.

Baker D. J. and Drew G. M. (1981) An investigation into the possible physiological roles of vascular α_1- and α_2-adrenoceptors. *J. Auton. Pharmacol.* **1,** 359–365.

Baker S. P. and Potter L. T. (1980) Effect of propranolol on beta-adrenoceptors in rat hearts. *Brit. J. Pharmacol.* **68,** 8–10.

Barnes P., Karliner J., Hamilton C., and Dollery C. (1979) Demonstration of alpha$_1$-adrenoceptors in guinea pig lung using ^{3}H-prazosin. *Life Sci.* **25,** 1207–1214.

Barnett D. B., Rugg E. L., and Nahorski S. R. (1978) Direct evidence for two types of beta-adrenoceptor binding site in lung tissue. *Nature* (Lond.) **273,** 166–168.

Barovsky K. and Brooker G. (1980) (-) ^{125}I-Iodopindolol, a new highly selective radioiodinated beta-adrenergic receptor antagonist: Measurement of beta-receptors on intact rat astrocytoma cells. *J. Cyclic Nucl. Res.* **6,** 294–307.

Berridge M. J. (1981) Phosphatidylinositol hydrolysis: A multifunctional transducing mechanism. *Mol. Cell. Endocrinol.* **24,** 115–140.

Berridge M. J. (1983) Rapid accumulation of inositol triophosphate reveals that agonists hydrolyze polyphosphoinositides instead of phosphatidylinositol. *Biochem. J.* **212,** 849–858.

Berthelsen S. and Pettinger W. A. (1977) A functional basis for classification of alpha-adrenergic receptors. *Life Sci.* **21,** 595–606.

Besse J. C. and Furchgott R. F. (1976) Dissociation constants and relative efficacies of agonists acting on alpha-adrenergic receptors in rabbit aorta. *J. Pharmacol. Exp. Ther.* **197,** 66–78.

Bird S. J. and Maguire M. E. (1978) The agonist-specific effect of magnesium ion on binding by beta-adrenergic receptors in S49 lymphoma cells: Interaction of GTP and magnesium in adenylate cyclase activation. *J. Biol. Chem.* **253,** 8826–8834.

Brittain R. T., Farmer J. B., Jack D., Marin L. E., and Simpson W. T. (1968) Alpha-[(*t*-butylamino)-methyl]-4-hydroxy-*m*-xylene-alpha^1-alpha^3-diol (AH 3365): A stimulant. *Nature* (Lond.) **219,** 862–863.

Brunello N., Barbaccia M. L., Chuang D.-M., and Costa E. (1982) Down-regulation of β-adrenergic receptors following repeated injections of desmethylimipramine: permissive role of serotonergic axons. *Neuropharmacol.* **21,** 1145–1149.

Burns J. J., Colville K. I., Lindsay L. A., and Salvador R. A. (1964) Blockade of some metabolic effects of catecholamines by *N*-isopropyl methoxamine (B. W. 61–43). *J. Pharmacol. Exp. Ther.* **144,** 163–171.

Cambridge D., Davey M. J., and Massingham R. (1977) Prazosin, a selective antagonist of postsynaptic alpha-adrenoceptors. *Brit. J. Pharmacol.* **59,** 514P.

Cannon W. B. and Rosenblueth A. (1937) *Autonomic Neuroeffector Systems.* New York, Macmillan.

Carlsson E. and Hedberg A. (1977) Are cardiac effects of noradrenaline and adrenaline mediated by different beta-adrenoceptors? *Acta Physiol. Scand.* (suppl. 44) **47.**

Cassel D. and Selinger Z. (1977) Mechanism of adenylate cyclase activation by cholera toxin: Inhibition of GTP hydrolysis at the regulatory site. *Proc. Natl. Sci. USA* **74,** 3307–3311.

Cheung Y-D., Barnett D. B., and Nahorski S. R. (1982) ^{3}H-Rauwolscine and ^{3}H-yohimbine binding to rat cerebral and human platelet membranes: Possible alpha$_2$-adrenoceptors. *Eur. J. Pharmacol.* **84,** 79–85.

Clark A. J. (1937) General Pharmacology, in *Heffter's Handbuch der experimentellen Pharmakologie, Erg. werk,* vol. 4, Springer, Berlin.

Cuatrecasas P., Tell G. P., Sica V., Parikh I., and Chang K-J. (1974) Noradrenaline binding and the search for catecholamine receptors. *Nature* (Lond.) **247,** 92–97.

Cubeddu L., Barnes E. M., Langer S. Z., and Weiner N. (1974) Release of norepinephrine and dopamine-beta-hydroxylase by nerve stimulation. I. Role of neuronal uptake and of alpha presynaptic receptors. *J. Pharmacol. Exp. Ther.* **190,** 431–450.

Dale H. H. (1906) On some physiological actions of ergot. *J. Physiol.* (Lond.) **34,** 163–206.

Davis J. N., Strittmatter W., Hoyler E., and Lefkowitz R. J. (1977) (^{3}H)-Dihydroergokryptine binding in rat brain. *Brain Res.* **132,** 327–336.

DeLean A., Stadel J. M., and Lefkowitz R. J. (1980) A ternary complex model explains the agonist specific binding properties of the adenylate cyclase-coupled β-adrenergic receptor. *J. Biol. Chem.* **255,** 7108–7117.

Docherty J. R. and McGrath J. C. (1980) A comparison of pre- and post-junctional potencies of several alpha-adrenoceptor agonists in the cardiovascular system and anococcygeus of the rat. Evidence for two types of post-junctional alpha-adrenoceptor. *Naunyn-Schmiedeberg's Arch. Pharmacol.* **312,** 107–116.

Dolphin A., Adrien J., Hamon M., and Bockaert J. (1979) Identity of ^{3}H-dihydroalprenolol binding sites and beta-adrenergic receptors coupled within the central nervous system: Pharmacological properties, distribution, and adaptive responsiveness. *Mol. Pharmacol.* **15,** 1–15.

Drew G. M. (1980) Postsynaptic alpha$_2$-adrenoceptors mediate pressor responses to 2-*N,N*,-dimethylamino-5,6-dihydroxy-1,2,3,4-tetrahydronapthalene (M-7). *Eur. J. Pharmacol.* **65,** 85–87.

Dunlop D. and Shanks R. G. (1968) Selective blockade of adrenoceptive beta-receptors in the heart. *Brit. J. Pharmacol.* **32,** 201–218.

Emmerson J. and Mackay D. (1981) A test of the null equation for functional antagonism. *Brit. J. Pharmacol.* **73,** 135–141.

Engel G. and Hoyer D. (1981) ^{125}I BE 2254, a new high affinity radioligand for alpha-1 adrenoceptors. *Eur. J. Pharmacol.* **73,** 221–224.

Engel G., Hoyer D., Berthold R., and Wagner H. (1981) (±)Iodocyanopindolol, a new ligand for beta-adrenoceptors: Identification and quantitation of subclasses of beta-adrenoceptors in the guinea pig. *Naunyn-Schmiedeberg's Arch. Pharmacol.* **317,** 277–285.

Fain J. N. and Garcia-Sainz J. A. (1980) Role of phosphatidylinositol turnover in alpha$_1$ and of adenylate cyclase inhibition in alpha$_2$ effects of catecholamines. *Science* **26,** 1183–1194.

Furchgott, R. F. (1966) The Use of Beta-Haloalkylamines in the Differentiation of Receptors and in the Determination of Dissociation Constants of Receptor-Agonist, in *Advances in Drug Research* vol. 3 (Harper N. J. and Simmonds A. B., eds.), pp. 21–55. Academic, London.

Furchgott R. F. (1972) The Classification of Adrenoceptors (Adrenergic Receptors). An Evaluation From the Standpoint of Receptor Theory, in *Handbook of Experimental Pharmacology* vol. 33, *Catecholamines,* Blaschko H. and Muscholl E., eds.), pp. 283–335, Springer-Verlag, Berlin.

Glaubiger G. and Lefkowitz, R. J. (1977) Elevated beta-adrenergic receptor number after chronic propranolol treatment. *Biochem. Biophys. Res. Comm.* **78,** 720–725.

Glaubiger G., Tsai B. S., Lefkowitz R. J., Weiss B., and Johnson E. M., Jr. (1978) Chronic guanethidine treatment increases cardiac beta-adrenergic receptors *Nature* (Lond.) **273,** 240–242.

Glossman H., Lubbecke F., and Bellemann P. (1981) ^{125}I-HEAT, a selective, high-affinity, high specific activity ligand for alpha$_1$-adrenoceptors. *Naunyn-Schmiedeberg's Arch. Pharmacol.* **318,** 1–9.

Goodhardt M., Ferry N., Geynet P., and Hanoune J. (1982) Hepatic α_1-adrenergic receptors show agonist-specific regulation by guanine nucleotides. *J. Biol. Chem.* **257,** 11,577–11,583.

Greenberg D. A., U'Prichard D. C., and Snyder S. H. (1976) α-Noradrenergic receptor binding in mammalian brain: Differential labeling of agonist and antagonist states. *Life Sci.* **19,** 69–76.

Greenglass P. and Bremner R. (1979) Binding characteristics of ^{3}H-prazosin to rat brain alpha-adrenergic receptors. *Eur. J. Pharmacol.* **55,** 323–326.

Haga R., Ross E. M., Anderson H. J., and Gilman A. G. (1977) Adenylate cyclase permanently uncoupled from hormone receptors in a novel variant of S49 mouse *Proc. Natl. Acad. Sci. USA* 74, 2016–2020.

Hancock A. A., DeLean A. L., and Lefkowitz R. J. (1979) Quantitative resolution of beta-adrenergic receptor subtypes by selective ligand binding: Application of fitting technique. *Mol. Pharmacol.* **16,** 1–9.

Harden T. K. (1983) Agonist-induced desensitization of the β-adrenergic receptor-linked adenylate cyclase. *Pharmacol. Rev.* **35,** 5–32.

Hawthorn M. H. and Broadley K. J. (1982) β-Adrenoceptor ligand binding and supersensitivity to isoprenaline of ventricular muscle after chronic reserpine pretreatment. *Naunyn-Schmiedeberg's Arch. Pharmacol.* **320,** 240–245.

Heidenreich K. A., Weiland G. A., and Molinoff P. B. (1980) Characterization of radioligand agonist binding to β-adrenergic receptors in mammalian tissues. *J. Cyclic Nucl. Res.* **6,** 217–230.

Hertel C., Miller P., Portenier M., and Staehelin M. (1983) Determination of the desensitization of β-adrenergic receptors by [^{3}H]-CHP-12177. *Biochem. J.* **216,** 669–674.

Hirata F. and Axelrod J. (1980) Phospholipid methylation and biological signal transmission. *Science* **209,** 1082–1090.

Hoffman B. B., Mullikin-Kilpatrick D., and Lefkowitz R. J. (1980) Heterogeneity of radioligand binding to alpha-adrenergic receptors: Analysis of guanine nucleotide regulation of agonist binding in relation to receptor subtypes. *J. Biol. Chem.* **255,** 4645–4652.

Hokfelt T., Fuxe K., Goldstein M., and Johansson O. (1974) Immunohistochemical evidence for the existence of adrenaline neurons in the rat brain. *Brain Res.* **66,** 235–251.

Homberger V., Lucas M., Rosenbaum E., Vassent G., and Bockaert J. (1981) Presence of both $beta_1$- and $beta_2$-adrenergic receptors in a single cell type. *Mol. Pharmacol.* **10,** 463–469.

Hornung R., Presek P., and Glossman H. (1979) Alpha adrenoceptors in rat brain: Direct identification with prazosin. *Naunyn-Schmiedeberg's Arch. Pharmacol.* **308,** 223–230.

Hyman A. L., Nandiwada P., Knight D. S., and Kadowitz P. J. (1981) Pulmonary vasodilator responses to catecholamines and sympathetic nerve stimulation in that vascular B-2 receptors are innervated. *Circ. Res.* **48,** 407–415.

Innes, I. R. (1962) Identification of the smooth muscle excitatory receptors for ergot alkaloids. *Brit. J. Pharmacol. Chemother.* **19,** 120–128.

Janowsky A., Okada F., Manier D., Hal A., Craig D., Sulser F., and Steranka L. (1982) Role of serotonergic input in the regulation of the β-adrenergic receptor-coupled adenylate cyclase system. *Science* **218,** 900–901.

Katada T., Bokoch G. M., Northrup J. K., Ui M., and Gilman A. G. (1984a) The inhibitory guanine nucleotide-binding regulatory component of adenylate cyclase: Properties and function of the purified protein. *J. Biol. Chem.* **259,** 3568–3577.

Katada T., Bockoch G. M., Smigel M. D., Ui M., and Gilman A. G. (1984b) The inhibitory guanine nucleotide regulatory component of adenylate cyclase: Subunit dissociation and the inhibition of adenylate cyclase in S49 lymphoma cyc- and wild type membranes. *J. Biol. Chem.* **259,** 3586–3595.

Kaumann A. J. (1982) Cultured heart cells as a model for β-adrenoceptors in a heart pacemaker. *Naunyn-Schmiedeberg's Arch. Pharmacol.* **320,** 119–129.

Kuo J. F. and Greengard P. (1969) Cyclic nucleotide-dependent protein kinases. IV. Widespread occurrence of adenosine 3′, 5′-monophosphate-dependent protein kinase in various tissues and phyla of the animal kingdom. *Proc. Natl. Acad. Sci. USA* **64,** 1348–1355.

Lands A. M., Arnold A., McAuliff J. P., Luduena F. P., and Brown T. G. (1967) Differentiation of receptor systems activated by sympathomimetic amines. *Nature* (Lond.) **214,** 597–598.

Langer S. Z. (1974) Presynaptic regulation of catecholamine release. *Biochem. Pharmacol.* **23,** 1793–1800.

Langer S. Z., Massingham R., and Shepperson N. B. (1980) Presence of postsynaptic α_2-adrenoceptors of predominantly extrasynaptic location in the vascular smooth muscle of the dog hind limb. *Clin. Sci.* **59,** 2255–2285.

Lefkowitz R. J. and Williams L. T. (1977) Catecholamine binding to the beta-adrenergic receptor. *Proc. Natl. Acad. Sci. USA* **74,** 515–519.

Lefkowitz R. J., Mukherjee C., Coverstone M., and Caron M. G. (1974) Stereospecific (^{3}H)-(-)-alprenolol binding sites, beta-adrenergic receptors and adenylate cyclase. *Biochem. Biophys. Res. Commun.* **60,** 703–709.

Lefkowitz R. J., Mullikin D., and Caron M. G. (1976) Regulation of beta adrenergic receptors by guanyl-5'-yl imidodiphosphate and other purine nucleotides. *J. Biol. Chem.* **251,** 4684–4692.

Levitzki A., Sevilia N., Atlas D., and Steer M. L. (1975) Ligand specificity and characteristics of the beta-adrenergic receptor in turkey erythrocyte plasma membranes. *J. Mol. Biol.* **97,** 35–53.

Levy B. (1966) The adrenergic blocking activity of *N*-tertbutylmethoxamine (butoxamine). *J. Pharmacol. Exp. Ther.* **151,** 413–422.

Limbird L. E., Speck J. L., and Smith S. K. (1982) Sodium ion modulates agonist and antagonists interactions with the human platelet alpha$_2$-adrenergic receptor in membrane and solubilized preparations. *Mol. Pharmacol.* **21,** 609–617.

Limbird L. E. and Lefkowitz R. J. (1978) Agonist-induced increase in apparent beta-adrenergic receptor size. *Proc. Natl. Acad. Sci. USA* **75,** 228–232.

Limbird L. E., Gill D. M., and Lefkowitz R. J. (1980) Agonist-promoted coupling of the beta-adrenergic receptor with the guanine nucleotide regulatory protein of the adenylate cyclase system. *Proc. Natl. Acad. Sci. USA* **77,** 775–779.

Maguire M. E., Van Arsdale P. M., and Gilman A. G. (1976) An agonist-specific effect of guanine nucleotides on binding to the beta adrenergic receptor. *Mol. Pharmacol.* **12,** 335–339.

Menkes D. B. and Aghajanian G. K. (1981) Alpha$_1$-adrenoceptor mediated responses in the lateral geniculate nucleus are enhanced by chronic antidepressant treatment. *Eur. J. Pharmacol.* **74,** 27–35.

Menkes D. B., Aghajanian G. K., and Gallager D. W. (1983) Chronic antidepressant treatment enhances agonist affinity of brain alpha$_1$-adrenoceptors. *Eur. J. Pharmacol.* **87,** 35–41.

Michell R. H. (1975) Inositol phospholipids and cell surface receptor function. *Biochim. Biophys. Acta* **415,** 81–147.

Minneman K. P., Dibner M. D., Wolfe B. B., and Molinoff, P. B. (1979a) Beta$_1$- and beta$_2$-adrenergic receptors in rat cerebral cortex are independently regulated. *Science* **204,** 866–868.

Minneman K. P., Hedberg A., and Molinoff P. B. (1979b) Comparison of beta adrenergic receptor subtypes in mammalian tissues. *J. Pharmacol. Exp. Ther.* **211,** 502–508.

Minneman K. P., Hegstrand L. R., and Molinoff P. B. (1979c) Simultaneous determination of beta-1 and beta-2 adrenergic receptors in tissues containing both receptor subtypes. *Mol. Pharmacol.* **16,** 34–46.

Minneman K. P., Weiland G. A., and Molinoff P. B. (1980) A comparison of the beta-adrenergic receptor of the turkey erythrocyte with mammalian beta$_1$ and beta$_2$ receptors. *Mol. Pharmacol.* **17,** 1–7.

Minneman K. P., Pittman R. N., and Molinoff P. B. (1981) Beta-adrenergic receptor subtypes: properties, distribution, and regulation. *Ann. Rev. Neurosci.* **4,** 419–461.

Minneman K. P., Fox A. W., and Abel P. W. (1983) Occupancy of alpha-1 adrenergic receptors and contraction of rat vas deferens. *Mol. Pharmacol.* **22,** 359–368.

Minneman K. P. and Abel P. W. (1984a) "Spare" alpha$_1$-adrenergic receptors and the potency of agonists in rat vas deferens. *Mol. Pharmacol.* **25,** 56–63.

Minneman K. P. and Abel P. W. (1984b) Relationship between α_1-adrenergic receptor density and functional response of rat vas deferens: Studies with phenoxybenzamine. *Naunyn-Schmiedeberg's Arch. Pharmacol.,* **329,** 238–246.

Moran N. C. and Perkins M. E. (1958) Adrenergic blockade of the mammalian heart by a dichloro analog of isoproterenol. *J. Pharmacol. Exp. Ther.* **124,** 223–237.

Morris M. J., Elghozi J-L., Dausse J-P., and Meyer P. (1981) Alpha$_1$- and alpha$_2$-adrenoceptors in rat cerebral cortex: Effect of frontal lobotomy. *Naunyn-Schmiedeberg's Arch. Pharmacol.* **316,** 42–44.

Nahorski S. R. (1976) Association of high affinity stereospecific binding of ^{3}H-propranolol to cerebral membranes with beta-adrenergic receptors. *Nature* (Lond.) **259,** 488–489.

Nahorski S. R. (1977) Altered responsiveness of cerebral beta adrenoceptors assessed by adenosine cyclic 3′,5′-monophosphate formation and ^{3}H-propranolol binding. *Mol. Pharmacol.* **13,** 679–689.

Nickerson M. (1949) The pharmacology of adrenergic blockade. *Pharmacol. Rev.* **1,** 27–101.

Nickerson M. (1956) Receptor occupancy and tissue response. *Nature* (Lond.) **178,** 697–698.

Nishizuka Y. (1984) The role of protein kinase C in cell surface signal transduction and tumor promotion. *Nature* (Lond.) **308,** 693–698.

O'Donnell S. R. and Wanstall J. C. (1979) The use of functional antagonism to determine whether β-adrenoceptor agonists must have a lower efficacy than isoprenaline to be trachea-atria selective in guinea pigs. *Brit. J. Pharmacol.* **60,** 255–262.

Olpe H. R. and Schellenberg A. (1980) Reduced sensitivity of neurons to noradrenaline after chronic treatment with antidepressant drugs. *Eur. J. Pharmacol.* **63,** 7–13.

Palmer G. C. (1972) Increased cyclic AMP response to norepinephrine in the rat brain following 6-hydroxydopamine. *Neuropharmacol.* **11,** 145–149.

Perry B. D. and U'Prichard D. C. (1981) ^{3}H-Rauwolscine (alpha-yohimbine): A specific antagonist radioligand for brain alpha-2-adrenergic receptors. *Eur. J. Pharmacol.* **76,** 461–464.

Pfeuffer T. (1977) GTP-Binding proteins in membranes and the control of adenylate cyclase activity. *J. Biol. Chem.* **252,** 7224–7234.

Pfeuffer T. and Helmreich E. J. M. (1975) Activation of pigeon erythrocyte membrane adenylate cyclase by guanylnucleotide analog and separation of a nucleotide binding protein. *J. Biol. Chem.* **250,** 867–876.

Reader T. A. and Briere R. (1982) Selective noradrenergic denervation and ^{3}H-prazosin binding sites in rat neocortex. *Brain. Res. Bull.* **10,** 155–158.

Rodbell M., Krans H. M. J., Pohl S. L., and Birnbaumer L. (1971) The glucagon-sensitive adenyl cyclase system in plasma membranes of rat liver. IV. Effects of guanyl nucleotides on binding of ^{125}I-glucagon. *J. Biol. Chem.* **246,** 1872–1876.

Rosell S. and Belfrage E. (1975) Adrenergic receptors in adipose tissue and their relation to adrenergic innervation. *Nature* (Lond.) **253,** 738–739.

Ross E. M. and Gilman A. G. (1980) Biochemical properties of hormone-sensitive adenylate cyclase. *Ann. Rev. Biochem.* **49,** 533–564.

Rouot B. R. and Snyder S. H. (1979) (^{3}H)-*Para*-amino-clonidine: A novel ligand which binds with high affinity to alpha-adrenergic receptors. *Life Sci.* **25,** 769–774.

Rouot B., Quennedey M. C., and Schwartz S. (1982) Characteristics of the ^{3}H-yohimbine binding on rat brain alpha$_2$-adrenoceptors. *Naunyn-Schmiedeberg's Arch. Pharmacol.* **321,** 253–259.

Ruffolo R. R., Jr. (1982) Important concepts of receptor theory. *J. Autonom. Pharmacol.* **2,** 277–295.

Russell M. P. and Moran N. C. (1980) Evidence for lack of innervation of beta-2 adrenoceptors in the blood vessels of the gracilis muscle of the dog. *Circ. Res.* **46,** 344–352.

Scatchard G. (1949) The attraction of proteins for small molecules and ions. *Ann. NY Acad. Sci.* **51,** 600–672.

Schoffelmeer A. N. M. and Mulder A. H. (1982) ^{3}*N*-Noradrenaline and ^{3}H-5-hydroxytryptamine release from rat brain slices and its presynaptic alpha-adrenergic modulation after long-term desipramine pretreatment. *Naunyn-Schmiedeberg's Arch. Pharmacol.* **318,** 173–180.

Sheys E. M. and Green R. D. (1972) A quantitative study of alpha adrenergic receptors in the spleen and aorta of the rabbit. *J. Pharmacol. Exp. Ther.* **180,** 317–325.

Slater I. H. and Powell C. E. (1957) Blockade of adrenergic inhibitory receptor sites by 1-(3′,4′-dichlorophenyl)-2-isopropylaminoethanol hydrochloride. *Fed. Proc.* **16,** 336.

Skolnick P., Stalvey L. P., Daly J. W., Hoyler E., and Davis J. N. (1978) Binding of alpha- and beta-adrenergic ligands to cerebral cortical membranes: Effect of 6-hydroxydopamine treatment and relationship to the responsiveness of cyclic-AMP generating systems in two rat strains. *Eur. J. Pharmacol.* **47,** 201–210.

Smith S. K. and Limbird L. E. (1981) Solubilization of human platelet alpha-adrenergic receptors: Evidence that agonist occupancy of the receptor stabilizes receptor–effector interactions. *Proc. Natl. Acad. Sci.* **78,** 4026–4030.

Snavely M. D. and Insel P. A. (1982) Characterization of alpha-adrenergic receptor subtypes in the rat renal cortex: Differential regulation of alpha$_1$-and alpha$_2$-adrenergic receptors by guanyl nucleotides and Na^+. *Mol. Pharmacol.* **22,** 532–546.

Sporn J. R., Harden T. K., Wolfe B. B., and Molinoff P. B. (1976) Beta-adrenergic receptor involvement in 6-hydroxydopamine-induced supersensitivity in rat cerebral cortex. *Science* **194,** 624–626.

Sporn J. R., Wolfe B. B., Harden T. K., and Molinoff P. B. (1977) Supersensitivity in rat cerebral cortex: Pre- and postsynaptic effects of 6-hydroxydopamine at noradrenergic synapses. *Mol. Pharmacol.* **13,** 1170–1180.

Starke K. (1981) Alpha-adrenoceptor subclassification. *Rev. Physiol. Biochem. Pharmacol.* **88,** 199–236.

Starke K., Borowski E., and Endo T. (1975) Preferential blockade of presynaptic alpha receptors by yohimbine. *Eur. J. Pharmacol.* **34,** 385–388.

Stephenson R. P. (1956) A modification of receptor theory. *Brit. J. Pharmacol.* **11,** 379–393.

Sternweis P. C., Northrup J. K., Smigel M. D., and Gilman A. G. (1981) The regulatory component of adenylate cyclase: purification and properties. *J. Biol. Chem.* **256,** 11517–11526.

Streb H., Irvine R. F., Berridge M. J., and Schulz I. (1983) Release of calcium from a non-mitochondrial intracellular store in pancreatic acinar cells by inositol 1,4,5-triphosphate. *Nature* (Lond.) **306,** 67–69.

Su Y-F., Harden T. K., and Perkins J. P. (1979) Isoproterenol-induced desensitization of adenylate cyclase in human astrocytoma cells: Relation of loss of hormonal responsiveness and decrement in beta-adrenergic receptors. *J. Biol. Chem.* **254,** 38–41.

Su T-F., Harden T. K., and Perkins J. P. (1980) Catecholamine-specific desensitization of adenylate cyclase: Evidence for a multi-step process. *J. Biol. Chem.* **255,** 7410–7419.

Sugrue M. F. (1982) A study of the sensitivity of rat brain $alpha_2$-adrenoceptors during chronic antidepressant treatments. *Naunyn-Schmiedeberg's Arch. Pharmacol.* **320,** 90–96.

Sugrue M. F. (1983) Chronic antidepressant therapy and associated changes in central monoaminergic receptor functioning. *Pharmacol. Ther.* **21,** 1–33.

Sutherland E. W. and Rall T. W. (1960) The relation of adenosine 3′,5′-monophosphate and phosphorylase to the actions of catecholamines and other hormones. *Pharmacol. Rev.* **12,** 265–299.

Terasaki W. L., Linden J., and Brooker G. (1979) Quantitative relationship between beta-adrenergic receptor number and physiologic responses as studied with a long-lasting beta-adrenergic antagonist. *Proc. Natl. Acad. Sci. USA* **76,** 6401–6405.

Tharp M. D., Hoffman B. B., and Lefkowitz R. J. (1981) Alpha-adrenergic receptors in human adipocyte membranes: Direct determination by ^{3}H-yohimbine binding. *J. Clin. Endocrinol. Metab.* **52,** 709–714.

Timmermans P. B. M. W. M., Kwa H. Y., and van Zwieten P. A. (1979) Possible subdivision of postsynaptic alpha drenoceptors mediating pressor responses in the pithed rat. *Naunyn-Schmiedeberg's Arch. Pharmacol.* **310,** 189–196.

Trendelenburg U. (1963) Supersensitivity and subsensitivity to sypathomimetic amines. *Pharmacol. Rev.* **15,** 225–276.

Tsai B. S. and Lefkowitz R. J. (1978a) Agonist-specific effects of monovalent and divalent cations on adenylate cyclase-coupled alpha-adrenergic receptors in rabbit platelets. *Mol. Pharmacol.* **14,** 540–548.

Tsai B. S. and Lefkowitz R. J. (1978b) ^{3}H-Dihydroergocryptine binding to alpha-adrenergic receptors in canine aortic membranes. *J. Pharmacol. Exp. Ther.* **204,** 606–614.

Tse J., Wrenn R. W., and Kuo J. F. (1980) Thyroxine-induced changes in characteristics and activities of beta-adrenergic receptors and adenosine 3′,5′-monophosphate and guanosine 3′,5′-monophosphate systems in the heart may be related to reputed catecholamine supersensitivity in hyperthyroidism. *Endocrinol.* **107,** 6–16.

U'Prichard D. C. and Snyder S. H. (1978) Guanyl nucleotide influences on ^{3}H-ligand binding to alpha-noradrenergic receptors in calf brain membranes. *J. Biol. Chem.* **253,** 3444–3452.

U'Prichard D. C., Bechtel W. D., Rouot B. M., and Snyder S. H. (1979) Multiple apparent alpha-noradrenergic receptor binding sites in rat brain: Effect of 6-hydroxydopamine. *Mol. Pharmacol.* **16,** 47–60.

U'Prichard D. C. and Snyder S. H. (1977) Binding of ^{3}H-catecholamines at alpha-noradrenergic receptor sites in calf brain. *J. Biol. Chem.* **252,** 6450–6453.

van den Brink F. G. (1973) The mode of functional interaction II. Experimental verification of a new model: The antagonism of beta-adrenoceptor stimulants and other agonists. *Eur. J. Pharmacol.* **22,** 279–286.

van Meel J. C. A., Qian T., Welffert B., DeJonge A., Timmermans P. B. M. W. M., and van Zwieten P. (1983) Effects of adrenalectomy and demedullation of the adrenals on vasopressor responses to stimulation of postjunctional α_2-adrenoceptors. *Eur. J. Pharmacol.* **91,** 171–179.

Venter J. C. (1979) High efficiency coupling between beta-adrenergic receptors and cardiac contractility: Direct evidence for "spare" beta-adrenergic receptors. *Mol. Pharmacol.* **16,** 429–440.

Vetulani J. and Sulser F. (1975) Action of various antidepressant treatments reduces reactivity of noradrenergic cyclic AMP generating systems in limbic forebrain. *Nature* (Lond.) **257,** 495–497.

Viveros O. H., Garlick D. G., and Renken E. M. (1968) Sympathetic beta adrenergic vasodilatation in skeletal muscle of the dog. *Am. J. Physiol.* **215,** 1218–1225.

Walsh D. A., Perkins J. P., and Krebs E. G. (1968) An adenosine 3′,5′-monophosphate-dependent protein kinase from rabbit skeletal muscle. *J. Biol. Chem.* **243,** 3763–3765.

Weiland G. A., Minneman K. P., and Molinoff P. B. (1979) Fundamental difference between the molecular interactions of agonists and antagonists with the beta-adrenergic receptor. *Nature* (Lond.) **281,** 114–117.

Wikberg J. E. S., Abero M., Caron M. G., and Hagen P. O. (1983) Norepinephrine-induced down regulation of alpha$_1$ adrenergic receptors in cultured rabbit aorta smooth muscle cells. *Life Sci.* **33,** 1409–1417.

Wilffert B., Timmermans P. B. M. W. M., and van Zwieten P. A. (1982) Extrasynaptic location of alpha-2 and noninnervated beta-2 adrenoceptors in the vascular system of the pithed normotensive rat. *J. Pharmacol. Exp. Ther.* **221,** 762–768.

Williams L. T., Mullikin D., and Lefkowitz R. J. (1976) Identification of alpha-adrenergic receptors in uterine smooth muscle membranes by ^{3}H-dihydroergokryptine binding. *J. Biol. Chem.* **251,** 6915–6923.

Williams L. T., Lefkowitz R. J., Watanabe A. M., Hathaway D. R., and Besch H. R., Jr. (1977) Thyroid hormone regulation of beta-adrenergic receptor number. *J. Biol. Chem.* **252,** 2787–2789.

Williams L. T., Mullikin D., and Lefkowitz R. J. (1978) Magnesium dependence of agonist binding to adenylate cyclase-coupled hormone receptors. *J. Biol. Chem.* **253,** 2984–2989.

Wolfe B. B., Zirrolli J. A., and Molinoff P. B. (1974) Binding of ^{3}H-*dl*-epinephrine to proteins of rat ventricular muscle: Nonidentity with beta-adrenergic receptors. *Mol. Pharmacol.* **10,** 582–596.

Wolfe B. B., Harden T. K., and Molinoff P. B. (1976) Beta-adrenergic receptors in rat liver: Effects of adrenalectomy. *Proc. Natl. Acad. Sci. USA* **73,** 1343–1347.

Wolfe B. B., Harden T. K., Sporn J. R., and Molinoff P. B. (1978) Presynaptic modulation of beta adrenergic receptors in rat cerebral cortex after treatment with antidepressants. *J. Pharmacol. Exp. Ther.* **207,** 446–457.

Wolfe B. B. and Harden T. K. (1981) Guanine nucleotides modulate the affinity of antagonists at beta-adrenergic receptors. *J. Cyclic Nucl. Res.* **7,** 303–312.

Yeh H. and Woodward P. J. (1983) Alterations in β-adrenergic physiological response characteristics after long-term treatment with desmethylimipramine. *J. Pharmacol. Exp. Ther.* **226,** 126–134.

Chapter 4
Serotonin Receptors

Stephen J. Peroutka

1. Introduction

Since the middle of the nineteenth century, physiologists have been aware of an endogenous vasoconstrictor substance. Often referred to as "vasotonin," the factor was present in the serum of clotted blood. Rapport et al. (1948) succeeded in isolating the compound and named it "serotonin." The chemical structure was soon found to be 5-hydroxytryptamine (5-HT). Simultaneously and independently, Italian scientists were studying a substance found in high concentrations in enterochromaffin cells of the intestinal mucosa that was also eventually found to be 5-HT (Erpsamer and Asero, 1952). The synthetic production of 5-HT (Hamlin and Fischer, 1951) led to an explosion of research into the physiologic function of this compound.

The initial discovery of 5-HT in the central nervous system was made by Twarog and Page (1953). Histochemical studies identified the raphe nuclei as the main 5-HT-containing cell bodies in the brain. Ascending fibers from the raphe enter the median forebrain bundle and provide diffuse innervation to all areas of the cortex. Descending axons from the caudal raphe nuclei project to the dorsal, lateral, and intermediate nuclei of the spinal cord. The characterization of specific synthesis, release, and storage mechanisms for 5-HT confirmed that the biogenic amine is a central neurotransmitter.

The concept of "receptors" for hormones and neurotransmitters is largely based on pharmacologic studies (Snyder, 1983). Prior to the use of molecular probes, pharmacologists could only measure the response of physiologic systems to pharmacologic agents. The actions of drugs in these systems were presumed to be mediated via specific recognition sites on cellular membranes. During the past decade, technical advances in the field of molecular pharmacology have allowed these recognition sites or "receptors" to be specifically analyzed.

However, evidence exists that not one, but multiple 5-HT receptors, exist in the central nervous system (Fillion, 1983). For example, at least two 5-HT binding sites in brain homogenates have been differentiated by ^{3}H-ligand techniques (Peroutka and Snyder, 1979; Peroutka and Snyder, 1983). Anatomic studies have confirmed that a variety of serotonergic recognition sites, with distinct regional localizations, exist in the central nervous system (Young and Kuhar, 1980; Palacios et al., 1983). Neurophysiologically, the variability of effects of 5-HT and related agents has also led to a hypothesis of multiple 5-HT receptors in the central nervous system (Rogawski and Aghajanian, 1981).

In this chapter, the effects of 5-HT in a number of physiologic systems will be reviewed. The focus will be on the use of radioligand binding studies to characterize 5-HT receptor subtypes in brain membranes. The data derived from these studies will then be compared with the effects of 5-HT and related agents as measured in a variety of physiologic systems. Thus, an attempt will be made to define functional correlates of 5-HT binding sites in order that these membrane recognition sites can be appropriately designated 5-HT "receptors."

2. Radioligand Binding Studies of Serotonin Receptors

Radioligand binding studies represent one of the most significant recent methodological breakthroughs in pharmacologic research (Snyder, 1984). The development of affinity chromatography, centrifugation separation, and, most importantly, rapid filtration methods enabled membrane receptors in the central nervous system to be studied by radioligand techniques. Radioreceptor assays can provide accurate information on receptor density, drug affinity, and kinetic constants. Perhaps most importantly, the assay allows for a large number of drugs to be simultaneously tested under the same experimental conditions.

2.1. Preliminary Binding Studies

Radioligand techniques in brain membranes have been used to characterize multiple neurotransmitter and drug receptor sites in brain homogenates (Yamamura et al., 1978; Snyder, 1984). The first successful radioligand analysis of 5-HT receptors was reported by Bennett and Aghajanian (1974). The binding of ^{3}H-*d*-lysergic acid diethylamide (LSD) was saturable, reversible, and displayed high affinity (K_d = 7.5 n*M*) for its membrane recognition site. In addition, the binding was stereoselective since *l*-LSD was more than three orders of magnitude weaker than *d*-LSD in competing for the radioligand binding site. The binding also displayed appropriate regional variations since brain regions with the highest density of receptors were areas known to receive a dense projection of 5-HT nerve terminals. These findings were soon extended and confirmed by other laboratories (Bennett and Snyder, 1975; Lovell and Freedman, 1976).

The second radioligand used to label 5-HT receptors was ^{3}H-5-HT (Bennett and Snyder, 1976). Like ^{3}H-LSD binding, the radiolabel bound with high affinity, and was saturable, stereospecific, and displayed appropriate regional variations. However, important discrepancies were noted between the binding of ^{3}H-LSD and ^{3}H-5-HT. First, agonists had greater affinity for ^{3}H-5-HT- than for ^{3}H-LSD-labeled sites. By contrast, classical antagonists such as cyproheptadine and cinanserin had markedly higher affinity for ^{3}H-LSD than ^{3}H-5-HT binding. Moreover, ^{3}H-LSD was invariably found to bind to a larger number of receptor sites than ^{3}H-5-HT. At the time, Bennett and Snyder (1976) suggested that ^{3}H-5-HT and ^{3}H-LSD did not label the same membrane recognition site, but rather two different "states" of the same receptor.

Radioligand analysis of 5-HT receptors was considerably advanced by the finding that ^{3}H-spiperone could also be used to label presumed 5-HT recognition sites (Leysen et al., 1978). Previously, ^{3}H-spiperone had been used as a label of dopamine receptors. However, in the rat frontal cortex where dopamine receptors are sparse, ^{3}H-spiperone was found to label a receptor that appeared to be "serotonergic" in the sense that 5-HT antagonists were the most potent displacers of the ligand. However, 5-HT and related tryptamines were extremely weak displacers of ^{3}H-spiperone. Nonetheless, a significant correlation was obtained between drug IC_{50} values for ^{3}H-LSD binding and ^{3}H-spiperone binding. The ability of ^{3}H-spiperone to label 5-HT receptors was quickly confirmed by other laboratories (Creese and Snyder, 1978; Quik et al., 1978).

2.2. Differentiation of 5-HT$_1$ and 5-HT$_2$ Receptors

Thus, subtle differences were noted between the binding characteristics of ^{3}H-5-HT, ^{3}H-LSD, and ^{3}H-spiperone (Peroutka and Snyder, 1979). If each ligand labeled the same membrane recogni-

Table 1
Drug Affinities for 5-HT_1 and 5-HT_2 Receptors[a]

Drug	K_i, nM		
	5-HT_1 ^{3}H-5-HT	5-HT_1 + 5-HT_2 ^{3}H-LSD	5-HT_2 ^{3}H-spiperone
Tryptamines			
5-Hydroxytryptamine	3.8	110	2700
5-Methoxytryptamine	11	210	2700
Bufotenine	37	220	840
5,6-Dihydroxytryptamine	300	3700	22000
Ergots			
D-LSD	10	7.3	13
Metergoline	12	9.0	3.9
Methysergide	88	8.4	2.6
Bromo-LSD	89	13	2.5
Antagonists			
Spiperone	730	18	0.51
Cyproheptadine	1500	90	2.0
Cinanserin	1800	130	18
Ketanserin	3000	—	2.1

[a]Rat frontal cortex membranes were incubated under standard assay conditions with 2.0 nM ^{3}H-5-HT, 4.2 nM ^{3}H-LSD, or 0.26 nM ^{3}H-spiperone, together with four concentrations of unlabeled drugs. K_i values were determined from IC_{50} values as calculated from log-probit analysis (from Peroutka and Snyder, 1979; Peroutka et al., 1981; Peroutka et al., 1983).

tion site, then unlabeled drugs should be equipotent in displacing ^{3}H-5-HT, ^{3}H-LSD, and ^{3}H-spiperone. This pattern is observed with LSD displacement of the three ligands. A K_i value of approximately 10 nM is obtained against each of the three ligands (Table 1). In marked contrast, 5-HT is approximately three orders of magnitude more potent in displacing ^{3}H-5-HT (3.8 nM) than ^{3}H-spiperone (2700 nM). Its apparent K_i for ^{3}H-LSD binding is 110 nM, a value that is intermediate between its affinity for ^{3}H-5-HT and ^{3}H-spiperone labeled sites. Furthermore, the Hill slope for 5-HT displacement of ^{3}H-LSD, but not of ^{3}H-5-HT or ^{3}H-spiperone, binding is significantly less than unity (Peroutka and Snyder, 1979). The converse pattern is observed with spiperone displacement of the three radioligands. For example, spiperone is extremely potent against ^{3}H-spiperone binding (0.51 n*M*), yet has fourteen hundred-fold less affinity for ^{3}H-5-HT labeled sites (730 n*M*). In addition, its apparent K_i for ^{3}H-LSD binding is 18 n*M*, but the displacement curve is biphasic with a "plateau" occurring between 10 and 30 n*M* spiperone.

All three ligands display monophasic saturation with a single binding component in evidence on Scatchard analysis (Peroutka and Snyder, 1979). However, whereas the maximal number of ^{3}H-5-HT and ^{3}H-spiperone binding sites is approximately 10 pmol/g wet wt tissue, the amount of specifically bound ^{3}H-LSD is slightly greater than 20 pmol/g wet wt tissue. Furthermore, the binding of ^{3}H-5-HT is completely blocked by 300 n*M* 5-HT, whereas ^{3}H-LSD binding is decreased by 50%. Similarly, 30 n*M* spiperone abolishes ^{3}H-spiperone binding, but only displaces half of ^{3}H-LSD binding. In both cases, saturation analysis by Scatchard plots reveals that the decrease in ^{3}H-LSD binding represents a decrease in the number of binding sites (B_{max}) with no change in the apparent affinity (K_D) of ^{3}H-LSD for the remaining sites.

Given the results outlined above and the fact that no correlation exists between drug potencies for ^{3}H-5-HT- and ^{3}H-spiperone-labeled "serotonergic" receptors, Peroutka and Snyder (1979) concluded that at least two distinct 5-HT membrane recognition sites exist in the central nervous system. The sites labeled by ^{3}H-5-HT were designated "5-HT$_1$ receptors" and those labeled by ^{3}H-spiperone were designated "5-HT$_2$ receptors." Since ^{3}H-LSD had equal affinity for both sites, it was proposed that this ligand could be used to label both 5-HT$_1$ and 5-HT$_2$ receptors.

The existence of 5-HT$_1$ and 5-HT$_2$ sites has been confirmed and extended by a number of laboratories. Each site has a number of unique characteristics (Table 2). Thus, the 5-HT$_1$ site can be labeled by ^{3}H-5-HT, ^{3}H-LSD, or ^{3}H-8-hydroxy-2-(di-*n*-propylamino)-tetralin (PAT) (Bennett and Snyder, 1976; Fillion et al., 1978; Nelson et al., 1978; Peroutka and Snyder, 1979; Gozlan et al., 1983). Although atypical displacement curves have suggested the possibility of 5-HT$_1$ subtypes (Pedigo et al., 1981), certain basic pharmacologic characteristics are shared by all 5-HT$_1$ receptors. For example, the 5-HT$_1$ site has nanomolar affinity for 5-HT, is saturable and stereospecific, and displays appropriate regional variations. The binding site has high affinity for 5-HT and related agonists, but comparatively low affinity for most classical 5-HT antagonists (Table 1). Only a few 5-HT antagonists, such as meter-

Table 2
Characteristics of $5\text{-}HT_1$ and $5\text{-}HT_2$ Receptors

	$5\text{-}HT_1$	$5\text{-}HT_2$
Labeled by	^3H-5-HT ^3H-LSD ^3H-PAT	^3H-Spiperone ^3H-LSD ^3H-Mianserin ^3H-Ketanserin ^{125}I-LSD ^3H-Mesulergine
5-HT affinity	Nanomolar	Micromolar
High-affinity antagonists	Metergoline Methiothepin	Metergoline Methiothepin Classical antagonists Neuroleptics Antidepressants
Effect of chronic antidepressants	None	Decreased number of receptors
Region of highest density	Hippocampus	Frontal cortex

goline and methiothepin, have nanomolar affinity for the $5\text{-}HT_1$ site. However, these drugs are also extremely potent at $5\text{-}HT_2$ sites. As a result, a selective, high-affinity $5\text{-}HT_1$ antagonist has not yet been described. In contrast, ^3H-spiperone, ^3H-LSD, ^3H-mianserin, ^3H-ketanserin, and ^{125}I-LSD can be used to label the $5\text{-}HT_2$ binding site (Leysen, 1981; Leysen et al., 1982; Peroutka and Snyder, 1983; Hartig et al., 1983; Engel et al., 1984). Serotonergic antagonists have high affinity for this site, whereas 5-HT and related tryptamines are markedly less potent. The number of $5\text{-}HT_2$, but not $5\text{-}HT_1$, binding sites can be decreased by chronic treatment with antidepressant drugs (Peroutka and Snyder, 1980a,b).

2.3. $5\text{-}HT_1$ Receptor Subtypes

Atypical displacement of ^3H-5-HT by spiperone led to the initial suggestion that $5\text{-}HT_1$ subtypes exist in the central nervous system (Pedigo et al., 1981). Sites with high affinity (K_i = 2–13 nM) for spiperone were designated 5-HT_{1A} sites, whereas sites with relatively low affinity for spiperone (K_i = 35 μM) were called 5-HT_{1B} sites. These two sites had different regional localizations, suggesting that at least two subtypes of 5-HT_1 receptors were present in rat brain. Similar results have been obtained in other species (Schnellmann et al., 1984). Autoradiographic studies (see below) have also confirmed the ability of spiperone to differentiate at least two types of sites labeled by ^{3}H-5-HT (Deshmukh et al., 1983).

More recently, it has been proposed that ^{3}H-PAT can specifically label a presynaptic 5-HT autoreceptor (Gozlan et al., 1983). In the hippocampus, the binding properties of ^{3}H-PAT correlated with ^{3}H-5-HT binding. However, striatal binding of ^{3}H-PAT did not correspond to ^{3}H-5-HT sites. The fact that 5,7-dihydroxytryptamine-induced lesions of the raphe system caused a loss of ^{3}H-PAT binding in the striatum, but not in the hippocampus, led the authors to hypothesize that ^{3}H-PAT labeled a presynaptic autoreceptor. It has recently been suggested that ^{3}H-PAT binding recognizes only the 5-HT_{1A} subtype of 5-HT receptor (Middlemiss and Fozard, 1983; Marcinkiewicz et al., 1984).

3. Regional Localization of 5-HT_1 and 5-HT_2 Receptors

3.1. Radioligand Binding Studies

Marked regional variations in both 5-HT_1 and 5-HT_2 binding sites have been noted (Peroutka and Snyder, 1981a). In general, 5-HT_1 sites predominate in all brain regions. The only exception is in the cerebral cortex, where equal numbers of both receptors are found. The hippocampus has the highest density of 5-HT_1 sites; the lowest density is in the cerebellum. In contrast, the highest level of 5-HT_2 binding is in the cerebral cortex and caudate, with all other brain regions having substantially fewer binding sites. 5-HT_2 regional studies have been performed using ^{3}H-spiperone, ^{3}H-LSD, ^{3}H-mianserin, and ^{3}H-ketanserin with excellent correlations between ligands (Peroutka and Snyder, 1981; Schotte et al., 1983).

3.2. Autoradiographic Analysis

Autoradiographic analysis of 5-HT receptors was first reported by Young and Kuhar (1980). Using ^{3}H-5-HT and ^{3}H-LSD, they found that 5-HT_1 receptor density was particularly high in the molecular layer of the dentate gyrus, CA3 and CA4 of the hippocampal formation, subiculum, and the median and dorsal raphe nuclei. A more detailed autoradiographic study of ^{3}H-LSD binding in the human brainstem confirmed the very high concentration of 5-HT$_1$ binding sites in the raphe nuclei (Palacios et al., 1983). In addition, autoradiographic analysis of ^{3}H-5-HT binding has also confirmed the high density of 5-HT$_1$ sites in the hippocampus (Biegon et al., 1982).

As previously mentioned, 5-HT$_1$ subtypes can also be labeled by autoradiographic techniques. Based on spiperone-displaceable ^{3}H-5-HT binding, 5-HT$_{1A}$ sites were found to predominate in the hippocampus, septum, and cerebral cortex (Deshmukh et al., 1983). On the other hand, the highest densities of 5-HT$_{1B}$ sites were found in the caudate nucleus, superior colliculus, lateral geniculate body, subiculum, and substantia nigra. 5-HT$_1$ sites can also be at least partially labeled by ^{3}H-PAT (Marcinkiewicz et al., 1984). In general, the distribution of specific ^{3}H-5-HT and ^{3}H-PAT binding was found to be very similar. In the choroid plexus, striatum, subiculum, and substantia nigra, however, very low densities of ^{3}H-PAT sites were identified despite the presence of high amounts of ^{3}H-5-HT binding. The specific binding in the choroid plexus has recently been designated the "5-HT$_{1C}$ site" (Cortes et al., 1984).

The 5-HT$_2$ site has been studied using ^{3}H-spiperone and ^{3}H-ketanserin. Using cinanserin-displaceable ^{3}H-spiperone binding as a label for 5-HT$_2$ sites, the highest density of specific binding was observed in lamina IV of the neocortex (Palacios et al., 1981). Similarly, a moderate amount of 5-HT$_2$ binding using ^{3}H-ketanserin was found in lamina IV of the cortex, whereas a high density of specific binding was identified in the striatum (Slater and Patel, 1983). The significance of the high-density ^{3}H-ketanserin binding in the caudate has not yet been determined.

4. Physiologic Correlates of 5-HT$_1$ and 5-HT$_2$ Receptors

Radioligand techniques provide a rapid and sensitive measure of drug potencies at specific membrane recognition sites. However, binding studies cannot differentiate the agonist vs antagonist properties of drugs, relative efficacy of agents, or drug interactions with other receptor sites involved in a specific physiologic response. This type of pharmacologic information can only be derived from physiologic experiments. Therefore, correlation of radioligand data with physiologic responses is necessary before a binding site can be considered a "receptor." However, physiologic studies are also limited in their ability to generate useful pharmacologic data. First, physiologic studies have usually analyzed only 5-HT and a few related agents. In order to correlate radioligand binding data with physiologic studies, an extensive analysis of both agonists and antagonists must be performed in both systems (Furchgott, 1978). Second, tissue preparations may differentially hinder pharmacologic agents from reaching the receptor site. The possible problems of drug absorption, metabolism, and tissue distribution are even more important in in vivo studies. Third, high concentrations of neurotransmitters such as 5-HT may act simultaneously at multiple sites. Attributing the total effect of an agonist to a single receptor site may be erroneous. The ideal physiologic system would therefore contain relatively "pure" populations of receptor subtypes.

In the following section, a number of physiologic models will be analyzed for possible relationships to 5-HT_1 and 5-HT_2 binding sites. As shown in Table 3, a relationship between various physiologic systems and radioligand binding sites has been proposed by a number of investigators. As will be described below, the

Table 3
Proposed Functional Correlates of 5-HT Binding Sites

<u>5-HT_1</u>
Adenylate cyclase stimulation
"Autoreceptor" control of neurotransmitter release
Contraction of canine basilar artery and saphenous vein
Serotonin behavioral syndrome
Contraction of rat fundus strip
Antinoceptive effects of 5-HT
Hypotensive effects of 5-HT
Neurophysiologic inhibition of neuronal firing
<u>5-HT_2</u>
5-Hydroxytryptophan- or mescaline-induced head twitches
Tryptamine-induced seizures
Contraction of vascular smooth muscle
Classical "D" receptors
Rat forepaw edema
Discriminative cue properties
Contraction of bronchial smooth muscle
Platelet shape changes and aggregation
Smooth muscle prostacyclin synthesis
Inositol phospholipid breakdown
Neurophysiologic excitation of neuronal firing

suggested relationships have been based on the similarities between the physiologic and radioligand potencies of 5-HT and related agents.

4.1. Serotonin-Sensitive Adenylate Cyclase

A 5-HT-sensitive adenylate cyclase in mammalian brain was first detected in newborn rats (Von Hungen et al., 1974; Von Hungen et al., 1975). Subsequent studies confirmed that half-maximal stimulation of the enzyme occurred at micromolar concentrations of 5-HT (Enjalbert et al., 1978a,b). Classical 5-HT antagonists also interacted with the enzyme at micromolar concentrations. An extensive analysis of the anatomic, developmental, and pharmacologic properties of this preparation has clearly shown that activation of this particular adenylate cyclase is not related to 5-HT_1 or 5-HT_2 binding sites (Nelson et al., 1980a,b; Leysen and Tollenaere, 1982).

However, a markedly different type of 5-HT sensitive adenylate cyclase has been detected by Fillion and coworkers (Fillion et al., 1979a,b, 1980, 1983). Using a radioimmunoassay technique that detects femtomole quantities of cyclic AMP, they have described a cyclase that is stimulated by nanomolar concentrations of 5-HT. The effect of 5-HT is antagonized by classical 5-HT antagonists, but only at micromolar or higher concentrations. The cyclase system is located primarily in purified synaptic membranes, in contrast to a low-affinity 5-HT cyclase system that is present in glial cell membranes and glial cell cultures (see below). The lesion studies also support an association between the 5-HT_1 binding site and the high-affinity 5-HT cyclase system (Fillion et al., 1979a, 1980, 1983). A similar type of 5-HT-sensitive adenylate cyclase has been analyzed in rat hippocampal membranes (Barbaccia et al., 1983), guinea pig hippocampal membranes (Shenker et al., 1983), and the NCB-20 hybrid clonal cell line (Berry-Kravis and Dawson, 1983).

Radioligand binding data also support an association between 5-HT_1 receptors and a 5-HT-sensitive adenylate cyclase. Regulation of neurotransmitter binding by guanine nucleotides often reflects an association of the agonist binding site with an adenylate cyclase (Maguire et al., 1977; Rodbell, 1980; Limbird, 1981). Thus, GTP and GDP, but not GMP, inhibit the binding of ^{3}H-5-HT to brain homogenates (Peroutka et al., 1979; Mallat and Hamon, 1982). In addition, guanine nucleotides significantly reduce agonist potencies for ^{3}H-5-HT binding sites, whereas antagonist potencies are not affected by nucleotides. Furthermore, guanine nucleotides have a profound synergistic effect on 5-HT-sensitive cyclase systems (Fillion et al., 1979b; Nelson et al., 1980b; Mallat and Hamon, 1982; Barbaccia et al., 1983). In summary, an association between the 5-HT_1 binding site and the high-affinity adenylate cyclase system is supported by both the pattern of pharmacological interactions at the cyclase in certain systems, as well as by the effects of guanine nucleotides on ^{3}H-5-HT binding.

However, it must be emphasized that this relationship is extremely dependent on the cyclase system being studied. For example, the 5-HT-sensitive cyclase located both in glial membranes (Fillion et al., 1980) and monkey cortex (Ahn and Makman, 1978) has an EC_{50} value in the micromolar range. The effect of 5-HT can be blocked by nanomolar concentrations of 5-HT antagonists and certain neuroleptics. In addition, under certain assay conditions, guanine nucleotides have been shown to affect the binding of ^{3}H-spiperone and ^{3}H-ketanserin to 5-HT_2 receptors (Kendall and Nahorski, 1983; Titeler et al., 1984). As a result, indirect evidence also supports a possible linkage of 5-HT_2 receptors, in certain systems, with an adenylate cyclase.

4.2. "Autoreceptor" Modulation of Neurotransmitter Release

Neurotransmitters are stored in synaptic vesicles located in the nerve terminal. Depolarization of the neuron by potassium or electrical stimulation causes release of the prestored neurotransmitter. The release can be inhibited by 5-HT and related agonists, presumably through a "presynaptic autoreceptor." Initial studies analyzed ^{3}H-5-HT release from rat hypothalamic synaptosomes or cortical slices (Farnebo and Hamberger, 1974; Cerrito and Raiteri, 1979; Gothert, 1980). In both systems, nanomolar concentrations of 5-HT inhibited ^{3}H-5-HT release. Slightly higher concentrations of methiothepin and metergoline were able to block the action of 5-HT. However, micromolar concentrations of cyproheptadine, methysergide, and mianserin had no effect on the action of exogenous 5-HT. Metergoline and methiothepin were also the most potent inhibitors of ^{3}H-5-HT release following potassium-evoked depolarization of rat raphe nuclei and hypothalamic slices (Ennis and Cox, 1982; Cox and Ennis, 1982).

Direct comparisons have been made between the pharmacologic interactions at "autoreceptors" and 5-HT_1 binding sites. Using rat hypothalamic synaptosomes, the inhibitory effect of nanomolar concentrations of 5-HT was blocked by methiothepin (IC_{50} = 3.8 n*M*) and quipazine (IC_{50} = 670 n*M*). Seven other 5-HT antagonists, including spiperone, cyproheptadine, and cinanserin, had no effect on the system at micromolar concentrations (Martin and Sanders-Bush, 1982a). In a separate study of eight agonists, a significant correlation ($r = 0.91$; $p < 0.01$) was found between potassium-evoked release of ^{3}H-5-HT from hypothalamic synaptosomes and agonist affinities for the 5-HT_1 binding site (Martin and Sanders-Bush, 1982b).

Engel and coworkers have compared the potencies of 14 indole derivatives at "autoreceptors" with their affinity for 5-HT_1 binding sites (Engel et al., 1983). A significant correlation between the 5-HT_1 binding site and the 5-HT "autoreceptor" was found using either electrical stimulation to induce release of ^{3}H-5-HT from rat cortical slices ($r = 0.91$), or overflow of ^{3}H-norepinephrine from strips of canine saphenous vein ($r = 0.78$). Furthermore, agonist interactions in each of these two systems was significantly correlated ($r = 0.83$). In addition, antagonist studies confirmed that methiothepin and metergoline were the most potent antagonists, whereas selective 5-HT_2 antagonists such as ketanserin, cyproheptadine, and cinanserin had no effect on the "autoreceptor" at micromolar concentrations. As a result, extensive pharmacologic analyses of 5-HT "autoreceptors" support the hypothesis that the receptor mediating release of 5-HT and certain other neurotransmitters from presumed nerve terminals is the 5-HT_1 binding site.

4.3. Smooth Muscle Contraction Studies

The effect of 5-HT on smooth muscle varies greatly depending on the specific organ being studied. This fact is particularly apparent with vascular smooth muscle. For example, reported $K_{ED_{50}}$ values for 5-HT range from 5.0 n*M* in the canine basilar artery (Allen et al., 1976) to 6000 n*M* in the rat aorta (Cohen et al., 1981). However, certain patterns of interactions have consistently been noted between serotonergic drugs and vasculature. First, intracranial vascular smooth muscle is invariably more sensitive to 5-HT than most extracranial vessels (Bohr et al., 1961; Toda and Fujita, 1973; Edvinsson and Hardebo, 1976; Edvinsson et al., 1978; Allen and Banghart, 1979; Forster and Whalley, 1982). Second, in vessels displaying nanomolar $K_{ED_{50}}$ values for 5-HT, inhibition by classical 5-HT antagonists is noncompetitive (Peroutka, 1984). Third, in vasculature with both high affinity for 5-HT and noncompetitive inhibition by antagonists, methysergide is often noted to act as a partial agonist (Allen et al., 1974; Curro et al., 1978; Apperley et al., 1980; Muller-Schweinitzer, 1980; Forster and Whalley, 1982). In contrast, methysergide and other classical 5-HT antagonists are potent competitive inhibitors in blood vessels in which the $K_{ED_{50}}$ value of 5-HT is greater than 100 n*M*.

Thus, the two distinct types of vascular responses may correlate with 5-HT_1 and 5-HT_2 binding sites (Peroutka, 1984). The canine basilar artery and saphenous vein, and the human basilar artery and pial vessels are all contracted by nanomolar concentrations of 5-HT (Peroutka et al., 1983). Antagonism with classical 5-HT antagonists and neuroleptics occurs at micromolar concentrations and is noncompetitive. In addition, autoradiographic analysis has shown that 5-HT_1, but not 5-HT_2, binding sites can be identified on the tunica media of the canine and human basilar arteries (Peroutka and Kuhar, 1984). In marked contrast, potent competitive antagonism of a number of vascular smooth muscle preparations has been documented. Specific correlations with 5-HT_2 binding sites have been claimed for the rabbit aorta, rat aorta, jugular vein, and caudal artery. In these vessels, approximately micromolar concentrations of 5-HT can be blocked by nanomolar concentrations of antagonists (Cohen et al., 1981; Humphrey et al., 1982; Leysen et al., 1982; Cohen et al., 1983; Feniuk et al., 1983; Maayani and Stollak, 1983). Furthermore, it has been suggested that the classical "D" receptor, as defined by Gaddum and Picarelli (1957), is synonomous with 5-HT_2 receptors (Humphrey et al., 1982; Bradley et al., 1983).

Correlates of 5-HT binding sites may also exist in nonvascular smooth muscle. For example, the rat stomach strip preparation can be contracted by 5-HT with $K_{ED_{50}}$ values in the nanomolar range (Wrigglesworth, 1983). Again, however, micromolar concentrations of potent 5-HT_2 antagonists, such as amitriptyline, trazodone, and cyproheptadine, are necessary to block the action of 5-HT in this preparation. Moreover, the most potent antagonists are metergoline and methysergide, whereas ketanserin and spiperone are orders of magnitude weaker (Leysen et al., 1982). A significant correlation ($r = 0.64$; $p < 0.01$) has been obtained between the rank order of drug interactions ($n = 19$) with the rat stomach strip and their affinity for 5-HT_1 sites (Leysen and Tollenaere, 1982). On the other hand, 5-HT-induced contractions of the bronchial smooth muscle located in the guinea pig trachea correlate significantly with 5-HT_2 receptor activation (Van Nueten et al., 1982).

4.4. Neurophysiologic Effects of 5-HT

The neurophysiologic effects of iontophoretically applied 5-HT and related agents have also led to the concept that multiple 5-HT receptors exist in the central nervous system (Roberts and Straughan, 1967; Haigler and Aghajanian, 1977; Rogawski and Aghajanian, 1981). 5-HT produces either inhibition or excitation of cell firing when applied microiontophoretically to single neurons. When 5-HT inhibits cell firing rates, the action is invariably mimicked by both LSD and methysergide. By contrast, the classical antagonists have no effect on cell inhibition produced by 5-HT. Only a single antagonist, metergoline, has been shown to antagonize the inhibitory effect of 5-HT (Sastry and Phillis, 1977). The comparatively higher potency of LSD vs 5-HT in inhibiting raphe neurons vs nonraphe cells has been cited as evidence that at least two types of inhibitory receptor are present in neurons. However, the ratio of potencies in the two neuronal groups is only 2.74 (Rogawski and Aghajanian, 1981).

5-HT-induced excitation, on the other hand, is invariably blocked by 5-HT antagonists such as cyproheptadine and cinanserin. In addition, both methysergide and LSD are antagonists of 5-HT-induced cellular excitation. Neither drug has any known 5-HT-mimetic action in terms of neuronal excitation. Furthermore, 5-HT-induced excitation is much less common than 5-HT-induced inhibition. Although direct comparisons with radioligand binding data have not been made, the pharmacologic characteristics of 5-HT-induced cellular inhibition and excitation appear to correspond with 5-HT_1 and 5-HT_2 binding sites, respectively (Peroutka et al., 1981).

4.5. Behavioral Studies

A complex behavioral hyperactivity results after central 5-HT stimulation with drugs such as 5-hydroxytryptophan (5-HTP) (Jacobs, 1976). The syndrome includes resting tremor, forepaw treading, hindlimb abduction, Straub tail, hyperreactivity, and head twitching. Attempts have been made to correlate many of the behavioral components with 5-HT_1 and/or 5-HT_2 binding sites. Drug antagonism of the "head-shake" or "head-twitch component" of the syndrome has been correlated with blockade of 5-HT_2 receptors (Peroutka et al., 1981; Leysen et al., 1982; Yap and Taylor, 1983; Colpaert and Janssen, 1983). In addition, tryptamine-induced seizure activity could be prevented by antagonists in a manner consistent with antagonism of the 5-HT_2 receptor (Leysen et al., 1978). In addition, Ortmann et al. (1982) found that antagonism of the head tremor and forepaw padding components of the syndrome also correlated with 5-HT_2 binding.

However, analysis of other behavioral components of the syndrome has shown that the entire set of behaviors could be blocked only by metergoline and methysergide (Lucki et al., 1984). In contrast, relatively high concentrations of ketanserin and pipamperone had no effect on 5-HTP-induced forepaw treading, head weaving, tremor, hindlimb abduction, extended posture, or Straub tail. These findings suggest that the more subtle components of the 5-HT behavioral "syndrome," as opposed to the "head-twitch," may be related to activation of 5-HT_1 receptors.

4.6. Other Systems

Methysergide and bromo-LSD, but not spiperone or ketanserin, inhibit the antinociceptive effects of intrathecal 5-HT. This finding was attributed to mediation of the spinal effects of 5-HT by the 5-HT_1 receptor (Schmauss et al., 1983). In a limited study of five tryptamine derivatives, the hypotensive potencies in pentobarbitone-anesthetized rats correlated with their affinity for 5-HT_1 sites ($r = 0.92$) (Kalkman et al., 1983b).

Antagonism of 5-HT-induced forepaw edema in the rat by 22 antagonists correlated with drug affinity for 5-HT_2 sites labeled by ^{3}H-spiperone (Ortmann et al., 1982). Two independent behavioral studies of the discriminative cue properties of 5-HT agonists have concluded that this behavioral response may be mediated by 5-HT_2 receptors (Glennon et al., 1983; Friedman et al., 1983). The dose-dependent pressor response that occurs after administration of 5-HT to pithed normotensive rat can also be blocked in a manner consistent with 5-HT_2 receptor mediation (Kalkman et al., 1983a). Induction of platelet shape changes and aggregation by 5-HT may also be mediated by 5-HT_2 receptors (DeClerck and Herman, 1983; Pletscher and Affolter, 1983; Geaney et al., 1984). Finally, preliminary analysis of 5-HT-induced prostacyclin synthesis in cultured bovine aortic smooth muscle cells (Coughlin et al., 1984) and inositol phospholipid hydroylysis in rat cortical slices (Kendall and Nahorski, 1984) suggests that these events may occur as a result of 5-HT_2 receptor activation.

5. Future Strategies for Further Differentiation of Serotonin Receptor Subtypes

In the present report, evidence is provided from a variety of biochemical, anatomic, and physiologic studies that multiple 5-HT receptors exist in the central nervous system. At the present time, the differentiation of 5-HT receptors into 5-HT_1 and 5-HT_2 subtypes appears to be the most relevant classification system. Indeed, a number of functional correlates of these 5-HT "binding sites" have been proposed (Table 3). For many of the suggested physiologic correlates, significant correlations have been documented between drug potencies in various physiologic systems and affinities for 5-HT_1 or 5-HT_2 binding sites (Tables 4 and 5). These correla-

Table 4
Physiologic Correlates of 5-HT_1 Binding Sites

Physiologic system	Number of drugs studied, *N*	Coorrelation coefficient vs ^{3}H-5-HT binding	Significance level, *p* value	References
Adenylate cyclase	10	0.92	0.01	Peroutka et al., 1981
"Autoreceptors" 1. Rat hypothalamic synaptosomes	8	0.91	0.01	Martin and Sanders-Bush, 1982b
2. Rat cortical slices	14	0.91	0.01	Engel et al., 1983
3. Canine saphenous vein	13	0.78	0.01	Engel et al., 1983
Canine basilar artery	11	0.84	0.01	Peroutka et al., 1983
Rat fundus strip	19	0.64	0.01	Leysen and Tollenaere, 1982
Hypotensive action of 5-HT	5	0.92	0.05	Kalkman et al., 1983

Table 5
Physiologic Correlates of $5\text{-}HT_2$ Binding Sites

Physiologic system	Number of drugs studied, *N*	Coorrelation coefficient vs ^3H-spiperone binding	Significance level, *p* value	References
Induced head twitch 1. 5-HTP 2. Mescaline	13	0.98	0.001	Peroutka et al., 1981
Tryptamine seizures	22	0.87	0.001	Leysen, 1981
Head-tremor and forepaw padding	22	0.80	0.01	Ortmann et al., 1982
Rat caudal artery contractions	24	0.82	0.001	Leysen, 1981
Forepaw edema	22	0.80	0.01	Ortmann et al., 1982
Guinea pig trachea contractions	23	0.62	0.01	Van Nueten et al., 1982
Inhibition of quipazine discrimination	10	0.87	0.01	Friedman et al., 1983a
Pressor response in pithed rats	11	0.89	0.01	Kalkman et al., 1983
Platelet shape changes	11	0.93	0.001	Geaney et al., 1984

tions allow the membrane recognition sites described by radioligand studies to be designated 5-HT "receptors."

To date, however, the lack of specific and selective pharmacologic probes has limited 5-HT research. Although the recent development of selective 5-HT_1 agonists, such as 8-OH-PAT, has facilitated the classification of 5-HT_1 receptor subtypes, a selective 5-HT_1 antagonist has not yet been described. Similarly, a potent and selective 5-HT_2 agonist has not yet been reported. The development of such a drug would allow for a simple and rational analysis of the effects of 5-HT in various physiologic systems. Eventually, such knowledge should lead to a better understanding of the role of 5-HT in the central nervous system.

Addendum

Since the submission of this chapter, rapid progress has been made in the differentiation and characterization of 5-HT_1 receptor subtypes. The 5-HT_{1A} site has been well characterized using ^{3}H-8-OH=DPAT (Hall et al., 1985), [^{3}H]-TVX Q 7821 (Dompert et al., 1985), and [^{3}H]-WB 4101 (Norman et al., 1985). The 5-HT_{1B} site has been more difficult to characterize, but progress has been made using [^{3}H]-5-HT (Sills et al., 1984; Peroutka, 1985) and 1^{125}-cyanopindolol (Pazos et al., 1985b). The 5-HT_{1C} site has been characterized in pig choroid plexus and cortex (Pazos et al., 1985a).

References

Ahn H. S. and Makman M. H. (1978) Serotonin sensitive adenylate cyclase activity in monkey anterior limbic cortex: antagonism by molindone and other antipsychotic drugs. *Life Sci.* **23,** 507–512.

Allen G. S. and Banghart S. B. (1979) Cerebral arterial spasm: Part 9. In vitro effects of nifedipine on serotonin-, phenylephrine-, and potassium-induced contractions of canine basilar and femoral arteries. *Neurosurg.* **4,** 37–42.

Allen G. S., Gross C. J., Henderson L. M., and Chou S. N. (1976) Cerebral arterial spasm. Part 4: In vitro effects of temperature, serotonin, analogues, large nonphysiologic concentrations of serotonin and extracellular calcium and magnesium on serotonin-induced contractions of the canine basilar artery. *J. Neurosurg.* **44,** 585–593.

Allen G. S., Henderson L. M., Chou S. N., and French L. A. (1974) Cerebral arterial spasm. Part 1: In vitro contractile activity of vasoactive agents on canine basilar and middle cerebral arteries. *J. Neurosurg.* **40,** 433–441.

Apperley E., Feniuk W., Humphrey P. P. A., and Levy G. P. (1980) Evidence for two types of excitatory receptor for 5-hydroxytryptamine in dog isolated vasculature. *Brit. J. Pharmacol.* **68,** 215–224.

Barbaccia M. L., Brunello N., Chuang D. M., and Costa E. (1983) Serotonin-elicited amplification of adenylate cyclase activity in hippocampal membranes from adult rat. *J. Neurochem.* **40,** 1671–1679.

Bennett J. L. and Aghajanian G. K. (1974) D-LSD binding to brain homogenates: possible relationship to serotonin receptors. *Life Sci.* **15,** 1935–1944.

Bennett Jr., J. P. and Snyder S. H. (1975) Stereospecific binding of D-lysergic acid diethylamide (LSD) to brain membranes: relationship to serotonin receptors. *Brain Res.* **94,** 523–544.

Bennett Jr., J. P. and Snyder S. H. (1976) Serotonin and lysergic acid diethylamide binding in rat brain membranes: relationship to postsynaptic serotonin receptors. *Mol. Pharmacol.* **12,** 373–389.

Berry-Kravis E. and Dawson G. (1983) Characterization of an adenylate cyclase-linked serotonin (5-HT_1) receptor in a neuroblastoma × brain explant hybrid cell line (NCB-20). *J. Neurochem.* **40,** 977–985.

Biegon A., Rainbow T. C., and McEwen B. S. (1982) Quantitative autoradiography of serotonin receptors in the rat brain. *Brain Res.* **242,** 197–204.

Bohr D. F., Goulet P. L., and Taquini A. C. (1961) Direct tension recording from smooth muscle of resistance vessels from various organs. *Angiology* **12,** 478–485.

Bradley P. B., Humphrey P. P. A., and Williams R. H. (1983) Are vascular 'D' and '5-HT_2' receptors for 5-hydroxytryptamine the same? *Brit. J. Pharmacol.* **79,** 295P.

Cerrito F. and Raiteri M. (1979) Serotonin release is modulated by presynaptic autoreceptors. *Eur. J. Pharmacol.* **57,** 427–430.

Cohen M. L., Mason N., Wiley K. S., and Fuller R. W. (1983) Further evidence that vascular serotonin receptors are of the 5-HT_2 type. *Biochem. Pharmacol.* **32,** 567–570.

Cohen M. L., Fuller R. W., and Wiley K. S. (1981) Evidence for 5-HT_2 receptors mediating contraction in vascular smooth muscle. *J. Pharmacol. Exp. Ther.* **218,** 421–425.

Colpaert F. C. and Janssen P. A. J. (1983) The head-twitch response to intraperitoneal injection of 5-hydroxytryptophan in the rat: antagonist effects of purported 5-hydroxytryptamine antagonists and of pirenperone, an LSD antagonist. *Neuropharmacol.* **22,** 993–1000.

Cortes R., Palacios J. M., and Pazos A. (1984) Visualisation of multiple serotonin receptors in the rat brain by autoradiography. *Brit. J. Pharmacol.* **82** (suppl), 202P.

Coughlin S. R., Moskowitz M. A., and Levine L. (1984) Identification of a serotonin type 2 receptor linked to prostacyclin synthesis in vascular smooth muscle cells. *Biochem. Pharmacol.* **33,** 692–694.

Cox B. and Ennis C. (1982) Characterization of 5-hydroxytryptaminergic autoreceptors in the rat hypothalamus. *J. Pharm. Pharmacol.* **34,** 438–441.

Creese I. and Snyder S. H. (1978) ^{3}H-Spiroperidol labels serotonin receptors in rat cerebral cortex and hippocampus. *Eur. J. Pharmacol.* **49,** 201–202.

Curro F. A., Greenberg S., Verbeuren T. J., and Vanhoutte P. M. (1978) Interaction between alpha adrenergic and serotonergic activation of canine saphenous veins. *J. Pharmacol. Exp. Ther.* **207,** 936–949.

DeClerck F. F. and Herman A. G. (1983) Hydroxytryptamine and platelet aggregation. *Fed. Proc.* **42,** 228–232.

Deshmukh P. P., Yamamura H. I., Woods L., and Nelson D. L. (1983) Computer-assisted autoradiographic localization of subtypes of $serotonin_1$ receptors in rat brain. *Brain Res.* **288,** 338–343.

Dompert W. U., Glaser T., and Traber J. (1985) ^{3}H-TVX Q 7821: Identification of 5-HT_1 binding sites as target for a novel putative anxiolytic. *Naunyn-Schmiedeberg's Arch. Pharmacol.* **328,** 467–470.

Edvinsson L. and Hardebo J. E. (1976) Characterization of serotonin receptors in intracranial and extracranial vessels. *Acta Physiol. Scand.* **97,** 523–525.

Edvinsson L., Hardebo J. E., and Owman C. (1978) Pharmacological analysis of 5-hydroxytryptamine receptors in isolated intracranial and extracranial vessels of cat and man. *Circ. Res.* **42,** 143–151.

Engel G., Gothert M., Muller-Schweinitzer E., Schlicker E., Sistonen L., and Stadler P. A. (1983) Evidence for common pharmacological properties of [^{3}H]5-hydroxytryptamine binding sites, presynaptic 5-hydroxytryptamine autoreceptors in CNS, and inhibitory presynaptic 5-hydroxytryptamine receptors on sympathetic nerves. *Naunyn Schmiedeberg's Arch. Pharmacol.* **324,** 116–124.

Engel G., Muller-Schweinitzer E., and Palacios J. M. (1984) 2-[^{125}Iodo] LSD, a new ligand for the characterisation and localisation of $5HT_2$ receptors. *Naunyn Schmiedeberg's Arch. Pharmacol.* **325,** 328–336.

Enjalbert A., Bourgoin S., Hamon M., Adrien J., and Bockaert J. (1978a) Postsynaptic serotonin-sensitive adenylate cyclase in the central nervous system. *Mol. Pharmacol.* **14,** 2–10.

Enjalbert A., Hamon M., Bourgoin S., and Bockaert J. (1978b) Postsynaptic serotonin-sensitive adenylate cyclase in the central nervous system. *Mol. Pharmacol.* **14,** 11–23.

Ennis C. and Cox B. (1982) Pharmacological evidence for the existence of two distinct serotonin receptors in rat brain. *Neuropharmacol.* **21,** 41–44.

Erpsamer V. and Asero B. (1952) Identification of enteramine, the specific hormone of the enterochromaffin cell system, as 5-hydroxytryptamine. *Nature* (Lond.) **168,** 800–801.

Farnebo L. O. and Hamberger B. (1974) Regulation of [^{3}H] 5-hydroxytryptamine release from rat brain slices. *J. Pharm. Pharmacol.* **26,** 642–644.

Feniuk W., Humphrey P. P. A., and Watts A. D. (1983) Further evidence for the heterogeneity of vascular receptors for 5-HT. *Brit. J. Pharmacol.* **79,** 296P.

Fillion G. (1983) 5-Hydroxytryptamine Receptors in Brain, in *Handbook of Psychopharmacology*, Vol. 17 (Iversen L. L., Iversen S. D., and Snyder S. H., eds.) pp. 139–166, Plenum, New York.

Fillion G. M. B., Rousselle J., Fillion M., Beaudoin D. M., Goiny M. R., Deniau J., and Jacob J. J. (1978) High-affinity binding of $[^3H]$5-Hydroxytryptamine to brain synaptosomal membranes: comparison with $[^3H]$lysergic acid diethylamide binding. *Mol. Pharmacol.* **14,** 50–59.

Fillion G., Beaudoin D., Rousselle J. C., Deniau J. M., Fillion M. P., Dray F., and Jacob J. (1979a) Decrease of $[^3H]$5-HT high-affinity binding and 5-HT adenylate cyclase activation after kainic acid lesion in rat brain striatum. *J. Neurochem.* **33,** 567–570.

Fillion G., Rousselle J. C., Beaudoin D., Pradelles P., Goiny M., Dray F., and Jacob, J. (1979b) Serotonin sensitive adenylate cyclase in horse brain synaptosomal membranes. *Life Sci.* **24,** 1813–1822.

Fillion G., Beaudoin D., Rousselle J. C., and Jacob J. (1980) $[^3H]$5-HT binding sites and 5-HT-sensitive adenylate cyclase in glial cell membrane fraction. *Brain Res.* **198,** 361–374.

Fillion G., Beaudoin D., Fillion M., Rousselle J. C., Robaut C., and Netter Y. (1983) 5-Hydroxytryptamine receptors in neurons and glia. *J. Neural Transmission,* Suppl. **18,** 307–317.

Forster C. and Whalley E. T. (1982) Analysis of the 5-hydroxytryptamine induced contraction of the human basilar arterial strip compared with the rat aortic strip in vitro. *Naunyn Schmiedeberg's Arch. Pharmacol.* **319,** 12–17.

Friedman R. L., Barrett R. J., and Sanders-Bush E. (1983) Discriminative cue properties of quipazine: mediation by serotonin-2 binding sites. *Soc. Neurosci. Abs.* **9,** 335.

Furchgott R. F. (1978) Pharmacological characterization of receptors: its relation to radioligand-binding studies. *Fed. Proc.* **37,** 115–120.

Gaddum J. H. and Picarelli Z. P. (1957) Two kinds of tryptamine receptor. *Brit. J. Pharmacol. Chemother.* **12,** 323–328.

Geaney D. P., Schachter M., Elliot J. M., and Grahame-Smith D. G. (1984) Characterisation of $[^3H]$lysergic acid diethylamide binding to a 5-hydroxytryptamine receptor on human platelet membranes. *Eur. J. Pharmacol.* **97,** 87–93.

Glennon R. A., Young R., and Rosecrans J. A. (1983) Antagonism of the effects of the hallucinogen DOM and the purported 5-HT agonist quipazine by 5-HT_2 antagonists. *Eur. J. Pharmacol.* **91,** 189–196.

Gothert M. (1980) Serotonin-receptor-mediated modulation of Ca^{2+}-

dependent 5-hydroxytryptamine release from neurons of the rat brain cortex. *Naunyn-Schmiedeberg's Arch. Pharmacol.* **314,** 223–230.

Gozlan H., El Mestikawy S., Pichat L., Glowinski J., and Hamon M. (1983) Identification of presynaptic serotonin autoreceptors using a new ligand: ^{3}H-PAT. *Nature* (Lond.) **305,** 140–142.

Haigler H. J. and Aghajanian G. K. (1977) Serotonin receptors in the brain. *Fed. Proc.* **36,** 2159–2164.

Hall M., El Mestikawy S., Emerit M., Pichat L., Hamon M., Gozlan H. (1985) ^{3}H-8-Hydroxy-2-(di-*n*-propylamino) tetralin binding to pre- and postsynaptic 5-hydroxytryptamine sites in various regions of the rat brain. *J. Neurochem.* **44,** 1686–1696.

Hamlin K. E. and Fischer F. E. (1951) The synthesis of 5-hydroxytryptamine. *J. Am. Chem. Soc.* **73,** 5007–5008.

Hartig P. R., Kadan M. J., Evans J. J., and Krohn A. M. (1983) ^{125}I-LSD: a high-sensitivity ligand for serotonin receptors. *Eur. J. Pharmacol.* **89,** 321–322.

Humphrey P. P. A., Feniuk W., and Watts A. D. (1982) Ketanserin—a novel antihypertensive drug?. *J. Pharm. Pharmacol.* **34,** 541.

Jacobs B. L. (1976) An animal behavioral model for studying central serotonergic synapses. *Life Sci.* **19,** 777–786.

Kalkman H. O., Batink H. D., Thoolen M. J. M. C., Timmermans P. B. M. W. M., and Van Zwieten P. A. (1983a) Correlation between the affinity for [^{3}H]mianserin-labeled receptors in brain and antagonism of the serotonin pressor response in pithed rats. *Biochem. Pharmacol.* **32,** 2111–2113.

Kalkman H. O., Boddeke H. W. G. M., Doods H. N., Timmermans P. B. M. W. M., and Van Zwieten P. A. (1983b) Hypotensive activity of serotonin receptor agonists in rats is related to their affinity for 5-HT_1 receptors. *Eur. J. Pharmacol.* **91,** 155–156.

Kendall D. A. and Nahorski S. R. (1983) Temperature-dependent 5-hydroxytryptamine (5-HT)-sensitive [^{3}H]spiperone binding to rat cortical membranes: regulation by guanine nucleotide and antidepressant treatment. *J. Pharmacol. Exp. Ther.* **227,** 429–434.

Kendall D. A. and Nahorski S. R. (1984) Suppression of 5-HT_2 receptor mediated inositol phospholipid breakdown in brain by chronic antidepressant treatment. *Brit. J. Pharmacol.* **82** (suppl), 206P.

Leysen J. E. (1981) Serotoninergic receptors in brain tissue: properties and identification of various ^{3}H-ligand binding studies in vitro. *J. Physiol.* (Paris) **77,** 351–362.

Leysen J. E. and Tollenaere J. P. (1982) Biochemical models for serotonin receptors. *Ann. Rev. Med. Chem.* **17,** 1–10.

Leysen J. E., Niemegeers C. J. E., Tollenaere J. P., and Laduron P. M. (1978) Serotonergic component of neuroleptic receptors. *Nature* (Lond.) **272,** 163–166.

Leysen J. E., Niemegers C. J. E., Van Nueten J. M., and Laduron P. M.

(1982) ^{3}H-Ketanserin (R 41 468), a selective ^{3}H-ligand for receptor binding sites. *Mol. Pharmacol.* **21,** 301–314.

Limbird L. E. (1981) Activation and attenuation of adenylate cyclase. *Biochem. J.* **195,** 1–13.

Lovell R. A. and Freedman D. X. (1976) Stereospecific receptor sites for D-lysergic acid diethylamide in rat brain: effects of neurotransmitters, amine antagonists, and other psychotropic drugs. *Mol. Pharmacol.* **12,** 620–630.

Lucki I., Nobler M. S., and Frazer A. (1984) Different actions of serotonin antagonists on two behavioral models of serotonin receptor activation in the rat. *J. Pharmacol. Exp. Ther.* **228,** 133–139.

Maguire M. E., Ross E. M., and Gilman A. G. (1977) β-Adrenergic receptor: ligand binding properties and the interaction with adenylyl cyclase. *Adv. Cyclic Nucleotide Res.* **8,** 1–83.

Maayani S. and Stollak J. (1983) 5-HT receptors in isolated rabbit aorta: Characterization by spiroperidol and other butyrophenones. *Fed. Proc.* **42,** 1150.

Mallat M. and Hamon M. (1982) Ca^{2+}-guanine nucleotide interactions in brain membranes. I. Modulation of central 5-hydroxytryptamine receptors in the rat. *J. Neurochem.* **38,** 151–161.

Marcinkiewicz M., Verge D., Gozlan H., Pichat L., and Hamon M. (1984) Autoradiographic evidence for the heterogeneity of 5-HT_1 sites in the rat brain. *Brain Res.* **291,** 159–163.

Martin L. L. and Sanders-Bush E. (1982a) The serotonin autoreceptor: antagonism by quipazine. *Neuropharmacol.* **21,** 445–450.

Martin L. L. and Sanders-Bush E. (1982b) Comparison of the pharmacological characteristics of 5 HT_1 and 5 HT_2 binding sites with those of serotonin autoreceptors which modulate serotonin release. *Naunyn-Schmiedeberg's Arch. Pharmacol.* **321,** 165–170.

Middlemiss D. N. and Fozard J. R. (1983) 8-hydroxy-2(DI-*n*-Propylamino)-tetralin discriminates between sybtypes of the 5-HT_1 recognition site. *Eur. J. Pharmacol.* **90,** 151–153.

Muller-Schweinitzer E. (1980) The mechanism of ergometrine-induced coronary arterial spasm: In vitro studies on canine arteries. *J. Cardiovas. Pharmacol.* **2,** 645–655.

Nelson D. L., Herbet A., Bourgoin S., Glowinski J., and Hamon M. (1978) Characteristics of central 5-HT receptors and their adaptive changes following intracerebral 5,7-dihydroxy-tryptamine administration in the rat. *Mol. Pharmacol.* **14,** 983–995.

Nelson D. L., Herbet A., Enjalbert A., Bockaert J., and Hamon M. (1980a) Serotonin-sensitive adenylate cyclase and [^{3}H]serotonin binding sites in the CNS of the rat. I. *Biochem. Pharmacol.* **29,** 2445–2453.

Nelson D. L., Herbet A., Adrien J., Bockaert J., and Hamon M. (1980b) Serotonin-sensitive adenylate cyclase and [^{3}H]serotonin binding sites in the CNS of the rat. II. *Biochem. Pharmacol.* **29,** 2455–2463.

Norman A. B., Battaglia G., Morrow A. L., and Creese I. (1985) ^{3}H-WB4101 labels S_1 serotonin receptors in rat cerebral cortex. *Eur. J. Pharmacol.* **106:** 461–462.

Ortmann R., Bischoff S., Radeke E., Buech O., and Delini-Stula A. (1982) Correlations between different measures of antiserotonin activity of drugs. *Naunyn-Schmiedeberg's Arch. Pharmacol.* **321,** 265–270.

Palacios J. M., Niehoff D. L., and Kuhar M. J. (1981) [^{3}H]Spiperone binding sites in brain: autoradiographic localization of multiple receptors. *Brain Res.* **213,** 277–289.

Palacios J. M., Probst A., and Cortes R. (1983) The distribution of serotonin receptors in the human brain: high density of [^{3}H]LSD binding sites in the raphe nuclei of the brainstem. *Brain Res.* **274,** 150–155.

Pazos A., Engel G., and Palacios J. (1985) Beta-adrenoceptor blocking agents recognize a subpopulation of serotonin receptors in brain. *Brain Res.* **343,** 403–408.

Pazos A., Hoyer D., and Palacios J. (1985) The binding of serotonergic ligands to the porcine choroid plexus: Characterization of a new type of serotonin recognition site. *Eur. J. Pharmacol.* **106,** 539–546.

Pedigo N. W., Yamamura H. I., and Nelson D. L. (1981) Discrimination of multiple [^{3}H]5-hydroxytryptamine binding sites by the neuroleptic spiperone in rat brain. *J. Neurochem.* **36,** 220–226.

Peroutka S. J. (1984) Vascular serotonin receptors: correlation with 5-HT_1 and 5-HT_2 binding sites. *Biochem. Pharmacol.* **33,** 2349–2353.

Peroutka S. J. (1985) Selective labeling of 5-HT_{1A} and 5-HT_{1B} binding sites in bovine brain. *Brain Res.* **344,** 167–171.

Peroutka S. J. and Kuhar M. J. (1984) Autoradiographic localization of 5-HT_1 receptors to human and canine basilar arteries. *Brain Res.* **310,** 193–196.

Peroutka S. J. and Snyder S. H. (1979) Multiple serotonin receptors: Differential binding of ^{3}H-serotonin, ^{3}H-lysergic acid diethylamide, and ^{3}H-spiroperidol. *Mol. Pharmacol.* **16,** 687–699.

Peroutka S. J. and Snyder S. H. (1980a) Long-term antidepressant treatment decreases spiroperidol-labeled serotonin receptor binding. *Science* **210,** 88–90.

Peroutka S. J. and Snyder S. H. (1980b) Regulation of $serotonin_2$ (5-HT_2) receptors labeled with ^{3}H-spiroperidol by chronic treatment with the antidepressant amitriptyline. *J. Pharmacol. Exp. Ther.* **215,** 582–587.

Peroutka S. J. and Snyder S. H. (1981) Two distinct serotonin receptors: Regional variations in receptor binding in mammalian brain. *Brain Res.* **208,** 339–347.

Peroutka S. J. and Snyder S.H. (1983) Multiple serotonin receptors and their physiological significance. *Fed. Proc.* **42,** 213–217.

Peroutka S. J., Lebovitz R. M., and Snyder S. H. (1979) Serotonin receptor binding affected differentially by guanine nucleotides. *Mol. Pharmacol.* **16,** 700–708.

Peroutka S. J., Lebovitz R. M., and Snyder S. H. (1981) Two distinct central serotonin receptors with different physiological functions. *Science* **212,** 827–829.

Peroutka S. J., Noguchi M., Tolner D. J., and Allen G. S. (1983) Serotonin induced contraction of canine basilar artery: mediation by 5-HT_1 receptors. *Brain Res.* **259,** 327–330.

Pletscher A. and Affolter H. (1983) The 5-hydroxytryptamine receptor of blood platelets. *J. Neural Transmission* **57,** 233–242.

Quik M., Iversen L. L., Lardner A., and Mackay A. V. P. (1978) Use of ADTN to define specific ^{3}H-spiperone binding to receptors in brain. *Nature* (Lond.) **274,** 513–514.

Rapport M. M., Green A. A., and Page I. H. (1948) Serum vasoconstrictor (serotonin). IV. Isolation and characterization. *J. Biol. Chem.* **176,** 1243–1251.

Roberts M. H. T. and Straughan D. W. (1967) Excitation and depression of cortical neurones by 5-hydroxytryptamine. *J. Physiol.* (Lond.) **193,** 269–294.

Rodbell M. (1980) The role of hormone receptors and GTP regulatory proteins in membrane transduction. *Nature* (Lond.) **284,** 17–21.

Rogawski M. A. and Aghajanian G. K. (1981) Serotonin autoreceptors on dorsal raphe neurons: structure–activity relationships of tryptamine analogs. *J. Neurosci.* **1,** 1148–1154.

Sastry B. S. R. and Phillis J. W. (1977) Metergoline as a selective 5-hydroxytryptamine antagonist in the cerebral cortex. *Can. J. Physiol. Pharmacol.* **55,** 130–133.

Schmauss C., Hammond D. L., Ochi J. W., and Yaksh T. L. (1983) Pharmacological antagonism of the antinociceptive effects of serotonin in the rat spinal cord. *Eur. J. Pharmacol.* **90,** 349–357.

Schnellmann R. G., Waters S. J., and Nelson D. L. (1984) [^{3}H]5-hydroxytryptamine binding sites: species and tissue variation. *J. Neurochem.* **42,** 65–70.

Schotte A., Maloteaux J. M., and Laduron P. M. (1983) Characterization and regional distribution of serotonin S_2-receptors in human brain. *Brain Res.* **276,** 231–235.

Shenker A., Maayani S., Weinstein H., and Green J. P. (1983) Characterization of a serotonin receptor coupled to adenylate cyclase in adult guinea pig hippocampus. *Soc. Neurosci. Abs.* **9,** 1152.

Sills M. A., Wolfe B. B., and Frazer A. (1984b) Determination of selective and nonselective compounds for the 5-HT_{1A} and 5-HT_{1B} receptor subtypes in rat frontal cortex. *J. Pharmacol. Exp. Ther.* **231,** 480–487.

Slater P. and Patel S. (1983) Autoradiographic distribution of serotonin$_2$ receptors in rat brain. *Eur. J. Pharmacol.* **92,** 297–298.

Snyder S. H. (1983) Molecular Aspects of Neurotransmitter Receptors: An Overview, in *Handbook of Psychopharmacology,* Vol. 17, (Iversen L. L., Iversen S. D., and Snyder S. H., eds.), pp. 1–12, Plenum, New York.

Snyder S. H. (1984) Drug and neurotransmitter receptors in the brain. *Science* **224,** 22–31.

Titeler M., Battaglia G., and Shannon M. (1984) Guanine nucleotides modulate cortical S_2 serotonin receptors. *Soc. Neurosci. Abs.* **9,** 334.

Toda N. and Fujita Y. (1973) Responsiveness of isolated cerebral and peripheral arteries to serotonin, norepinephrine, and transmural electrical stimulation. *Circ. Res.* **33,** 98–104.

Twarog B. M. and Page I. H. (1953) Serotonin content of some mammalian tissues and urine and a method for its determination. *Am. J. Psychiat.* **175,** 157–161.

Van Nueten J. M., Leysen J. E., Vanhoutte P. M., and Janssen P. A. (1982) Serotonergic responses in vascular and nonvascular tissues. *Arch. Int. Pharmacodyn.* **256,** 331–334.

Von Hungen K., Roberts S., and Hill D. F. (1974) Developmental and regional variations in neurotransmitter-sensitive adenylate cyclase systems in cell-free preparations from rat brain. *J. Neurochem.* **22,** 811–819.

Von Hungen K., Roberts S., and Hill D. F. (1975) Serotonin-sensitive adenylate cyclase activity in immature rat brain. *Brain Res.* **84,** 257–267.

Wrigglesworth S. J. (1983) Heterogeneity of 5-hydroxytryptamine receptors in the rat uterus and stomach strip. *Brit. J. Pharmacol.* **80,** 691–697.

Yamamura H. I., Enna S. J., and Kuhar M. J. (eds.) (1978) *Neurotransmitter Receptor Binding.* Raven Press, New York.

Yap C. Y. and Taylor D. A. (1983) Involvement of 5-HT_2 receptors in the wet-dog shake behavior induced by 5-hydroxytryptophan in the rat. *Neuropharmacol.* **22,** 801–804.

Young W. S., III and Kuhar M. J. (1980) Serotonin receptor localization in rat brain by light microscopic autoradiography. *Eur. J. Pharmacol.* **62,** 237–239.

Chapter 5
Tryptamine and Phenylethylamine Recognition Sites in Brain

Kenneth J. Kellar and Caren S. Cascio

1. Introduction

Tryptamine (T) and 2-phenylethylamine (PE) can be formed in vivo by the decarboxylation of the precursor amino acids L-tryptophan and L-phenylalanine, respectively. As would be expected of immediate products of amino acid decarboxylation, both of these amines are phylogenetically very old. They have been detected in invertebrate, as well as mammalian, nervous systems (*see* Boulton and Juorio, 1982, for review). These amines can be synthesized locally by mammalian brain (Saavedra and Axelrod, 1973; Snodgrass and Horn, 1973; Snodgrass, 1974), and because they cross the blood–brain barrier, they can enter the brain from the circulation (Marsden and Curzon, 1978; Randrup and Munkvad, 1966). Both T and PE have effects in the CNS after systemic injection.

When injected after inhibition of monoamine oxidase (MAO), T produces several centrally mediated effects, including increased locomotor activity in rats (Foldes and Costa, 1975; Marsden and Curzon, 1978), hyperthermia and forepaw clonus in rabbits (Quock and Weick, 1978), and antagonism of 5-hydroxytryptophan-induced head twitches in mice (Jones, 1981). Foldes and Costa (1975) and Marsden and Curzon (1978) have suggested that part of the so-called "serotonin behavioral syndrome" seen in rats following administration of a monoamine oxidase inhibitor and L-tryptophan could be related to increases in brain tryptamine, as well as to increases in serotonin (5-HT). In humans, the effects of infused T were reported to be similar to those of LSD (Martin and Sloan, 1970).

The effects of PE after systemic injection into rats resemble those of amphetamine (*see* Wyatt et al., 1977, for review). In fact, it has been speculated that PE may act as an endogenous amphetamine, and that it could be involved in schizophrenia (Sandler and Reynolds, 1976; Potkin et al., 1969). Whereas the methyl substitution on the alpha-carbon of amphetamine renders it resistant to oxidation by MAO, PE is an excellent substrate for the MAO type B enzyme (Yang and Neff, 1973), and it is quickly converted to phenylacetaldehyde and then to phenylacetic acid (Wu and Boulton, 1975). Thus, the actions of PE are abbreviated.

It is not known whether T and PE act principally by modifying 5-HT and catecholamine neurotransmission processes, and thus play possible roles as neuromodulators, or whether they act through distinct systems of their own, or both. For example, iontophoretic application of T modified the firing rate of rat cerebral cortical neurons in response to 5-HT (Jones and Boulton, 1980a), but it also produced postsynaptic responses (depression of firing rate) on cortical neurons that appeared to be independent of 5-HT systems (Jones, 1982a). In addition, administration of 5-HT to chicks produced behavioral depression and EEG synchronization, whereas T produced opposite behavioral and EEG responses (Dewhurst and Marley, 1965; Jones, 1981). In rats, T and 5-HT produced opposite effects on rat body temperature when injected into the preoptic area of the hypothalamus (Cox et al., 1981).

Iontophoretically applied PE reduced the firing rate of neurons in the striatum and cerebral cortex in a manner similar to dopamine and norepinephrine (Henwood et al., 1979). It also potentiated the decrease in firing rate produced by dopamine (Jones and Boulton, 1980b).

To postulate that the effects of T and/or PE might be independent of their actions on the more established biogenic amine neurotransmitters, 5-HT, and the catecholamines, and to propose T or PE as candidate neurotransmitters or neurohormones in the CNS, certain minimum criteria must be met. These include the presence of the amine and a route of metabolism, release of the amine, a pharmacological analysis that is consistent with an independent role for the amine, and a postsynaptic response that is dependent on the amine.

Obviously, both T and PE are present in the CNS, and there exist metabolic pathways for their synthesis and degradation—both locally in the CNS and in the periphery, from which the synthesized amines can be carried to the CNS by the circulation. However, degradation by MAO in brain presumes that T and PE would have access to the enzyme, which is intracellular. This might be accomplished by simple diffusion of these amines from the synaptic space into neurons and glia; or, alternatively, they might enter neurons via uptake pumps—either their own or those of 5-HT and catecholamine neurons. Evidence for T uptake into brain slices has been reported by Jones and Broadbent (1982); however, it remains to be determined whether this uptake is saturable and has the properties expected of a specific uptake system for T.

In an elegant experiment, Jones (1982b) found that stimulation of the median raphe nucleus could produce responses in cortical neurons that were characteristic of ionotophoretically applied T. Further, there were differences in the ability of drugs to block the responses to raphe stimulation and between iontophoretically applied 5-HT and T. This suggests that stimulation of raphe neurons can lead to release of T.

At the present time, a stronger case for an independent neurohormone action can be made for T than for PE. This is largely attributable to the studies by Boulton and Jones and their colleagues. However, high-affinity binding sites for both [^{3}H]-tryptamine ([^{3}H]-T) and [^{3}H]-phenylethylamine ([^{3}H]-PE) have been found in brain, and if these binding sites can be shown to be specific receptors for T and PE, it would considerably strengthen the case for neurohormonal actions of these amines. The remainder of this chapter is aimed at reviewing the data concerning these high-affinity recognition sites for [^{3}H]-T and [^{3}H]-PE.

2. [^{3}H]-Tryptamine Recognition Sites

High-affinity binding sites for [^{3}H]-T have been found in rat and human brain (Kellar and Cascio, 1982; Cascio and Kellar, 1982, 1983; Brüning and Rommelspacher, 1984; Wood et al., 1984). These sites have kinetic and pharmacological characteristics of distinct T recognition sites in rat brain (Cascio and Kellar, 1983; Wood et al., 1984), and a site with similar characteristics is present in peripheral tissue (Brüning and Rommelspacher, 1984; Cascio and Kellar, unpublished).

The [^{3}H]-T binding assay utilizing high-specific radioactivity [^{3}H]-T (20–40 Ci/mmol; New England Nuclear) has been carried out under similar conditions in three laboratories, and there is good agreement on the characteristics of the site (Kellar and Cascio, 1982; Cascio and Kellar, 1983; Brüning and Rommelspacher, 1984; Wood et al., 1984).

In binding studies utilizing labeled agonists, unknown competition by the endogenous agonist remaining in the tissue is possible. In the case of many biogenic monoamines, this problem can be reduced by initially homogenizing the tissue in buffer that does not contain a MAO inhibitor and then, after two centrifugation steps that wash the tissue free of soluble constituents, resuspending it in the incubation buffer that does contain a MAO inhibitor. In the case of [^{3}H]-T binding, competition by endogenous T did not appear to be a problem, probably because of the very low concentrations of endogenous T. It is important, however, to include a MAO inhibitor in the final preparation of tissue suspension and to allow a few minutes at least for the inhibitor to inactivate the enzyme. In our laboratory we have used 10 μ*M* pargyline in the final preparation of tissue and in the incubation buffer. We have routinely preincubated the tissue for 5 min at 37°C with the buffer containing pargyline before addition of the [^{3}H]-T and drugs, although this 37°C preincubation step does not appear to be critical. However, the presence of pargyline or another suitable MAO inhibitor is critical. In experiments in which pargyline was omitted, specific binding of [^{3}H]-T was reduced (Cascio and Kellar, 1983). This probably reflects rapid oxidation of [^{3}H]-T even at the reduced temperature of the binding assay. Another reason for including a MAO inhibitor in the assay is to be sure that the binding site being measured is not monoamine oxidase enzyme itself.

2.1. Kinetics of [^{3}H]-Tryptamine Binding Sites

Studies of [^{3}H]-T binding in brain have been carried out in a simple Tris-HCl buffer (50 m*M*; pH 7.4 at 0°C) containing pargyline and ascorbic acid to prevent oxidation of [^{3}H]-T and added drugs. Nonspecific binding has been defined as the binding in the presence of 10 μ*M* unlabeled tryptamine and has been subtracted from total binding to yield specific binding. Initial studies were carried out at a reduced temperature (0°C) and without added ions because these conditions usually favor the binding of agonists. However, both temperature and sodium ion have important effects on [^{3}H]-T binding (Cascio and Kellar, 1983; also, see below).

Saturation analyses of [^{3}H]-T binding in cerebral cortex indicated that the binding capacity (B_{max}) is 25–30 pmol/g tissue (approximately 450 fmol/mg protein) and that the equilibrium dissociation constant (K_d) is approximately 3 n*M* (Fig. 1). Hill plots of binding saturation data have slopes of 1, indicating that over the [^{3}H]-T concentration range utilized in these studies (0.5–20 n*M*), binding is to a single class of sites (Cascio and Kellar, 1983).

At 0°C, the half-time ($t_{1/2}$) for association is approximately 7 min and the $t_{1/2}$ for dissociation is approximately 13 min. The K_d

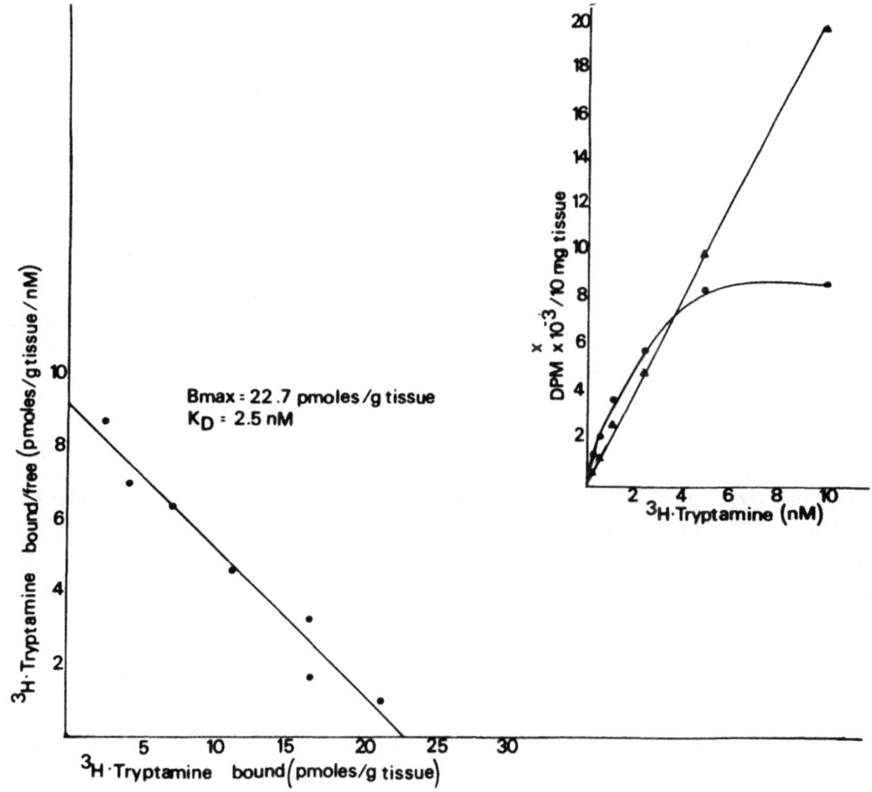

Fig. 1. [^{3}H]-Tryptamine binding in rat cerebral cortex. Right panel: Saturation analysis of [^{3}H]-tryptamine binding to rat cortical homogenates. The tissue homogenate was incubated with varying concentrations of [^{3}H]-tryptamine at 0°C for 60 min. Nonspecific binding (▲) was determined in the presence of 10 μM unlabeled tryptamine. Specific binding (●) is the difference between total binding and nonspecific binding. Left panel: Scatchard analysis of the specific binding of [^{3}H]-tryptamine. K_d and B_{max} were determined by least squares linear regression analysis. These data are representative of 10 saturation analyses (from Kellar and Cascio, 1982).

calculated from the kinetic rate constants (k_{-1}/k_1) is 1.7 nM, which is in reasonable agreement with that found in saturation studies (Cascio and Kellar, 1983). Rate constant studies and saturation analysis of [^{3}H]-T binding to peripheral tissue have yielded similar K_d values (Brüning and Rommelspacher, 1984). At 37°C, however, [^{3}H]-T dissociates much more rapidly, with a $t_{1/2}$ of less than 1 min. This rapid dissociation results in much lower affinity (higher K_d values) in studies done at 37°C than at 0°C (Cascio and Kellar, 1983).

TABLE 1
Effect of Various Ions on [^{3}H]-Tryptamine Binding[a]

Ion	IC_{50}, μM
NaCl	49.0 ± 15.9
LiCl	24.7 ± 2.9
KCl	24.7 ± 6.9
$MgCl_2$	>100

[a]Cortical homogenates were incubated with 2 n*M* [^{3}H]-tryptamine and 4–6 concentrations of the ion. The ion concentration that inhibited specific binding by 50% (IC_{50}) was determined graphically (from Cascio and Kellar, 1983).

Ions have important effects on the [^{3}H]-T binding site. The monovalent cations inhibit binding, but the divalent cations do not (Table 1). Of the cations tested, however, only sodium significantly inhibits binding within its physiological concentration range. This effect appears to be caused mainly by a reduction in the affinity of the site for [^{3}H]-T.

2.2. Subcellular Localization

Boulton and Baker (1975) found that T in brain is located primarily in the synaptosomal fraction, suggesting that it originated, or at least is stored, in neurons. Consistent with a neuronal function of T, subcellular localization studies of the [^{3}H]-T binding site have found most of the sites in the synaptosomal fraction of brain (Table 2). But a significant portion of the sites appeared to sediment with the mitochondrial fraction. This may be caused by limitations of the methods of subcellular fractionation used (sucrose density gradient centrifugation), as well as to heterogeneity with regard to size and isopycnic density among the synaptosomes.

2.3. Pharmacology of [^{3}H]-Tryptamine Binding Sites

The studies described above served to establish that the [^{3}H]-T binding site has certain properties that are consistent with it being the recognition site of a receptor. Further identification relied heavily on pharmacological studies. In drug competition studies (Table 3), several analogs of T with a single substitution on the ring or methylation of the amine group have high affinity for the [^{3}H]-T binding site (none as high as T itself). Thus, *N*-methyltryptamine

TABLE 2
Subcellular Distribution of [^{3}H]-Tryptamine Binding in Rat Cerebral Cortex[a]

Fraction	Specific binding, fmol/mg protein	Specific binding, fmol/fraction
S_1 Total homogenate	339 ± 63	3378 ± 147
Synaptosomal	345 ± 33	1804 ± 194
Myelin	50 ± 16	91 ± 36
Mitochondrial	317 ± 160	181 ± 96

[a]The binding of 2 n*M* [^{3}H]-tryptamine in total homogenate (S_1 fraction), synaptosomal, mitochondrial, and myelin fractions were isolated using a sucrose gradient centrifugation. The data are reported as means ± SEM obtained from three animals (from Cascio and Kellar, 1983).

retains relatively high affinity for the [^{3}H]-T site, and even the hallucinogenic *N*,*N*-dimethyltryptamine has low micromolar potency in competing for the site. But more extensive substitutions result in much lower affinity; thus, the 5-methoxy-*N*,*N*-dimethyl derivative and melatonin, a pineal hormone, have low affinity for the site (Table 3).

Although 5-HT has relatively high affinity for the [^{3}H]-T binding site, other drugs with high affinity for 5-HT-1 and 5-HT-2 receptors—including LSD, methysergide, and spiperone—are ineffective in competing for the site (Table 3). Thus, there appears to be a clear distinction between [^{3}H]-T recognition sites and either of the 5-HT binding sites. Compared to methysergide, two other 5-HT antagonists, methiothepin and metergoline, have much higher affinity for the [^{3}H]-T site (Table 3). Interestingly, both of these antagonists are much more potent than methysergide in blocking T-induced hyperthermia (Cox et al., 1981). The affinities of a series of drugs in several other classes distinguish the [^{3}H]-T binding site from other known receptor binding sites in brain (Table 3). Thus, [^{3}H]-T appears to bind to a specific recognition site for T.

Members of two chemical classes, the phenylethylamines and the beta-carbolines, have been found to have high affinity for the [^{3}H]-T recognition site. Interestingly, both classes contain members that are endogenous to brain and peripheral tissues. Phenylethylamine itself has high affinity for the [^{3}H]-T site. Substitutions on the ring, as in tyramine, or on the side chain, as in the amphetamines, decrease the affinity by 10–30-fold. Nevertheless, the [^{3}H]-

TABLE 3
Competition for [^{3}H]-Tryptamine Binding Sites by Various Drugs[a]

	K_i, nM
Tryptamine derivatives	
Tryptamine	2.4 ± 0.5
5-Methyltryptamine	16 ± 2
N-Methyltryptamine	150 ± 36
5-Methoxytryptamine	220 ± 60
N,N-Dimethyltryptamine	2109 ± 350
5-Methoxy-*N,N*-dimethyltryptamine	>50,000
5-Methoxy-*N*-acetyltryptamine (melatonin)	>50,000
Serotonin drugs	
5-HT	280 ± 50
Quipazine	1570 ± 190
Methiothepin	12900 ± 3640
Metergoline	30000 ± 6700
LSD	>50,000
Methysergide	>50,000
Mianserin	>50,000
Spiperone	>50,000
Dopamine drugs	
Dopamine	470 ± 120
Haloperidol	14,300 ± 3400
Chlorpromazine	14,900
Apomorphine	>50,000
Butaclamol	>50,000
Tetrahydro-β-carbolines	
Tetrahydro-β-carboline	28 ± 8
6-Methoxy-THBC	2160 ± 170
6-Hydroxy-1-methyl-THBC	11,020 ± 5520
6-Hydroxy-THBC	18,250 ± 8040

[a]Cortical homogenates were incubated with 2 nM [^{3}H]-tryptamine and 5–7 concentrations of each competing drug. The concentration of each drug that inhibited binding by 50% (IC_{50}) was determined graphically and converted to an inhibitory constant (K_i) by the equation $K_i = IC_{50}/1 + ([^3\text{H-tryptamine}]/K_d)$. The K_i values reported are the mean ± SEM of 3–5 determinations, except for chlorpromazine, which is the average of two determinations (from Cascio and Kellar, 1983).

(continued on next page)

TABLE 3 *(continued)*

	K_i, nM
Phenylethylamines	
Phenylethylamine	87 ± 27
p-Chloroamphetamine	570 ± 310
Tyramine	940 ± 40
d-Amphetamine	1000 ± 61
Methamphetamine	3800 ± 2600
l-Amphetamine	10,080 ± 1200
Fenfluramine	>50,000
Other drugs	
Iprindole	15,000 ± 1370
Amitriptyline	21,400 ± 1900
Noradrenaline	22,400 ± 8200
Psilocybin	>50,000
GABA	>500,000

T recognition site displays relatively marked stereospecificity for the pharmacologically more active D-amphetamine; and, as a class, the amphetamines retain affinities high enough to make actions at [^{3}H]-T recognition sites a realistic possibility (Table 3; Cascio and Kellar, 1983; Wood et al., 1984). It is noteworthy that PE and the three amphetamine analogs that have relatively high affinity for the [^{3}H]-T recognition site are psychomotor stimulants (Cox and Maickel, 1972), and that in contrast, fenfluramine, an amphetamine analog that has low affinity for the site, is a psychomotor depressant.

Tetrahydro-beta-carbolines may be formed by condensation reactions between indoleamines and aldehydes. For examples, the unsubstituted tetrahydro-beta-carboline (THBC) is the product of condensation between T and formaldehyde, and 6-hydroxy-1-methyl-THBC is the product formed by 5-HT and acetaldehyde. The tetrahydro-beta-carbolines may be formed in vivo (Rommelspacher and Greiner, 1978; Barker et al., 1979; Faull et al., 1982), and they have effects on several neurotransmission processes (for a review of the chemistry, detection, and pharmacology of the beta-carbolines, *see* Bloom et al., 1982). For example, they competitively inhibit the high-affinity uptake of 5-HT into brain homogenates (Kellar et al., 1976; Rommelspacher et al., 1976; Buckholtz and

Boggan, 1976). Their structural similarity to T and pharmacological effects suggested that the tetrahydro-beta-carbolines might also act at [^{3}H]-T recognition sites. Among a series of tetrahydro-beta-carbolines that we have examined, the unsubstituted THBC has the highest affinity for the [^{3}H]-T recognition site (Table 3). The affinity of THBC for [^{3}H]-T sites (28 n*M*) is 200 and 400 times higher than its affinity for 5-HT-1 and 5-HT-2 binding sites, respectively (Table 4). In fact, all of the tetrahydro-beta-carbolines that we have examined have higher affinity for [^{3}H]-T sites than for either of the 5-HT binding sites (Table 4). In contrast, LSD has much higher affinity for 5-HT-1 and 5-HT-2 binding sites than for [^{3}H]-T sites (Table 4). The affinity of THBC for the [^{3}H]-T recognition site is higher than any nontryptamine we have tested. However, both Wood et al. (1984) and Brüning and Rommelspacher (1984) have reported that certain other beta-carbolines, including norharmane (an aromatic beta-carboline) and harmaline (a dihydro-beta-carboline) have equal or even higher affinities for the [^{3}H]-T recognition site.

Binding studies provide important but limited information. They do not provide direct information about the functions of T, or whether drugs with affinity for the [^{3}H]-T recognition site are

TABLE 4
Binding Inhibition Constants (K_i) of Four Tetrahydro-β-carbolines and LSD at [^{3}H]-Tryptamine, [^{3}H]-5-HT, and [^{3}H]-Spiperone Binding Sites in Cerebral Cortex[a]

	Binding site K_i, μM		
Drug	[^{3}H]-Tryptamine	[^{3}H]-5-HT	[^{3}H]-Spiperone
THBC	0.028 ± 0.008	6.03 ± 1.58	12.03 ± 3.03
6-Methoxy-THBC	2.16 ± 0.17	17.84 ± 4.44	12.43 ± 4.64
6-Hydroxy-1-methyl-THBC	11.02 ± 5.52	36.62 ± 1.88	>50
6-Hydroxy-THBC	18.25 ± 8.04	29.02 ± 5.78	>50
LSD	>50	0.013 ± 0.001	0.008 ± 0.001

[a]K_i's were calculated from the concentration of drugs that decreased binding by 50% (IC_{50}) using the equation $K_i = IC_{50}/1 + ([^3\text{H-ligand}]/K_d)$. Values are mean ± SEM of 3–5 experiments (from Cascio and Kellar, 1982).

agonists or antagonists. Two studies bear on these questions and, as is often the case, both studies examined peripheral tissue responses. In one study, T-induced increases in diastolic blood pressure in pithed rats was investigated, and this effect could be distinguished from 5-HT and alpha-adrenergic vasopressor effects by the much lower potencies of the antagonists methysergide and phentolamine, respectively, in blocking the T-induced responses (Feniuk et al., 1982). In another study, Hicks and Langer (1983) found that both T and 5-HT produced contraction of the rat tail artery in vitro. Methysergide was a much more potent antagonist of 5-HT-induced contractions than of T-induced contractions, and, in contrast, THBC was a more potent antagonist of T-induced contractions than of those induced by 5-HT.

2.4. Distribution of [^{3}H]-Tryptamine Binding Sites in Brain

Studies of the regional distribution of the [^{3}H]-T binding site in brain indicated that the site is widely but unevenly distributed in rat brain. At a subsaturating concentration of [^{3}H]-T, the highest binding was found in the cortex, hippocampus, and striatum and the lowest binding was found in the pons, medulla, and spinal cord (Table 5). Saturation analyses in several rat brain regions indicated that among these regions the affinity of the binding sites for [^{3}H]-T is similar, but that the density of sites varies (Table 6). In human cerebral cortex, [^{3}H]-T bound with an affinity of approximately 4 n*M* to one-third to one-half the number of sites found in rat cortex (Table 6). These brain regional studies revealed that the [^{3}H]-T binding sites are at least as dense as either 5-HT-1 or 5-HT-2 binding sites in most brain regions. Autoradiographic localization of [^{3}H]-T sites in brain and spinal cord sections agrees closely with the results from these homogenate binding studies (Perry et al., 1982; McCormack et al., 1983).

We examined the possibility that [^{3}H]-T binding sites might be located on 5-HT or catecholamine axons by injecting 5,7-dihydroxytryptamine (5,7-DHT) or 6-hydroxydopamine (6-OHDA), respectively, into the lateral ventricles of rats. At the time of sacrifice, 30 d after injection of the neurotoxins, the high-affinity uptake of 5-HT and the catecholamines (noradrenaline and dopamine) in several brain areas was decreased by at least 50% by 5,7-DHT and 6-OHDA, respectively. Despite the widespread degeneration of 5-HT and catecholamine axons produced by the neurotoxin lesions, no evidence of a decrease in [^{3}H]-T binding sites was found in any of four areas of rat brain (Table 7). Thus, the [^{3}H]-T recognition site does not appear to be located on 5-HT or catecholamine axons.

TABLE 5
Regional Distribution of [^{3}H]-Tryptamine Binding[a]

Region	Specific binding, pmol/g tissue
Striatum (5)	11.8 ± 0.8
Cortex (5)	11.0 ± 0.3
Hippocampus (4)	10.3 ± 0.7
Olfactory cortex (3)	6.9 ± 0.6
Hypothalamus (3)	5.0 ± 0.2
Cerebellum (5)	5.0 ± 0.5
Midbrain (6)	4.0 ± 0.7
Medulla (6)	3.5 ± 0.6
Pons (6)	2.9 ± 0.3
Spinal cord (3)	2.0 ± 0.3

[a]The specific binding of 2 n*M* [^{3}H]-tryptamine to various areas of rat brain was determined. The binding is expressed as pmol/g tissue ± SEM for the number of animals, indicated in parentheses (from Cascio and Kellar, 1983).

Brüing and Rommelspacher (1984) found [^{3}H]-T binding sites widely distributed among peripheral tissues, with the highest number of binding sites in the liver and heart, and the lowest number in the lung. The K_d in each of the tissues was similar to that in brain, and a detailed comparison of the binding sites in kidney

TABLE 6
Binding Constants of [^{3}H]-Tryptamine in Various Brain Regions[a]

Brain region	B_{max}, pmol/g tissue	K_d, n*M*
Hippocampus (3)	39.9 ± 2.0	3.7 ± 0.8
Striatum (3)	33.3 ± 3.9	3.0 ± 0.7
Cortex (5)	26.6 ± 3.4	2.8 ± 0.2
Cerebellum (3)	15.7 ± 0.2	2.7 ± 0.1
Human cortex (4)	11.7 ± 4.0	4.5 ± 1.0

[a]Tissue homogenates were incubated with varying concentrations of [^{3}H]-tryptamine as described in the text. K_d and B_{max} were determined using Scatchard analyses. The number of determinations is indicated in parentheses. Values are reported as the mean ± SEM of the mean (from Cascio and Kellar, 1983).

TABLE 7
The Effects of Lesions of Catecholamine and Serotonin Axons on [^{3}H]-Tryptamine (1 n*M*) Binding in Rat Brain[a]

	[^{3}H]-Tryptamine (1 n*M*) specifically bound, pmol/g tissue		
	Control	6-Hydroxy-dopamine	5,7-Dihydroxy-tryptamine
Cortex	7.6 ± 0.2	7.4 ± 0.2	7.9 ± 0.2
Striatum	7.8 ± 0.3	7.7 ± 0.6	8.2 ± 0.5
Hypothalamus	7.8 ± 0.5	7.7 ± 0.2	7.2 ± 0.3
Hippocampus	7.1 ± 0.8	7.7 ± 0.5	7.4 ± 0.7

[a]Rats were anesthetized with pentobarbital and injected intraventricularly with 6-hydroxydopamine (200 μg) or 5,7-dihydroxytryptamine (200 μg). The rats were sacrificed 30 d after injection, and [^{3}H]-tryptamine binding sites were measured in the cortex, striatum, hypothalamus, and hippocampus. The data are expressed as mean ± SEM of 4–8 animals per group.

and cerebral cortex indicated that the [^{3}H]-T sites in the two tissues were very similar (Brüning and Rommelspacher, 1984).

2.5. Regulation of [^{3}H]-Tryptamine Binding Sites In Vivo

In many neurotransmitter and hormone systems, the density of receptors is influenced by the concentration of agonist available to them (for review, *see* Creese and Sibley, 1981). Two drugs that could potentially alter T concentrations in brain are reserpine and pargyline. Reserpine disrupts neuronal storage vesicles for 5-HT and the catecholamines; however, the issue of whether T is stored in reserpine-sensitive vesicles is still open. Some studies have indicated that reserpine treatment decreases endogenous T in brain (Snodgrass and Horn, 1973), whereas other studies have indicated that reserpine treatment does not affect T content (Saavedra and Axelrod, 1973; Marsden and Curzon, 1974). Some of the differences may be related to the difficulty of accurately measuring T (Jones, 1982c). In contrast, the effects of treatment with MAO inhibitors on T concentration in brain are much clearer. Increases of up to 300-fold have been reported in brain from several species following treatment with different MAO inhibitors (Marsden and Curzon, 1974; Philips and Boulton, 1979; Boulton et al., 1975; Tabakoff et al., 1977). In rat brain, treatment with pargyline produces marked increases in T concentration (Warsh et al., 1977).

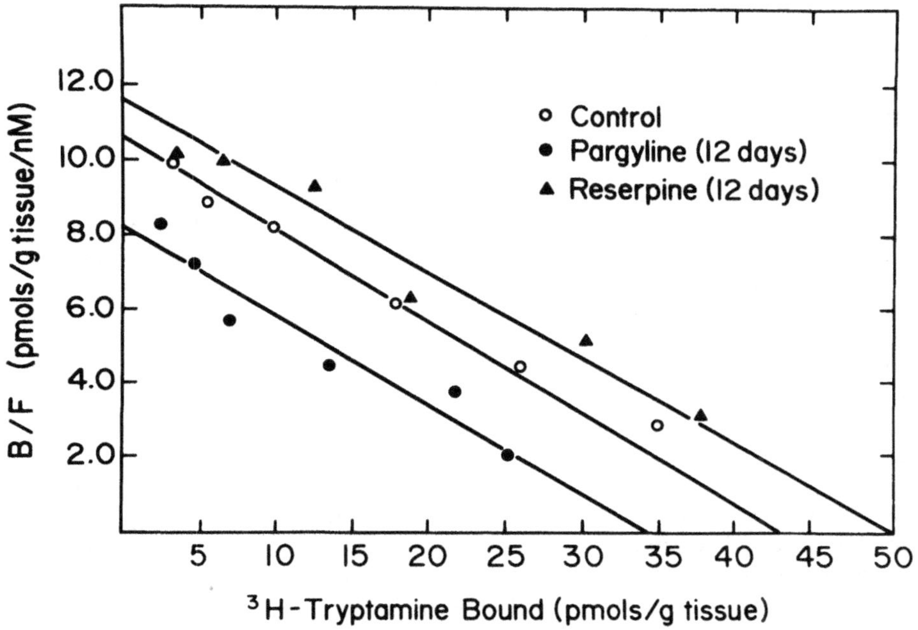

Fig. 2. Scatchard plots of [^{3}H]-tryptamine binding in cerebral cortex from rats treated for 12 d with saline (control, ○); reserpine (▲); or pargyline (●). Binding was measured using 0.5–16 n*M* [^{3}H]-tryptamine.

To determine whether [^{3}H]-T binding sites in the brain could be regulated in vivo, rats were treated for 12 d with either reserpine or pargyline. Reserpine treatment produced a statistically significant, but unimpressive, increase in [^{3}H]-T binding in the cerebral cortex and hippocampus, but not in the striatum or hypothalamus (Table 8). The increase in [^{3}H]-T binding in the cortex appeared to be caused by an increase in the density of binding sites (Fig. 2). This small and variable increase in [^{3}H]-T binding sites in different brain areas following reserpine treatment could be a reflection of high but variable rates of T turnover in different brain areas. Nevertheless, it suggests that in some brain areas, T might be associated with a reserpine-sensitive storage vesicle.

The effects of MAO inhibition were much more consistent. Pargyline treatment decreased [^{3}H]-T binding in all four brain areas examined (Table 8). The largest decrease was seen in the hypothalamus. Scatchard analyses in the cortex indicated that pargyline treatment decreased the number of [^{3}H]-T binding sites, but did not change the apparent affinity of the sites (Fig. 2). Thus, chronic treatment with pargyline down-regulates [^{3}H]-T binding sites.

In other studies, we found no change in [^{3}H]-T binding follow-

TABLE 8
[^{3}H]-Tryptamine (3 nM) Binding in Rat Brain Areas[a]

	Control	Reserpine, pmol/g tissue	Pargyline
Cortex	14.0 ± 0.5	16.6 ± 0.4[b]	10.4 ± 0.4[b]
Striatum	12.8 ± 0.9	13.6 ± 0.9	9.4 ± 0.8[b]
Hippocampus	13.9 ± 0.9	15.5 ± 0.4[b]	10.6 ± 1.2[b]
Hypothalamus	9.0 ± 0.5	8.5 ± 0.7	5.0 ± 0.3[b]

[a]Binding following 12 d treatment with reserpine (2.5 mg/kg, days 1 and 2; 0.5 mg/kg, days 3–12), or pargyline (10 mg/kg, 12 d). The rats were sacrificed 24 h after the last injection. Values are mean ± SEM of 4–8 rats.

[b]$p < 0.05$ compared to control.

ing either 4 wk of treatment with lithium in the diet (serum Li^+ levels of 0.9 mEq/L), 12 d of electroconvulsive shock (once a day), or 12 d treatment with amphetamine.

3. [^{3}H]-Phenylethylamine Recognition Sites

Binding sites for [^{3}H]-PE in rat brain have been found by Hauger et al. (1982). These sites have been measured by incubating [^{3}H]-PE (40–48 Ci/mmol; New England Nuclear) with brain homogenates in Tris-HCl buffer at 0–4°C for 60 min. Binding in the presence of 100 μM nonradioactive PE was used to define nonspecific binding. Under these conditions, [^{3}H]-PE was found to bind to forebrain homogenates with a K_d of 55 nM and a B_{max} of approximately 1000 fmol/mg protein. The site is sensitive to heat and to trypsin and chymotrypsin digestion (Hauger et al., 1982).

In preliminary studies, Hauger et al. (1982) found that the highest [^{3}H]-PE binding in brain was in the hypothalamus and striatum, areas where the endogenous levels of PE are reported to be high (Willner et al., 1974; Philips et al., 1978; Karoum et al., 1981). In drug competition studies, the [^{3}H]-PE recognition site was reported to have a profile distinct from catecholamine or cholinergic receptors; and neither neuroleptic drugs, tricyclic antidepressants, benzodiazepines, nor amphetamine competed effectively for the binding site (Hauger et al., 1982).

These studies suggest that a specific recognition site for PE might exist in brain. However, the studies were carried out in the absence of any MAO inhibitor; thus, the possibility that the binding site is monoamine oxidase itself can not be excluded at this time.

4. Summary

Specific recognition sites [^{3}H]-T have been found in rat and human brain and in rat peripheral tissues. These sites have the characteristics expected of a distinct receptor for T. Electrophysiological and body temperature responses to T suggest that this amine might function as a neurotransmitter or neuromodulator in the CNS. Studies of the vascular responses to T indicate that it contracts arterial smooth muscle by mechanisms that can be distinguished from those of other pressor agents, such as 5-HT and alpha-adrenergic agonists. The pressor response to T is blocked by THBC, a potent competitor for the [^{3}H]-T recognition site. Thus, both the biochemical characteristics of the [^{3}H]-T binding site and the physiological responses to T suggest the possibility for independent functions of this amine mediated by the [^{3}H]-T recognition site.

The [^{3}H]-PE binding site that has been found in brain is potentially interesting, but further characterization of the site and the actions of PE are required before considering the [^{3}H]-PE binding site to be a potential specific mediator of PE action.

References

Barker S. E., Harrison R. E., Brown G. B., and Christian S. T. (1979) Gas chromatographic/mass spectrometric evidence for the identification of 1,2,3,4-tetrahydro-beta-carboline as an *in vivo* constituent of rat brain. *Biochem. Biophys. Res. Commun.* **87,** 146–154.

Bloom F., Barchas J., Sandler M., and Usdin E. (eds.) (1982) *Beta-Carbolines and Tetrahydroisoquinolines.* Alan R. Liss, New York.

Boulton A. A. and Baker G. B. (1975) The subcellular distribution of beta-phenylethylamine, *p*-tyramine, and tryptamine in rat brain. *J. Neurochem.* **25,** 477–481.

Boulton A. A. and Juorio A. V. (1982) Brain Trace Amines, in *Handbook of Neurochemistry,* Vol. 1 (Lajtha A., ed)., pp. 189–222, Plenum, New York.

Boulton A. A., Juorio A. V., Philips S. R., and Wu P. H. (1975) Some arylalkylamines in rabbit brain. *Brain Res.* **96,** 212–216.

Brüning G. and Rommelspacher H. (1984) High affinity [^{3}H]-tryptamine binding sites in various organs of the rat. *Life Sci.* **34,** 1441–1446.

Buckholtz N. S. and Boggan W. O. (1976) Effects of tetrahydro-beta-carbolines on monoamine oxidase and serotonin reuptake in mouse brain. *Biochem. Pharmacol.* **25,** 2319–2321.

Cascio C. S. and Kellar K. J. (1982) Tetrahydro-beta-carbolines: Affinities for tryptamine and serotonergic binding sites. *Neuropharmacology* **21,** 1219–1221.

Cascio C. S. and Kellar K. J. (1983) Characterization of [^{3}H]-tryptamine binding sites in brain. *Eur. J. Pharmacol.* **95,** 31–39.

Cox R. H. and Maickel R. P. (1972) Comparison of anorexigenic and behavioral potency of phenylethylamines. *J. Pharmacol. Exp. Ther.* **181,** 1–9.

Cox B., Lee T. F., and Martin D. (1981) Different hypothalamic receptors mediate 5-hydroxytryptamine- and tryptamine-induced core temperature change in the rat. *Brit. J. Pharmacol.* **72,** 477–482.

Creese I. and Sibley D. R. (1981) Receptor adaptations to centrally acting drugs. *Ann. Rev. Pharmacol. Toxicol.* **21,** 357–391.

Dewhurst W. G. and Marley E. (1965) Action of sympathomimetic and allied amines on central nervous system of the chicken. *Brit. J. Pharmacol.* **25,** 705–727.

Faull K., Holman R. B., Elliot G. R., and Barchas J. D. (1982) Tryptolines: Artifact or Reality. A New Method of Analysis Using GC/MS, in *Beta-Carbolines and Tetrahydroisoquinolines* (Bloom F., Barchas J., Sandler M., and Usdin E., eds.), pp. 135–154, Alan R. Liss, New York.

Feniuk W., Humphrey P. P. A., and Perren M. J. (1982) Tryptamine-induced vasopressor responses in pithed rats are not predominantly mediated via 5-hydroxytryptamine receptors. *Brit. J. Pharmacol.* **77,** 521P.

Foldes A. and Costa E. (1975) Relationship of brain monoamine and locomotor activity in rats. *Biochem. Pharmacol.* **24,** 1617–1621.

Hauger R. L., Skolnick P., and Paul S. M. (1982) Specific [^{3}H]-beta-phenylethylamine binding sites in rat brain. *Eur. J. Pharmacol.* **83,** 147–148.

Henwood R. W., Boulton A. A., and Phillis J. W. (1979) Iontophoretic studies of some trace amines in the mammalian CNS. *Brain Res.* **164,** 347–351.

Hicks P. E. and Langer S. Z. (1983) Tetrahydro-beta-carboline (THBC), an antagonist of tryptamine-induced contractions in perfused rat tail artery. *Brit. J. Pharmacol.* **80,** 649P.

Jones R. S. G. (1981) *In vivo* pharmacological studies on the interactions between tryptamine and 5-hydroxytryptamine. *Brit. J. Pharmacol.* **73,** 485–493.

Jones R. S. G. (1982a) A comparison of the responses of cortical neurones to iontophoretically applied tryptamine and serotonin in the rat. *Neuropharmacology* **21,** 209–214.

Jones R. S. G. (1982b) Responses of cortical neurons to stimulation of the nucleus raphe medianus: a pharmacological analysis of the role of indoleamines. *Neuropharmacology* **21,** 511–520.

Jones R. S. G. (1982c) Tryptamine: A neuromodulator or neurotransmitter in mammalian brain? *Prog. Neurobiol.* **19,** 117–139.

Jones R. S. G. and Boulton A. A. (1980a) Tryptamine and 5-hydroxytryptamine: Actions and interactions on cortical neurons in the rat. *Life Sci.* **27,** 1949–1956.

Jones R. S. G. and Boulton A. A. (1980b) Interactions between *p*-tyramine, *m*-tyramine, or beta-phenylethylamine and dopamine on single neurones in the cortex and caudate nucleus of the rat. *Can. J. Physiol. Pharmacol.* **58,** 222–227.

Jones R. S. G. and Broadbent J. (1982) Differential effects of fluoxetine and zimelidine on the uptake of 5-hydroxytryptamine and tryptamine by cortical slices and on responses of cortical neurons to stimulation of the nucleus raphe medianus. *Eur. J. Pharmacol.* **81,** 681–685.

Karoum F., Chuang L. W., and Wyatt R. J. (1981) Presence and distribution of phenylethylamine in rat spinal cord. *Brain Res.* **225,** 442–446.

Kellar K. J. and Cascio C. S. (1982) [^{3}H]-Tryptamine: High-affinity binding sites in rat brain. *Eur. J. Pharmacol.* **78,** 475–478.

Kellar K. J., Elliott G. R., Holman R. B., Danellis J. V., and Barchas J. D. (1976) Tryptoline inhibition of serotonin uptake in rat forebrain homogenates. *J. Pharmacol. Exp. Ther.* **198,** 619–625.

Marsden C. A. and Curzon G. (1974) Effects of lesions and drugs on brain tryptamine. *J. Neurochem.* **23,** 1171–1176.

Marsden C. A. and Curzon G. (1978) The contribution of tryptamine to the behavioral effects of l-tryptophan in tranylcypromine-treated rats. *Psychopharmacology* **57,** 71–76.

Martin W. R. and Sloan J. W. (1970) Effects of infused tryptamine in man. *Psychopharmacology* **18,** 231–237.

McCormack J. K., Beitz A. J., and Larson A. A. (1983) [^{3}H]-Tryptamine binding in the brain and spinal cord: *In vitro* autoradiography. *Soc. Neurosci. Abst.* **9,** 333.

Perry D. C., Manning D. C., and Snyder S. H. (1982) *In vitro* autoradiographic localization of [^{3}H]-tryptamine binding sites in rat brain. *Soc. Neurosci. Abst.* **8,** 783.

Philips S. R. and Boulton A. A. (1979) The effect of monoamine oxidase inhibitors on some arylalkylamines in rat striatum. *J. Neurochem.* **33,** 159–167.

Philips S. R., Rozdilsky B., and Boulton A. A. (1978) Evidence for the presence of *m*-tyramine, tryptamine, and phenylethylamine in rat brain and several areas of human brain. *Biol. Psychiatry* **13,** 51–57.

Potkin S., Karoum F., Chuang L. W., Cannon-Spoor H. E., Phillips I., and Wyatt R. J. (1979) Phenylethylamine in paranoid schizophrenia. *Science* **206,** 470–471.

Quock R. M. and Weick B. G. (1978) Tryptamine-induced drug effects insensitive to serotonergic antagonists: Evidence of specific tryptaminergic receptor stimulation? *J. Pharm. Pharmacol.* **30,** 280–283.

Randrup A. and Munkvad I. (1966) Dopa and other naturally occurring substances as causes of stereotypy in rats. *Acta Psychiat. Scand.* (Suppl.) **191,** 193–199.

Rommelspacher H., Bade P., Coper H., and Kossmehl G. (1976) Inhibition of the reuptake of serotonin by tryptoline. *Naunyn Schmiedebergs Arch. Pharmacol.* **292,** 93–95.

Rommelspacher H. and Greiner B. (1978) Formation of beta-carbolines: A new metabolic pathway of indoles in the brain of rats. *Naunyn Schmiedebergs Arch. Pharmacol.* **302,** R60.

Saavedra J. M. and Axelrod J. (1973) Effect of drugs on the tryptamine content of rat tissues. *J. Pharmacol. Exp. Ther.* **185,** 523–529.

Sandler M. and Reynolds G. P. (1976) Does phenylethylamine cause schizophrenia? *Lancet* **i,** 76.

Snodgrass S. R. (1974) Phenylalanine, brain phenethylamine, and motor activity in the rat. *J. Pharm. Pharmacol.* **26,** 931–936.

Snodgrass S. R. and Horn A. (1973) An assay procedure for tryptamine in brain and spinal cord using [^{3}H]-dansyl derivative. *J. Neurochem.* **21,** 687–696.

Tabakoff B., Moses F., Philips S. R., and Boulton A. A. (1977) Effects of tranylcypromine and pargyline on brain tryptamine. *Experentia* **33,** 380–381.

Warsh J. J., Godse D. D., Stancer H. C., Chan P. W., and Coscina D. V. (1977) Brain tryptamine in rats by a new gas chromatographic mass fragmentographic method. *Biochem. Med.* **18,** 10–20.

Willner J., LeFevre H., and Costa E. (1974) Assay by multiple ion detection of beta-phenylethylamine and phenylethanolamine in rat brain. *J. Neurochem.* **23,** 857–859.

Wood P. L., Pilapil C., LaFaille F., Nair N. P. V., and Glennon R. A. (1984) Unique [^{3}H]-tryptamine binding sites in rat brain: Distribution and pharmacology. *Arch. Int. Pharmacodyn.* **268,** 194–201.

Wu P. H. and Boulton A. A. (1975) Metabolism, distribution and disappearance of injected beta-phenylethylamine in the rat. *Can. J. Biochem.* **53,** 42–50.

Wyatt R. J., Gillin J. C., Stoff D. M., Majo E. A., and Tinklenberg J. R. (1977) Beta-phenylethylamine and Neuropsychiatric Disturbances, in *Neuroregulators and Psychiatric Disorders* (Usdin E., Hamburg D. A., and Barchas J. D., eds.), pp. 31–44, Oxford University Press, New York.

Yang H. Y. T. and Neff N. H. (1973) Beta-phenylethylamine: A specific substrate for type B monoamine oxidase in brain. *J. Pharmacol. Exp. Ther.* **187,** 365–371.

Chapter 6
Muscarinic Cholinergic Receptors

Andrew P. Braun and Prakash V. Sulakhe

1. Introduction

In the central nervous system (CNS), modulation of neuronal activity is brought about by the actions of endogenous "neurotransmitters" that specifically interact with receptor proteins located in the neuronal cell membrane. In the cholinergic system, acetylcholine is the primary neurotransmitter released in the CNS, and appears to alter neuronal excitability by affecting such processes as cGMP production and K^+ flux (Heilbronn and Bartfai, 1978). In contrast to the periphery, in which the cholinergic receptor population is divided between nicotinic and muscarinic subclasses, the CNS-associated cholinergic receptor appears to be predominantly of the muscarinic type, although there is evidence that the spinal cord contains substantial amounts of nicotinic receptors (Heilbronn and Bartfai, 1978). Historically, the initial characterization of cholinergic actions was provided by A. S. V. Burgen. Although these early studies, using smooth muscle tissue (Burgen and Spero, 1968; Burgen et al., 1974), provided initial clues about the peripheral muscarinic receptors, since then there have been numerous investigations, especially following the advent of radioligand binding assays, from which have come many novel observations of physiological significance. Among these, the findings that muscarinic cholinergic receptors can be categorized as either "M_1" or "M_2" types, as well as the dominance of M_1 subtype in certain areas of the CNS and of M_2 in various peripheral tissues, have provided excitement for the present research. It is this aspect that provides the major framework of the following article.

The concept of receptor "subtypes" to explain the subtle differences in the actions of hormones in various tissues of the body has become an increasingly useful approach. Presently, evidence for this type of subclassification is available for other receptor systems [e. g., adrenergic, (Stiles and Lefkowitz, 1984); histaminergic, (Ash and Schild, 1966; Black et al., 1972)] in which the specific receptors are subdivided primarily on the basis of their interactions (i. e., rank order, affinity) with a select series of ligands (both agonists and antagonists). This line of thinking is being extended to the muscarinic cholinergic system, for which evidence describing the existence of at least two receptor subtypes for cholinergic action is now available; the study of Goyal and Rattan (1978) on the muscarinic cholinergic receptor in the upper GI tract provided the basis for the nomenclature "M_1" and "M_2." Until now, this type of differentiation has been difficult because classical antagonists, such as atropine, show similar affinities for both subtypes of the receptor (Luthin and Wolfe, 1984). However, with the emergence of the selective antagonist pirenzepine (Hammer et al., 1980) and agonist McN-A-343 (Birdsall et al., 1983), identification of the M_1 receptor by ligand interaction now appears to be possible. These recent findings have lead to further investigations that show the distribution of the putative M_1 and M_2 receptors is not homogeneous in the CNS, but follows distinct anatomical patterns. This novel observation has brought with it considerable new research, focusing primarily on the M_1 receptor in the CNS, select regions of which are thought to be the major location of this subtype. This differs from the M_2 subtype, which is predominant in the periphery, although it is also known to exist in the CNS. Thus, it will be the aim of this chapter to examine some of the present information emerging from this rapidly growing area, and to determine what effect, if any, it may have on earlier studies dealing with CNS-related muscarinic cholinergic receptors.

2. Pirenzepine

Perhaps the single most important factor contributing to the identification of the putative M_1 receptor has been the selective muscarinic antagonist pirenzepine. This drug, a tricyclic compound with an endocyclic amide group (Trummlitz et al., 1984), recognized for its antiulcer activity, has recently become invaluable as a tool for receptor subtype classification. Based on its selective inhibition, it was discovered (Hammer et al., 1980) that pirenzepine could specifically antagonize a subpopulation of muscarinic receptors subserving certain functions (e. g., gastric secretion) while, at the same time, have only minimal effects on others (e.g., bradycardia). These differences in action, revealed primarily by Hammer and associates (1980, 1982), strongly supported the hypothesis of distinct muscarinic receptor subtypes. Recent work (Hammer et al., 1980) has, in part, shown the differential effects of pirenzepine—primarily the higher affinity of this compound for muscarinic receptors in the CNS, leading to its selective action at lower concentrations. In comparison, the "classical" antagonist atropine does not show this differential affinity for receptors in the CNS and periphery.

Using this selectivity as a basis, pirenzepine has thus become known as an "M_1 antagonist" because its actions appear to be limited to a select group of receptors we now refer to as the "M_1" population. The importance of this compound to the present discussion will be borne out in latter sections.

3. McN-A-343

The advent of pirenzepine as an M_1 selective antagonist was closely followed by the introduction of an agonist described as having greater selectivity for the M_1 receptor. This compound, 3-(*m*-chlorophenylcarbamoyloxy)-2-butynyltrimethylammonium (McN-A-343), has been shown to stimulate M_1 receptors (Hammer and Giachetti, 1982) and appears to act differentially on M_1 and M_2 sites (Birdsall et al., 1983). In this respect, it is known that various cholinergic actions, such as negative chronotropy and inotropy (Roberts and Konjovic, 1969), show differential sensitivity to stimulation by the agonist acetylcholine. Additionally, muscarinic agonists are known to have differing affinities for the modulation of various effector systems [e.g., cAMP production and phosphatidylinositol hydrolysis (Brown and Brown-Masters, 1984)], which further illustrates the concept of "agonist selectivity." However, whereas McN-A-343 shows a "preference" for the M_1 subtype, a preferential agonist for the M_2 receptor has yet to be reported.

4. Current Methodology

Much of the present information on the M_1 and M_2 subtypes of cholinergic receptors has come from the in vitro studies involving radioligand binding methods. These radiolabeled compounds are commonly cholinergic antagonists or agonists with high radiospecific activity. Agonist interaction with the receptors was previously studied through the use of various radiolabeled antagonists and more recently with radiolabeled agonists. Although most of the earlier studies described agonist interaction with the receptor, they nevertheless suffered because they utilized radiolabeled antagonists to examine this interaction with the receptor. For example, these studies provided only "indirect" information on agonist binding to the receptor, and in particular, to a high-affinity site. Thus, more of the recent work utilizes both antagonist and agonist binding techniques to the membrane fraction in question, thereby providing more complete information regarding agonist/receptor interaction. A brief overview of the previous work, combined with a discussion of recent data may therefore be helpful.

4.1. Experiments Involving Radiolabeled Antagonists

4.1.1. Displacement of Radiolabeled Antagonists

In these types of experiments, radiolabeled antagonists were displaced by both "cold" agonists and antagonists, which allowed one to look at the interaction of the receptor with competing ligands. The assumption was made in these experiments that competition occurred at the same site on the receptor as identified by the radiolabel. Although the data of numerous studies using antagonists and agonists confirm this view, certain exceptions have been encountered. For example, gallamine, a neuromuscular blocking agent, appears to interact at the ligand binding site in addition to another site, thought to be allosteric in nature (Dunlap and Brown, 1983). In these studies, one of two commonly used techniques is seen: (a) rapid filtration (Yamamura and Snyder, 1974), in which the binding reaction is stopped by passing the assay contents over a membrane filter under mild suction, and (b) centrifugation (Birdsall et al., 1978), in which the reaction is halted by quick centrifugation of the assay tubes at a gravitational force of 14,000 in order to remove the tissue fraction from the reaction medium. The radiolabels used frequently have been antagonists, such as [^{3}H]-propylbenzilylcholine, [^{3}H]-*N*-methylscopolamine (NMS) (Hulme et al., 1978), [^{3}H]-*N*-methyl-4-piperidyl benzilate (4NMPB) (Kloog and Sokolovsky, 1978), and, perhaps the most common, [^{3}H]-quinuclidinyl benzilate (QNB) (Yamamura and Snyder, 1974). Historically, radiolabeled antagonists were primarily used because: (a) they are known to interact with high affinity, (b) the receptor concentration in most tissues is believed to be extremely low, and (c) radiolabeled antagonists at low concentrations are expected to interact almost exclusively with the receptor. Assessment of the nonspecifically bound radioactivity, frequently referred to as "nonspecific binding," is generally done by including high, saturating concentrations of a nonradiolabeled, potent antagonist that is thought to displace specifically bound radioactivity; that is, specific binding is the difference between the radioactivity retained on the membrane filter when the assays are devoid of, or contain a saturating concentration of a potent antagonist.

An early report that described the centrifugation technique was from Birdsall et al. (1978), in which both tritiated agonists and antagonists were utilized to examine muscarinic cholinergic receptors in rat cerebral cortex. Briefly, centrifugation was used to terminate the binding reaction in which incubations of the tissue fraction, radiolabeled and optional competing ligands, were done in microcentrifugation tubes at 30°C for 15 min. Following this interval, assay tubes were centrifuged for either 30 or 60 s at 14,000*g*, and the resulting pellet washed with buffer several times, then dissolved in a scintillation medium and counted by liquid scintillation spectrometry. Variations of this method are presently used by numerous laboratories.

The method of rapid filtration was originally described by Yamamura and Snyder (1974) and is based on the separation of protein-bound radioactivity from the remainder of the radioactivity by filtering the incubated medium through a glass fiber filter (generally with a pore size of less than 1–2 μm) by application of negative pressure. In this way, membrane protein-bound radioactivity is trapped and can then be quantitated by liquid scintillation spectrometry. Although minor modifications of the original technique have been introduced, it still appears to be the most commonly used, based on its simplicity, speed, and convenience.

Determining nonspecific binding by the manner described carries with it the assumption that the radioactivity displaced by the competing antagonist (or agonist) is primarily receptor-associated. However, despite this assumption, it seems likely that this procedure may inherently over- or under-estimate specific binding by resulting in displacement of either too much or too little radioactivity from the tissue fraction. Therefore, caution must be applied in this regard since "nonspecific" binding may vary appreciably depending upon the manner in which it is "defined experimentally."

4.2. Experiments With Radiolabeled Agonists

Experimentally, a procedure similar to that outlined above for the radiolabeled antagonists is followed, with the exception that radiolabeled agonists are used in the binding assay. However, because these radiolabeled agonists can only be used in very low concentrations (e.g., p*M*–n*M*), they label primarily high-affinity receptor sites. Thus, these experiments provide complementary and perhaps "direct" assessment of agonist interaction with the high-affinity site of the receptor. It should be noted that tritiated antagonists bind with similar affinity to both high- and low-affinity sites and, therefore, the total receptor number estimated from radiolabeled antagonist binding is always greater than that estimated from radiolabeled agonist binding, because the latter only provides an estimation of high-affinity agonist sites on the receptor. Radiolabeled agonists used so far in the study of brain cholinergic subtypes have been [^{3}H]-*cis*-dioxolane (Vickroy et al., 1983) and [^{3}H]-oxotremorine-M(OXO) (Birdsall et al., 1978). However, those who do employ such radiolabels have found it necessary to introduce technical "tricks" to avoid intolerable artifactual errors that seem to be inherent to the use of these newly available agonists. Just as the strong hydrophobic interactions of some radiolabeled antagonists with the tissue fraction presented a problem, so it seems that the hydrophilic nature of tritiated agonists has also created a complication. Fortunately, these difficulties seem to have been overcome in part (Harden et al., 1983). Published work to date with these radiolabeled agonists has been done primarily using filtration assays, although the centrifugation method has also been employed (Birdsall et al., 1978).

4.3. Use of M_1-Directed Radiolabeled Pirenzepine

The introduction of [^{3}H]-pirenzepine (PRZ) as a preferential radioligand for the M_1 subtype has meant that the binding properties of the M_1 receptor can now be studied directly. This development has thus aided M_1 receptor studies that earlier had to be done with radiolabels apparently unable to discriminate between M_1 and M_2 subtypes. The procedure for using [^{3}H]-PRZ has been recently refined (Watson et al., 1983), such that many of the problems previously associated with its use appear to have been minimized. Presently, [^{3}H]-PRZ binding studies commonly utilize rapid filtration techniques, although there have been attempts to employ centrifugation methods as well (Watson et al., 1982).

4.4. Analysis

There are several criteria that are commonly applied to assess the interaction of ligand with receptor. Some of these appear below:

(a) receptor affinity for either radiolabeled or nonradiolabeled ligands,
(b) saturability of the receptor sites,
(c) the theoretical maximum number of receptor binding sites, and
(d) kinetic rates of ligand dissociation/association, which could be used to assess affinity constants; the latter can also be derived from saturation kinetics.

For these purposes, the types of data analyses most frequently used have been the Hill (Birdsall et al., 1979a), Scatchard (1949), and Schild (Arunlakshana and Schild, 1959) plots, which are helpful for the interpretation of data with respect to the presence of one, two, or more binding sites, as well as the dissociation constants (K_d) and theoretical maximal binding values (B_{max}). In general, results to date have shown that antagonist binding follows classical Langmuir isotherms, and the interactions can often be explained by simple mass-action kinetics (Birdsall et al., 1979a). However, deviation from this type of antagonist interaction may be observed under certain conditions. Expected data for these interactions are seen in the form of sigmoidally shaped displacement curves and linear Scatchard plots. Evidence for nonsimple interaction by a ligand can be observed by its competitive displacement of radiolabeled antagonists, and is illustrated by flattened equilibrium competition curves, nonlinear Scatchard plots, and Hill coefficients less than unity (Birdsall et al., 1978). The mathematical basis for many of these interpretations is presented in a review by Birdsall et al. (1979a). Using this approach, Birdsall et al. (1978) were the first to suggest the presence of two major binding sites (high- and low-affinity) and a third minor site (superhigh-affinity) in the rat cerebral cortex, all of which showed similar antagonist affinities. More recently, Bevan (1984), using [^{3}H]-OXO-M (an agonist) in the rat brain, has reported superhigh- and high-affinity populations of muscarinic cholinergic receptors, as shown by nonlinear Scatchard analysis. The potential relevence that these types of data may have with respect to the M_1 subtype will be discussed at a later point.

Currently, much of the analysis performed in radioligand binding studies is done with the aid of computers, as a result of their greater accessibility and ease of use. Analytical procedures such as those just described are now being performed by laboratory computers more efficiently and accurately than earlier methods. Further discussion on this topic is provided in a later section.

4.5. Autoradiography

In addition to the two procedures just described, characterization of brain muscarinic cholinergic receptors has also been performed using autoradiographic techniques. It is generally believed that the information derived from autoradiographic studies is more pertinent to the in vivo properties of the receptor than that obtained from "pure" binding studies carried out using cell free preparations. Although several autoradiographic studies have looked at CNS-associated muscarinic receptors (Wamsley et al., 1980; Eckelman et al., 1984; Nonaka and Moroji, 1984), only recently has this technique been used to focus on the localization of the "M_1 vs M_2" subtype in the brain (Potter et al., 1984; Wamsley et al., 1984). Simply, this procedure is as follows: Whole or dissected brain regions are "snap frozen" and then sliced (10–20 μm thick) using a cryostat. The slices are then thaw-mounted on slides, which can then be incubated in a buffer solution containing an appropriate radiolabel; this step is followed by brief washes. Nonspecific binding of the radioligand in the slice can be determined by including a saturating (e.g., 1 μ*M*) concentration of a competing antagonist (e.g., atropine) in the incubation medium. After incubation is complete, slices can be dried, and the bound radioactivity determined by long exposure of the slices to tritium-sensitive film (e.g., LKB Ultrofilm), followed by densitometric scanning. This protocol is essentially the same as that utilized by Wamsley et al. (1984), who used both tritiated PRZ and (−)QNB to differentiate specific M_1 sites in brain slices. Their data clearly show that [^{3}H]-PRZ distinguishes a subpopulation of muscarinic cholinergic receptors in the brain, because [^{3}H]-(−)QNB binding not only overlapped that of [^{3}H]-PRZ, but in addition, recognized receptors in other regions devoid of [^{3}H]-PRZ, binding. This indicated not only the existence of muscarinic cholinergic receptor heterogeneity, but also an unequal distribution of the two subtypes that appear to be region specific.

4.6. Receptor Solubilization

There are many reports in the literature that focus on the solubilization of brain muscarinic cholinergic receptors. Based on apparent similarities between the M_1 and M_2 subtypes, these data may thus be able to provide certain clues about the properties of both subtypes, despite the fact that an M1/M2 distinction was not considered experimentally.

The procedures outlined by those who have attempted solubilization of brain muscarinic receptors have generally followed similar protocols (Aronstam et al., 1978b; Gorissen et al.,

1978; Repke and Matthies, 1980; Potter et al., 1984). This usually involves treatment of a partially purified membrane fraction with a detergent, ionic or nonionic, followed by high-speed centrifugation. Soluble receptors are then recovered in the resulting supernatant fluid.

This technique appears to work best when using digitonin as the solubilization detergent (Aronstam et al., 1978b; Repke and Matthies, 1980). Other detergents utilized, both ionic (for example, sodium dodecyl sulfate) (Hurko, 1978) sodium deoxycholate, sodium cholate (Aronstam et al., 1978b; Baron and Abood, 1984), and CHAPS (Shirakawa et al., 1983) and nonionic, e. g., Triton X-100, Tween, and Lubrol (Hurko, 1978) are also capable of receptor solubilization, but in some cases, were found to be less effective in comparison to digitonin. In addition, salt extraction techniques, particularly with NaCl, have been employed by Alberts and Bartfai (1976). However, this procedure was found by both Hurko (1978) and Repke and Matthies (1980) to be substantially inferior to solubilization by digitonin. The reason for this discrepancy is unclear.

Although digitonin appears to be the best suited of the available detergents for this purpose, it does carry with it some disadvantages as well. As outlined by Aronstam et al. (1978b), solubilization with digitonin appears to increase nonspecific binding of the receptor preparation, which apparently decreases in the presence of 1*M* NaCl. In addition, the solubilizing action of digitonin was found to be inconsistent, in that it varied with different batches of the compound. These problems were addressed by Repke and Matthies, who suggested that the detergent in high enough concentrations is able to encapsulate the receptor, forming a micelle-type structure. This may thus explain, in part, the observed increase in nonspecific binding following solubilization. The latter point (variable activity) would appear to result from differences in the amount of gitonin present in digitonin. Thus, Repke and Matthies reported that a 3:2 proportion of digitonin to gitonin was optimal for the solubilization process. Differences in the amounts of these components in the commercially available preparations may therefore, in part, account for instances of "ineffective" digitonin solutions (Aronstam et al., 1978b).

4.7. Properties

4.7.1. Membrane-Bound Receptors

Much of the in vitro characterization of membrane-bound receptors has been carried out in cell-free preparations. Using radiolabeled antagonists, the following observations have been commonly made (for review, *see* Sulakhe et al., 1985):

(a) antagonists bind to a single, high-affinity receptor site,
(b) binding by antagonists is apparently unaffected by the actions of guanine nucleotides or divalent cations,
(c) Scatchard plots are generally linear and Hill coefficients are close to unity for antagonist binding, and
(d) alkylation of sulfhydryl groups by NEM does not appear to significantly affect antagonist binding.

These well-reported observations have been challenged in part by additional studies that suggest the interaction of antagonist with the membrane-bound receptor may not always be so "clear-cut." Pretreatment of tissue fractions, differences in assay temperature, and use of chemically different antagonists have been utilized to demonstrate that the interaction between receptor and antagonist can be altered from its expected pattern.

Additional properties of membrane-bound receptors have been provided by studies focused on the competitive displacement of radiolabeled antagonists by unlabeled agonists. In contrast to antagonists, characterization of agonist interaction with the receptor has shown the following (for review, *see* Sulakhe et al., 1985):

(a) agonists bind to receptors with two distinct affinities,
(b) agonist high-affinity binding is converted to low-affinity by the addition of guanine nucleotides (GTP),
(c) both monovalent and divalent cations are observed to alter high-affinity agonist binding,
(d) Scatchard plots are generally nonlinear, and Hill coefficients are significantly less than one, and
(e) sulfhydryl groups appear to be functionally important in the regulation of high-affinity agonist binding sites.

These general properties of membrane-bound receptors, defined by both agonist and antagonist binding, can be further divided on the basis of the M_1/M_2 receptor distinction. Recent studies have indicated that agonist binding to both M_1 and M_2 sites can be differentially affected by those conditions already mentioned. Specifically, guanine nucleotides do not appear to alter M_1 high-affinity agonist binding in the same fashion as observed for the M_2 site; alkylation of sulfhydryl groups by NEM is reported to enhance agonist binding in the M_1 receptor, opposite to that for the M_2 agonist high-affinity site; divalent cations have little effect on the M_1 receptor, whereas significant decreases in high-affinity agonist binding are reported for the M_2 receptor. Thus, membrane bound receptors display properties that are specific for both antagonist and agonist binding, as well as more subtle distinctions related to the M_1/M_2 subtype separation.

4.7.2. Solubilized Receptors

Receptors solubilized with digitonin appear to retain most of the characteristics reported for those in membrane-bound preparations. These include stereospecificity for antagonists (Gorissen et al., 1978; Hurko, 1978), heterogeneous agonist displacement of radiolabeled antagonists (Hurko, 1978; Baron and Abood, 1984), as well as temperature sensitivity (Baron and Abood, 1984). This latter point, however, may not be applicable to the M_1 subtype, as suggested by reinterpretation of existing data. Furthermore, Potter et al. (1984) have reported that association/dissociation rates differ only slightly for both M_1 and M_2 subtypes as compared to membrane-bound receptors. Pirenzepine displacement of [^{3}H]-(−)QNB was also similar to that seen with membrane preparations, suggesting that the M_1/M_2 distinction was not imparted by the lipid environment. In addition, the solubilization process may include another protein that is functionally important for the M_2 subtype; thus, a receptor complex may result during such partial purification.

4.8. Receptor Size

Unfortunately, no evidence so far exists that clearly shows the specific size and structural characteristics of the M_1 and M_2 subtypes. However, this apparent difficulty can be dealt with by assuming that the data may reflect, in part, M_1 receptor qualities, based on the regions examined. Using this approach, then, Amitai et al. (1982) found the molecular mass of the muscarinic receptor in the rat cerebral cortex to be approximately 86,000 daltons, as identified by a covalently bound, highly specific radioligand marker. The procedure utilized photosensitive, tritiated ligands, created by incorporating an azido group into the benzilic moiety of two high-affinity antagonists, 3-QNB and 4NMPB. Binding of these compounds to the receptors was then made irreversible by brief exposure of the radioligand-membrane media to UV light, followed by polyacrylamide gel electrophoresis. From their results, radioactivity was found to be associated primarily with a single protein band, having a molecular mass of 86,000 daltons. This protein band was thought to contain the monomeric muscarinic receptor site or to consist exclusively of the receptor itself. Since no other radiolabeled band was detected in this study, it was also thought that the muscarinic cholinergic receptors from the CNS so detected may not be oligomeric protein molecules. Additionally, Venter (1983) has examined this question, and has reported a range of 78,000–82,000 daltons for the brain muscarinic receptor, depending upon the species and method used. In his work, SDS/polyacrylamide gel electrophoresis, as well as radiation inactivation-target analysis, were used to estimate receptor size. This latter technique is based on measuring the progressive loss of enzymatic activity by increasingly stronger doses of radiation. Graphically, a linear relationship is observed for activity loss vs radiation. When two enzymes of known molecular mass (one predetermined as a standard) are thus inactivated, the ratio of the slopes of the lines for activity loss vs radiation (referred to as the inactivation ratio) is found to be proportional to the ratio of the molecular mass of the enzyme and that of the standard. Plotting the inactivation ratios against the molecular mass ratios (using a common enzyme standard) for various enzymes reveals a linear relationship, which can serve as a standard curve. For the muscarinic receptor, the inactivation ratio can be determined by measuring the loss of binding sites (as indicated by [^{3}H]-QNB) produced by doses of radiation, and this can then be compared to the standard curve of inactivation ratios vs molecular mass ratios. The unknown molecular mass of the muscarinic receptor can thus be calculated. A broader range has come from Potter et al. (1984) who reported a molecular mass of 80,000–100,000 daltons. Although these molecular masses appear to be similar not only to each other, as well as to reports by others using neural (Birdsall et al., 1979b; Haga, 1980) and nonneural tissues (Herron and Schimerlik, 1984), it is important to remember that these reported masses surely reflect those of heterogenous muscarinic-cholinergic receptor populations. One can thus only assume that the molecular mass of the M_1 subtype may fall somewhere within this range, based on the observation that the various CNS regions examined are now judged to contain considerable populations of M_1 receptors.

4.9. Specific Brain Locations

Much of the evidence supporting the existence, as well as the selective enrichment, of the M_1 vs M_2 receptors has come from studies employing mainly radiolabeled pirenzepine and QNB. This evidence has also come from data based on work with radiolabeled agonists and autoradiography, the results of which tend to agree. The major portion of this work has come from Yamamura and associates, whose research has contributed considerably to the elucidation of the M_1 subtype in the CNS. By using [^{3}H]-pirenzepine, several areas of the brain have been established as containing substantial subpopulations of M_1 receptors. Other regions in which the M_2 subtype predominates have also been recognized by similar means. These are indicated in Table 1.

5. Reinterpretation of Published Data in Terms of Receptor Subtypes

With the aid of recently available cholinergic agents demonstrating M_1/M_2 preferential binding or selectivity, reevaluation of previously published results in light of these more recent developments may be considered. For example, concepts such as receptor isomerization (Henis and Sokolovsky, 1983), binding rate kinetics (Jarv et al., 1980), and so on, have previously been used to explain the variety of data generated by receptor/ligand interaction studies. However, more recent work allows the introduction of an additional explanation. What this leads to is the possible reinterpretation of data in terms of M_1/M_2 characteristics as opposed to previous interpretations in which the receptor population was considered to be homogeneous, that is, consisting of a single type of protein molecule. Thus, the reported observation of agonist binding with superhigh- and high-affinity, as initially demonstrated by Birdsall et al. (1978) and later by Bevan (1984), may reflect binding to two distinct receptors (M_1 and M_2) present in the cortex. In this latter work, it was shown that high-affinity binding, but not super-high binding, was insensitive to guanine nucleotides, a characteristic that now appears to be that of the M_1 receptor subtype (Watson et al., 1983). Early observations of guanine nucleotide insensitivity were provided by Sokolovsky et al. (1980) using murine cerebral cortex and hippocampus, two regions now thought to contain substantial M_1 populations. Similar extrapolation can account for the data of Nukada et al. (1983) using the caudate putamen. Here, a subpopulation of receptors showing insensitivity to guanine nucleotides (M_1-associated?) also demonstrated heat stability; this finding would thus appear to suggest that the M_1 subtype may also be heat stable in comparison

TABLE 1
Distribution of M_1 and M_2 Sites in the Brain

Areas found to contain M_1 subtypes	Areas predominately with M_2 subtypes	Radioligand(s) used	References
Corpus striatum > cerebral cortex ≥ hippocampus	Cerebellum	[^{3}H]-PRZ/QNB	Watson et al., 1983
Corpus striatum > cerebral cortex ≥ hippocampus >> pons-medulla	Cerebellum	[^{3}H]-PRZ/QNB	Luthin and Wolfe, 1984
Caudate putamen ≥ hippocampus ≥ dentate gyrus > cerebral cortex > dorsal horn (spinal cord) > pons-medulla >> cerebellum[a]	Hindbrain nuclei, ventral horn (spinal cord)[a]	[^{3}H]-PRZ/QNB	Wamsley et al., 1984
Hippocampus > striatum > cerebral cortex	Brain stem + superior colliculus	[^{3}H]-(−)QNB	Potter et al., 1984

[a]Refer to article for detailed description of specific nuclei.

to its M_2 counterpart. From these few, yet pertinent, examples, it is apparent that the recent work done with pirenzepine and other subtype-specific agents (for example, McN-A-343) has provided the opportunity for reexamination of both past and present studies. Although interpretation of the manner provided here is speculative, it is hoped it will prove to be quite valuable in that it may stimulate novel experimental approaches to future work.

6. In Vitro Characterization of M_1 and M_2 Receptor Subtypes

It is evident from previous sections that the properties of M_1 and M_2 subtypes differ. For example, only M_1 receptors are apparently guanine-nucleotide insensitive (Vickroy et al., 1983); that is, their affinities for agonists are not altered in the presence of GTP and its analogs. In addition, the high-affinity state of the M_1 subtype appears to be heat stable (authors' interpretation), whereas the M_2 high-affinity state shows sensitivity to temperature (Sulakhe et al., 1985). The actions of *N*-ethylmaleimide (NEM) are different with respect to agonist interaction with either subtype, which may be indicative of differing roles of sulfhydryls in regulating agonist binding to the two receptors. Effects of divalent metals (e. g., Mg^{2+}) are also not evident in the M_1/agonist interaction (Vickroy et al., 1983), which differs again from that of the M_2 subtype (Harden et al., 1983). Each of these apparent differences between the two subtypes deserves further comment.

The insensitivity of M_1 receptors to GTP, which has been observed by Vickroy et al., (1983), may be indicative of the lack of influence of a regulatory coupling protein (Ni) on this receptor in contrast to the now-recognized role of Ni in the M_2 receptor-linked adenylate cyclase system (Sulakhe et al., 1985). It would indeed be interesting in future work to determine whether in an exclusively M_1-containing tissue, there exists cholinergic, agonist-induced inhibition of adenylate cyclase activity. This finding will provide one of the necessary evidences for the previously stated view. This by no means excludes the role of a GTP binding protein (Ns) similar to Ni in the regulation of effector systems other than adenylate cyclase by M_1 receptor/agonist interaction. Additionally, it is conceivable that M_1-dependent effector regulation may occur by mechanisms not involving a Ni- or Ns-like intermediate protein.

The idea of differing heat sensitivity can be thought of in one of two ways. First, the differences exist in the receptor molecules themselves, which could account for the maintenance of the high-agonist binding affinity of M_1 receptors at higher temperatures. Alternatively, the high-affinity state of the M_2 subtype is in fact dependent upon the stability of the Ni protein, thus implying it is this component that imparts to the receptor the observed heat lability. There is now evidence that the Ni protein is more heat labile than the GTP binding protein (Ns), which has been implicated in the stimulation of adenylate cyclase by various hormones (Sulakhe et al., 1985).

The importance of the sulfhydryl groups in cholinergic receptor-ligand interaction was borne out very early on by Aronstam et al. (1978a) in their study on the rat brain cortical muscarinic receptors. The recent observations that NEM increases high-affinity agonist binding to the M_1 receptor and inhibits such binding to the M_2 sites suggests: (a) differences in the sulfhydryl residues regulating agonist binding, and (b) differences in the roles of sulfhydryls in events distal to the receptors. It has been considered recently that the coupling protein (Ni) may possess a sulfhydryl group(s) (Sulakhe et al, 1985). Regardless of the above, it is attractive to consider that sulfhydryl groups of M_1 and M_2 receptor-effector systems regulate agonist binding in an opposite manner; that is, reductive alkylation of M_1 receptor-associated sulfhydryl residue(s) inhibits the inhibitory influence of this group on agonist binding. This possibility is consistent with the speculative interpretation provided by Nukada and coworkers (1983).

It has been well recognized that divalent cations are able to modify considerably the interaction of agonist with the M_2 subtype (Sulakhe et al., 1985). However, recent data indicate that these actions by various metals (e.g., Mg^{2+}, Co^{2+}, Ni^{2+}) are minimal or absent in the M_1 subtype (Vickroy et al., 1983; Nukada et al., 1983). This observed dissimilarity in cation action may thus suggest that the two receptor subtypes differ structurally from each other, such that the M_1 subtype is insensitive to the effects of these divalent metals, or that the site of action of the cation may in fact be located on the Ni protein, so that the metal influences the interaction of receptor and agonist through its action on the regulatory protein, which so far would appear to be only M_2-related. At present, further resolution of this issue does not appear possible.

7. What Constitutes the Muscarinic Receptor High-Affinity Site?

Although it is well established that agonists can bind to a "high-affinity" state of the receptor, this phenomenon is conceptually unresolved because it has yet to be ascertained exactly what state of the receptor (conformational, coupled, or otherwise) defines "high-affinity." In the case of the M_2 subtype, the possibilities that the high-affinity conformational state may be defined by the receptor protein alone, in its native conformation, or by the dimeric existence of the receptor with the coupling protein, Ni (Potter et al., 1984), still remain. Furthermore, one may ask whether the two subtypes adhere to only a single, common high-affinity "mechanism" (high-affinity represented by the receptor molecule alone), or alternatively, whether the M_2 subtype can have two high-affinity conformations (coupled to Ni or as a functional monomer in the membrane), whereas the M_1 high-affinity state is defined by only the latter of these two possible conformations.

Recently, binding constants for the high-affinity state of the M_1 subtype have been reported in the literature. Dissociation constants (K_d) for [^{3}H]-pirenzepine range from approximately 8 to 18 n*M* (Watson et al., 1982, 1983; Wamsley et al., 1984; Luthin and Wolfe, 1984), and K_i values for unlabeled pirenzepine are reported from 15 to 50 n*M* (Hammer et al., 1980; Watson et al., 1983; Luthin and Wolfe, 1984). These values for the high-affinity M_1 site appear to overlap somewhat with those reported for the CNS-associated M_2 receptors.

8. Likely Effectors Regulated by M_1 and M_2 Receptors

Although several receptor systems have been linked to the M_2 subtype (e. g., adenylate cyclase) (Sulakhe et al., 1985), whether any of these are also acted on by M_1 receptors has yet to be discerned. Clues as to the identity of possible effectors are given in part by the observation that the M_1 population appears to be excitatory in nature; that is, M_1 receptor stimulation leads to depolarization (Hammer and Giachetti, 1982; Wilfert et al., 1983; Potter et al., 1984). Therefore, even though functional coupling to an effector like adenylate cyclase seems unlikely based on previous discussion, M_1 receptors may be associated with the regulation of other effector systems, such as phosphatidylinositol turnover or K^+ channel, which may not necessarily require the participation of a GTP-binding regulatory protein. Clearly, evidence in this regard is lacking.

9. Methods and Materials

(a) Radioligands:
  (i) L-[benzilic-4,4'-^{3}H(*N*)] quinuclidinyl benzilate, mol wt = 337.4, specific activity = 30–60 Ci/mmol (New England Nuclear, Boston, MA),
  (ii) [methyl-^{3}H] oxotremorine-M acetate, mol wt = 254.2; specific activity = 70–90 Ci/mmol (New England Nuclear, Boston, MA),
  (iii) L-(+)-CIS-[2-methyl-^{3}H] dioxolane, mol wt = 239.98; specific activity = 38.1 Ci/mmol (New England Nuclear, Boston, MA),
  (iv) [*N*-methyl-^{3}H]-pirenzepine, mol wt = 351.4; specific activity = 70–87 Ci/mmol (New England Nuclear, Boston, MA), and
  (v) [*N*-methyl-^{3}H] scopolamine methyl chloride, mol wt = 380.3; specific activity = 50–90 Ci/mmol (New England Nuclear, Boston, MA)

(b) Atropine sulfate, anhydrous, mol wt = 678.8 (Sigma)
(c) Bovine serum albumin, A6003 (Sigma)
(d) Carbamylcholine chloride (carbachol), mol wt = 182.65 (Sigma)
(e) 5'-guanylylimidodiphosphate, disodium [Gpp(NH)p], mol wt = 545.9 (ICN)
(f) Magnesium chloride mol wt = 203.3 (Sigma)
(g) *N*-ethylmaleimide (NEM) mol wt = 125.1 (Sigma)
(h) 1,4-*Bis*-[2,5-phenoxyloxazolyl] benzene; 2,2'-phenylene-bis-(5-phenyloxazole) (POPOP) mol wt = 364.4 (Sigma)
(i) 2,5-diphenyloxazole (PPO) mol wt = 221.3 (Sigma)
(j) Toluene, scintillation grade, mol wt = 92.14 (BDH)
(k) Tris (hydroxymethyl) aminomethane (TRIS) mol wt = 121.1 (Sigma)
(l) Whatman GF/C glass microfiber filters (1.2 μm)
(m) Prosil-28 (organosilane concentrate surface-treating agent) (Special Chemicals, Gainesville, FL)
(n) Polyethylenimine (Sigma)
(o) Mercuric chloride, granular, mol wt = 271.52 (Mallinckrodt, St. Louis, MO)
(p) Bray's solution (New England Nuclear, Boston, MA)
(q) Vacuum filtration manifold (VFM1 System, Amicon Canada Ltd., Oakville, Ontario, Canada). This cup-seal vacuum filtration manifold has 10 sample wells that adapt to either 25- or 13-mm diameter filters

and a tank that collects the radioactive filtrate. Vacuum connection (vacuum pump, General Electric Co., USA) is made to this tank through one of the two external ports. The bulk filtrate is constantly aspirated by a connection made to the lower port. Each sample well contains a plain white, mixed cellulose ester membrane filter disk of pore diameter 0.8 μm and a sample cup that provides vacuum when it is seated in place.

9.1. Reagents and Solutions

(a) Radioligands are diluted (usually 10:1) to a working stock solution by addition of ethanol/DDW (double distilled water) (7:3). Thus, for an original stock (82.0 Ci/mmol,1 mCi/1.4 mL) solution of [^{3}H]-oxotremorine-M (8.711 μ*M*,0.7143 μCi/μL), 100 μL can be added to 900 μL of ethanol/DDW to prepare a 1 mL solution of radiolabel with a concentration of 871.1 n*M*, or 71.43 nCi/μL. From this working stock, assay standards can be prepared by further dilution with DDW only. For example, 20 μL of this working solution can be diluted 34.84 × by addition of 678.8 μL of DDW to get a final concentration of 25 n*M*. To check the accuracy of the dilution, the specific radioactivity of such a concentration can be determined by counting a small (e. g., 15 μL) sample of this solution. Theoretically, the radioactivity present in this sample should be:

$$0.7143\ \mu Ci/\mu L \times 100\ \mu L/1000\ \mu L \times 20\ \mu L/696.8\ \mu L \times 15\ \mu L = 0.03075\ \mu Ci;$$

Using the conversion factor of 2.2×10^6 dpm/μCi, we can calculate the theoretical number of dpm that this 15 μL volume should give (N.B.: dpm, as opposed to cpm, appears to be a preferable method for measuring radioactivity, because one is then able to correct for the counting efficiency of the liquid scintillation counter). Also, the theoretical conversion factor (see above) is known. Thus,

$$0.03075\ \mu Ci \times 2.2 \times 10^6\ dpm/\mu Ci = 67656.2\ dpm$$

Thus, when the standard is counted, the number of dpm/15 μL can be converted to give the actual concentration of the solution, using the conversion factor. For example, if the sample gave 66,000 dpm, then:

$$66{,}000 \text{ dpm} \times 1\ \mu\text{Ci}/2.2 \times 10^6 \text{ dpm} = 0.030\ \mu\text{Ci}$$

Referring to the specific activity of the original stock (82.0 Ci/mmol), the concentration can be calculated based on the radioactivity:

$$0.030\ \mu\text{Ci} \times 1 \text{ nmol}/82.0\ \mu\text{Ci} = 0.0003658 \text{ nmol} = 365.8 \text{ fmol}$$

Thus, 365.8 fmol of [^{3}H]-oxotremorine-M is present per 15 µL, which makes the concentration 24.39 n*M*. Preparation of other radioligand solutions can be made in similar fashions.

(b) Tris-HCl buffer (150 m*M*, pH 7.4–7.6): Dissolve 1.8165 g of Tris base in 75 mL of DDW, and adjust to desired pH using 1*N* HCl. Add sufficient DDW to make final volume 100 mL.

(c) Magnesium chloride (60 m*M*): Dissolve 0.6099 g of $MgCl_2$-$6H_2$) in about 30 mL of DDW, and adjust volume to 50 mL with more DDW. This stock can then be further diluted if required, or used as is.

(d) *N*-ethylmaleimide (30 m*M*): Dissolve 75.06 mg in about 15 mL of DDW. Adjust to pH 7.5 with 1*M* Tris base and make up to final volume of 20 mL with DDW.

(e) Atropine sulfate (15 m*M*): Dissolve 20.30 mg of atropine sulfate in 2 mL DDW. Further diluted concentrations can be prepared from this stock.

(f) Carbachol chloride (15 m*M*): Dissolve 5.48 mg carbachol chloride in 2 mL of DDW. Further concentrations can be prepared from this stock by dilution with DDW.

(g) 5'-Gpp(NH)p (1.5 mM): Dissolve 3.28 mg Gpp(NH)p in 2 mL of DDW. More dilute concentrations can be prepared by addition of DDW.

(h) BSA (10 mg/mL): Dissolve 100 mg BSA in 10 mL DDW. Dilute 1 mL of this stock in 9 mL DDW to obtain a solution of 1 mg/mL. This latter concentration can be checked spectrophotometrically, $A^{280} = 0.667$.

(i) Scintillation fluid: Dissolve 16 g PPO and 200 mg POPOP in 4 L of toluene, mix thoroughly. Preparation of this solution in a fume hood is recommended.

(j) 1% Prosil-28—0.75% BSA: Amount prepared depends on number of filters to be used. Typically, to make 100 mL of solution, dissolve 0.75 g BSA in 99 mL of DDW, and add 1 mL of concentrated Prosil-28, mix well. (*Note:* **Prosil-28 is an irritant; breathing of vapors and skin contact should be avoided.**)

(k) 0.1% Polyethylenimine: Amount prepared depends on number of filters to be used. To prepare 100 mL, add 0.1 mL (disposable syringe) of the reagent to about 100 mL of DDW, mix thoroughly. (*Note:* **Polyethylenimine is irritating to skin; avoid direct contact.**)

9.2. Buffers

A variety of buffers (for example, HEPES, $NaKPO_4$, $NaPO_4$, Tris, modified Krebs-Henseleit, and so on) is generally used for receptor binding experiments, and most often, without problem. However, it has been found that Tris buffer, in higher concentrations (>25 m*M*) can have deleterious effects on both antagonist and agonist interactions (Hosey, 1983; Murphy and Sastre, 1983). Thus, it is recommended that final concentrations of this buffer should not exceed 20 m*M*. Switching from a 50- to a 10-m*M* final concentration of $NaKPO_4$ buffer was also observed to optimize binding of [^{3}H]-*cis*-dioxolane (Watson et al., 1983). Thus, it is recommended that the chosen buffer be "tested" with the desired radioligand so that complications arising from the buffer be avoided.

9.3. Stability of Solutions

The buffer solutions can be stored conveniently at 4°C and are stable for several months. Drug solutions and radioligands are distributed in small aliquots and stored at −20°C. Carbachol and atropine solutions are prepared fresh monthly. Scintillation fluid is used at laboratory temperature.

9.4. Binding Assay Procedure

9.4.1. Rapid Filtration Technique

(a) Additions: All additions are to be made at 4°C (i.e., test-tube rack placed on ice), with the final concentrations as given below; volume is then made up to 400 μL using DDW.

Reagents	*Final concentration*
Tris-HCl, pH 7.5	10 m*M*
$MgCl_2$	variable
Gpp(NH)p	100 μ*M* (variable)
Drugs	0.1n*M*–1 m*M*
Radioligands	0.05–5 n*M*

(b) Incubation: Start reaction binding by the addition of 50 μL of the tissue fraction (50–200 μg protein/50 μL aliquot) and vortex thoroughly. Tubes can be incubated at room temperature (20–30°C) for 1–2 h, or for longer periods (12–24 h) at colder (0–5°C) temperatures. The final reaction volume should now be 450 μL.

(c) Filtration: After the incubation period is complete, stop the reaction by addition of 5 mL of ice-cold buffer (10 m*M* Tris-HCl, pH 7.5). Test tube contents should then be quickly poured over the glass fiber filter that is placed on the filter disk seat of the vacuum manifold, under mild suction, about 10–15 psi (this lower pressure is desirable because it allows the wash volume to completely cover the filter to assure proper washing). Once tube contents are filtered, the tube is rinsed with 5 mL of buffer, which is also filtered. The GF/C filter itself is then washed with 2 × 5 mL of buffer. This entire process of filtration should not take more than 15 s.

An alternative to filtering the entire assay contents is to remove a fixed volume only, which can then be filtered directly and washed as previously described (Wei and Sulakhe, 1979). Either method appears to work equally well.

(d) After the washing of a filter is complete, the vacuum cup can be quickly removed and the filter taken off with blunt forceps (Millipore) and placed in a scintillation vial (Fisher, 20-mL capacity). Once the filtration step is complete for all assay tubes, the filters should be dried at 60°C for approximately 2 h.

(e) Total radioactivity of each sample is determined by addition of 9 mL of scintillation fluid to the vial followed with counting by liquid scintillation spectrometry (typically, 40–45% counting efficiency for tritium).

(f) The amount of protein added to the assays can be determined from the original tissue fraction by the method of Lowry et al. (1951), using BSA (1 mg/mL) as the standard.

(g) Radioligand standards can be prepared by adding an accurately measured volume (e. g., 25 μL) of the same radioligand solution as that used in the assay to dry GF/C filters placed in scintillation vials. These vials can be dried along with the others.

(h) Assay blanks can be prepared exactly as the other tubes (i.e., same additions) except that no tissue fraction is added. In this way, the amount of radiolabel bound only to the filter can be determined; the fraction volume should be replaced with DDW.

(i) Specific and nonspecific binding: The amount of radioactivity present on the filter is the sum of the radioligand bound specifically to the receptors and the amount bound nonspecifically to other tissue protein. Thus, nonspecific binding of the radioligand is defined as the radioactivity bound to the fraction in the presence of 1 μ*M* (final) atropine sulfate (or other potent antagonists) and usually does not exceed more than 20%. The specific binding of the radioligand is then calculated from the difference in total counts (dpm) in the absence and presence of atropine. The concentration of free ligand is then corrected by subtracting the amount bound from the total added.

(j) Calculations: The specific activity of the radioligand should first be corrected for radioactive decay. Tritium has a half-life of 12.3 yr, so by using the following relationship, loss of radioactivity from the time of production can be calculated:

$$N/No = (0.5)^{x/147.6}$$

where N = present radioactivity, No = original radioactivity, x = time elapsed since production (mo), 147.6 = half-life (mo)

Making this correction is also important when checking the concentration of a prepared radioligand solution, because the specific activity is required to convert from dpm to Ci to moles of substance. Thus, for a 1-yr-old sample of [^{3}H]-(−)QNB (33.1 Ci/mmol = original), the new radiospecificity would be:

$$N/33.1 \text{ Ci/mmol} = (0.5)^{12/147.6}$$
$$= 31.286 \text{ Ci/mmol}$$

Calculation of [^{3}H]-(−)-QNB specific binding (fmol/mg protein) can be done as follows, using the radiospecificity just calculated:

total radioactivity counted (− atropine) − total radioactivity counted (+ atropine) = radioactivity bound specifically

specific binding (dpm) × 1 fmol/31.286 pCi × 1 pCi/2.2 dpm × mg protein = fmol of [^{3}H]-(−)-QNB specifically bound per mg protein

If the binding of a radioligand is determined over a range of concentrations, then a plot of the amount specifically bound against the free concentration of radiolabeled represents a saturation or equilibrium binding isotherm, which is commonly hyperbolic in shape. Once at this point, Scatchard analysis of the same data can be performed. Very simply, this is a plot of the ratio of the amount of specifically bound radiolabel vs its "free" concentration in the assay (ordinate) against the amount specifically bound (abscissa). Graphically, this relationship of "bound/free vs bound" is often a linear one (Wei and Sulakhe, 1979), with a negative slope and extrapolated x intercept. This point where the curve intersects the x axis is referred to as the theoretical binding maximum, or B_{max}, whereas the negative inverse of the slope of the curve is representative of the dissociation constant (K_d) for the measured ligand. It should be recognized that this type of analysis is only applicable to binding isotherms or experimental data in which the degree of binding by a radiolabel (or unlabeled competing ligand) at different concentrations was determined. An example of this type of calculation is provided below:

If 10 fmol/mg protein of [^{3}H]-(−)QNB were specifically bound at a free radiolabel concentration of 0.1 n*M*, then the *y* coordinate value would be the ratio of bound/free, or 100 fmol [^{3}H]-QNB/mg protein/n*M*, and the *x* coordinate value would simply be 10 fmol [^{3}H]-QNB/mg protein. Thus, the first point on the graph would be (10,100). Similar calculations can then be made for all data points, followed by graphical presentation. Linear or nonlinear regression analysis can be used to draw the curve through the plotted points.

Once this analysis has been completed, it allows the calculation of the Hill coefficient (n_H), which is derived from the Hill plot (Birdsall et al., 1979a). Essentially, the set of values for the ordinate, or *y* axis, is determined by the following relationship: quantity of ligand specifically bound at the concentration used (B) divided by the difference between the theoretical binding maximum and the amount specifically bound $\{B/(B_{max} - B)\}$. The concentrations of free ligand are then plotted on the abscissa. Thus, the value of B_{max} is required for this set of calculations, indicating that Scatchard analysis must be done prior to this analytical procedure. A sample calculation for this procedure is given below:

using the same values as given above and an imaginary B_{max} of 1000 fmol/mg protein, the *y* coordinate would be

$$10 \text{ fmol}/(1000 \text{ fmol} - 10 \text{ fmol}) = 0.0101$$

The corresponding *x* coordinate will simply be the free ligand concentration, which in this example is 0.1 n*M* (assuming no correction is required).

This calculation can then be done for the different values of bound ligand that were determined at each separate ligand concentration used in the assay. For the actual plotting of these points, it is suggested that both the *x* and *y* coordinates be expressed in log units, because this has been found to nicely accomodate the wide range of ligand concentrations that are used experimentally. Once plotted, a linear relationship should be observed; taking the slope of this line then yields the Hill coefficient, or n_H. Refer to the text for the significance of this calculated value.

(k) Special requirements: Intolerably high filter blank values are commonly observed with [^{3}H]-oxotremorine-M and [^{3}H]-pirenzepine binding experiments, and these obscure absolute binding values. To reduce the amount of radioactivity retained by the filter, 1% Prosil-28 – 0.75% BSA solution (2–3 mL per filter) should be used for treating GF/C filters (minimum 2 h) when [^{3}H]-oxotremorine-M is used as the radioligand. Doing so reduces the number of dpm retained by the filter to about 0.5% of total dpm filtered (Braun, unpublished observations; Harden et al., 1983). For [^{3}H]-pirenzepine, treatment of filters in a solution of 0.1% polyethylenimine for about 1 h is required (Pretreatment in 0.05% polyethylenimine is also recommended for binding studies with [^{3}H]-*cis*-dioxolane). Vacuum pressure for these two radiolabels should be 10–12 psi (allows the wash volume to completely cover the filter), and the filters should be allowed to remain under this suction for at least 2 min (allows the bulk of the moisture to be removed) following filtration and washing. (*Note:* Using 3–4 vacuum wells simultaneously for filtering allows successive samples to be filtered and the filters then removed at 3–4 tube intervals later. [^{3}H]-pirenzepine and oxotremorine-M are also somewhat photolabile; to avoid this instability from affecting results, binding experiments should be carried out using dim, indirect lighting (Braun, unpublished observations).

9.5. Data Analysis

Binding data, once calculated, can be analyzed by several methods. As mentioned, the theory behind this analysis has been described by Birdsall et al. (1979a), along with the use of such techniques as Scatchard and Hill plots. Presently, these types of analyses are being applied through the use of computer-driven software, utilizing programs that have been written to provide curve-fitting (both linear and nonlinear), statistical analysis, and fundamental calculation of dissociation constants, Hill coefficients, Scatchard analysis, and so on. Not only do these application programs drastically reduce the workload, they also provide increased accuracy and speed of analysis. Currently, availability and variety of software tailored to the needs of ligand binding analysis are steadily growing, as evidenced by such systems as LIGAND (Munson and Rodbard, 1980), EBDA (McPherson, 1983), and BMDPAR (Health Science Computing Faculty, UCLA).

9.6. Additional Comments

Assays using radiolabeled agonists require much more care than those with radiolabeled antagonists. We recommend that the user carry out a preliminary study of the conditions (e.g., varying the assay reagents, incubation times, temperature, reaction volume, and so on) prior to adaption of the agonist binding assay described here.

Besides the rapid filtration assay, a centrifugation assay for muscarinic receptor binding can also be used. In this case, binding assays can be done in microcentrifugation tubes and, following incubation, the reactions are stopped by centrifugation (100,000*g* for 1 min). Pellets are then carefully washed two or three times with buffer, and finally dissolved using 0.1*N* NaOH, 1% SDS, or commercially available tissue solubilizers. The dissociated pellet can then be removed, placed in a vial, and counted (Bray's solution) by liquid scintillation spectrometry.

10. Conclusions

From all the work examined, it still appears quite evident that our present understanding of the putative "M_1 subtype" is very limited, most certainly because of the novelty of this receptor subtype. Although important inroads have been made through the use of pirenzepine, current knowledge would seem to represent only the tip of the "M_1 receptor iceberg." In spite of this, present data have still provided some basic conceptual clues by which future work can be guided. Affinity states, sulfhydryl groups, receptor properties, effector coupling, and regulatory proteins are only some of the major issues confronting researchers in this field. One positive point that has emerged is the greater confidence with which the M_1 receptor story can be related. The work described above appears to provide more than circumstantial evidence for the existence of a separate subpopulation of muscarinic cholinergic receptors in the CNS, distinguishable both functionally and anatomically. However, it is obvious from the present state of knowledge that the majority of work in this field remains to be done.

Acknowledgments

This work was supported by a grant from the Saskatchewan Heart Foundation. APB is a recipient of a graduate student scholarship from the University of Saskatchewan.

References

Alberts P. and Bartfai T. (1976) Muscarinic acetylcholine receptor from rat brain. Partial purification and characterization. *J. Biol. Chem.* **251,** 1543–1547.

Amitai G., Avissar S., Balderman M., and Sokolovsky M. (1982) Affinity labeling of muscarinic receptors in rat cerebral cortex with a photolabile antagonist. *Proc. Natl. Acad. Sci. USA* **79,** 243–247.

Aronstam R. S., Abood L. G., and Hoss W. (1978a) Influence of sulfhydral reagents and heavy metals on the functional state of the muscarinic acetylcholine receptor in rat brain. *Mol. Pharmacol.* **14,** 575–586.

Aronstam R. S., Schuessler Jr. D. C., and Eldefrawi M. E. (1978b) Solubilization of muscarinic acetylcholine receptors of bovine brain. *Life Sci.* **23,** 1377–1382.

Arunlakshana O. and Schild H. O. (1959) Some quantitative uses of drug antagonists. *Br. J. Pharmacol.* **14,** 48–58.

Ash A. S. F. and Schild H. O. (1966) Receptors mediating some actions of histamine. *Br. J. Pharmacol.* **27,** 427–43.

Baron B. and Abood L. G. (1984) Solubilization and characterization of muscarinic receptors from bovine brain. *Life Sci.* **35,** 2407–2414.

Bevan P. (1984) [^{3}H] Oxotremorine-M binding to membranes prepared from rat brain and heart. Evidence for subtypes of muscarinic receptors. *Eur. J. Pharmacol.* **101,** 101–110.

Birdsall N. J. M., Burgen A. S. V., and Hulme E. C. (1978) The binding of agonists to brain muscarinic receptors. *Mol. Pharmacol.* **14,** 723–736.

Birdsall N. J. M., Burgen A. S. V., Hulme E. C. (1979a) Muscarinic receptors: Biochemical and Binding Studies, in *Recent Advances in Receptor Chemistry*. (Gualtieri F., Giannella M., and Malchiorre C., eds.) Elsevier/North Holland Biomedical, Amsterdam.

Birdsall N. J. M., Burgen A. S. V., and Hulme E. C. (1979b) A study of the muscarinic receptor by gel electrophoresis. *Br. J. Pharmacol.* **66,** 337–342.

Birdsall N. J. M., Burgen A. S. V., Hulme E. C., Stockton J. M., and Zigmond M. J. (1983) The effect of McN-A-343 on muscarinic receptors in the cerebral cortex and heart. *Br. J. Pharmacol.* **78,** 257–259.

Black J. W., Duncan W. A., Durant C. J., Ganellin C. R., and Parsons E. M. (1972) Definition and antagonism of histamine H2-receptors. *Nature* (Lond.) **236,** 385–390.

Brown J. H. and Brown-Masters S. (1984) Muscarinic regulation of phosphatidylinositol turnover and cyclic nucleotide metabolism in the heart. *Fed. Proc.* **43,** 2613–2617.

Burgen A. S. V. and Spero L. (1968) The action of acetylcholine and other drugs on the efflux of potassium and rubidium from smooth muscle of the guinea pig ileum. *Br. J. Pharmacol.* **34,** 99–115.

Burgen A. S. V., Hiley C. R., and Young J. M. (1974) Binding of [^{3}H]-propylbenzilylcholine mustard by longitudinal muscle strips from guinea pig small intestine. *Br. J. Pharmacol.* **50,** 145–151.

Dunlap J. and Brown J. H. (1983) Heterogeneity of binding sites on cardiac muscarinic receptors induced by the neuromuscular blocking agents gallamine and pancuronium. *Mol. Pharmacol.* **24,** 15–22.

Eckelman W. C., Reba R. C., Rzeszotarski W. J., Gibson R. E., Hill T., Holman B. L., Budinger T., Conklin J. J., Eng R., and Grissom M. P. (1984) External imaging of cerebral muscarinic acetylcholine receptors. *Science* **223,** 291–293.

Gorissen H., Aerts G., and Laduron P. (1978) Characterization of digitonin-solubilized muscarinic receptor from rat brain. *FEBS Lett.* **96,** 64–68.

Goyal R. K. and Rattan S. (1978) Neurohumoral, humoral, and drug receptors for the lower esophageal sphincter. *Gastroenterology* **74,** 598–619.

Haga T. (1980) Molecular size of muscarinic acetylcholine receptors of rat brain. *FEBS Lett.* **113,** 68–72.

Hammer R., Berrie C. P., Birdsall N. J. M., Burgen A. S. V., and Hulme E. C. (1980) Pirenzepine distinguishes between different subclasses of muscarinic receptors. *Nature* (Lond.) **283,** 90–92.

Hammer R. and Giachetti A. (1982) Muscarinic receptor subtypes: M_1 and M_2 biochemical and functional characterization. *Life Sci.* **31,** 2991–2998.

Harden T. K., Meeker R. B., and Martin M. W. (1983) Interaction of a radiolabeled agonist with cardiac muscarinic cholinergic receptors. *J. Pharmacol. Exp. Ther.* **227,** 570–578.

Heilbronn E. and Bartfai T. (1978) Muscarinic acetylcholine receptor. *Prog. Neurobiol.* **11,** 177–188.

Henis Y. I. and Sokolovsky M. (1983) Muscarinic antagonists induce different receptor conformations in rat adenohypophysis. *Mol. Pharmacol.* **24,** 357–365.

Herron G. S. and Schimerlik M. I. (1984) Protein composition of the atrial muscarinic acetylcholine receptor partially purified by wheat germ agglutinin affinity chromatography. *Arch. Biochem. Biophys.* **230,** 533–542.

Hosey M. M. (1983) Regulation of ligand binding to cardiac muscarinic receptors by ammonium ion and guanine nucleotides. *Biochim. Biophys. Acta.* **757,** 119–127.

Hulme E. C., Birdsall N. J. M., Burgen A. S. V., and Mehta P. (1978) The binding of antagonists to brain muscarinic receptors. *Mol. Pharmacol.* **14,** 737–750.

Hurko H. (1978) Specific [^{3}H] quinuclidinyl benzilate binding activity in digitonin-solubilized preparations from bovine brain. *Arch. Biochem. Biophys.* **190,** 434–445.

Jarv J., Hedlund B., and Bartfai T. (1980) Kinetic studies on muscarinic antagonist-agonist competition. *J. Biol. Chem.* **255,** 2649–2651.

Kloog Y. and Sokolovsky M. (1978) Studies on muscarinic acetylcholine receptors from mouse brain. Characterization of the interaction with antagonists. *Brain Res.* **144,** 31–48.

Lowry O. H., Rosebrough N. J., Farr A. L., and Randall R. J. (1951) Protein measurements with the Folin phenol reagent. *J. Biol. Chem.* **183,** 265–275.

Luthin G. R. and Wolfe B. B. (1984) Comparison of [^{3}H] pirenzepine and [^{3}H] quinuclidinylbenzilate binding to muscarinic cholinergic receptors in rat brain. *J. Pharmacol. Exp. Ther.* **228,** 648–655.

McPherson G. A. (1983) A practical computer-based approach to the analysis of radioligand binding experiments. *Comput. Programs Biomed.* **17,** 107–113.

Munson P. J. and Rodbard D. (1980) LIGAND: A versatile computerized approach for the characterization of ligand-binding systems. *Anal. Biochem.* **107,** 220–239.

Murphy K. M. and Sastre K. (1983) Obligatory role of a Tris/choline allosteric site in guanine nucleotide regulation of [^{3}H]-L-QNB binding muscarinic acetylcholine receptors. *Biochem. Biophys. Res. Commun.* **113,** 280–285.

Nonaka R. and Moroji T. (1984) Quantitative autoradiography of muscarinic cholinergic receptors in the rat brain. *Brain Res.* **296,** 295–303.

Nukada T., Haga T., and Ichiyama A. (1983) Muscarinic receptors in porcine caudate nucleus (II). *Mol. Pharmacol.* **24,** 374–379.

Potter L. J., Flynn D. D., Hanchett H. E., Kalinoski D. L., Luber-Narod J., and Mash D. C. (1984) Independent M_1 and M_2 receptors: Ligands, autoradiography and functions. *Trends in Pharmacol. Sci.* (Suppl.) 22–31.

Repke H. and Matthies H. (1980) Biochemical characterization of solubilized muscarinic acetylcholine receptors. *Brain Res. Bull.* **5,** 703–709.

Roberts, C. M. and Konjovic, J. (1969) Differences in the chronotropic and inotropic response of the rat atrium to choline esters, cholinesterase inhibiters, and certain blocking agents. *J. Pharmacol. Expt. Therap.* **169,** 109–119.

Scatchard G. (1949) The attraction of proteins for small molecules and ions. *Ann. NY Acad. Sci.* **51,** 660–672.

Shirakawa O., Kuno T., and Tanaka C. (1983) The glycoprotein nature of solubilized muscarinic acetylcholine receptors from bovine cerebral cortex. *Biochem. Biophys. Res. Commun.* **115,** 814–819.

Sokolovsky M., Gurwitz D., and Galron R. (1980) Muscarinic receptor binding in mouse brain: Regulation by guanine nucleotides. *Biochem. Biophys. Res. Commun.* **94,** 487–492.

Stiles G. L. and Lefkowitz R. J. (1984) Cardiac adrenergic receptors. *Ann. Rev. Med.* **35,** 149–164.

Sulakhe P. V., Jagadeesh G., and Braun A. P. (1985) Cardiac Autonomic Receptors and Adenylate Cyclase in Health and Science, in *Regulation of Heart Function—Basic Concepts and Clinical Applications* (Rupp H. ed.), Thieme-Stratton, New York.

Trummlitz. G., Schmidt G., Wagner H.-U., Luger P. (1984) Confirmational analysis of the antiulcer drug pirenzepine. *Arzneim.-Forsch./Drug Res.* **34,** 849–859.

Venter J. C. (1983) Muscarinic cholinergic receptor structure. *J. Biol. Chem.* **258,** 4842–4848.

Vickroy T. W., Yamamura H. I., and Roeske W. R. (1983) Differential regulation of high affinity agonist binding to muscarinic sites in the rat heart, cerebellum, and cerebral cortex. *Biochem. Biophys. Res. Commun.* **116,** 284–290.

Wamsley J. K., Zarbin M. A., Birdsall N. J. M., and Kuhar M. J. (1980) Muscarinic cholinergic receptors: Autoradiographic localization of high and low affinity agonist binding sites. *Brain Res.* **200,** 1–12.

Wamsley J. K., Gehlert D. R., Roeske W. R., Yamamura H. I. (1984) Muscarinic antagonist binding site heterogeneity as evidenced by autoradiography after direct labeling with $[^3H](-)$QNB and $[^3H]$ pirenzepine. *Life Sci.* **34,** 1395–1402.

Watson M., Roeske W. R., and Yamamura H. I. (1982) $[^3H]$ pirenzepine selectively identifies a high affinity population of muscarinic cholinergic receptors in the rat cerebral cortex. *Life Sci.* **31,** 2019–2023.

Watson M., Yamamura H. I., and Roeske W. R. (1983) A unique regulatory profile and regional distribution of $[^3H]$ pirenzepine binding in rat provide evidence for distinct M_1 and M_2 muscarinic receptor subtypes. *Life Sci.* **32,** 3001–3011.

Wei J.-W. and Sulakhe P. V. (1979) Properties of muscarinic cholinergic receptors in rat atrium. *Naunyn Schmiedebergs Arch. Pharmacol.* **309,** 259–269.

Wilfert B., Davidesko D., Timmermans P., and Van Zwieten P. A. 1983 Differential role of M_1 and M_2 receptors in sympathetic ganglia of the pithed normotensive rat in alpha-adrenoceptor mediated vasoconstriction. *J. Pharmacol. Exp. Ther.* **226,** 855–860.

Yamamura H. I. and Snyder S. H. (1974) Muscarinic cholinergic binding in rat brain. *Proc. Natl. Acad. Sci. USA* **71,** 1725–1729.

Chapter 7
Excitatory Amino Acid Receptors

John W. Ferkany and Joseph T. Coyle

1. Introduction

The dicarboxylic amino acids L-glutamate (Glu) and L-aspartate (Asp) are found throughout the mammalian central nervous system. Although it was initially believed that these amino acids participated solely in the metabolic activities of the brain, evidence that Glu and Asp are excitatory neurotransmitters is now substantial (reviewed in Curtis et al., 1972; Davies et al., 1979; Davies et al., 1980; Coyle, 1980; Watkins and Evans, 1981). Beyond a fundamental role as chemical mediators of synaptic transmission, there is an increasing suspicion that alterations in the neurochemistry of excitatory amino acids are involved in the etiology or symptomatic presentation of several neuropsychiatric and neurodegenerative disorders (Olney and Gubareff, 1978; Nadler, 1981; Coyle, 1982a).

Both electrophysiological and biochemical evidence suggest that multiple types of synaptic receptors for excitatory amino acids are present in brain (Foster and Fagg, 1984) (Table 1). Defined by the potency of antagonists to selectively inhibit agonist-induced neuroexcitation, these receptors have been divided into three categories and include (a) receptors activated by Glu and the conformationally restricted Glu analog, quisqualic acid (Quis), and antagonized by glutamic acid diethylester (GDEE); (b) receptors responsive to the synthetic analog of Asp, *N*-methyl-D-aspartic acid (NMDA), the neurotoxin, ibotenic acid (Ibo), and possibly to Asp itself, and antagonized by Mg^{2+}, D-(−)-2-amino-5-phosphonopentanoic acid (AP5), and D-(−)-2-amino-7-phosphonoheptanoic acid (APH); and (c) receptors activated by the pyrrolidine neurotoxin, kainic acid (KA). No selective antagonists for the neuroexcitotoxic actions of KA have been identified (however, see Krogsgaard-Larsen and Honore, 1983).

TABLE 1
Major Receptor Categories for Excitatory Amino Acids in the Mammalian Central Nervous System

Agonists	Antagonists
Glutamate/quisqualate	
L-Glutamic acid	GDEE
Quisqualic acid	
AMPA	
N-Acetylaspartyl glutamate	DL-(±)-APB
N-Methyl-D-aspartic acid	
N-Methyl-D-aspartic acid	ADA
Ibotenic acid	D-(−)-AP5
L-Homocysteic acid	D-(−)-APH
Quinolinic acid	
Aspartic acid (?)	
Kainic acid	
Kainic acid	
Domoic acid	

The focus of the following chapter is twofold. First, we undertake to review briefly the current understanding of receptors for excitatory amino acids as defined by in vitro ligand binding techniques. Second, a discussion of some of the practical aspects of receptor binding assays for excitatory amino acid receptor ligands is presented. Because of the limitations of space, our review of excitatory amino acids is necessarily confined to what we feel are the key issues that must be appreciated prior to attempting receptor binding assays with these ligands. Readers interested in the electrophysiological aspects of excitatory amino acids or a more comprehensive discussion of the pharmacology and neurochemistry of these compounds are referred to any of the excellent reviews cited above. Details of the theoretical considerations of receptor binding methods can be found in the first chapter of this volume by Hrinda and elsewhere (Cuatrecasas and Hollenberg, 1976; Bylund, 1980).

2. Receptors for Excitatory Amino Acids

Both Asp and Glu are present in neuronal and glial elements of the brain and can be detected at high concentrations (up to 10 μmol/g) at birth and throughout the life of the animal (Altman, 1972; Bradford, 1976; Currie and Kelly, 1981). Synthetic and degradative enzymes specific for the neurotransmitter pools of Asp and Glu have not been unequivocally identified, and the use of these markers to detect pathways that release Glu or Asp has proven difficult (Skaggs et al., 1983). Although excitatory amino acids are accumulated in brain by high-affinity transport systems, such processes do not appear to be restricted to neuronal components of the CNS (Wofsey et al., 1971; Bennett et al., 1972; Roberts and Watkins, 1975; Weiler et al., 1978; Hamberger et al., 1979; Currie and Kelly, 1981). Thus, the standard assumptions and methodologies that have proven useful in characterizing other neurotransmitters, including the identification and localization of transport systems and the localization of the synthetic or catabolic enzymes or of the transmitter itself, have been of limited value in delineating excitatory amino acid neuronal systems.

With the advent of in vitro ligand binding techniques came the hope that these procedures would afford a simple and direct avenue to characterizing selected aspects of excitatory amino acid neurotransmission. Although initial expectations have not been fully realized, the application of ligand binding methods has allowed the tentative identification of the three major categories of excitatory amino acid neurotransmitter receptors in brain. However, as will become apparent in the discussion that follows, a variety of factors, including the method of tissue preparation, tissue storage, and the composition of the incubation buffer in which the binding assay is performed, can have dramatic effects on the characteristics of the ligand–receptor interaction. Additionally, many of the excitatory amino acid agonists and antagonists that by electrophysiological criteria are considered to be selective for a particular category of receptor appear to be less receptor-selective in ligand binding assays in vitro. In order to facilitate the discussion of various binding assays and receptor categories, we have divided the following sections broadly, according to whether compounds exhibit agonist or antagonist properties in vivo and secondarily, according to the individual ligand under consideration.

2.1. Agonists

2.1.1. L-Glutamate

Partly because Glu was among the first of the acidic amino acids recognized to possess neuroexcitatory properties in brain (Curtis et al., 1972), it has become one of the most extensively studied compounds of its class. Both electrophysiological and biochemical investigations have provided substantial evidence that Glu may be the "common" excitatory neurotransmitter in brain. Neurons in virtually all regions of the CNS undergo depolarization following iontophoretic application of Glu (Davies et al., 1979; Davies et al., 1980; Watkins and Evans, 1981). Release of Glu from tissue stores can be evoked in vivo and in vitro by elevated extracellular potassium or electrical stimulation (Reubi, 1980; Keller et al., 1981; Ferkany and Coyle, 1983a). Stimulus-induced release is largely dependent on the presence of Ca^{2+}, suggesting that releasable Glu is coupled to a neuronal pool (Potashner, 1978; Foster and Roberts, 1980; Ferkany and Coyle, 1983a,c); and lesion or ablation of putative glutamatergic pathways is accompanied by a decline in the uptake and evoked release of the amino acid in distant loci innervated by the interrupted tract (McGeer et al., 1977; White et al., 1977; Storm-Mathisen and Upsahl, 1978; Zaczek et al., 1979; Walaas and Fonnum, 1980).

Efforts to identify a synaptic Glu receptor in brain using ligand binding techniques were pioneered by Roberts (1974) using [^{14}C]-Glu. In cerebral synaptic membranes, Roberts identified two saturable binding sites for the ligand that were distinguished by both the density of sites (B_{max}) and the requirement that Na^+ be present in the incubation medium to detect one but not the other component of binding. This latter phenomenon, referred to as "sodium-dependency," is frequently encountered in ligand binding studies when the neurotransmitter itself is used as a tracer and generally reflects the interaction of the ligand with sodium-dependent transport sites rather than the synaptic receptor (Baudry and Lynch, 1981; Sharif and Roberts, 1981a).

The Na^+-independent component of [^{14}C]-Glu binding (K_d = 8.3 μM) observed by Roberts possessed some of the characteristics predicted for a synaptic Glu receptor. Binding was stereospecific and L-Glu, the form of Glu found in brain, was a more potent inhibitor of specific binding than the D-isomer. Specific binding of the ligand was also inhibited by high concentrations of the putative Glu receptor antagonist, GDEE.

After the introduction of [^{3}H]-Glu of high specific radioactivity, several laboratories began to explore the biochemical, pharmacological, and regulatory aspects of Glu receptors in the CNS. Among the first of the reports to appear was that of Foster and Roberts (1978), who examined [^{3}H]-Glu binding to freshly prepared cerebellar membranes incubated in Tris-HCl buffer (37°C). Expanding upon Roberts' earlier study, these authors demonstrated a single, saturable binding component for [^{3}H]-Glu of moderate affinity (K_d = 0.7 μM) and high density (B_{max} = 73 pmol/mg protein). Specific binding was competetively inhibited by several excitatory amino acids and excitatory amino acid analogs, including Quis, Ibo, and homocysteic acid (K_i s = 3–5 μ*M*), as well as by D- and L-Asp, the excitatory amino acid antagonist D,L(±) amino adipic acid (ADA), and D,L(±)APB (K_i s = 12–20 μ*M*). Neither NMDA nor KA displayed significant inhibitory activity at the [^{3}H]-Glu binding site (Table 2).

Other early investigation of specific [^{3}H]-Glu binding to brain revealed different characteristics for the ligand–receptor interaction (Biziere et al., 1980). Using previously frozen brain membranes as a source of receptors and conducting incubations at 2°C in Tris-citrate buffer, two pharmacologically similar but kinetically distinct (K_d s = 11; 85 n*M*) binding sites for [^{3}H]-Glu were detected. With the exception of membranes prepared from heart muscle, specific binding of the ligand was confined to the CNS, and although a marked regional difference in receptor density was detected (frontal cortex > hippocampus > striatum > thalamus > cerebellum > hypothalamus), the ratio of high-affinity to low-affinity receptors (1:9) was constant throughout the regions examined. Kainic acid-induced lesions, (*see* section 2.1.4) of the corpus striatum, but not the destruction of excitatory cortico/striatal efferents, resulted in a significant decrease in the binding of [^{3}H]-Glu to striatal membranes, consistent with the suggestion that the receptors being examined were located on striatal interneurons and not associated with Glu transport sites present on the axonal projections of cortical neurons. Of the compounds examined, the excitatory amino acids Glu, Asp, cysteine sulfinic acid, homocysteic acid, and cyclopentylglutamic acid were found to be potent inhibitors of specific binding, as was the NMDA-type antagonist, ADA (Table 3). When tested at 100 μ*M*, NMDA, KA, Quis, Ibo, and AP5 were found to be ineffective inhibitors of ligand binding.

The reports of Foster and Roberts (1978) and Biziere et al. (1980) serve to illustrate the complexities of examining Glu receptors in brain using in vitro ligand binding techniques. As becomes apparent in the following discussion, methodological differences most likely account for the disparate results of these studies. Indeed, it is now recognized that specific binding of [^{3}H]-Glu to brain membranes is exquisitely sensitive to both the conditions of the binding assay, as well as the methods of tissue preparation and storage (*see* Foster and Fagg, 1984, for detailed discussion). Thus, at least two major subdivisions of receptors for Glu can be detected in vitro, including (a) recognition sites that requires the presence of Cl^- and is labile to prior freezing of the tissue, and (b) a Glu receptor that is resistant to freezing and can be detected independent of the addition of Cl^- to the incubation (Foster and Fagg, 1984).

TABLE 2
Pharmacologic Characteristics of Chloride- and Calcium-Dependent Specific [^{3}H]-Glutamate Binding to Brain[a]

	Inhibition constants$_{50}$, μM				
Compound	A	B[b]	C	D	E
<u>Agonists</u>					
L-Quisqualic acid	0.6	0.2	0.1	0.24	8.4
L-Glutamic acid	1.0	0.8	0.4	0.21	5
Ibotenic acid	3.1	1.2	0.5	0.4	8.1
L-Cysteine sulfinic acid	7.1	4.7			
L-Homocysteic acid		1.6			10
D-Glutamic acid	15	10	9.0	2.6	29
L-Aspartic acid	28	6.0	11	2.1	
D-Aspartic acid	29	50	27		140
AMPA				28	
Cyclopentylglutamate	10				
<u>Antagonists</u>					
D,L-(±)-AP3				>500	
D,L-(±)-APB	4.6	4.2		1.2	26
D-(−)-APB			75		
L-(+)-APB			5		
D,L-(±)-AP5			39	4.4	
D-(−)-AP5	14				
L-(+)-AP5	16				
D,L-(±)-APHex	>100		5		
D,L-(±)-APH	13				
D-(−)-APH			48		
L-(+)-APH			12		
D,L-(±)-ADA		1.3			26
D-(−)-ADA	2.4		0.7		
L-(+)-ADA	3.4				

[a]Compounds that fail to inhibit specific binding of [^{3}H]-L-glutamate include: NMDA, domoic acid, kainic acid, dihydrokainic acid, GDEE, gamma-D-glutamylglycine, and alpha-methyl-L-glutamic acid.

Conditions and references:

A. Cerebellum; Tris-HCL (0.05*M*, pH 7.1, 37°C); Slevin et al. (1982).
B. Hippocampus; Tris-HCL (0.05*M*; pH 7.4, 37°C); Baudry and Lynch (1981).
C. Whole forebrain (plasma membranes) HEPES-KOH with $CaCl_2$ (2.5 m*M*) Fagg et al. (1983a).
D. Hippocampus; Tris-HCL (0.05*M*; pH 6.9, 37°C); Werling et al. (1983).
E. Cerebellum; Tris-HCL (0.05*M*, pH 7.4, 37°C); Foster and Roberts, 1978.

[b]Reported for "Glu_A" receptors only.

TABLE 3
Pharmacologic Characteristics of Calcium- and Chloride-Independent Specific [^{3}H]-L-Glutamate Binding to Brain[a]

Compound	Inhibition constants$_{50}$, μM	
	A[b]	B[c]
Agonists		
L-Glutamic acid	0.1	0.4
D-Glutamic acid	3.9	244
L-Aspartic acid	1.8	0.8
D-Aspartic acid	>100	270
Cysteine sulfinic acid	3.4	
Homocysteic acid	2.0	59
Cyclopentyl glutamic acid	4.0	
Antagonists		
D-(−)-ADA		>100
D,L-(±)-ADA	30	
D,L-(±)-APB	100	
D,L-(±)-AP5		7
D-(−)-APH		6

[a]Inactive compounds include: KA, NMDA, Ibo, and GDEE.
[b]A, From Biziere et al., 1980.
[c]B, From Fagg et al., 1983a.

In the presence of Cl^- (as Tris-HCl), multiple but apparently noninteracting binding sites for [^{3}H]-Glu can be discriminated in cerebellar membranes (Slevin et al., 1983). Although [^{3}H]-Glu itself appears to interact with a single population of recognition sites, several analogs of Glu display biphasic patterns to inhibit ligand binding. Such behavior is not unprecedented in binding studies. For instance, [^{3}H]-quinuclidinyl benzilate, a muscarinic cholinergic receptor antagonist, labels an apparently homogeneous class of receptors in brain that can be shown to be multiple binding sites according to the ability of various muscarinic agonists and antagonists to inhibit binding (Birdsall et al., 1978). In cerebellar membranes a variety of Glu analogs, including Ibo, Asp, cyclopentyl glutamate, ADA, aminopimelic acid, and the 4,5- and 7-carbon α-amino-phosphonocarboxylic acids, inhibited [^{3}H]-Glu binding in a biphasic manner. The higher affinity component of [^{3}H]-Glu binding (K_i for inhibitors = 0.5–6.0 μM) was found to represent 63% of the total receptor complement (Table 2). Consistent with other reports, neither KA nor any of its analogs, NMDA or GDEE, displaced specifically bound [^{3}H]-Glu.

The influence of ions other than Cl^- on [^{3}H]-Glu binding to brain membranes has been examined. Notably, Baudry and coworkers (Baudry et al., 1981a,b, 1983; Baudry and Lynch, 1981) have explored the effects of Ca^{2+} on Glu binding to brain, and although methodological differences between these and other studies are extensive (in particular, Baudry et al. terminate the binding reaction by vacuum filtration rather than centrifugation, as described in section 4.2.1), the pharmacologic profile of binding reported by Baudry and Lynch (1981) is consistent with the identification of a Cl^--dependent Glu receptor (Table 2). Specific binding of the ligand is inhibited by μM concentrations of Quis, Ibo, ADA, APB, and Asp (Baudry and Lynch, 1981) (Table 2). Addition of Ca^{2+} to the incubation resulted in a 2–3-fold increase in the apparent number of binding sites (Baudry et al., 1981a,b) when membranes were prepared from either the cortex or hippocampus. The influence of Ca^{2+} was temperature-dependent (not observed at 2°C) and showed a marked developmental profile. However, it did not appear to be generalized since in cerebellum, the effect of Ca^{2+} was greatly reduced or absent (Baudry et al., 1981a,b). Increases in ligand binding were caused by an increase in receptor number and were irreversible, since extensive washing of the membranes following exposure to Ca^{2+} did not result in a decline in receptor density to control values (Baudry et al., 1981b, 1983). Following coincubation of membranes with Ca^{2+}, fragments of a high molecular weight protein have been detected, and it has been suggested that Ca^{2+} initiates a protease-mediated event that removes a protein, hypothesized to be fodrin, that occludes Glu receptors in the membranes (Baudry and Lynch, 1980; Baudry et al., 1983). Because the effect of Ca^{2+} is reduced or absent in some brain regions, Glu receptors may be regulated at the molecular level by different processess throughout the brain. If true, this suggests that it may be possible to develop pharmacological strategies that selectively modify glutamatergic neurotransmission in a particular area of brain.

The effects of calcium alone and the interaction of Ca^{2+} with Cl^- on specific [^{3}H]-Glu binding to purified synaptic plasma membranes have been extensively explored by others (Foster and Fagg, 1984). Whereas little influence of Ca^{2+} (as calcium acetate) on specific binding of [^{3}H]-Glu was detected in synaptic plasma membranes, Cl^- ($EC_{50} = 4\ \mu M$) had a marked stimulatory effect characterized by a 2–3-fold increase in the apparent number of binding sites for the ligand (Mena et al., 1982; Fagg et al., 1982a,b, 1983a).

Moreover, the combination of Ca^{2+} (2.5 m*M*) with Cl^- (5 m*M*) induced an additional 2-fold increase in receptor number (Fagg et al., 1983a). Other divalent cations, including Sr^{2+}, Mn^{2+}, and Mg^{2+}, mimick the effect of Ca^{2+} in the presence of Cl^-, although these are markedly less potent. In high concentrations, Br^- and NO_3^- can also mimick the effect of Cl^- (Mena et al., 1982). The specific binding of [^{3}H]-Glu in the presence of Ca^{2+}/Cl^- displayed kinetic features indicative of a homogeneous population of receptors. Detailed examination of the pharmacological characteristics of binding in the presence of Ca^{2+}/Cl^-, revealed two classes of Glu recognition sites (Fagg et al., 1982b, 1983a), the majority (75–80%) of which were sensitive to inhibition by micromolar concentrations of APB, Ibo, Quis, AP5, APH, and the ω-sulphonic amino acids serine-O-sulfate, homocysteic acid, and cysteine sulfinic acid (Fagg et al., 1982b, 1983a) (Table 2). Inactive compounds at this population of receptors included NMDA, KA, and GDEE. In agreement with the observations of Biziere et al. (1980), in the absence of added Ca^{2+}/Cl^- or following freezing of the tissue, APB ($K_i > 1$ m*M*) and Quis were much less potent inhibitors of specific [^{3}H]-Glu binding (Fagg et al., 1982a, 1983b) (Table 3).

Based on the aggregate of pharmacological information currently available, it has been suggested that the Ca^{2+}/Cl^--dependent [^{3}H]-Glu binding site in brain most likely represents the excitatory amino acid-like receptor activated by an unidentified transmitter that is antagonized by L-(−)-APB (Foster and Fagg, 1984; Collins, 1982; Koerner and Cotman, 1981). Because Asp, AP5, APH, and to a lesser extent Ibo are among the most potent inhibitors of [^{3}H]-Glu binding at the Ca^{2+}/Cl^--independent receptor, this latter component of binding may represent an NMDA-like recognition site (Foster and Fagg, 1984; Fagg et al., 1983a). Supporting this hypothesis, Fagg has recently demonstrated the presence of Ca^{2+}/Cl^--independent [^{3}H]-Glu binding sites on isolated postsynaptic densities that are inhibited (K_i = 12 μ*M*) by NMDA itself (Foster and Fagg, 1984; Fagg and Matus, 1985). NMDA has also been shown to inhibit [^{3}H]-Glu binding to brain slices in vitro (Monaghan et al., 1983b).

Multiple Cl^--dependent [^{3}H]-Glu binding sites have also been reported to be present on hippocampal synaptic membranes, and these receptors can be differentiated according to the potencies of Quis and Ibo to inhibit binding (Werling et al., 1983). Termed Glu_A and Glu_B receptors, the relationship of these binding sites to previously identified Glu recognition sites is unclear. Whereas the Glu_A site is pharmacologically similar to the Glu/Quis receptor defined above, the Glu_B receptor can not be classified by current excitatory amino acid terminology, since both Quis and NMDA fail to affect the ligand–receptor interaction (Table 2).

In addition to an exquisite sensitivity to the ionic composition of the incubation buffer and to tissue preparation, the specific binding of [^{3}H]-Glu to brain homogenates can be modified by a variety of agents, including guanine nucleotides (Sharif and Roberts, 1981a) and KA (Fuxe et al., 1983), both of which alter the affinity of the receptor for Glu and the total number of Glu receptors. More recently, Ferkany et al. (1984a,b) and Zaczek et al. (1985) have demonstrated that [^{3}H]-Glu binding in Cl^--containing buffers is enhanced (2–4-fold) in the presence of a structurally related series of dipeptides containing carboxyterminal glutamate residues. As is the case for Ca^{2+}/Cl^-, dipeptides (represented by L-phenylalanyl-L-glutamate) increase the apparent number of [^{3}H]-Glu binding sites and alter the pharmacology of the ligand–receptor interaction. Although the physiological significance of this effect remains under investigation, the response has been demonstrated to be reversible and has a substantial regional heterogeneity and developmental profile (Ferkany and Coyle, submitted; Ferkany, 1984a). Since the action of L-phenylalanyl-L-glutamate is most likely allosteric, it is possible that receptors for excitatory amino acids are subject to regulation by endogenous organic substances, as well as by the ionic composition of the membrane environment.

Isolation of a Glu-binding protein(s) from brain has proven difficult. Michaelis et al. (1983) have purified to apparent homogeneity a low molecular weight (mol wt = 14,000 daltons), acidic, and highly hydrophobic substance from rat synaptic plasma membranes that binds [^{3}H]-Glu in a complex manner determined by tissue concentration. Specific binding of [^{3}H]-Glu to this protein is inhibited by Glu, Asp, and cysteine sulfinic acid, whereas Quis, Ibo, and ADA are only weakly active.

In summary, both physiological and biochemical studies have provided evidence that Glu may be an excitatory neurotransmitter in brain. Although it is well documented that receptors for Glu are present in the CNS, the binding of [^{3}H]-Glu to brain membranes in vitro is complex. Currently, at least three and possibly as many as five receptors for Glu have been identified, including a Na^+-dependent uptake site, a Ca^{2+}/Cl^--dependent receptor site, and a Ca^{2+}/Cl^--independent binding site. These latter two components can each be further subdivided using kinetic and pharmacological criteria into two discrete receptors. To varying degrees, these recognition sites possess characteristics predicted for the synaptic Glu receptor involved in neurotransmission. However, binding at all sites is inhibited by compounds thought to act in vivo at other excitatory amino acid receptor subtypes. Thus, the final determination of the validity of in vitro [^{3}H]-Glu binding assays to identify physiologically relevant receptors awaits the introduction of more selective agonist and antagonist drugs.

2.1.2. AMPA

Since [^{3}H]-Glu labels multiple receptors on brain membranes in vitro and the characteristics of the ligand–receptor interaction vary according to the conditions of the assay, it would be advantageous to develop ligands more selective for the various Glu receptor subtypes. One compound of potential interest in this regard is AMPA (R,S-amino-3-hydroxy-5-methyl-isoxazole-4-propionic acid), a rigid synthetic analog of Glu derived from Ibo (Honore et al., 1981). When applied to mammalian spinal neurons, AMPA is a potent neuroexcitant, and the action is antagonized by GDEE, suggesting that Quis/Glu-type receptors may mediate this response in the CNS (Hosli et al., 1983b).

Tritiated AMPA binds to brain membranes and cultured brain tissues in vitro in a saturable and reversible manner (Honore et al., 1982; Hosli et al., 1983a). Scatchard analysis of [^{3}H]-AMPA binding yielded biphasic saturation isotherms with high affinity, low capacity (K_d = 9 nM; B_{max} = 0.1 pmol/mg protein) and low affinity, high capacity (K_d = 2.4 μM; B_{max} = 7.0 pmol/mg protein) components. In contrast to Ca^{2+}/Cl^--dependent [^{3}H]-Glu binding, binding of [^{3}H]-AMPA is optimal in Tris-citrate buffer (pH = 7.4); prior freezing of the tissue increases the amount of specifically bound ligand by 60%. The effects of freezing and thawing the tissue prior to assay are complex, and after repeated freezing–thawing, only a single binding site of intermediate affinity and capacity is detected (Honore et al., 1982).

Specific binding of [^{3}H]-AMPA is inhibited in order of potency by AMPA and Quis (K_i = 0.3 μM), Glu, and several analogs of Ibo (K_i = 1.3–2.5 μM). Intriguingly, KA (K_i = 6.0 μM) is a moderately potent inhibitor of binding, whereas Ibo, Asp, D-Glu, and GABA are all without effect when tested at a concentration of 100 μM. In the presence of chloride ions, AMPA is a very weak inhibitor of the specific binding of [^{3}H]-Glu to brain membranes.

AMPA is a potent neuroexcitant and in vitro binding studies using AMPA as a ligand reveal a recognition site possessing characteristics indicative of an excitatory amino acid receptor. Classification of [^{3}H]-AMPA binding according to the recognized scheme for these receptors will require additional investigation.

This should be forthcoming now that AMPA is commercially available in tritiated form.

2.1.3. L-Aspartate; N-Methyl-D-aspartic Acid

As is the case for Glu, sufficient electrophysiological and biochemical evidence exists to warrant the consideration of Asp as an excitatory amino acid neurotransmitter in brain (Watkins and Evans, 1981; Foster and Fagg, 1984). However, the difficulties that accompany the examination of the role of Glu in neurotransmission apply equally well to the study of Asp. Although Asp is a potent neuroexcitant, Asp-induced neuronal depolarization can be demonstrated throughout the brain, and documentation that neurons of a particular terminal field are sensitive to this amino acid is not adequate to suggest aspartergic innervation (Watkins and Evans, 1981; Davies et al., 1979, 1980). Furthermore, Asp is involved in intermediate metabolism in the CNS, and although Asp is accumulated in synaptosomes by a high-affinity carrier system, this transport process is not restricted to neuronal elements (Wofsey et al., 1971; Bennett et al., 1972; Logan and Snyder, 1972; Balcar et al., 1977). Indeed, one of the confusing aspects of excitatory mino acid neurochemistry is that both Asp and Glu appear to be transported by a similar or identical carrier in brain and it is not clear how selective neurotransmitter pools of these amino acids arise.

Receptor binding studies employing [^{3}H]-Asp are less numerous than those that have utilized [^{3}H]-Glu. Consistent with the hypothesis that receptors for [^{3}H]-Asp are involved in synaptic neurotransmission, specific binding of the ligand to brain is enriched 9-fold in synaptic junctional membranes compared to whole particulate fractions of brain (Fagg et al., 1982a; Foster et al., 1981b). Although Asp inhibits the binding of Glu (and vice versa), and Asp and Glu have similar affinities for their recognition sites (200–800 n*M*), when measured in the same tissue, the number of binding sites for the latter compound is 2–5-fold greater than for [^{3}H]-Asp. Binding sites for [^{3}H]-Glu can be further differentiated by the sensitivity of the ligand–receptor interactions to sodium, calcium, magnesium, and chloride ions (Fagg et al., 1982a; Sharif and Roberts, 1981b). Thus, it appears likely that the majority of recognition sites for [^{3}H]-Asp in brain are distinct from those for Glu.

Both Shariff and Roberts (1981b) and Di Lauro et al. (1982) have explored the general pharmacological and kinetic properties of [^{3}H]-Asp binding to cerebellar and whole brain preparations and have reached similar conclusions. Binding of the ligand is saturable, reversible, and occurs to a single, noninteracting population of receptors in the absence of sodium ions; in the presence of sodium, an additional population of recognition sites most likely corresponding to the Asp transport site can be defined. Specific binding of [^{3}H]-Asp is inhibited by a restricted number of compounds, including Glu, Ibo, D,L-α-aminosuberic acid and the NMDA-type antagonist HA966 (Table 4). The inhibitors of the Na^+-dependent Asp transport system, threo-3-hydroxyaspartate and aspartate-β-hydroxymate, are devoid of displacing activity, as are Quis, GDEE, APB, KA, ADA, and NMDA. Freezing the membranes prior to assay has been reported to result in a precipitous decline in the specific binding, although the effects of this tissue treatment remain in dispute (Sharif and Roberts, 1981b; Di Lauro et al., 1982). Also, $CaCl_2$ has been reported to either enhance or have no effect on ligand binding (Fagg et al., 1982a; Sharif and Roberts, 1981b).

Summarizing available pharmacological, biochemical, and kinetic data, Foster and Fagg (1984) have suggested that [^{3}H]-Asp and Ca^{2+}/Cl^--independent [^{3}H]-Glu binding to brain in vitro occur at a similar or identical receptor. Such a hypothesis is interesting, since other evidence suggests that this receptor may represent an NMDA-type recognition site in brain (Foster and Fagg, 1984).

The synthetic Asp analog NMDA is approximately 10-fold more potent than Asp and Glu as a neuroexcitant, but substantially less potent than either KA or Ibo as a neurotoxin (Zaczek et al., 1981). The neuroexcitatory actions of NMDA in vivo and in vitro are selectively antagonized by the 5- and 7-carbon α-amino-ω-phosphonoamino acids, as well as by low concentrations of the divalent cation Mg^{2+} (Perkins et al., 1981; Evans et al., 1982; Davies and Watkins, 1982; Lehmann and Scatton, 1982). The interactions between NMDA, AP5, APH, and Mg^{2+} are considered highly specific and form the basis for defining the NMDA-subtype of excitatory amino acid receptor in brain.

Unfortunately, with the exception of a single report (Snodgrass, 1979), attempts to identify an NMDA-like receptor using ligand binding methods with the parent compound have been unsuccessful, although NMDA has been available labeled to a high specific activity for some years. Although the reasons for this are not clear, it should be noted that the potency of NMDA as a neuroexcitant is based on electrophysiological studies in which direct determination of the concentration of drug is difficult. Even the use of in vitro systems to determine the relative potency of excitatory amino acids to induce neurochemical alterations may

TABLE 4
Pharmacologic Characteristics of Specific
[^{3}H]-Aspartic Acid Binding to Brain[a]

Compound	Inhibition constants$_{50}$, μM	
	A[b]	B[c]
L-Aspartic acid	5.0	0.3
D-Aspartic acid	457	100
L-Glutamic acid	2.0	5.0
Ibotenic acid	10	
(±)-Aminosuberic acid	10	
Cysteine sulfinic acid	125	
D,L-(±)-ADA	360	

[a]Inactive compounds include: KA, NMDA, Quis, D,L-(±)-APB, and D-glutamic acid.

[b]A, Sharif and Roberts, 1981b.

[c]B, Di Lauro et al., 1982.

overestimate the potency of NMDA vis-a-vis amino acids like Glu and Asp, since the latter compounds are accumulated by high-affinity transport systems in brain. Thus, it is possible that the affinity of NMDA for its receptor may be substantially lower than previously thought. Furthermore, if the Ca^{2+}/Cl^--independent Glu binding site in brain represents an NMDA-type receptor, then the number of these sites would be expected to be small. This combination of low affinity and low capacity might account for the failure to detect [^{3}H]-NMDA binding in vitro. This notwithstanding, the current consensus holds that NMDA itself is an inappropriate ligand for labeling NMDA receptors; attempts to identify an NMDA-type recognition site using labeled NMDA antagonists are discussed in section 2.2.2.

To summarize, specific binding sites for the putative excitatory amino acid neurotransmitter Asp have been identified in brain and appear to be distinct from Ca^{2+}/Cl^--dependent [^{3}H]-Glu binding sites. The significance of these sites, however, must await further investigation. Attempts to label an NMDA receptor in brain with the parent compound have been unsuccessful.

2.1.4. *Kainic Acid*

Kainic acid was isolated and identified by Takemato (1978) during a search for the antiascarial principle of the seaweed, *Di-*

genea simplex, and, unlike Glu and Asp, KA is not endogenous to brain. As initially demonstrated by Shinozaki and Konish (1970), iontophoretic application of KA to neurons is followed by pronounced neuronal depolarization. KA is also among the most potent axon-sparing neuroexcitotoxins known, and following intracerebral application, neurons with cell bodies near the area of injection degenerate, whereas axons of passage or dendritic terminal fields remain unaffected (Coyle and Schwarcz, 1976; Schwarcz and Coyle, 1977; Schwarcz et al., 1978; Coyle et al., 1978, 1981a). Because of the obvious structural similarities between KA and Glu, it was initially thought that the neuroexcitotoxic actions of KA were a consequence of the prolonged activation of Glu receptors. More recent investigations have not substantiated this hypothesis; as noted in section 2.1.1 (Foster and Fagg, 1984), KA has negligible affinity ($K_i > 1$ mM) for Ca^{2+}/Cl^--dependent and -independent Glu binding sites in brain.

That KA represents an archetype to form a third category of receptors for excitatory amino acids is substantiated by physiological investigations (reviewed in Watkins and Evans, 1981; Foster and Fagg, 1984). Antagonists of both Quis/Glu- and NMDA-type responses are ineffective in attenuating the excitatory or neurotoxic actions of KA. Unfortunately, a precise understanding of the role that KA receptors play in neurotransmission is difficult to obtain, since selective antagonists at these receptors have not been identified (Krogsgaard-Larsen and Honore, 1983). However, interest in KA remains high, since in animals, both KA-induced lesions and epileptiform activity bear some resemblance to those observed in human neuropsychiatric disorders (McGeer et al., 1978; Coyle et al., 1981a; Nadler, 1981; London et al., 1981).

Receptors for [^{3}H]-KA were first reported to be present in brain by Simon et al. (1976) and shortly thereafter by Vincent and McGeer (1979) and London and Coyle (1979). Although most extensively studied in rat brain, specific binding sites for [^{3}H]-KA have been detected in nervous tissue from a number of species, including hydra, crayfish, dogfish, frog, chicken, pigeon, and man (London et al., 1979; Henke, 1979; Henke et al., 1981). In both rat and human brain, two kinetically distinct but pharmacologically similar receptors have been observed, one having an apparent affinity of 20–50 nM and found throughout the brain, and a second, higher affinity site (K_d = 2 nM) that is restricted to forebrain structures (London and Coyle, 1979; Slevin et al., 1983; O'Brien and Ferkany, unpublished). Low-affinity binding sites can account for up to 70% of the total receptor population (London and Coyle, 1979; London et al., 1979). Receptors for KA are enriched in synaptic plasma membranes (Foster et al., 1981a), suggesting a neuronal localization. Receptor density is highest in telencephalic regions, caudate nucleus, nucleus accumbens, and hippocampus (London and Coyle, 1979; Henke et al., 1981; Monaghan and Cotman, 1982). The relationship of high- and low-affinity [^{3}H]-KA binding sites to the neurotoxic and neuroexcitatory actions of KA is not clear, but KA is neurotoxic in the cerebellum, an area of brain reported to contain only the low-affinity binding component; and this receptor type may be the important species for the induction of cell death (London and Coyle, 1979; Roberts et al., 1981; Slevin et al., 1983). That the receptors identified in ligand binding studies are involved in neurotoxicity is further supported by the observation that regions of brain having low densities of these receptors are resistant to the neurotoxic effects of the drug (Coyle, 1982b), whereas in regions of brain sensitive to the neurotoxic influence of KA, KA-induced lesions result in a substantial depletion of specific [^{3}H]-KA binding sites (London and Coyle, 1979; Biziere and Coyle, 1979; Henke et al., 1981; Coyle et al., 1981b; Biziere et al., 1982).

Only limited reports are available concerning the influence of ions and detergents on the specific binding of [^{3}H]-KA to brain, and these indicate that both mono- and divalent cations have no effect or slightly depress binding (Beaumont et al., 1979). Likewise, both fresh and frozen membrane preparations have been employed in various studies without apparent effect on either the high- or low-affinity binding component (O'Brien and Ferkany, unpublished; Slevin et al., 1983; Beaumont et al., 1979). Lack of information concerning the influences of different conditions on KA binding may be explained in part by the difficulties encountered in detecting subtle changes in the characteristics of [^{3}H]-KA binding, since this assay has been difficult to perform because of the low-specific radioactivities of previously available ligands. The recent introduction of [^{3}H]-KA of high-specific radioactivity should permit more detailed investigations into the effects of various conditions on KA receptors, and this is an area in need of additional investigation.

The pharmacological characteristics of the KA recognition site are distinct from those predicted for either the Quis/Glu or NMDA receptor subtypes (London and Coyle, 1979; Coyle, 1980; Coyle et al., 1981a,b; Slevin et al., 1983; Foster and Fagg, 1984). Although both L-Glu and Quis inhibit binding, these compounds are 100- and 30-fold less potent than KA itself (Table 5). Specific binding is insensitive to 100 μM concentrations of the Glu-type antagonists D-(−)-APB and GDEE, and the Quis agonist AMPA. Compounds

TABLE 5
Pharmacologic Characteristics of Specific [^{3}H]-Kainic Acid Binding to Rat Brain Membranes[a]

	Inhibition constants$_{50}$, n*M*			
Compound	A[b]	B[c]	C[d]	D[e]
Domoic acid			13	
Kainic acid	19	1.5	21	25
L-Glutamic acid	722	63	1500	1200
Quisqualic acid	255	23	610	75
alpha-Keto kainic acid			430	
alpha-allo-Keto kainic acid			5700	
Dihydrokainic acid	31500	6100	7400	
Ibotenic acid	36000	3400	9400	
Homocysteic acid	87000	71000		

[a]The following compounds failed to inhibit specific binding when tested at 10^4 n*M*: -D-glutamic acid, GDEE, L-(+)-ADA, D,L-(±)-APB, D,L-(±)-AP5, D,L-(±) APHex, D,L-(±)-APH, gamma-D-glutamylglycine, (±)-cis-2,3-piperdine dicarboxylic acid.

[b]A, From London and Coyle, 1979, low affinity site.

[c]B, From London and Coyle, 1979, high affinity site.

[d]C, From Slevin et al., 1983.

[e]D, From Simon et al., 1976.

thought to function via NMDA-like receptors, including ADA, AP5, APH, and NMDA, only weakly inhibit the binding of [^{3}H]-KA when tested at high concentrations, although the putative NMDA receptor agonist Ibo inhibits binding with an IC_{50} of 10 μ*M* (London and Coyle, 1979; Slevin et al., 1983).

Of the compounds that are effective in inhibiting [^{3}H]-KA binding in vitro, only domoic acid, a structural analog of KA, in which the isopropylene side chain is modified, is as potent (K_i = 13 n*M*) as KA itself. Alpha-keto-KA, in which the isopropylene side chain has been replaced by a ketone group, displays inhibitory activity but is 10–20-fold less potent than the parent compound. The analogs dihydro-KA, alpha-allo-KA, and kainic acid diethylester have 100–3000- fold lower affinities for the [^{3}H]-KA recognition site, and the putative KA-type receptor antagonists (±)-*cis*-2,3-piperdine dicarboxylic acid and gamma-D-glutamylglycine are also inactive (Slevin et al., 1983) (Table 5). Both domoic acid and alpha-keto-KA possess neuroexcitotoxic properties in brain and have potencies relative to KA that are similar to those observed in binding studies (Coyle et al., 1981b).

The weight of both pharmacological and physiological evidence indicates that receptors for KA are distinct from receptors for Glu or NMDA. Because KA receptors may be involved in certain human neuropsychiatric disorders, it is not surprising that attempts have been made to identify an endogenous KA-like factor in brain. Recently, Ruck et al (1980) have reported that derivatives of folic acid are potent, competitive inhibitors of [^{3}H]-KA binding in vitro. Although this observation generated initial excitement, attempts to confirm it have been unsuccessful (Ferkany et al., 1982; Roberts et al., 1981; Morgan and El-Lakany, 1982). Several dipeptides, including the lead compound phenylalanyl-L-glutamate, have been found to be modestly potent (K_i = 5–50 μM) inhibitors of [^{3}H]-KA binding to brain (Ferkany et al., 1984). Since numerous similar compounds are known to exist in brain, these findings are of considerable interest with regard to the possible identity of KA-like neurotransmitters (Ferkany et al., 1984).

2.2. Antagonists

From the foregoing discussion, it is clear that attempts to identify excitatory amino acid receptors with radiolabeled agonists as probes have encountered difficulties. Not surprisingly, efforts have progressed to exploit excitatory amino acid antagonists for this purpose. From previous experience with other, well-characterized receptors, the use of receptor antagonists as ligands offers distinct advantages.

First, antagonists may discriminate among subpopulations of receptors responsive to a single agonist. Notably, in ligand binding studies agonists such as GABA, dopamine, and serotonin may recognize several kinetically or pharmacologically distinct populations of binding sites. In the case of [^{3}H]-GABA, for instance, multiple receptors can be identified and only a portion of these are sensitive to inhibition by the GABA-receptor antagonist bicuculline (reviewed in Enna and Gallagher, 1983). In contrast, binding at bicuculline-insensitive receptors is selectively inhibited by the antispastic agent baclofen (Lioresal ™) (Hill and Bowery, 1981; Wilkin et al., 1982). Also, two classes of muscarinic cholinergic receptors labeled by the antagonists [^{3}H]-quinuclidinyl benzilate and [^{3}H]-pirenzepine can be differentiated according to the ability of various agonists to inhibit specific binding, as well as by the tissue and brain distribution of ligand binding sites (Luthin and Wolfe, 1984; Watson et al., 1983).

The second reason for selecting antagonists rather than agonists to label receptors is that the former compounds often have a higher affinity for the receptor. Thus, the K_d for spiroperidol at dopamine receptors in brain is 30-fold higher than for dopamine itself (Burt et al., 1975; Fields et al., 1977). Finally, antagonists of synaptic receptors generally show little affinity for neurotransmitter uptake sites, and the probability that these receptors will be labeled by the ligand is reduced. This combination of selectivity and sensitivity often makes antagonists more robust probes for receptor binding assays. The recent availability of tritium-labeled excitatory amino acid antagonists has provided a promising new avenue to examine the properties of synaptic receptors for these neurotransmitters.

2.2.1. 2-Amino-4-phosphonobutyric Acid (APB)

As noted in section 2.1.1 (Foster and Fagg, 1984), [^{3}H]-Glu labels a heterogeneous population of recognition sites on brain membranes and the characteristics of the sites identified by [^{3}H]-Glu in vitro are markedly dependent on the presence of certain ions and the methods of tissue preparation. Notably, in the presence of low concentrations of Ca^{2+} and Cl^-, and with fresh tissue, specific binding of [^{3}H]-Glu occurs to at least two populations of receptors that can be discriminated according to the ability of L-(+)-APB to inhibit ligand binding (Foster and Fagg, 1984). To explore more directly the nature of the APB-sensitive Glu binding site, two groups have examined the characteristics of specific [^{3}H]-APB binding to brain (Butcher et al., 1983; Monaghan et al., 1983a). Although assay conditions and tissue preparations differed substantially in these studies, overall results are not markedly at odds. Thus, binding of the ligand is saturable with an apparent affinity of 1–5 μM. The 2-fold higher density of binding sites reported by Monaghan vis-a-vis Butcher most probably reflects the use of synaptic plasma membranes as opposed to crude mitochondrial preparations, since the former preparation would be expected to concentrate specific receptors relative to other membrane-bound proteins. Although Butcher et al. reported a biphasic association of the ligand with the receptor, both groups demonstrated that specific binding achieves a stable equilibrium by 60 min of incubation. Binding of [^{3}H]-APB in both studies was optimal at 37°C and fully reversible; slight differences in the effects of pH were observed and our experience indicates that the optimal pH for this assay is between 6.8 and 7.1 (Ferkany, unpublished).

As predicted from the effects of APB on Ca^{2+}/Cl^--dependent [^{3}H]-Glu binding, the specific binding of [^{3}H]-APB to brain is strictly dependent on the presence Ca^{2+}/Cl^- ions in the incubation buffers. Specific binding is absent when the tissue preparations are frozen for 24 h prior to assay.

The pharmacological profile of [^{3}H]-APB binding to brain membranes is consistent with the identification of a recognition site for excitatory amino acid neurotransmitters (Table 6). Of the compounds examined, Quis (IC_{50} = 0.4 μM) was the most potent displacer followed by Glu > Asp > D-Glu > D-Asp. L-homocysteate, cysteine sulfinic acid, and serine-*O*-sulfate were also effective inhibitors of binding, whereas neither NMDA, KA, nor the rigid NMDA analog quinolinic acid were active at 100 μM. Among antagonists of excitatory amino acids, the α-amino-ω-phosphonoamino acids all displayed inhibitory effects at the [^{3}H]-APB binding site with K_i's in the 1–30 μM range and, as predicted from previous studies employing [^{3}H]-Glu (Fagg et al., 1982b, 1983a), the L-(+) forms of these compounds were consistently more potent (10-fold) than the corresponding D-(−) isomers. Equipotent to the APB were the electrophysiologically less selective excitatory amino acid antagonists D,L-aminosuberate, D,L-α-aminoadipate and D,L-α-aminopimelate.

As noted by Butcher et al. (1983) and by one of us (Ferkany, unpublished), care should be taken in determining the potency of various compounds to compete at the [^{3}H]-APB binding site, since several compounds including D and L-APB, as well as Quis and the sulfur-containing analogs of cysteine, displace significantly more bound ligand than do other amino acid analogs. Thus, determination of appropriate values for the nonspecific binding component of ligand–receptor interaction is difficult. The use of the racemic mixture of [^{3}H]-APB for binding studies may confound interpretation of data, since the possibility exists that these isomers identify distinct receptor entities. Nevertheless, the use of D,L-(±)-[^{3}H]-APB as a ligand to explore receptors for excitatory amino acids is promising and, in one regard, presents distinct advantages over the use of [^{3}H]-Glu. [^{3}H]-APB does not bind to transport sites for Glu; therefore, the possibility that this class of recognition sites will contribute to the pattern of ligand binding is eliminated.

2.2.2. 2-Amino-5-phosphonopentanoic Acid (AP5)

The α-amino-ω-phosphonoamino acid, AP5 is the higher homolog of APB. Unlike the latter compound, however, the physiological effects of AP5 are limited primarily to antagonism of NMDA-type responses and the compound is much less potent at putative Glu or KA receptors in brain. Thus, it has been suggested that AP5 would provide a useful ligand with which to examine NMDA-specific receptors. Although [^{3}H]-AP5 is not commercially

TABLE 6
Pharmacologic Characteristics of Specific D,L-[^{3}H]-2-Amino-4-phosphonobutyric Acid Binding to Brain[a]

Compound	Inhibition constants$_{50}$, μM	
	A	B
Agonists		
Quisqualic acid	0.38	0.4
L-Glutamic acid	0.80	1.70
Cysteine sulfinic acid		7.1
Ibotenic acid	0.54	10
L-Aspartic acid	5.35	7.0
Homocysteic acid		1.21
D-Glutamic acid	6.9	18
D-Aspartic acid	15	50
Antagonists		
L-(+)-ADA		3
D-(−)-ADA		16
D,L-(±)-APB		3.6
L-(+)-APB	1.45	2.2
D-(−)-APB	10	30
D,L-(±)-AP5	5.8	40
L-(+)-AP5		20
D-(−)-AP5		126
D,L-(±)-APHex	2.8	14
D,L-(±)-APH	19	32
L-(+)-APH		19
D-(−)-APH		100

[a]Inactive compounds included NMDA, KA, serine-*O*-sulfate, quinolinic acid, and D,L-(±)-APP.

[b]A, Monaghan et al., 1983a.

[c]B, Butcher et al., 1983.

available, the compound has been tritiated in Watkin's laboratory and a preliminary report of the binding of [^{3}H]-AP5 to brain has appeared (Olverman et al., 1984).

Specific binding of D,L-(±)-[^{3}H]-AP5 to extensively washed brain membranes is optimal when incubations are performed at 37°C and at physiological pH. Binding of the ligand is saturable and of comparable affinity (K_d = 0.5 μM) to that determined for other excitatory amino acids. The interaction of [^{3}H]-AP5 with brain membranes displays monophasic saturation isotherms, suggesting that the ligand identifies only a single class of receptors. An 8-fold variation in the amount of specific binding is observed among membranes prepared from different regions of brain (hippocampus > cortex ⩾ striatum ⩾ midbrain > cerebellum > medulla).

In contrast to other excitatory amino acid ligands, specific binding of [^{3}H]-AP5 to brain membranes is inhibited by both NMDA and L-Asp, and these compounds are only 20-fold less potent in inhibiting binding than D-(−)-AP5 itself (Table 7). In agreement with electrophysiological evidence, L-(+)-AP5 in vivo is a much less potent displacer of [^{3}H]-AP5 than is the D-(−) isomer, and KA, Quis, and D,L-(±)-APB are also only weak inhibitors of specific binding. The pharmacological profile is consistent with the suggestion that D-(−)-[^{3}H]-AP5 labels an NMDA-like receptor in vitro. However, the most potent inhibitor of specific binding is Glu (K_i = 0.04 μ*M*), followed by homocysteic and cysteine sulfinic acids (Table 6). Additionally, the KA antagonists gamma-D-glutamyl glycine and (±)-*cis*-2,3-piperdine dicarboxylic acid are only 4- and 7-fold less effective, respectively, than NMDA in inhibiting the interaction, whereas the endogenous NMDA-like compound quinolinic acid (Foster et al., 1983) is 35-fold less potent in this regard. The influence of ions and other treatments on [^{3}H]-AP5 binding to brain has not been reported.

2.2.3. 2-Amino-7-phosphonoheptanoic Acid (APH)

Like AP5, APH is an antagonist of NMDA-type responses in physiological test systems (Davies and Watkins, 1982; Evans et al., 1982; Lehmann et al., 1983). Attempts to label NMDA-like receptors in vitro using D,L-(±)-[^{3}H]-APH, however, have yielded ambiguous results (Ferkany and Coyle, 1983b). Although binding of the ligand to brain occurs at a single population of receptors at 37°C (K_d = 3.0 μ*M*), assays are best conducted using previously frozen, extensively washed, and preincubated membrane preparations (Ferkany and Coyle, 1984; Ferkany and Coyle, submitted). Unlike the situation with [^{3}H]-Glu and [^{3}H]-APB binding, Cl^- (50 m*M*) attenuates the specific binding of [^{3}H]-APH, and Ca^{2+} has little apparent effect (Ferkany and Coyle, submitted).

The interaction of [^{3}H]-APH with its recognition site in brain is inhibited most potently by Quis, followed by Ibo, Glu, serine-*O*-sulfate, and aminoadipic acid (Table 7). D,L-(±)-APB and D,L-(±)-AP5 are equipotent, although these are 5-fold less potent than D,L-(±)-APH. Interestingly, the L-(+) isomers of the α-amino-ω-

TABLE 7
Pharmacologic Characteristics of Specific D,L-[^{3}H]-2-Amino-5-phosphonopentanoic Acid and D,L-[^{3}H]-2-Amino-7-phosphonoheptanoic Acid Binding to Brain Membranes

Compound	Inhibition constants$_{50}$, μM	
	AP5[a]	APH[b]
Agonists		
Quisqualic acid	250	0.6
L-Glutamic acid	0.90	5.6
Cysteine sulfinic acid	13	22
Ibotenic acid		3.0
Homocysteic acid	3.9	14
NMDA	11	> 100
Serine-*O*-sufate	25	1.1
Quinolinic acid	350	110
Kainic acid	870	>1000
Antagonists		
D,L-(±)-AP3	560	> 100
D,L-(±)-APB	260	17
D,L-(±)-AP5	1.7	21
D-(−)-AP5	0.6	
L-(+)-AP5	55	
D,L-(±)-APHex	42	
D,L-(±)-APH	2.4	3.6
D-(−)-ADA		5.8
gamma-D-Glutamylglycine	40	≫ 100
(±)-*cis*-2,3-Piperdine dicarboxylic acid	78	≫ 100

[a]AP5, From Olverman et al., 1984.
[b]APH, From Ferkany and Coyle, 1983b.

phosphono compounds are more effective displacers of ligand binding than the D-(−) isomers, although this may reflect the use of the racemic mixture of [^{3}H]-APH (Ferkany and Coyle, submitted). Neither NMDA, Asp, nor KA and its analogs attenuate the binding of [^{3}H]-APH to brain, and quinolinic acid is only a weakly effective inhibitor (K_i = 200 μ*M*). As with the binding of [^{3}H]-AP5, recognition sites for APH are most concentrated in the hippocampus, with fewer sites detected in the cerebellum and midbrain (Ferkany and Coyle, 1983b).

Both D-(−)-[^{3}H]-AP5 and D,L-(±)-[^{3}H]-APH represent promising ligands with which to examine the characteristics of excitatory amino acid-like binding sites. However, as with most other excitatory amino acid-like ligands, the characteristics of the binding of [^{3}H]-AP5 and [^{3}H]-APH to brain membranes in vitro is not entirely consistent with those predicted for NMDA-type receptors from studies using physiological approaches. Thus, the final judgment as to the usefulness of these compounds to examine a particular population of receptors for excitatory amino acids must await the results of further investigations.

3. Assay Methods

3.1. General Considerations

Although similarities exist among binding assays that employ ligands specific for excitatory amino acid receptors, sufficient differences are present to warrant careful attention to detail in every instance. For example, although assays for [^{3}H]-KA can be performed over a broad range of pH s (6.8–7.7) (O'Brien and Ferkany, unpublished observation), the assay for [^{3}H]-APB has been reported to be optimal using more acidic buffers (pH 6.8; Monaghan et al., 1983a). Specific binding of [^{3}H]-APH to brain membranes should always be examined at a rigidly controlled pH of 7.3, since shifts in the hydrogen ion concentration of as little as 0.2 units can reduce specific binding of the ligand by as much as 30% (Ferkany and Coyle, submitted). Also, methods of tissue preparation and storage, buffer composition, the length of time required for the ligand–receptor interaction to reach equilibrium, acceptable limits on both ligand and receptor concentrations, methods of separating bound from free ligand, the choice of appropriate unlabeled compounds, and the concentration of these compounds necessary to determine nonspecific binding of the ligand to the membranes (Butcher et al., 1983; Ferkany, unpublished observations) all vary depending upon the ligand being employed and the goals of the particular experiment.

One common requirement for ligand binding assays that examine excitatory amino acid receptors is the refrigerated centrifuge. Whereas in many procedures the centrifuge is used to terminate the binding reaction by separating the tissue containing the bound from free radioactivity, it is also an essential piece of equipment necessary for the preparation of tissue samples. Because the centrifuge is used for both applications, it should be equipped with rotors capable of handling a number of small (7–15 mL) reaction tubes, as well as a lesser number of large preparatory tubes (40–50 mL). Additionally, some procedures (Butcher et al., 1983; Olvermann et al., 1984) describe assays performed in microtubes (1.5 mL) and it is convenient to have available a high-speed microcentrifuge that achieves maximum speeds in a short period of time and brakes rapidly. Specific binding of [^{3}H]-Glu to brain membranes has been performed using vacuum filtration techniques to separate bound from free ligand (Baudry et al., 1981a,b, 1983), and this procedure may be applicable to study the binding of [^{3}H]-KA to brain when [^{3}H]-KA of high specific activity is employed (Zaczek et al., unpublished). Thus, a manifold capable of maintaining multiple filters under reduced pressure is useful. However, both rapidity and uniformity of aspiration is essential to ensure consistent results, especially with affinities in the high n*M* range typical of these receptors. Several different types of commercially supplied machines are available for this purpose.

The other essential pieces of equipment necessary to perform ligand binding assays successfully are common to any laboratory routinely engaged in biochemical research. These include refrigerators and freezers for the storage of tissue, drugs, and ligands, preparative equipment including polytrons, homogenizers, and sonic tissue disrupters, and a liquid scintillation counter for the determination of radioactivity. Since many assays are conducted at elevated temperatures, a waterbath capable of maintaining a set range of temperatures (25–37°C) is essential. Recent advances in microcircuits have made the small computer one of the least expensive additions to the laboratory and these instruments are useful for both the storage and reduction of data. Moreover, programs to assess all facets of receptor binding studies are now commercially available and are invaluable for data reduction, particularly when the ligand labels two or more pharmacologically or kinetically distinct receptors.

All chemicals and reagents necessary to perform excitatory amino acid binding assays can be obtained from the standard suppliers. Many of the drugs employed to probe the pharmacological characteristics of excitatory amino acid receptors can also be obtained at a reasonable cost in milligram quantities. Notably, these include the entire series of α-amino-ω-phosphonocarboxylic and piperidine-*X,Y*-dicarboxylic acids, Ibo, Quis, KA, cysteine sulfinic acid, NMDA, serine-*O*-sulfate, and the older, prototypical agents, such as GDEE and ADA (for example, *see* Sigma Chemical Company, St. Louis, MO, and Cambridge Research Biochemicals, Atlantic Beach, NY).

Ligands of interest, including tritiated KA, Glu, AMPA, Asp, APB, and APH can be obtained from either Amersham, Inc., Chicago, IL, or New England Nuclear, Boston, MA, among other suppliers. Although ligands are all of high-specific radioactivity (sp.act. > 20 Ci/m*M*) and initial purity, investigators should be prepared to reassess the radiochemical purity of tracers upon receipt and at periodic intervals thereafter, since in some instances the stability of the ligand in solution has not been determined.

3.2. Tissue Preparation

The fundamental goals of tissue preparation for receptor binding assays are 2-fold: (a) to remove as much endogenous substances present in the tissue that might interfere with the ligand–receptor interaction as possible, and (b) to increase the density of receptors specific for the ligand relative to other tissue components. With respect to the first of these objectives, it should be remembered that the brain content of acidic amino acids such as Glu is remarkably high, e.g., 1–10 μmol/g tissue. Thus, since the concentration of Glu in brain is 3–10 m*M*, and since the affinity of excitatory amino acid ligands for their receptors is usually in the high n*M* range, it is apparent that specific binding of the ligand would not be detected without prior removal or dilution of endogenous inhibitors. For this reason, extensive washing of the tissue is required. This is easily achieved by repeated centrifugation of the tissue with intermittent resuspension of the pellet in a large volume (1:20 w/v) of fresh buffer.

The second consideration of tissue preparation, to concentrate the receptors of interest, reflects the fact that recognition sites for neurotransmitters constitute only a minute fraction of the structural elements of brain and, at most, may be present at concentrations of only pmol/mg protein. In contrast, nonspecific (nonreceptor) binding sites for a ligand that arise because of hydrogen bonding or ionic interactions of the tracer with tissue are virtually infinite. By concentrating the receptors of interest and removing other tissue elements, it is possible to increase the ratio of specific to nonspecific binding and, thus, enhance the precision of the binding assay (Ferkany and Enna, 1981).

One of the most popular methods for preparing brain for excitatory amino acid binding assays was developed by Enna and Snyder (1977), and is frequently referred to as the "buffy coat" method. In this procedure (performed at 2°C), fresh brain tissue is collected and immediately homogenized in 20 vol (w/v) of isotonic sucrose (0.32*M*; 109.5 g/L) by using a motor-driven Teflon pestle fitted to a glass tube. The resulting suspension is centrifuged (1000*g* for 10 min), and the supernatant is decanted to a clean tube and recentrifuged at 15,000*g* for 20 min, a step that yields the crude mitochondrial fraction (P_2), a fraction moderately enriched in synaptosomes and dendrosomes (Cotman, 1974). The second supernatant is gently decanted and an equal volume of glass-distilled water is added to the tube containing the pellet. Using either sonication or a mechanical homogenizer (Brinkman Polytron), the pellet is lysed and resuspended in the water, and the mixture is recentrifuged at 8000*g* for 20 min to produce a heterogeneous pellet composed of a dense, tightly adhering core surrounded by a loosely packed "buffy coat." The latter is dispersed by gently shaking the tube and, along with the supernatant, is collected by an additional centrifugation (50,000*g* for 15 min) to produce the initial material for assay. It has been our experience that this final step and all additional washes of the tissue are best performed following the addition of a small amount (10–20% v/v) of assay buffer to the supernatant, since this produces a more stable pellet resistant to the rigors of repeated handling. For binding assays in which tissue can be stored frozen, batch preparation of membranes is possible. In other assays, notably those employing [^{3}H]-APB, freezing the tissue dramatically reduces the amount of specifically bound ligand and fresh tissue prepared as described above should be used immediately (Fagg et al., 1982a, 1983b; Butcher et al., 1983; Monaghan et al., 1983a).

Alternative techniques are available to concentrate receptors of interest for excitatory amino acid binding assays. For example, Cotman's laboratory has made extensive use of a method that yields semipurified junctional synaptic plasma membranes (Foster et al., 1981b; Fagg et al., 1982b, 1983a). Although this procedure employs differential centrifugation similar to that described above, additional steps using discontinuous gradient centrifugation are required. Because the technique does not produce a homogeneous preparation of receptors, the increases achieved in receptor concentration should be considered in terms of the complexity of the method and the additional time required for the routine use of this procedure.

Regardless of the method of tissue preparation that is selected once the initial material is obtained, additional handling of the tissue varies considerably among binding assays for excitatory amino acid receptors. This is illustrated in the following section in which representative protocols for examining the specific binding of tritiated Glu, APH, and KA are discussed.

3.3. Protocols

As noted in section 3.2, individual laboratories may have different methods of preparing tissue for excitatory amino acid binding assays. However, it has been our experience that the "buffy coat" preparation provides a suitable starting material to explore the pharmacological or biochemical characteristics of excitatory amino acid receptors. For the following descriptions of experimental protocols, the assumption has been made that the "buffy coat" technique has been adopted.

3.3.1. [^{3}H]-L-Glutamate

Several groups have reported that [^{3}]-Glu labels a heterogeneous population of receptors on brain membranes in in vitro binding assays (section 2.1), and a portion of these receptors appear to be labile to freezing and thawing the tissue (Fagg et al., 1982b, 1983b). To examine the total population of recognition sites for [^{3}]-Glu, it is necessary to employ freshly prepared brain membranes and, equally important prior to conducting the assay, endogenous Glu must be removed from the tissue. To achieve this latter goal, the initial membrane preparation described in section 3.2 is washed an additional three times by centrifugation and resuspension in fresh buffer (20 vol w/v). Following the third centrifugation, the tissue is resuspended a final time in sufficient fresh assay buffer to yield a protein concentration of 250–500 μg/mL. One-milliliter aliquots of the suspension are added directly to polyethylene scintillation/centrifugation tubes (Beckman Mini-Poly Q™) containing 1 mL of buffer, [^{3}H]-Glu (final assay concentration ≤0.300 μ*M*), and the compounds of interest, each of which are assayed in triplicate. Routinely, nonspecific binding is defined as residual radioactivity bound to the tissue in the presence of 1000 μ*M* unlabeled L-Glu. The incubation is continued for 30 min at 37°C and the reaction terminated by centrifugation (50,000*g* for 10 min). The supernatant is decanted, and the pellet washed rapidly and superficially with 2 × 3 mL of ice-cold buffer. Following solubilization of the pellet in a suitable solvent (e.g., Protosol™, New England Nuclear, Boston, MA), fluor is added and radioactivity determined by liquid scintillation counting.

If economy of tissue, ligand, or drug is of critical importance, it is possible to reduce the final volume of the [^{3}H]-Glu binding assay considerably. As with other binding assays, this is done by making parallel reductions in the volume of all additions. The absolute reductions in volumes that can be achieved in practice are limited by the amount of radioactivity that is specifically bound to the tissue. Fortunately, the high-specific radioactivity of commercial preparations of [^{3}H]-Glu combined with the moderately low affinity of the ligand–receptor interaction ($K_d \geq 0.5\ \mu M$) means that a substantial concentration of [^{3}H]-Glu (0.1 μM) can be added to microassays without infringing upon the theoretical considerations of ligand binding techniques (e.g., that the concentration of ligand routinely employed should be at a concentration less than that of the K_d of the reaction). A convenient way to perform the miniaturized [^{3}H]-Glu binding assay is to conduct the incubations in 1.5 mL conical centrifuge tubes and terminate the reaction by rapid centrifugation at room temperature using a high-speed microcentrifuge. In this instance, the supernatant is removed by careful aspiration and the pellet washed only a single time with 1 mL of cold buffer. The bottom of the tube is cut off and the tissue treated as described above.

To this point no mention has been made of buffers to use in conducting the [^{3}H]-Glu binding assay since, as noted previously, the characteristics of specific binding of [^{3}H]-Glu to brain membranes is radically altered according to the ionic composition of the incubation medium (section 2.1.1). In particular, calcium and chloride, alone or in combination, reveal subsets of Glu receptors that are both pharmacologically and kinetically distinct from those observed in the absence of these ions. Thus, to a large extent the choice of buffers will depend upon the goal of the particular investigation.

To examine the Ca^{2+}/Cl^--independent binding of [^{3}H]-Glu to brain, HEPES/KOH, Tris-acetate, or Tris-citrate buffers (0.05M, pH 6.8–7.1, 37°C) may be selected. The first of these possesses a large buffering capacity at physiological pH, but suffers from the presence of K^+ ions in the buffer, which, like Na^+ at sufficient concentrations, can affect the characteristics of [^{3}H]-Glu binding (Baudry and Lynch, 1981). HEPES may be titrated with Tris base to the desired pH, but this offers little advatage over the Tris-acetate itself and HEPES is reletively expensive. Tris-citrate has less buffering capacity at the desired pH, but provides the advantage of weakly chelating divalent cations in solution and may adsorb these as they are released from tissue stores. Chloride-dependent specific [^{3}H]-Glu binding is best studied using Tris-HCl buffer or Tris-acetate supplemented with ammonium chloride at the desired concentration. Likewise, the influence of calcium on the ligand–receptor interaction can be examined using the latter buffer in combination with calcium acetate or following the addition of $CaCl_2$ to Tris-HCl. Sodium, preferably the acetate salt, can be added to any of these buffers as desired.

3.3.2. [^{3}H]-Kainic Acid

Kainic acid binds to a population of receptors in brain that by several criteria are distinct from sites labeled by other excitatory amino acid ligands (reviewed in Coyle, 1980; Watkins and Evans, 1981; Foster and Fagg, 1984) (section 2.1.4). Unlike [^{3}H]-Glu binding, the specific binding of [^{3}H]-KA to brain membranes is not labile to freezing, and batch preparation of tissue is possible. Previous assays using [^{3}H]-KA have employed centrifugation to terminate the reaction; however, this technique may be outmoded since ligand of high specific activity (50–60 Ci/mmol) is now available and the high-affinity component of [^{3}H]-KA binding (K_d = 1–2 n*M*) can be examined using rapid filtration techniques to separate bound from free ligand (Zaczek et al., in preparation). Examination of the low-affinity component of [^{3}H]-KA binding, however, is best done using centrifugation procedures to terminate the reaction (O'Brien and Ferkany, unpublished).

Tissue preparation and the final dilutions of the tissue suspension for the [^{3}H]-KA binding assay are identical to that described in section 3.2. Incubations are conducted in triplicate for 60 min at 2°C, a temperature at which the high-affinity component of the interaction is best revealed. The final assay volume is 2 mL and either KA ($10^{-5}M$) or Glu ($10^{-3}M$) is used to define nonspecific binding. The assay does not appear to be markedly affected by the choice of buffers, and Tris-HCl (pH 7.1; 0.05*M*) may be routinely used (O'Brien and Ferkany, unpublished).

Selective examination of the high-affinity [^{3}H]-KA recognition site is achieved by maintaining a ligand concentration of 2 n*M* or less; at higher ligand concentrations, binding to the low-affinity recognition site becomes apparent. Utilization of membranes prepared from the cerebellum affords some advantages both practically and with regard to interpretation of data, since this region of brain contains only low-affinity receptors and ligand concentrations of 20 n*M* or more can be used routinely (London and Coyle, 1979). Miniaturization of the [^{3}H]-KA binding assay has not been characterized, since attempts to use this approach with the previously available batches of ligand (sp. act. = 4–5 Ci/mmol) yielded very low amounts of specificially bound radioactivity. Since the specific activity of the ligand is no longer an impediment, reducing the overall volume of the [^{3}H]-KA binding assay may be possible.

3.3.3. [^{3}H]-2-Amino-7-phosphonoheptanoic Acid

"Buffy coat" preparations have also been used successfully to examine the specific binding of the NMDA-type antagonist [^{3}H]-

APH to brain (Ferkany and Coyle, 1983b; Ferkany et al., 1984; Ferkany and Coyle, submitted). In contrast to the assay for [^{3}H]-Glu, it is recommended that membranes be stored frozen at −80°C for at least 24 h, since this procedure elicits a substantial increase in the amount of ligand specifically bound to the tissue. Once prepared, frozen tissue may be stored for at least 30 d without loss of specific binding activity.

On the day of the assay, the frozen membrane preparation is thawed, resuspended in 20 vol (w/v) of Tris-citrate buffer (pH 7.3, 0.05*M*, 37°C), and washed four times by centrifugation. Prior to initiating the binding reaction, the tissue (200–300 μg protein/mL) is preincubated for 30 min at 37°C, a step that induces an additional 45% increase in the apparent number of binding sites for the ligand (Ferkany and Coyle, submitted). Following the preincubation, 1-mL aliquots of the tissue are added in triplicate to tubes containing 1 mL of buffer, [^{3}H]-APH (routinely 500 n*M*), and the compounds of interest. The incubation is continued for 90 min at 37°C and the reaction is terminated by centrifugation (50,000*g* for 10 min, 2°C). The supernatant is decanted and the pellet is washed rapidly and superficially with ice-cold buffer and processed in a routine fashion.

Unlike the assays for [^{3}H]-Glu and [^{3}H]-KA, the specific binding of [^{3}H]-APH to brain is not amenable to examination using rapid filtration techniques (Ferkany, unpublished), since this procedure results in a complete loss of measurable specifically bound radioactivity. Also, specific binding of [^{3}H]-APH is reduced in the presence of concentrations of chloride, such as those found in 50 m*M* Tris-HCl. Calcium has only a modest effect on the interaction; however, as little as 1 m*M* Na^+ can dramatically reduce the specific binding of [^{3}H]-APH to brain membranes (Ferkany and Coyle, 1983b; Ferkany and Coyle, submitted).

In summary, receptor binding assays for three major excitatory amino acid ligands are similar with regard to both technical and practical features. Because of their simplicity, these methods are easily mastered by technicians and students. Once one method has been successfully learned, adapting the technique for additional ligands of interest is routine.

4. Applications

The fundamental use of in vitro ligand binding assays is to dissect the biochemical and pharmacological characteristics of neurotransmitter receptors. Although this goal is of intrinsic merit, it is ultimately hoped that a better knowledge of receptor properties will provide a conceptual framework in which to understand the molecular events of neurotransmission in both the normal and diseased brain, and in the latter instance, provide a rational basis for the development of disease-specific pharmacological interventions.

Perhaps one of the better examples of the utility of receptor binding assays to inspect the relationship of drugs and neurotransmission at the molecular level is provided by our current understanding of the mechanism of action of the anxiolytic benzodiazepines and sedative/hypnotic barbiturates. Electrophysiological evidence has for some time indicated that both barbiturates and benzodiazepines are capable of facilitating the neuronal hyperpolarization induced by the inhibitory neurotransmitter, gamma-aminobutyric acid (reviewed in Olsen, 1981; Enna and Gallagher, 1983). Using in vitro ligand binding assays for GABA and the benzodiazepines, several laboratories have established that bidirectional interactions occur between these agents and that each potentiates the binding of the other at their respective recognition sites on brain membranes (Tallman et al., 1978, 1979; Karobath and Sperk, 1979; Skerritt et al, 1982a,b; Skerritt and Johnston, 1983). In contrast, picrotoxinin is a potent antagonist of the effects of GABA at the physiological level (Enna and Gallagher, 1983). Using in vitro techniques, it is evident that picrotoxin-mediated antagonism of GABA is not the result of a direct interaction with the postsynaptic GABA recognition site, but rather the result of an interaction with a closely associated molecular complex (possibly, the Cl^- ionophore linked to the GABA receptor) (Ticku et al., 1978; Squires et al., 1983). Like the benzodiazepines, certain barbiturates can enhance the specific binding of [^{3}H]-GABA in vitro (Willow and Johnston, 1981a, 1981b), and these same barbiturates competitively inhibit specific binding of ligands at the picrotoxinin recognition site while allosterically influencing benzodiazepine binding (Leeb-lundberg et al., 1981; Olsen and Leeb-lundberg, 1981; Skolnick et al., 1981, 1982; Ticku, 1981; Leeb-lundberg and Olsen, 1982). The present conceptualization of the postsynaptic GABA receptor, based largely upon the results of in vitro binding studies, depicts at least three distinct but interacting molecules, including a recognition site for GABA, the benzodiazepines, picrotoxin, and the barbiturates (Andrews and Johnston, 1979; Enna and Gallagher, 1983). Additional investigations, notably those of Guidotti and coworkers and Enna et al. (Enna and Snyder, 1977; Toffano et al., 1978; Massotti et al., 1981; Guidotti et al., 1982), further suggest the presence of a low molecular weight protein coupled to the GABA receptor that regulates the affinity of the receptor for GABA itself.

Although the model of excitatory amino acid receptors is less complete than that for the GABA receptor complex, substantial progress in this area is being achieved. The observations that calcium and chloride regulate Glu binding in vitro (section 2.1.1, Foster and Fagg, 1984) suggest that these ions may be intimately involved in the physiology of excitatory neurotransmission and make it interesting to speculate that subtle changes in the ionic mileau of the neuron, and not in the receptor or in the process of neurotransmission itself, may contribute to the symptoms of epilepsy or other disorders in which excitatory amino acid neurotransmission may be involved. In addition, the finding that certain dipeptides influence excitatory amino acid ligand binding to brain membranes in vitro provides the first evidence that receptors for these substances may consist of multiple, interacting components, perhaps similar to the GABA/benzodiazepine/picrotoxin receptor complex (Ferkany et al., 1984; Ferkany and Coyle, 1984; Ferkany and Coyle, submitted; Zaczek et al., 1985).

Because the action of many of the currently available, therapeutically useful drugs are mediated through specific receptors, there is a growing appreciation of the utility of ligand binding assays to rapidly screen large numbers of compounds for further testing. Indeed, because of the simplicity of this method, it is routinely possible to examine 35–50 compounds in a single day in any particular assay. Although in vitro binding assays do not generally allow discrimination of agonist or antagonist properties of receptor-active agents, binding assays do provide a useful primary screen for the selection of compounds; once activity is observed in a receptor binding assay, agents can be examined for biological activity in more selective but tedious physiological test systems.

One drawback to receptor binding methodology is that compounds that are active in vitro may lack potency or efficacy in vivo because of rapid metabolism or, particularly in the case of the CNS, failure to reach the target organ following peripheral administration. This notwithstanding, receptor binding assays can provide valuable clues concerning the structural characteristics necessary for compounds to interact with a particular receptor, and such information can be used to further modify the parent compound or to identify additional compounds of interest.

Given the suspected link between excitatory amino acid neurotransmission and selected neurological disorders, there is growing interest in developing compounds capable of antagonizing the physiological actions of these neurotransmitters. Although efforts in this area are in an early stage, useful and potentially exciting developments are already apparent. Thus, the α-amino-ω-phosphonoamino acid, APH, is an antagonist of NMDA-induced neuronal responses in vivo and, inhibits the specific binding of tritiated Glu, AP5, APB, and APH to brain membranes in vitro (sections 2.1.4, 2.2). Systemic administration of APH provides some protection against seizures in animals (Croucher et al., 1982; Czuczwar and Meldrum, 1982; Meldrum et al., 1983; Peterson et al., 1983). As a prototypical agent, APH represents a useful structure for the further development of antagonists of excitatory amino acid neurotransmission.

5. Directions

As is clear from the foregoing discussions, substantial evidence indicates that diacidic amino acids may be widespread excitatory neurotransmitters in brain. Nevertheless, several difficulties with this hypothesis are apparent. The excitatory amino acids are intimately involved in a variety of metabolic functions and neither discrete neurotransmitter pools nor neurotransmitter-related transport processes have been conclusively identified. Virtually all neurons in the CNS are depolarized by excitatory amino acids, and this lack of selectivity in vivo and in vitro makes it difficult to assign transmitter-specific receptors to these agents. Of the substances for which adequate evidence exists to suggest specific receptors, KA is not endogenous to brain, and recognition sites for NMDA, also a synthetic compound, have not been conclusively identified in vitro. Not surprisingly, the hypothesis that Asp and Glu are excitatory neurotransmitters has come under increasing scrutiny, and efforts to identify alternative neurotransmitter candidates are extensive. In addition, the search for an endogenous KA- and NMDA-like substance in brain has gained increasing significance since evidence has accrued that such substances may be involved in neurological disorders.

Recently, Zaczek and coworkers (1983) have isolated from brain a peptide, *N*-acetylaspartyl-L-glutamate (NAAG), which is equipotent with Quis in inhibiting the specific binding of [^{3}H]-Glu to brain membranes in vitro (Koller and Coyle, 1984b). As with other excitatory amino acid analogs, NAAG inhibits binding at only a portion of the total sites labeled by [^{3}H]-Glu, an observation confirmed by the finding that there are fewer specific [^{3}H]-NAAG than [^{3}H]-Glu binding sites in brain (Zaczek et al., 1983; Koller and Coyle, 1984b; Koller et al., in press). Differences in both the ontogenetic and regional distribution of receptors for NAAG and Glu confirm that the former compound identifies a subpopulation of excitatory amino acid receptors (Koller and Coyle, 1984a,b; Koller et al., in press). More exciting, NAAG elicits excitatory responses in the lateral olfactory tract, an area of brain in which evidence suggests that Glu may not be the endogenous excitatory neurotransmitter (ff-French-mullen et al., in press). Other investigations have revealed that a limited number of Glu-containing dipeptides may indirectly modulate the affinity and number of excitatory amino acid receptors in vitro (Ferkany et al., 1984; Ferkany and Coyle, submitted; Zaczek et al., 1985). Taken together, these findings provide evidence that Glu may not be the only mediator of synaptic events mediated by excitatory amino acid-like compounds in brain.

Efforts by Schwarz and Foster to identify an NMDA-like substance in brain have also been promising (Foster et al., 1983; Schwarcz et al., 1983). Thus, the endogenous tryptophan metabolite quinolinic acid (QUIN) has been shown to be present in brain and is a potent neuroexcitant (Moroni et al., 1984). Based on the antagonist profile of the action of QUIN, excitation appears to be mediated by NMDA-type receptors since it is effectively attenuated by AP5 and APH (Stone and Perkins, 1981; Foster et al., 1983; Perkins and Stone, 1983a,b; Lehmann et al., 1983). Of great interest, QUIN possesses neurotoxic properties of the axon-sparing type when injected into brain (Schwarcz et al., 1983; Foster et al., 1983). Although these findings remain preliminary, they suggest a mechanism by which dysfunction of excitatory neurotransmission may be involved in neurodegenerative disorders. Furthermore, since the actions of QUIN on brain are blocked by NMDA-type antagonists, potential therapeutic approaches to disorders that may involve QUIN are apparent.

Efforts to identify a KA-like agonist or antagonist native to the brain have been fruitless to date. Nevertheless, the findings discussed above provide hope for the success of these endeavors. Undoubtedly, as more compounds are extracted from brain and examined both in vivo and in vitro for biological activity, the list of substances that are candidates to replace Glu and Asp as the sole excitatory amino acid-like neurotransmitters in brain will also expand.

References

Altman J. J. (1972) Postnatal development of the cerebellar cortex in the rat. III. Maturation of the components of the granular layer. *J. Comp. Neurol.* **145,** 465–514.

Andrews P. R. and Johnston G. A. R. (1979) GABA agonists and antagonists. *Biochem. Pharmacol.* **28,** 2697–2702.

Balcar V. J., Borg J., and Mandel P. (1977) High affinity uptake of L-glutamate and L-aspartate by glial cells. *J. Neurochem.* **28,** 87–93.

Baudry M. and Lynch G. (1980) Regulation of hippocampal glutamate receptors: Evidence for the involvement of a calcium-activated protease. *Proc. Natl. Acad. Sci. USA* **77,** 2298–2302.

Baudry M. and Lynch G. (1981) Characterization of two [^{3}H]glutamate binding sites in rat hippocampal membranes. *J. Neurochem.* **36,** 811–820.

Baudry M., Arst D., Oliver M., and Lynch G. (1981a) Development of glutamate binding sites and their regulation by calcium in the rat hippocampus. *Dev. Brain Res.* **1,** 37–48.

Baudry M., Smith E., and Lynch G. (1981b) Influence of temperature, detergents and enzymes on glutamate receptor binding and its regulation by calcium in rat hippocampal membranes. *Mol. Pharmacol.* **20,** 280–268.

Baudry M., Siman R., Smith E., and Lynch G. (1983) Regulation by calcium ions of glutamate receptor binding in hippocampal slices. *Eur. J. Pharmacol.* **90,** 161–168.

Beaumont K., Maurin Y., Reisine T. D., Fields J. Z., Spokes E., Bird E. D., and Yamamura H. I. (1979) Huntington's disease and its animal model: Alterations in kainic acid binding. *Life Sci.* **24,** 809–816.

Bennett J. P., Logan W. J., and Snyder S. H. (1972) Amino acid neurotransmitter candidates: Sodium-dependent high affinity uptake by unique synaptosomal fractions. *Science* **178,** 997–999.

Birdsall N. J. M., Burger A. S., and Hulme E. C. (1978) The binding of agonists to brain muscarinic receptors. *Mol. Pharmacol.* **14,** 723–736.

Biziere K. and Coyle J. T. (1979) Localization of receptors for kainic acid on the inner nuclear layer of retina. *Neuropharmacology* **18,** 409–413.

Biziere K., Thompson H., and Coyle J. T. (1980) Characterization of specific high-affinity binding sites for L-[^{3}H]glutamic acid in rat brain membranes. *Brain Res.* **183,** 421–433.

Biziere K., Slevin J. T., Zaczek R., Collins J. C., and Coyle J. T. (1982) Kainic Acid: Neurotoxicity and Receptor Interactions, in *Advances in Pharmacology and Therapeutics II* (Yoshida H., Hagihara Y., and Ebashi S., eds.), pp. 271–276, Peragamon, New York.

Bradford H. F. (1976) Metabolism and Transmitter Function of Amino Acids in the Nervous System, in *Biologically Active Substances: Exploration and Exploitation* (Henns D. A., ed.), pp. 266–280. Wiley, London, UK.

Burt D., Enna S. J., Creese I., and Snyder S. H. (1975) Dopamine receptor binding in the corpus striatum of mammalian brain. *Proc. Natl. Acad. Sci. USA* **72,** 4655–4659.

Butcher S. P., Collins J. F., and Roberts P. J. (1983) Characterization of the binding of D,L-[^{3}H]2-amino-4-phosphonobutyrate to L-glutamate-sensitive sites on rat brain synaptic membranes. *Brit. J. Pharmacol.* **80,** 355–364.

Bylund D. (1980) Analysis of receptor binding data. *Soc. Neurosci. Short Course,* 70–99.

Collins G. G. S. (1982) Some effects of excitatory amino acid receptor antagonists on synaptic transmission in the rat olfactory cortex slice. *Brain Res.* **244,** 311–318.

Cotman C. W. (1974) Isolation of Synaptosomal and Synaptic Plasma Membrane Fractions, in *Methods in Enzymology* (Fleisher L. and Palker L., eds.), pp. 445–452, Academic, New York.

Coyle J. T. (1980) Excitatory Amino Acid Receptors, in *Neurotransmitter Receptors* (Enna S. J. and Yamamura H. I., eds.), pp. 5–40, Chapman and Hall, London, UK.

Coyle J. T. (1982a) Neurotoxic amino acids in human degenerative disorders. *Trends Neurosci.* **5,** 287–288.

Coyle J. T. (1982b) Excitatory Amino Acid Neurotoxins, in *Handbook of Psychopharmacology* (Iversen, L. L., Iversen S. D., and Snyder S. H., eds.), pp. 237–269, Plenum, New York.

Coyle J. T. and Schwarcz R. (1976) Lesion of striatal neurons with kainic acid provides a model for Huntington's chorea. *Nature* (Lond.) **263,** 244–246.

Coyle J. T., Bird S. J., Evans R., Gulley R., Nadler J. V., Nicklas W. J., and Olney J. W. (1981a) Excitatory amino acid neurotoxins: selectivity, specificity and mechanism of action. *Neurosciences Res. Prog. Bull.* **19.**

Coyle J. T., Molliver M., and Kuhar M. J. (1978) *In situ* injection of kainic acid: A new method for selectively lesioning neuronal cell bodies while sparing axons of passage. *J. Comp. Neurol.* **180,** 301–324.

Coyle J. T., Zaczek R., Slevin J., and Collins J. C. (1981b) Neuronal Receptor Sites for Kainic Acid: Correlations with Neurotoxicity, in *Glutamate as a Neurotransmitter* (Di Chiara G. and Gessa G. L., eds.), pp. 337–346, Raven, New York.

Croucher M. J., Collins J. F., and Meldrum B. S. (1982) Anticonvulsant action of excitatory amino acid antagonists. *Science* **216,** 899–901.

Cuatrecasas P. and Hollenberg M. D. (1976) Membrane receptors and hormone action. *Adv. Protein Chem.* **30,** 251–451.

Currie D. N. and Kelly J. S. (1981) Glial versus neuronal uptake of glutamate. *J. Exp. Biol.* **95,** 181–193.

Curtis D. R., Duggan A. W., Felix D., Johnston G. A. R., Tebecis A. K., and Watkins J. C. (1972) Excitation of mammalian central neurons by acidic amino acids. *Brain Res.* **41,** 283–301.

Czuczwar S. J. and Meldrum B. (1982) Protection against chemically induced seizures by 2-amino-7-phosphonoheptanoic acid. *Eur. J. Pharmacol.* **83,** 335–338.

Davies J. and Watkins J. C. (1982) Actions of the D- and L-forms of 2-amino-5-phosphonovalerate and 2-amino-4-phosphonobutyrate in the cat spinal cord. *Brain Res.* **235,** 3789–386.

Davies J., Evans R. H., Francis A. A., Jones A. W., and Watkins J. C. (1980) Excitatory Amino Acid Receptors in Vertebrate Central Nervous System, in *Neurotransmitters and Their Receptors* (Littauer U., Silman I., Teichberg V., and Vogel Z., eds.), pp. 333–347, Wiley, London, UK.

Davies J., Evans R. H., Francis A. A., and Watkins J. C. (1979) Excitatory amino acid receptors and synaptic excitation in the mammalian central nervous system. *J. Physiol.* (Paris) **75,** 641–654.

Di Lauro A., Meek J. L., and Costa E. (1982) Specific high affinity binding of L-[^{3}H] aspartate to rat brain membranes. *J. Neurochem.* **38,** 1261–1267.

Enna S. J. and Gallagher J. P. (1983) Biochemical and electrophysiological characteristics of mammalian GABA receptors. *Int. Rev. Neurobiol.* **24,** 181–212.

Enna S. J. and Snyder S. H. (1977) Influence of ions, enzymes and detergents on gamma-aminobutyric acid receptor binding in synaptic membranes of rat brain. *Mol. Pharmacol.* **13,** 442–453.

Evans R. H., Francis A. A., Jones A. W., Smith D. A., and Watkins J. C. (1982) The effect of a series of ω-phosphono-2-carboxylic amino acids on electrically evoked and excitant amino acid-induced responses in isolated spinal cord preparations. *Brit. J. Pharmacol.* **75,** 65–75.

Fagg G. E. and Matus A. (1985) Selective association of *N*-methyl aspartate and quisqualate types of L-glutamate receptor with brain postsynaptic densities. *Proc. Natl. Acad. Sci. USA* **81,** 6876–6880.

Fagg G. E., Mena E. E., Monaghan D. T., and Cotman C. W. (1982a) Acidic amino acid receptor populations in synaptic membranes: Separation of Cl^-/Ca^{2+}-dependent and -independent sites by freezing. *Soc. Neurosci. Abs.* **8,** 878.

Fagg G. E., Foster A. C., Mena E. E., and Cotman C. W. (1982b) Chloride and calcium ions reveal a pharmacologically distinct population of L-glutamate binding sites in synaptic membranes: Correspondence between biochemical and electrophysiological data. *J. Neurosci.* **2,** 958–965.

Fagg G. E., Foster A. C., Mena E. E., and Cotman C. W. (1983a) Chloride and calcium ions separate L-glutamate receptor populations in synaptic membranes. *Eur. J. Pharmacol.* **88,** 105–110.

Fagg G., Mena E., Monaghan T., and Cotman C. W. (1983b) Freezing eliminates a specific populations of L-glutamate receptors in synaptic membranes. *Neurosci. Lett.* **38,** 157–162.

Ferkany J. W. and Coyle J. T. (1983a) Evoked release of aspartate and glutamate: Disparities between prelabeling and direct measurement. *Brain Res.* **278,** 279–282.

Ferkany J. W. and Coyle J. T. (1983b) Specific binding of [^{3}H]2-amino-7-

phosphonoheptanoic acid to rat brain membranes *in vitro*. *Life Sci.* **33,** 1295–1305.

Ferkany J. W. and Coyle J. T. (1983c) Kainic acid selectively stimulates the release of endogenous excitatory amino acids from brain slices *in vitro*. *J. Pharmacol. Exp. Ther.* **225,** 399–406.

Ferkany J. W. and Enna S. J. (1982) Radioreceptor Assays, in *Analysis of Biogenic Amines* (Baker G. B. and Coutts R. T., eds.), pp. 267–284, Elsevier/North Holland, Amsterdam.

Ferkany J. W., Slevin J. T., Zaczek R., and Coyle J. T. (1982) Failure of folic acid derivatives to mimic the actions of kainic acid in brain *in vitro* or *in vivo*. *Neurobehav. Tox. Teratol.* **4,** 573–579.

Ferkany J. W. and Coyle J. T. (1984a) Influence of various conditions and age on the specific binding of [^{3}H]2-amino-7-phosphonoheptanoic acid to rat brain in the presence and absence of phenylalanylglutamate. *Soc. Neurosci. Abs.* **10,** 225.

Ferkany J. W., Zaczek R., Markl A., and Coyle J. T. (1984b) Glutamate-containing dipeptides enhance specific binding at glutamate receptors and inhibit specific binding at kainate receptors in rat brain. *Neurosci. Lett.* **44,** 281–286.

Fields J. Z., Reisine T. D., and Yamamura H. I. (1977) Biochemical demonstration of dopaminergic receptors in rat and human brain using [^{3}H]spiroperidol. *Brain Res.* **136,** 5778–5584.

Foster A. C. and Fagg G. E. (1984) Acidic amino acid binding in mammalian neuronal membranes: Their characteristics and relationship to synaptic receptors. *Brain Res. Rev.* **7,** 103–184.

Foster A. C. and Roberts P. J. (1978) High affinity L-[^{3}H]glutamate binding to postsynaptic sites on rat cerebellar membranes. *J. Neurochem.* **31,** 1467–1477.

Foster A. C. and Roberts P. J. (1980) Endogenous amino acid release from rat cerebellum *in vitro*. *J. Neurochem.* **35,** 517–519.

Foster A. C., Collins J. F., and Schwarcz R. (1983) On the exitotoxic properties of quinolinic acid, 2,3-piperidine dicarboxylic acids and structurally related compounds. *Neuropharmacology* **22,** 1331–1341.

Foster A. C., Mena E. E., Monaghan D. T., and Cotman C. W. (1981a) Synaptic localization of kainic acid binding sites. *Nature* (Lond.) **289,** 73–75.

Foster A. C., Mena E. E., Fagg G. E., and Cotman C. W. (1981b) Glutamate and aspartate binding sites are enriched in synaptic junctions isolated from rat brain. *J. Neurosci.* **1,** 620–625.

ff-French-mullen J. M. H., Koller K. J., Zaczek R., Hori L., Coyle J. T., and Carpenter D. O. *N*-acetylaspartyglutamate: Possible role as the neurotransmitter of the lateral olfactory tract. *Proc. Natl. Acad. Sci. USA* (in press).

Fuxe K., Agnati L. F., and Celani M. F. (1983) Evidence for interactions between [^{3}H]glutamate and [^{3}H]kainic acid binding sites in rat striatal membranes. Possible relevance for kainic acid neurotoxicity. *Neurosci. Lett.* **35,** 233–238.

Guidotti A., Konkel D. R., Ebstein B., Corda M. G., Wise B. C., Krutzsch H., Meek J. L., and Costa E. (1982) Isolation, characterization and purification to homogeneity of a rat brain protein (GABA-modulin). *Proc. Natl. Acad. Sci. USA* **79,** 6084–6088.

Hamberger A. C., Chiang G. H., Nylen E. S., Scheff S. W., and Cotman C. W. (1979) Glutamate as a CNS transmitter. I. Evaluation of glucose and glutamine as precursors for the synthesis of preferentially released glutamate. *Brain Res.* **168,** 515–530.

Henke H. (1979) Kainic acid binding in human caudate nucleus: Effect of Huntington's disease. *Neurosci. Lett.* **14,** 247–251.

Henke H., Beaudet A., and Cuenod M. (1981) Autoradiographic localization of specific kainic acid binding sites in pigeon and rat cerebellum. *Brain Res.* **219,** 95–105.

Hill D. R. and Bowery M. (1981) [^{3}H]Baclofen and [^{3}H]GABA bind to bicuculline-insensitive $GABA_b$ sites in rat brain. *Nature* (Lond.) **290,** 149–152.

Honore T., Lauridsen J. and Krogsgaard-Larsen P. (1981) Ibotenic acid analogues as inhibitors of [^{3}H]glutamic acid binding to cerebellar membranes. *J. Neurochem.* **36,** 1302–1304.

Honore T., Lauridsen J., and Krogsgaard-Larsen P. (1982) The binding of [^{3}H]AMPA, a structural analogue of glutamic acid, to rat brain membranes. *J. Neurochem.* **389,** 173–178.

Hosli E., Krogsgaard-Larsen P., and Hosli L. (1983a) Binding sites for the glutamate analogue [^{3}H]AMPA in cultured rat brainstem and spinal cord. *Brain Res.* **268,** 177–180.

Hosli L., Hosli E., Lehmann R., and Eng P. (1983b) Effect of the glutamate analogue AMPA and its interaction with antagonists on cultured rat spinal and brain stem neurons. *Neurosci. Lett.* **36,** 59–62.

Karobath M. and Sperk G. (1979) Stimulation of benzodiazepine receptor binding by gamma-aminobutyric acid. *Proc. Natl. Acad. Sci. (USA)* **76,** 1004–1006.

Keller E., Davis J. L., Tachiki K. H., Cummins J. T., and Baxter C. F. (1981) L-proline inhibition of glutamate release. *J. Neurochem.* **37,** 1335–1337.

Koerner J. F. and Cotman C. W. (1981) Micromolar L-2-amino-4-phosphonobutyric acid selectively inhibits perforant path synapses from the lateral entorhinal cortex. *Brain Res.* **216,** 192–198.

Koller K. J. and Coyle J. T. (1984a) Ontogenesis of *N*-acetylaspartate and *N*-acetylaspartylglutamate in rat brain. *Develop. Brain Res.* **317,** 137–140,

Koller K. J. and Coyle J. T. (1984b) Characterization of the interactions of *N*-acetylaspartylglutamate with [^{3}H]L-glutamate receptors. *Eur. J. Pharmacol.* **98,** 193–199.

Koller K. J. and Coyle J. T. Specific labelling of brain receptors by [^{3}H]*N*-acetylaspartylglutamate. *Eur. J. Pharmacol.* (in press).

Koller K. J., Zaczek R., and Coyle J. T. *N*-acetylaspartylglutamate: Regional levels in rat brain and the effects of brain lesions as determined by a new HPLC method. *J. Neurochem.* (in press).

Krogsgaard-Larsen P. and Honore T. (1983) Glutamate receptors and new glutamate agonists. *Trends Pharmacol. Sci.* **4,** 31–33.

Leeb-lundberg F., Napios C., and Olsen R. W. (1981) Dihydropicrotoxinin binding sites in mammalian brain: Interaction with convulsant and depressant benzodiazepines. *Brain Res.* **216,** 399–408.

Leeb-lundberg F. and Olsen R. W. (1982) Interactions of barbiturates of various pharmacological categories with benzodiazepine receptors. *Mol. Pharmacol.* **121,** 320–328.

Lehmann J. and Scatton B. (1982) Characterization of the excitatory amino acid receptor-mediated release of [^{3}H]acetylcholine from rat striatal slices. *Brain Res.* **252,** 77–85.

Lehmann J., Schaeffer P., Ferkany J. W., and Coyle J. T. (1983) Quinolinic acid evokes [^{3}H]acetylcholine release in striatal slices: Mediation by NMDA-type excitatory amino acid receptors. *Eur. J. Pharmacol.* **96,** 111–115.

Logan W. J. and Snyder S. H. (1972) High affinity uptake systems for glycine, glutamic and aspartic acid in synaptosomes of rat central nervous system. *Brain Res.* **42,** 413–431.

London E. D. and Coyle J. T. (1979) Specific binding of [^{3}H]kainic acid to receptor sites in rat brain. *Mol. Pharmacol.* **15,** 492–505.

London E. D., Kelman N., and Coyle J. T. (1979) Phylogenetic distribution of [^{3}H]kainic acid receptor binding sites in neuronal tissue. *Br. Res.* **192,** 463–476.

London E. D., Yamamura H. I., Bird E. D., and Coyle J. T. (1981) Decreased receptor binding sites for kainic acid in brains of patients with Huntington's disease. *Biol. Psychiatry* **16,** 155–162.

Luthin G. R. and Wolfe B. B. (1984) Comparison of [^{3}H]pirenzepine and [^{3}H]quinuclidinyl benzilate binding to muscarinic cholinergic receptors in rat brain. *J. Pharmacol. Exp. Ther.* **228,** 648–655.

Massotti M., Guidotti A., and Costa E. (1981) Characterization of benzodiazepine and gamma-aminobutyric recognition sites and their endogenous modulators. *J. Neurosci.* **1,** 409–418.

McGeer P. L., McGeer E. G., Scherer U., and Singh K. (1977) A glutamatergic corticostriatal path? *Brain Res.* **128,** 369–373.

McGeer E. G., Olney J. W., and McGeer P. L., eds. (1978) *Kainic Acid as a Tool in Neurobiology,* Raven, New York.

Meldrum B. S., Croucher M. J., Badman G., and Collins J. (1983) Antiepileptic actions of excitatory amino acid antagonists in the photosensitive baboon, *Papio papio. Neurosci. Lett.* **39,** 101–104.

Mena E. E., Fagg G. E., and Cotman C. W. (1982) Chloride ions enhance L-glutamate binding to rat synaptic membranes. *Brain Res.* **243,** 378–381.

Michaelis E. K., Michaelis M. L., Stormann T. M., Leroychiterlander W., and Grubbs R. D. (1983) Purification and molecular characterization of the brain synaptic membrane glutamate-binding protein. *J. Neurochem.* **40,** 1742–1753.

Monaghan D. T. and Cotman C. W. (1982) The distribution of [^{3}H]kainic acid binding sites in rat CNS as determined by autoradiography. *Brain Res.* **252,** 91–100.

Monaghan P., McMillis M. D., Chamberlin A. R., and Cotman C. W. (1983a) Synthesis of [^{3}H]-2-amino-4-phosphonobutyric acid and characterization of its binding to rat brain membranes: A selective ligand for the chloride/calcium-dependent class of L-glutamate binding sites. *Brain Res.* **278,** 137–144.

Monaghan D. T., Holets V. R., Toy P. W., and Cotman C. W. (1983b) Anatomical distribution of four pharmacologically distinct [^{3}H]-L-glutamate binding sites. *Nature* (Lond.) **306,** 176–178.

Morgan I. G. and El-Lakany S. (1982) Folic acid derivatives do not reproduce the neurotoxic effects of kainic acid on chicken retina. *Neurosci. Lett.* **34,** 69–73.

Moroni F., Lombardi G., Carla V., and Moneti G. (1984) The excitotoxin quinolinic acid is present and unevenly distributed in the rat brain. *Brain Res.* **295,** 352–355.

Nadler J. V. (1981) Kainic acid as a tool for the study of temporal lobe epilepsy. *Life Sci.* **29,** 2031–2042.

Olney J. W. and Gubareff T. (1978) Glutamate neurotoxicity and Huntington's chorea. *Nature* (Lond.) **271,** 557–559.

Olsen R. W. (1981) GABA, benzodiazepine–barbiturate receptor interactions. *J. Neurochem.* **37,** 1–13.

Olsen R. W. and Leeb-lundberg F. (1981) Convulsant and Anticonvulsant Drug Binding Sites Related to the GABA Receptor/Ionophore System, in *Neurotransmitters, Seizures and Epilepsy* (Morselli P., Lloyd K., Loscher W., Meldrum B., Chir B., and Reynolds E., eds.), 151–164, Raven Press, New York.

Olverman H. J., Jones A. W., and Watkins J. C. (1984) L-Glutamate has higher affinity than other amino acids for [^{3}H]D-AP5 binding sites in rat brain membranes. *Nature* (Lond.) **307,** 460–462.

Perkins M. N. and Stone T. W. (1983a) Pharmacology and regional variations of quinolinic acid-evoked excitations in the rat central nervous system. *J. Pharmacol. Exp. Ther.* **226,** 551–561.

Perkins M. N. and Stone T. W. (1983b) On the interaction of 2-amino-7-phosphonoheptanoic acid and quinolinic acid in mice. *Eur. J. Pharmacol.* **89,** 297–300.

Perkins N. M., Stone T. W., Collins J. E., and Curry K. (1981) Phosphonate analogues of carboxylic acids as amino acid antagonists on rat cortical neurons. *Neurosci. Lett.* **23,** 333–336.

Peterson D. W., Collins J. F., and Bradford H. F. (1983) The kindled amygdala model of epilepsy: Anticonvulsant action of amino acid antagonists. *Brain Res.* **275,** 169–172.

Potashner S. J. (1978) The spontaneous and electrically evoked release from slices of guinea pig cerebral cortex of endogenous amino acids labelled via metabolism of D[^{14}C]glucose. *J. Neurochem.* **34,** 177–186.

Reubi J. (1980) Comparative study of the release of glutamate and GABA, newly synthesized from glutamine in various regions of the central nervous system. *Neuroscience* **5,** 2145–2150.

Roberts P. (1974) Glutamate receptors in the rat central nervous system. *Nature* (Lond.) **232,** 399–401.

Roberts P. J. and Watkins J. C. (1975) Structural requirements for the inhibition for L-glutamate uptake by glia and nerve endings. *Brain Res.* **85,** 120–125.

Roberts P. J., Foster G. A., and Thomas E. M. (1981) Neurotoxic action of methyltetrahydrofolate in rat cerebellum unrelated to direct activation of kainate receptors. *Nature* (Lond.) **243,** 654–655.

Ruck A., Kramer S., Metz J., and Brennan M. J. W. (1980) Methyltetrahydrofolate is a potent and selective agonist for kainic acid receptors. *Nature* (Lond.) **287,** 852–853.

Schwarcz R. and Coyle J. T. (1977) Striatal lesions with kainic acid; neurochemical characteristics. *Brain Res.* **127,** 235–249.

Schwarcz R., Whetsell W. O., and Mangano R. M. (1983) Quinolinic acid, an endogenous metabolite, can produce axon-sparing lesions in rat brain. *Science,* **219,** 316–318.

Schwarcz R., Zaczek R., and Coyle J. T. (1978) Microinjection of kainic acid into the rat hippocampus. *Eur. J. Pharmacol.* **50,** 209–220.

Sharif N. A. and Roberts P. J. (1981a) Regulation of cerebellar L-[^{3}H]glutamate binding: Influence of guanine nucleotides and Na^+ ions. *Biochem. Pharmacol.* **30,** 3019–3022.

Sharif N. A. and Roberts P. J. (1981b) L-Aspartate binding sites in rat cerebellum: A comparison of the binding of L-[^{3}H]aspartate and [^{3}H]glutamate to synaptic membranes. *Brain Res.* **211,** 293–303.

Shinozaki H. and Konish S. (1970) Actions of several antihelminthics and insecticides on rat cortical neurons. *Brain Res.* **24,** 368–371.

Simon J. R., Contrera J. F., and Kuhar M. J. (1976) Binding of [^{3}H]kainic acid, an analogue of L-glutamate, to brain membranes. *J. Neurochem.* **26,** 141–147.

Skaggs K. K., Wenthold R. J., and Altschuler R. A. (1983) Immunocytochemical distribution of aspartate aminotransferase and glutaminase in the cerebellum. *Soc. Neurosci. Abs.* **9,** 1090.

Skerritt J. H. and Johnston G. A. R. (1983) Enhancement of GABA binding by benzodiazepines and related compounds. *Eur. J. Pharmacol.* **89,** 193–198.

Skerritt J. H., Willow M., and Johnston G. A. R. (1982a) Diazepam enhancement of low affinity GABA binding to rat brain membranes. *Neurosci. Lett.* **29,** 63–66.

Skerritt J. H., Chen Chow S., and Johnston G. A. R. (1982b) Differences in the interaction between GABA and benzodiazepines. *Neurosci. Lett.* **33,** 173–178.

Skolnick P., Moncada V., Barker J. L., and Paul S. M. (1981) Pentobarbital: Dual actions to increase brain benzodiazepine receptor affinity. *Science* **211,** 1448–1450.

Skolnick P., Rice K. C., Barker J. L., and Paul S. M. (1982) Interaction of barbiturates with benzodiazepine receptors in the central nervous system. *Brain Res.* **233,** 143–156.

Slevin J. T., Collins J. F., and Coyle J. T. (1983) Analogue interactions with the brain receptor labeled by [^{3}H]kainic acid. *Brain Res.* **265,** 169–172.

Slevin J., Collins J., Lindsley K., and Coyle J. T. (1982) Specific binding of [^{3}H]L-Glutamate to cerebellar membranes: Evidence for recognition site heterogeneity. *Brain Res.* **249,** 353–360.

Snodgrass S. R. (1979) *In vitro* binding studies with ^{3}H-*N*-methylaspartate. *Soc. Neurosci. Abs.* **5,** 572.

Squires R. F., Casida J. E., Richardson M., and Saederup E. (1983) [^{35}S]-*t*-butylbicyclophosphorothionate binds with high affinity to brain specific sites coupled to gamma-aminobutyric acid-A and ion recognition sites. *Mol. Pharmacol.* **23,** 326–336.

Stone T. W. and Perkins M. N. (1981) Quinolinic acid: A potent endogenous excitant at amino acid receptors in the CNS. *Eur. J. Pharmacol.* **72,** 411–412.

Storm-Mathisen J. and Upsahl M. W. (1978) Aspartate and/or glutamate may be transmitters in hippocampal efferents to septum and hypothalamus. *Neurosci. Lett.* **9,** 65–70.

Takemoto T. (1978) Isolation and Structural Identification of Naturally Occurring Excitatory Amino Acids, in *Kainic Acid as a Tool in Neurobiology* (McGeer E. G., Olney J. W., and McGeer P. L., eds.), pp. 1–15, Raven, New York.

Tallman J. F. and Gallager D. W. (1979) Modulation of benzodiazepine binding site sensitivity. *Pharmacol. Biochem. Behav.* **10,** 809–813.

Tallman J. F., Thomas J. W., and Gallager D. W. (1978) GABAergic modulation of benzodiazepine binding site sensitivity. *Nature* (Lond.) **274,** 383–388.

Ticku M. J. (1981) Interaction of depressant, convulsant and anticonvulsant barbiturates with the [^{3}H]diazepam binding site of the benzodiazepine GABA-receptor-ionophore complex. *Biochem. Pharmacol.* **30,** 1573–1579.

Ticku M. K., Van Ness P. C., Haycock J. W., Levy W. B., and Olsen R. W. (1978) Dihydropicrotoxinin binding sites in rat brain: Comparison to GABA receptors. *Brain Res.* **150,** 642–647.

Toffano H., Guidotti A., and Costa E. (1978) Purification of an endogenous protein inhibitor of the high affinity binding of gamma-aminobutyric acid to synaptic membranes of rat brain. *Proc. Natl. Acad. Sci. USA* **75,** 4024–4028.

Vincent S. R. and McGeer E. G. (1979) Kainic acid binding to brain membranes of striatal neurons. *Life Sci.* **24,** 265–270.

Walaas I. and Fonnum F. (1980) Biochemical evidence for glutamate as a transmitter in hippocampal efferents to the basal forebrain and hypothalamus in the rat brain. *Neuroscience* **5,** 1691–1698.

Watkins J. C. and Evans R. H. (1981) Excitatory amino acid transmitters. *Ann. Rev. Pharmacol. Toxicol.* **21,** 165–204.

Watson M., Roeske W. R., and Yamamura H. I. (1983) [^{3}H]Pirenzepine selectively identifies a high affinity population of muscarinic cholinergic receptors in the rat cerebral cortex. *Life Sci.* **31,** 2019–2023.

Weiler C. T., Nystrom B., and Hamberger A. (1978) Characteristics of glutamine versus glutamate transport in isolated glia and synaptosomes. *J. Neurochem.* **32,** 559–565.

Werling L. L., Doman K. A., and Nadler J. V. (1983) L-[^{3}H]Glutamate binding to hippocampal synaptic membranes: Two binding sites discriminated by differing affinities for quisqualate. *J. Neurochem.* **41,** 586–593.

White W. F., Nadler J. V., Hamberger A., Cotman C. W., and Cummins J. T. (1977) Glutamate as a transmitter of hippocampal perforant path. *Nature* (Lond.) **270,** 356–357.

Wilkin G. E., Hudsan A. L., Hill D. R., and Bowery N. G. (1982) Autoradiographic localization of $GABA_B$ receptors in rat cerebellum. *Nature* (Lond.) **294,** 584–589.

Willow M. and Johnston G. A. R. (1981a) Enhancement by anesthetic and convulsant barbituates of GABA binding to rat brain synaptosomal membranes. *J. Neurosci.* **1,** 364–367.

Willow M. and Johnston G. A. R. (1981b) Pentobarbitone slows the dissociation of GABA from rat brain synaptosomal binding sites. *Neurosci. Lett.* **23,** 71–74.

Wofsey A. R., Kuhar M. J., and Snyder S. H. (1971) A unique synaptosomal fraction which accumulates glutamic and aspartic acids in brain tissue. *Proc. Natl. Acad. Sci. USA* **6,** 1102–1106.

Zaczek R., Collins J., and Coyle J. T. (1981) *N*-Methyl-D-aspartic acid: A convulsant with weak neurotoxic properties. *Neurosci. Lett.* **24,** 181–186.

Zaczek R., Hedreen J. C., and Coyle J. T. (1979) Evidence for a hippocampal–septal glutamatergic pathway in rat brain. *Exp. Neurol.* **65,** 145–156.

Zaczek R., Koller K., Cotter R., Heller D., and Coyle J. T. (1983) *N*-Acetylaspartyl-glutamate: An endogenous peptide with high affinity for a brain "glutamate" receptor. *Proc. Natl. Acad. Sci. USA* **80,** 1116–1119.

Zaczek R., Arlis S., Murphy T., and Coyle J. T. (1985) Effects of acidic dipeptides on brain glutamate receptors. *Trans. Am. Soc. Neurochem.* **16,** 303.

Chapter 8
GABA Receptor Binding

Kenneth George Lloyd

1. Introduction

GABA (γ-aminobutyric acid) receptors are among the most ubiquitous in the brain, having been identified (neurophysiologically) in all regions and at all levels (e.g., Iversen, 1978). All neuronal cells respond to the iontophoretic application of GABA, usually with a decrease in firing rate associated with hyperpolarization (Curtis, 1979). However, well-known instances of GABA-induced membrane depolarization occur, and the same cells may respond with a local depolarization or hyperpolarization, depending on where the GABA is applied (dendrite vs axon hillock vs cell body, and so on) (Andersen et al., 1980). This electrophysiological evidence indicates that all cells possess GABA receptors (most likely, $GABA_A$ receptors, see below), but it does not at all provide evidence for the existence of GABA synapses at these same cells. Indeed, the dorsal root ganglion (rat or cat) possesses $GABA_A$ receptors linked to chloride ion channels, activation of which results in membrane depolarization; there is, however, no GABA-ergic input to the dorsal root ganglion. This is therefore an example of an extrasynaptic GABA receptor (Desarmenien et al., 1980; Gallagher et al., 1983).

Although neurophysiological techniques identify GABA receptors, these techniques are difficult and time-consuming to perform. Furthermore, a quantitative estimate of GABA receptor density is usually not possible and identification of receptor subtypes is laborious. It is such problems, together with the need to make the GABA receptor more available for general study, that led to the development and massive acceptance of binding techniques for the study of GABA receptors.

Early attempts at GABA binding (Gottesfeld and Elliott, 1971; Sano and Roberts, 1963; DeFeudis, 1973) were not convincing that the binding sites identified represented GABA receptors; indeed they probably labeled uptake sites rather than binding sites. In 1974, Zukin et al., and in 1975, Enna and Snyder published a technique that with only a few modifications has become the standard assay for measuring the Na^+-independent $GABA_A$ receptor. In 1981, further advances were made when Hill and Bowery demonstrated a completely different Ca^{2+}-dependent $GABA_B$ receptor.

These $GABA_A$ and $GABA_B$ receptors are presently considered to be the major GABA receptors, or at least the important Na^+-independent [^{3}H]-GABA binding sites in nervous tissue. Although the technique is more rapid and efficient than the neurophysiological approach, [^{3}H]-GABA binding cannot easily distinguish synaptic vs extrasynaptic receptors or physiologically relevant vs pharmacologically relevant sites.

The present review concentrates on the methodological aspects of [^{3}H]-radioligand binding assays for $GABA_A$ and $GABA_B$ receptors. Although the pharmacological characterization of these different sites is considered, their distribution, pathology, and interactions with recognition sites for other ligands (e.g., benzodiazepines, barbiturates) is outside the scope of this article.

2. $GABA_A$ Receptors

The neurophysiological definition of the $GABA_A$ receptor includes bicuculline sensitivity and linkage to a chloride ion channel. In terms of binding criteria, the requirement for bicuculline sensitivity is maintained since bicuculline acts directly at the GABA recognition site (Johnston, 1976). However, the demonstration of chloride ion channel linkage by binding methods is much more difficult because this site is distal to the recognition site within the $GABA_A$ receptor macromolecular complex. Under certain conditions, the $GABA_A$ site can be shown to be part of a larger GABA-benzodiazepine chloride ion channel macromolecular complex (Möhler et al., 1981); however, for most experiments it has not been verified whether or not the $GABA_A$ recognition site studied is part of this complex. Furthermore, even in those regions in which the macromolecular complex is demonstrated, there is no evidence that all $GABA_A$ recognition sites are combined in a complex with benzodiazepine recognition sites.

2.1. Membrane Preparation for $GABA_A$ Receptor Binding

Membrane preparation is the critical aspect, much more so than the actual binding assay, for the GABA receptor assay. For the present purposes we will consider the preparation of membranes from brain tissue; it is possible to prepare membranes from peripheral tissues, but each tissue needs individual modification of the technique and the relative recovery of binding sites between tissues is unknown. In the membrane preparation of brain tissue for $GABA_A$ binding, two variables generally occur: (1) the use of fresh or frozen tissue, and (2) the use or nonuse of the nonionic detergent Triton-X-100.

The use of fresh or frozen tissue depends to a large extent on the experimental design. If one is preparing large amounts of membranes to serve as a homogenous source for a series of pharmacological or physicochemical experiments, then the use of fresh tissue as the starting material is preferable, since the partial purification of synaptosomes results in a higher density of recognition sites in the final preparation. However, for *ex vivo* experiments, regional comparison studies, and human postmortem tissue, it is often impossible to prepare fresh membranes from all of the different regions to be studied. In these instances, dissection of the regions and immediate freezing, or freezing of the whole brain and later dissection of specific neuroanatomical areas from coronal sites, is an acceptable procedure.

Triton-X-100 increases the affinity and density of the recognition site for $GABA_A$ receptors, as indicated by a lower K_d (or IC_{50} for displacement of unlabeled GABA) (Enna and Snyder, 1977) (Table 1). This occurs when either fresh or frozen tissue is used as starting material. There are several possibly different mechanisms involved in the action of Triton-X-100. First, the detergent may remove endogenous GABA that remains bound to the membranes. In order to accurately estimate the kinetics of ^{3}H-ligand binding to the recognition site, it is essential that the ligand concentration be known exactly. If endogenous GABA remains in the membrane preparation, an overestimation of K_d and underestimation of B_{max} will result. However, in the absence of Triton-X-100, multiple tissue washes are sufficient to remove all endogenous GABA (Greenlee et al., 1978) and similar inhibitors of GABA binding.

Second, in its role as a detergent Triton-X-100 removes membrane phospholipids. This results in an altered configuration of the GABA receptor, making the recognition site more accessible to the ligand. The removal of membrane phospholipids is probably a major component of the Triton-X-100 effect, since treatment of brain membranes with phospholipase C reproduces the action of Triton-X-100 (Lloyd and Beaumont, 1980; Giambalvo and Rosenberg, 1976).

Third, the use of Triton-X-100 may change the relationship of the $GABA_A$ site to that of the other components of the GABA–

Table 1
Kinetic Characteristics of $GABA_A$ Recognition Sites in Rat Cerebral and Human Cerebellar Membranes

Ligand conditions	[^{3}H]-GABA, n*M*	[^{3}H]-Muscimol, n*M*	[^{3}H]-Isoguvacine, n*M*	[^{3}H]-Bicuculline, n*M*
IC_{50} for GABA				
Rodent brain				
No Triton	200–400[a–e]	8[l], 24[m], 38[o], 130[p], 200[m]	690[r]	400–600[v,w]
Plus Triton	20–60[b,f,u]	8[u] 27[o] 200[n]	100[s], 75[t], 100[a]	
Human cerebellum				
No Triton	70–215[c,e,g]			
K_d of ligand				
Rodent brain				
No Triton	K_d = 370–511[a,e]	K_{d1} = 2–3[l,n,o], 9[o]	K_{d1} = 6–10[t]	K_{d1} = 85[v]; 35[w]
		K_{d2} = 30[o], 44[p]	K_{d2} = 63–90[t]	K_{d2} = 362[v]
		K_{d2} = 60[l,n]		
	K_{d1} = 4–16	K_d = 2–3[l,m]	76[s]	
	K_{d2} = 27–175[b,f,r]	K_{d1} = 0.9		
		K_{d2} = 5.8		
		K_{d3} = 32		

Human cerebellum	
No Triton	$K_{d1} = 6–23$
	$K_{d2} = 110–192$[h]
	$K_d = 101–335$[i,j]
Plus Triton	$K_d = 38$[j]

[a]From Enna and Snyder, 1975.
[b]From Enna and Snyder, 1977.
[c]From Lloyd and Dreksler, 1979.
[d]From Greenlee et al., 1978.
[e]From Lloyd et al., 1977a.
[f]From Wong and Horng, 1977.
[g]From Iversen et al., 1979.
[h]From Van Ness and Olsen, 1979.
[i]From Lloyd et al., 1977b.
[j]From Lloyd and Davidson, 1979.
[k]From Krogsgaard-Larsen et al., 1983.
[l]From Beaumont et al., 1978.
[m]From Snodgrass, 1978.
[n]From Wang et al., 1979.
[o]From Williams and Risley, 1979.
[p]From Leach and Wilson, 1978.
[q]From Johnston et al., 1979.
[r]From Collins et al., 1980.
[s]From White and Snodgrass, 1983.
[t]From Morin and Wasterlain, 1980.
[u]From Lloyd et al., 1982.
[v]From Möhler, 1979.
[w]From Olsen and Snowman, 1983.

benzodiazepine–chloride ionphore macromolecular complex. This may involve the denuding of the complex of endogenous compounds (GABA-modulin, GABArins) that normally maintain the configuration of the $GABA_A$ receptor, determined by the state of activation of the benzodiazepine recognition site (Johnston and Skerritt, 1984).

The merits of membrane treatment with Triton-X-100 may be questionable in some experimental designs. In the case of the establishment of a structure–activity relationship of different chemical series, it is highly desirable to use Triton-X-100-pretreated membranes (usually 0.05%). This gives the most active, homogeneous membrane preparation with reduced problems from phospholipids, endogenous inhibitors, and so on. However, if one is studying the regional distribution of "physiologically responsive" GABA recognition sites under different conditions, then Triton-X-100 pretreatment may give misleading results, since the activation of GABA binding by the detergent is not constant between brain regions (Enna et al., 1979; Lloyd and Dreksler, 1979). Similarly, when pathological conditions are under investigation, especially those involving lipid abnormalities, the use of Triton-X-100 may obliterate or greatly reduce meaningful differences between pathological and control tissue (Lloyd and Davidson, 1979).

The actual membrane preparation for [^{3}H]-GABA binding (or for other ligands for the $GABA_A$ recognition site) has varied very little since the original descriptions (Zukin et al., 1974; Enna and Snyder, 1975, 1977; Lloyd et al., 1977a; Wong and Horng, 1977). When using fresh rat brain as the starting material, the procedure is as follows: The brains are homogenized (glass-Teflon homogenizer) in 15 vol of ice-cold, 0.32*M* sucrose and centrifuged at 1000*g* for 10 min, the pellet is discarded, and the supernatant recentrifuged at 20,000*g* for 20 min. This pellet is resuspended (Polytron, 30 s, midsetting) in distilled water (50 mL/g of original brain weight) and centrifuged at 8000*g* for 20 min. The supernatant together with the buffy coat (upper bilayer of the pellet, which is very rich in GABA receptors), are centrifuged for 20 min at 48,000*g*. The resultant pellet is then frozen on dry ice for at least 2 h (preferably overnight at –80°C). The frozen pellet is resuspended (Polytron) in deionized water, left at room temperature for 30 min, and then centrifuged for 10 min at 48,000*g*. The pellets are then resuspended in buffer (either 0.05*M* Tris-citrate, pH 7.1, or 0.05*M* potassium phosphate, pH 7.2) for use in the binding assay.

It should be noted that variation in collection of the "buffy coat" will result in parallel charges in the density (but not the affinity) of $GABA_A$ sites in the final membrane preparation.

If frozen tissue is used as the starting material, the tissue is homogenized (Polytron) in cold, deionized water (50 mL/g tissue). The suspension is left at room temperature for 30 min and then centrifuged at 48,000*g* for 10 min. The pellet is washed twice with deionized water and the final pellet resuspended in buffer so that a concentration of 40 mg starting tissue/mL buffer is obtained (approximately 2 mg protein/mL).

If Triton-X-100 is to be used, the procedures are modified in the following manner: When starting from fresh tissue, the frozen pellet is resuspended and incubated at 37°C for 30 min in buffer containing 0.05% Triton-X-100. The suspension is centrifuged at 48,000*g* for 10 min and then resuspended for use in the binding assay. If frozen brain tissue is the starting material, then the detergent is added to the buffer (0.05%) in the initial homogenization step. It should be noted that although Triton-X-100 increases the affinity, the number of binding sites, and the specific binding (to 92–95% from 75–80% using frozen human cerebellum) (Lloyd and Dreksler, 1979), the total amount of protein recovered is greatly decreased. When using very small amounts of frozen tissue (e.g., substantia nigra), this can result in the loss of the pellet.

2.2. Binding to the $GABA_A$ Recognition Site: Choice of Ligand

Once the membranes have been prepared, the binding assay is technically very simple. However, certain precautions must be taken in order to ensure that the specific binding measured is actually to the $GABA_A$ recognition site, and not to other GABA recognition sites in the membrane, e.g., $GABA_B$ or other receptors, uptake sites, sites on enzymes, and so on. This is possible by carefully controlling the incubation conditions, respecting the laws of receptor interactions, and by the use of "specific" ligands.

In terms of incubation conditions, the essential factor is a sodium-free medium. Under these conditions, GABA uptake sites are inactive, a state that can be further guaranteed by performing the incubation in an icebath (0–4°C). Any binding to non$GABA_A$ receptor sites (nonspecific binding) is accounted for by duplicate tubes containing a high concentration of unlabeled GABA (usually 1 m*M*).

For the $GABA_A$ recognition site, the following conditions should also be met: saturability, bicuculline-sensitivity, and a pharmacology parallel to that observed in neurophysiological experiments with identified $GABA_A$ agonists.

The choice of the ^3H-ligand for $GABA_A$ binding is a more complex matter. There are several available, and each has its advantages and disadvantages.

2.2.1. [^{3}H]-GABA

This ligand is the personal choice of the author, since by definition it is recognized by the $GABA_A$ site. Changes observed in pathological conditions (disease, lesion, and so on) can be related logically, and with a large degree of certainty, to alterations in $GABA_A$ receptor function. Similarly, displacement of [^{3}H]-$GABA_A$ binding by a given compound is indicative of an action at the $GABA_A$ site. The disadvantage of using [^{3}H]-GABA as the ligand is that these observations are valid only if all other recognition sites for GABA (see above) are inactivated or otherwise accounted for, which is possible under most circumstances. However, the use of nonendogenous ligands circumvents this problem.

2.2.2. [^{3}H]-Muscimol

In addition to [^{3}H]-GABA, this ligand is widely used for the measurement of $GABA_A$ receptor binding. The compound is a $GABA_A$ agonist, as identified neurophysiologically (Johnston, 1976; Krogsgaard-Larsen et al., 1983), with little activity at $GABA_B$ receptors (Hill and Bowery, 1981) (see below) or on GABA metabolizing enzymes. Furthermore, the affinity of muscimol for the $GABA_A$ recognition site is much greater than for the uptake recognition site (Brehm et al., 1979). Together with its very high affinity for the $GABA_A$ site (see below), this would appear to make it an ideal ligand. However, the disadvantages of muscimol are the following: (1) it may not label all $GABA_A$ sites, especially those in a low-affinity configuration. This is of potential importance since it is not clear which affinity state is of the greatest physiological relevance; and (2) the in vivo profile of muscimol is not that of a wide-spectrum GABA agonist (Worms et al., 1982; Lloyd et al., 1983).

2.2.3. [^{3}H]-Isoguvacine

Isoguvacine is a GABA agonist, as identified neurophysiologically (Krogsgaard-Larsen et al., 1979), and has been proposed as a useful ^{3}H-ligand for $GABA_A$ receptors. The major advantage of this compound is that it is devoid of activity at GABA uptake (Brehm et al., 1979) and $GABA_B$ recognition sites (see below). However, an uptake and release system for isoguvacine itself apparently exists (White and Snodgrass, 1983). The major disadvantage is that it behaves differently from GABA itself under certain conditions. Notably, isoguvacine is relatively weak in enhancing [^{3}H]-diazepam binding, which is a bicuculline-sensitive action of GABA itself, muscimol, and other $GABA_A$ agonists (Krogsgaard-Larsen and Arnt, 1980; Krogsgaard-Larsen et al., 1984). Although several publications exist (e.g., Collins et al., 1980; Morin and Wasterlain, 1980; White and Snodgrass, 1983; Lloyd et al., 1982), [^{3}H]-isoguvacine has not been generally used as a ligand.

2.2.4. [^{3}H]-Piperidine-4-Sulfonic Acid (P4S)

It has been proposed that [^{3}H]-P4S binds to $GABA_A$ receptors, identifying both high- and low-affinity recognition sites, although the kinetic parameters of [^{3}H]-P4S binding are somewhat different from those of [^{3}H]-GABA binding (Krogsgaard-Larsen et al., 1981). Not enough literature data are available to properly assess this ligand. However, it should be noted that P4S actually decreases [^{3}H]-diazepam binding, rather than enhancing it, as do most GABA agonists (Krogsgaard-Larsen and Arnt, 1980; Krogsgaard-Larsen et al., 1984).

2.2.5. [^{3}H]-THIP

THIP (4,5,6,7-tetrahydroisoxazolo[5,4-c]pyridin-3-o1) is a specific $GABA_A$ receptor agonist, as identified neurophysiologically (Krogsgaard-Larsen et al., 1975, 1979), that has been used as a radioligand to identify $GABA_A$ sites (Falch and Krogsgaard-Larsen, 1982). However, [^{3}H]-THIP binding identifies fewer sites than does [^{3}H]-$GABA_A$ binding, and the regional distribution of these ligands is different (Falch and Krogsgaard-Larsen, 1982). Unlike GABA itself, THIP is at most a weak enhancer of [^{3}H]-diazepam binding (Krogsgaard-Larsen and Arnt, 1980; Krogsgaard-Larsen et al., 1984). [^{3}H]-THIP binding has not been reported extensively enough in the literature to be able to critically assess it as a ligand.

2.2.6. [^{3}H]-Bicuculline

Bicuculline is a prototypic $GABA_A$ antagonist (Johnston, 1976), and has been proposed as a highly specific ligand for the $GABA_A$ recognition site (Collins and Cryer, 1978; Möhler and Okada, 1978; Möhler, 1979). Since one of the criteria for $GABA_A$ recognition site identification is bicuculline-sensitivity, this would appear to be a most suitable choice. Unfortunately, bicuculline is also an inhibitor of acetylcholinesterase (Svenneby and Roberts, 1973), and therefore part of its binding may be to this enzyme. A practical hindrance to the use of [^{3}H]-bicuculline is the low degree of specific binding obtained (16–50%). This may be a result of different agonist and antagonist configurations of the $GABA_A$ receptor. [^{3}H]-Bicuculline binding has not been widely reported in the literature.

2.3. Assay Conditions for Binding to the $GABA_A$ Recognition Site

2.3.1. [^{3}H]-GABA

Since the conditions used in the studies on [^{3}H]-GABA binding to either rat and/or human membranes have been maintained over the past 6–10 yr, only the original papers are presently cited (e.g., Zukin et al., 1974; Enna and Snyder, 1975, 1977; Lloyd et al., 1977a,b; Wong and Horng, 1977; Greenlee et al., 1978).

For the binding assay, the incubation volumes may vary from 0.5 to 2.5 mL; the buffer is either 0.05*M* Tris-citrate (pH 7.1) or 0.05*M* potassium phosphate (pH 7.2). The optimum pH range in both buffers is rather broad—between 7.0 and 7.4, or even slightly greater (e.g., Enna and Snyder, 1977; Lloyd and Dreksler, 1979). The incubation is normally performed in an icebath for 5 min, although higher temperatures (up to 35°C, Enna and Snyder, 1977) may be used. Binding is already maximal at less than 5 min. For displacement studies, [^{3}H]-GABA concentrations of 8–25 n*M* are common, the lower concentrations usually being used for Triton-X-100-treated membranes. For kinetic analysis, [^{3}H]-GABA concentrations of 1–750 n*M* are usual. Nonspecific binding is usually assessed by parallel tubes containing 1 m*M* GABA, although 0.1 m*M* bicuculline may also be used. The binding assay is terminated by centrifugation (48,000*g*, 10 min). The fluid is decanted and the pellet rinsed rapidly and *superficially* with 10 mL of ice-cold deionized water, which is decanted after each rinse. Any resuspension of the pellet results in a variable loss of binding. Any droplets remaining on the tube wall are absorbed with a cotton swab. The final pellet is dissolved using a tissue solubilizer (e.g., Soluene, Protosol, and so on), acidified if necessary (to remove chemiluminescence), and scintillation fluid added for estimation of radioactivity.

Under these conditions, specific binding is 45–65% in the rat brain (Enna and Snyder, 1975, 1977; Greenlee et al., 1978) and 75–95% for human cerebellar preparations (Lloyd and Dreksler,

1979; Greenlee et al., 1978). The only significant deviation from this procedure was described by Lester and Peck (1979), who used 10 m*M* HEPES buffer (pH 7.4) and filtration to terminate the assay. This technique has not found wide acceptance.

Using the standard centrifugation technique and [^{3}H]-GABA as the ligand, the characterization of the $GABA_A$ recognition site is highly reproducible between laboratories (Table 1). In the absence of Triton-X-100 pretreatment, the rat brain and human cerebellum have IC_{50} levels for GABA in the same range. Scatchard analysis usually indicates one site, but some authors have reported two or three (e.g., Falch and Krogsgaard-Larsen, 1982). In Triton-X-100-treated human cerebellar material, a high-affinity K_d is always observed, but a lower-affinity component is somewhat variable.

2.3.2. [^{3}H]-Muscimol

Although the use of [^{3}H]-muscimol as a ligand for $GABA_A$ receptors is more recent (Beaumont et al., 1978; Snodgrass, 1978) and has been studied somewhat less frequently, the variation in techniques is greater than that for the use of GABA itself. However, most authors (e.g., Beaumont et al., 1978; Williams and Risley, 1979) follow the same procedure. Membrane preparation is basically that of Zukin et al. (1974) and Enna and Snyder (1975), with variations including different buffers for resuspension (Snodgrass, 1978; Wang et al., 1979) and sources other than rat brain (DeFeudis, 1981).

The buffer used is uniformly Tris in nature, usually 0.05*M* Tris-citrate, pH 7.1 (Beaumont et al., 1978; Williams and Risley, 1979), although 0.04*M* HEPE-Tris, pH 7.0 (Snodgrass, 1978), and 0.05*M* Tris-HCl, pH 7.3 (Wang et al., 1979), have also been used. However, it is often better to avoid the incorporation of chloride ions, since they may interact with the recognition complex and mask certain effects (Madtes, 1984). The incubation volume varies from 0.6 (Beaumont et al., 1978) to 1.05 mL (Snodgrass, 1978).

The concentration (for displacement studies) of [^{3}H]-muscimol is usually 2 n*M*, although a range of 1–5 n*M* has been utilized without obviously affecting the results. In almost all assay systems, the incubation temperature is 0–4°C (icebath), and an incubation time of 30 min is used (Beaumont et al., 1978; Wang et al., 1979; William and Risley, 1979). An incubation time of 10 min has also been reported (Snodgrass, 1978), but this appears to be rather short, since equilibrium takes 10–25 min (e.g., Beaumont et al., 1978; Williams and Risley, 1979). Bowery et al. (1979) have used a 10-min preincubation with test agents (e.g., other agonists, and so on), and then a 10-min incubation with [^{3}H]-muscimol. There do not appear to be any clear advantages to this latter procedure. Nonspecific binding is assessed in parallel tubes by means of high concentrations of GABA (10–200 μ*M*) or muscimol (1–1000 μ*M*).

In contrast to [^{3}H]-GABA binding, the slow equilibrium time of muscimol allows filtration to be used as a means of assay termination. In this case, Whatman GFB glass filters are used in a Millipore 1225 filter apparatus. The membranes are rapidly rinsed twice (total procedure less then 15 s) with either icewater or buffer. The protein is then dissolved and assayed for radioactivity (Wang et al., 1979; Williams and Risley, 1979; Leach and Wilson, 1978). Centrifugation is also a valid means of terminating the assay, and is especially convenient when using small volumes in 1.5-mL centrifuge tubes. In the case of centrifugation, the pellet is usually rinsed once or twice and recentrifuged. As for [^{3}H]-GABA binding, the pellet is solubilized then assayed for radioactivity using acidified scintillation fluid (Beaumont et al., 1978; Snodgrass, 1978; Wang et al., 1979; Bowery et al., 1979). Under these assay conditions, specific binding is 80–90% of total counts (Beaumont et al., 1978; Wang et al., 1979; Williams and Risley, 1979).

The major exception to the above procedures is the report of Maurer (1979), who used a dialyzed membrane preparation instead of freeze-thawing or Triton-X-100 treatment. In addition, he reported an incubation pH of 6.7 (0.05*M* Tris-citrate) and an incubation temperature of 20°C. Non-specific binding was assessed by parallel tubes with 1 μ*M* muscimol (incubation concentration, 2 n*M* [^{3}H]-muscimol), and filtration was used to terminate the assay. Using these conditions, the muscimol binding site identified differs somewhat from that of other authors (see below).

As can be seen from Table 1, the results from different [^{3}H]-muscimol binding assays are much more variable than for [^{3}H]-GABA binding to the $GABA_A$ site. Thus, in the absence of Triton pretreatment, a high-affinity site of 2–3 n*M* is generally observed, with a lower affinity site of 30–70 n*M*, although one report indicates the presence of only one low-affinity site (Leach and Wilson, 1978) and another reports variable sites (one or two), depending on the particular assay (Wang et al., 1979). Triton-X-100 pretreatment of the membranes generally results in a single affinity site of 2–3 n*M*; however, one report, using a very high concentration of detergent (0.5%), demonstrated three binding sites for [^{3}H]-muscimol, but only two for [^{3}H]-GABA under the same conditions (Johnston et al., 1979).

The different [^{3}H]-muscimol binding sites also have different affinities for GABA (as indicated by the IC_{50}, Table 1). In the absence or presence of Triton-X-100 treatment, the IC_{50} for GABA varied from 8 to 200 n*M*! These activities seem very similar to the high- and low-affinity sites for [^{3}H]-GABA, suggesting that in the [^{3}H]-muscimol assay there are variable states of the recognition site that respond to GABA and muscimol differently. In general, Triton-X-100 has much less effect on [^{3}H]-muscimol binding than on [^{3}H]-GABA binding, and one report (Wang et al., 1979) could not demonstrate any action of Triton-X-100 on [^{3}H]-muscimol binding.

2.3.3. [^{3}H]-Isoguvacine

The membrane preparation used in [^{3}H]-isoguvacine binding to the $GABA_A$ recognition site is exactly the same as that used for [^{3}H]-GABA (e.g., Zukin et al., 1974; Enna and Snyder, 1975) or [^{3}H]-muscimol binding (Snodgrass, 1978). The only major point of contention is the use of Triton-X-100. Collins et al. (1980) reported that use of this detergent resulted in a loss of binding activity, whereas Morin and Wasterlain (1980) reported an increase in binding capacity without a change in affinity after treatment of membranes with 0.01% Triton-X-100. Similarly, White and Snodgrass (1983) reported that 0.05% Triton-X-100 was essential for reproducible [^{3}H]-isoguvacine binding with a high percent of specific binding. If Triton-X-100 is not used, it would appear that a large number of membrane washes (up to six, Morin and Wasterlain, 1980) are necessary in order to expose all of the [^{3}H]-isoguvacine binding sites.

The binding assay is the same as that used for [^{3}H]-GABA (Enna and Snyder, 1975) or [^{3}H]-muscimol (Snodgrass, 1978). Tris-citrate buffer (0.05*M*, pH 7.1) at 4°C has routinely been used with an incubation time of 5–10 min. Concentrations of 5, 8, and 10 n*M* [^{3}H]-isoguvacine give similar results. Nonspecific binding is assessed by parallel incubation tubes containing 0.5–1 m*M* GABA or 0.01–0.1 m*M* isoguvacine. Termination is by centrifugation, since filtration has been reported to result in a lower specific binding (Morin and Wasterlain, 1980). Under these conditions, specific binding has been reported to be about 50% at a concentration of 5 n*M* [^{3}H]-isoguvacine.

In the absence of Triton-X-100, the IC_{50} for GABA is reported to be rather high (690 n*M*), whereas with Triton-X-100-treated membranes, the values vary from 75 to 100 n*M* (Table 1). Untreated membranes exhibit two binding sites. In one study (Morin and Wasterlain, 1980), Triton-X-100 was reported to increase the binding capacity without an alteration in affinity. However, using Triton-X-100-treated membranes, White and Snodgrass (1983) could only demonstrate low-affinity (K_d, 76 n*M*) [^{3}H]-isoguvacine binding sites.

It is likely that at least some of these discrepancies are a result of the number of times the membranes are washed (i.e., a variable removal of GABA and other endogenous agents that interfere with binding), since the binding characteristics alter upon successive washes for at least six repetitions (Morin and Wasterlain, 1980).

2.3.4. [^{3}H]-Bicuculline

Of the ligands used to identify the $GABA_A$ recognition site, bicuculline is the only antagonist. Although this ligand has rather low affinity for displacing agonist ligands (see below), the possibility that the $GABA_A$ site exists in both agonist and antagonist states cannot be excluded. Thus, bicuculline is of at least theoretical interest as a ligand to assess this hypothesis.

For [^{3}H]-bicuculline binding, the membranes are prepared in the same manner as for [^{3}H]-GABA binding (Zukin et al., 1974; Greenlee et al., 1978). Triton-X-100 is not used in the preparation, since this detergent eliminates [^{3}H]-bicuculline binding (Möhler and Okada, 1978; Möhler, 1979; Collins and Cryer, 1978). In addition, it has been reported that freezing the tissue reduces the specific binding (Olsen and Snowman, 1983).

For the actual binding, three different forms of [^{3}H]-bicuculline have been reported—the methobromide (Collins and Cryer, 1978), the methoiodide (Möhler, 1979), and the methochloride (Olsen and Snowman, 1983); this may be a source of variation in the binding parameters.

As for agonist binding, the medium should be sodium-free, using Tris-citrate (pH 7.1), Tris-HCl (pH 7.4), or potassium phosphate (pH 7.5) buffer (50 n*M*). The specific activity is enhanced by using a chaotropic agent, usually SCN^- or ClO_4^-, the latter being more efficient (Möhler, 1979). The incubation temperature may be either 4 or 20°C, and the incubation time varies from 10 to 15 min, without altering the results. The ligand concentration is 2–5 n*M* and a protein concentration of about 2 mg/2 mL assay is desirable. Nonspecific binding is accounted for by the presence of either unlabeled GABA (0.1–1 m*M*) or bicuculline (0.1 m*M*). The use of GABA would seem to be preferable, since bicuculline may have displaceable recognition sites other than the $GABA_A$ receptor (e.g., acetylcholinesterase).

Termination of the assay is by centrifugation (8000–50,000*g* for 10 min). The pellet is superficially rinsed, then solubilized, acidified, and assayed for radioactivity in the presence of liquid scintillation fluid.

If a chaotropic agent is not used in the assay, specific binding is 16–20%; the use of SCN^- results in 30–50% specific binding (the higher value occurs when fresh tissue is used); ClO_4^- is reported to result in 60% specific binding (85–90% specific binding for [^{3}H]-GABA under similar conditions) (Möhler and Okada, 1978; Collins and Cryer, 1978; Möhler, 1979; Olsen and Snowman, 1983).

Using these conditions (no Triton, presence of a chaotropic agent) one or two binding sites has been described (Table 1). For the single binding site, the slope of the Hill plot was 0.7, indicating the possible presence of more than one site (Olsen and Snowman, 1983). The IC_{50} for GABA is much higher than when either [^{3}H]-GABA or [^{3}H]-muscimol are used as ligands, but is similar to that of [^{3}H]-isoguvacine in the absence of Triton-X-100 (Table 1).

2.4. In Vitro Pharmacological Characterization of the $GABA_A$ Recognition Site by Binding Techniques

The classical pharmacological definition of $GABA_A$ receptor function has been a bicuculline-dependent alteration in chloride ion flux. This is a technically sophisticated and time-consuming task that is not available to all laboratories wishing to investigate $GABA_A$ receptors. Although neurophysiological testing remains the fundamental criterion of synaptic pharmacology, binding techniques have provided a tool that allows all laboratories to investigate $GABA_A$ recognition sites. With respect to the pharmacological characterization of the $GABA_A$ site, the binding technique has both advantages and disadvantages compared to the neurophysiological technique.

The major disadvantage of the binding technique is, at least for $GABA_A$ recognition sites, that it is difficult to differentiate between an agonist and an antagonist; both displace the [^{3}H]-ligand. By comparing the effects of an unknown compound on agonist (e.g., GABA) vs antagonist (bicuculline) binding, a strong indication may be obtained. However, in addition, such a hypothesis should be verified by neurophysiological means.

The advantage of the binding techniques (in addition to rapidity and relative simplicity) is the qualitative aspect, although it must be remembered that all binding sites identified are probably not physiologically relevant receptors. This allows an estimation of IC_{50} values and thus an accurate comparison between different agonists or antagonists. Furthermore, an estimation of an increase (up-regulation) or decrease (down-regulation) of receptor number and affinity after pharmacological manipulations or in pathological conditions is possible using binding techniques.

The pharmacological profile of the $GABA_A$ recognition site in both the rodent and human brain is well defined by the affinities (reflected by the IC_{50} for the displacement of the radioligand) of different agonists and antagonists (most of which have been identified also by neurophysiology) (Krogsgaard-Larsen et al., 1979; Johnston, 1976) for different ligands under different conditions.

Table 2 provides selected information on the most routinely used displacers and ligands studied for the rodent brain. Only GABA itself and muscimol have been tested for all ligands in the different testing conditions. Other displacing agents that have been widely utilized include 3-aminopropane sulfonic acid, imidazoleacetic acid, THIP, isoguvacine, and bicuculline. It is of note that the antagonist bicuculline displaces radioligands for the $GABA_A$ site, and that its affinity increases considerably when either [^{3}H]-bicuculline is used as the ligand or else when a chatropic agent is present for [^{3}H]-GABA binding. Antagonists are the most sensitive displacers of [^{3}H]-bicuculline binding (Möhler, 1979; Möhler and Okada, 1978; Olsen and Snowman, 1983).

A much more limited set of data is available for the pharmacological characterization of the $GABA_A$ site in the human brain (Table 3). As can be seen using [^{3}H]-GABA as the ligand, the pharmacology of the $GABA_A$ recognition site is parallel in rodent and human brains. The IC_{50} values for the major displacers (e.g., GABA, muscimol, 3-aminopropanesulfonic acid, isoguvacine) fall within the same ranges and the rank ordering of the displacing agents is very similar.

In addition to the displacement of $GABA_A$ binding by pharmacological agents, there are endogenous agents other than GABA itself that decrease GABA binding (termed GABArins, by Johnston and Skerritt, 1984). Among these, homocarnosine (cf. Table 3) will interfere with the displacement assay for GABA in tissue extracts or fluids (e.g., CSF, blood) unless steps are taken to remove it. Phospholipids also decrease GABA binding; phosphatidylethanolamine is the most active in this regard (Johnston and Kennedy, 1978). The removal of these phospholipids is likely the mechanism of the activation of $GABA_A$ binding by phospholipase C (see below).

Another major endogenous inhibitor of GABA binding is GABA-modulin, a 16-kdalton peptide that inhibits GABA binding noncompetitively with an affinity of 0.5–1 μM (cf., Toffano et al., 1978; Toffano, 1983; Guidotti et al., 1984, for further details). It is proposed that diazepam enhances GABA binding by uncoupling GABA-modulin from the GABA recognition site.

As indicated above, $GABA_A$ binding can be enhanced as well as displaced. The first instance of increased $GABA_A$ binding in vitro was observed for diazepam (Guidotti et al., 1978) and resulted in a 20–30% increase at concentrations of diazepam ranging from 1–20 μ*M* (Guidotti et al., 1978, 1984; Lloyd and Dreksler, 1979; Toffano, 1983; Johnston and Skerritt, 1984). This appears to be caused by a change from a low- to a high-affinity state of the receptor, and is reflected by an enhanced neurophysiological response in vivo (cf., Haefely et al., 1981). Although [^{3}H]-GABA and [^{3}H]-muscimol binding are both enhanced by diazepam, [^{3}H]-THIP binding is not (Olsen and Snowman, 1982). Other benzodiazepines also enhance [^{3}H]-GABA binding (Guidotti et al., 1978; Johnston and Skerritt, 1984).

The increase of binding following Triton-X-100 treatment of the membranes has already been extensively discussed. This is likely caused by several factors: removal of endogenous GABA; removal of phospolipids; removal of GABA-modulin.

Phospholipase C has a marked effect on $GABA_A$ binding, resulting in a decrease in K_i or IC_{50} (Giambalvo and Rosenberg, 1976; Lloyd and Davidson, 1979; Lloyd and Beaumont, 1980; Toffano, 1983). In tissue previously treated with Triton-X-100, phospholipase C no longer enhances GABA binding (Lloyd and Davidson, 1979; Lloyd and Beaumont, 1980). Phospholipases "A" or "D" do not enhance $GABA_A$ binding (Enna and Snyder, 1977).

Other compounds that increase GABA binding include choatropic agents (Enna and Snyder, 1977; Möhler, 1979; Olsen and Snowman, 1983), barbiturates (Olsen and Snowman, 1982), and S-adenosylmethonine (Di Perri et al., 1983).

2.5. In Vivo Binding Studies of the $GABA_A$ Receptor

Administration of [^{3}H]-muscimol in vivo followed by ex vivo analysis of different brain regions for saturable or displacable binding has been reported (Snodgrass, 1978; Guidotti and Ferrero, 1985) (Table 4). Both studies used similar doses of muscimol, and despite differences in species and methodology, the results are similar. Thus, [^{3}H]-muscimol binding in vivo is displaceable and the distribution in different brain areas parallels that observed for in vitro binding. In addition, Guidotti and Ferrero report that the K_d for muscimol in vivo is 2–3 n*M*, that the binding is displaced by other $GABA_A$ agonists such as THIP and progabide, and that [^{3}H]-muscimol binding in vivo is enhanced by diazepam. The displacement of in vivo [^{3}H]-muscimol binding by progabide or muscimol parallels their anticonvulsant activity against the onset of isoniazed-induced seizures.

The major points of contention between the two reports is the muscimol remaining in the brain at the time of sacrifice, and the time at which sacrifice should occur. However, these two sets of

TABLE 2
Pharmacological Definition of the $GABA_A$ Receptor: Rodent Brain

	[³H]-GABA			[³H]-Muscimol		[³H]-Bicuculline	[³H]-Isoguvacine		[³H]-THIP	[³H]-P4S
Displacer, IC_{50}, nM	No triton[a–h,m,w]	Triton-X-100[d–k,m,w,y]	SCN ClO_4[l,m]	No triton[n,o]	Triton-X-100[j,k,n,p]	SCN ClO_4[l,m,q]	No triton[r]	Triton-X-100[j,s,t]	No triton[u]	No triton[v]
GABA	200–370	12–60	29–40	38,200	8, 27	430–600	690	27–100	14	30
Muscimol	24–50	3–9	7	10, 20	5–8	58–100	58	5–32	5	4
3-Aminopropane-sulfonic acid	70–250	4–80	50	30, 100	5	500–2000	—	—	—	40
Imidazoleacetic acid	70–300	4–80	350	30, 100	5	—				200
Piperidine-4-sulfonic acid	90	34	—	—	—	1000	—	20	18	50
trans-4-Aminocrotonic acid	150–430	1	—	51	7	2000	—	—	—	60
Isoguvacine	150–1400	37–120	70	250	—	1000	51	14, 41	10	30
THIP	630–1300	80–110	—	6300	—	—	—	51	20	200
Isonipecotic acid	—	330	—	—	—	—	—	—	110	—
β-OH-GABA	—	700	—	—	—	—	—	—	—	—
SL 75102	—	860–1400	—	—	900	—	—	500	—	—
5-Aminovaleric acid	5000	—	—	—	—	—	—	—	—	—

(+)Bicuculline	2000 –4000	5000 –10,000	110	6000	2000	10–70	—	—	—	2000
(±)Bicuculline	—	10,000	—	—	2000	—	—	36,000	—	—
(±)Bicuculline methiodide	—	—	570	—	—	30–222	—	97,000	—	—
N-Me-Muscimol	—	4200	—	—	—	—	—	—	—	—
GABAmide	—	17,000	—	—	—	—	—	—	—	—
Progabide	—	34,000 –35,000	—	—	8000	—	—	8000	—	—
Taurine	44,000	—	—	—	—	—	—	—	—	—
N-Me-GABA	—	67,000	—	—	—	—	—	—	—	—
d-Tubocurarine	85,000	—	—	—	—	—	—	—	—	—
Baclofen	$>10^6$	55,000–100,000	—	—	—	—	—	—	—	—

[a]From Olsen et al., 1979.
[b]From Olsen et al., 1978.
[c]From Greenlee et al., 1978.
[d]From Johnston and Kennedy, 1978.
[e]From Enna and Snyder, 1975.
[f]From Enna and Snyder, 1977.
[g]From Krogsgaard-Larsen et al., 1983.
[h]From Bartholini et al., 1979.
[i]From Horng and Wong, 1979.
[j]From Lloyd et al., 1982.
[k]From Enna et al., 1978.
[l]From Möhler, 1979.
[m]From Olsen and Snowman, 1983.
[n]From Williams and Risley, 1979.
[o]From Wang et al., 1979.
[p]From Beaumont et al., 1978.
[q]From Möhler and Okada, 1978.
[r]From Collins et al., 1980.
[s]From Krogsgaard-Larsen and Arnt, 1980.
[t]From Morin and Wasterlain, 1980.
[u]From Falch and Krogsgaard-Larsen, 1982.
[v]From Krogsgaard-Larsen et al., 1981.
[w]From Olsen et al., 1978.
[x]From Krogsgaard-Larsen et al., 1979.
[y]From Bowery et al., 1982.

TABLE 3
Pharmacological Definition of the $GABA_A$ Receptor in Human Cerebellum[a]

Displacer, IC_{50}, n*M*	[^{3}H]-GABA Binding: No Triton	[^{3}H]-GABA Binding: Triton-X-100
GABA	70–250	14
Muscimol	14–45	3
3-Aminopropane sulfonic acid	83	40
Imidazoleacetic acid	500–1420	90
Isoguvacine	650	35
THIP	1300	80
Piperidine-4-sulfonic acid	—	13,500
β-OH-GABA	2921	—
SL 75102	800–830	—
5-aminovaleric acid	2730	—
(+)-Bicuculline	—	4000
(±)-Bicuculline	17,700 –18,000	—
Progabide	39,000	—
Homocarnosine	43,300	—
Baclofen	$>10^6$	360,000

[a]From Lloyd et al., 1982; Iversen et al., 1979; Enna et al., 1979; Lloyd and Dreksler, 1979; Bartholini et al., 1979.

TABLE 4
In Vivo Binding to $GABA_A$ Receptors Using [^{3}H]-Muscimol as Ligand

Parameter	Snodgrass, 1978	Guidotti and Ferrero, 1985
Species	Rat	Mouse
Dose of [^{3}H]-muscimol	2.5 μg/kg, iv	8.5 nmol/kg, iv
Time of sacrifice	45 min	10 min
Displacer for "specific" binding"	Bicuculline	Muscimol
Specific binding	35–57%	35–40%
Muscimol as % of brain radioactivity	>90%	10–20%

preliminary data indicate that after methodological improvements, the in vivo binding pharmacology of $GABA_A$ receptors may be feasible and practical.

3. $GABA_B$ Receptor Binding

A second sodium-independent membrane recognition site for GABA has been recently identified, and, is associated with a physiological, biochemical, and neuropharmacological spectrum of activity that is different from that of $GABA_A$ receptors. The recognition site has been termed the $GABA_B$ receptor by Hill and Bowery (1981), and this terminology has been generally accepted.

The prototypic agonist for $GABA_B$ sites is (–)-baclofen, as well as GABA itself. The $GABA_B$ site is bicuculline-insensitive and is not associated with a chloride ionophore, but rather with a calcium channel. In this regard, $GABA_B$ receptor binding is Ca^{2+}-dependent (Hill and Bowery, 1981; Bowery et al., 1984). In terms of second-messenger effects, $GABA_B$ receptor stimulation *per se* does not increase cyclic nucleotide production, and under certain conditions may inhibit adenylate cyclase activity (Wojcïk and Neff, 1984). However, $GABA_B$ agonists greatly potentiate cAMP production as a result of the activation of different neurotransmitter receptors, e.g., noradrenaline, dopamine (Hill, 1985; Enna and Karbon, 1984; Karbon et al., 1984). In contrast, $GABA_A$ agonists have a direct effect, reducing cerebellar cGMP levels (Mao et al., 1975).

One of the present problems associated with the study of $GABA_B$ receptor function and classification is the lack of a specific antagonist. Two antagonists have been described, 5-aminovaleric acid (Muhyaddin et al., 1982), and homotaurine (Giotti et al., 1983). However, both are $GABA_A$ receptor agonists (*see* above and Luzzi et al., 1985) with similar affinities for both $GABA_A$ and $GABA_B$ sites. This lack of specificity considerably reduces the interest in these compounds for neurochemical and neuropharmacological studies.

Although the availability of a specific antagonist would be useful, the conditions for in vitro binding can be manipulated so that the specific binding of [^{3}H]-GABA to membranes is almost exclusively to $GABA_B$ sites, although baclofen- and bicuculline-insensitive $GABA_C$ sites have recently been proposed (Johnston and Allan, 1984).

3.1. Membrane Preparation for $GABA_B$ Binding

The literature is quite uniform with regard to the preparation of rat brain membranes for $GABA_B$ binding studies, using the procedure as originally described by Hill and Bowery (1981). This method is, in effect, very similar to the preparation of membranes for $GABA_A$ binding as described by Enna and Snyder (1975) and Zukin et al. (1974). Membranes may be prepared from either fresh synaptosomes, or from frozen and thawed material. It should be noted that if frozen brain material is to be used, membranes must be prepared within 72 h of freezing (at –80°C), since after this time a decrease in $GABA_B$ binding occurs. However, once the membranes are prepared, they may be frozen for lengthy periods of time.

Before use in the $GABA_B$ binding assay, the membranes must be washed much more thoroughly than for $GABA_A$ binding (Hill and Bowery, 1981; Bowery et al., 1983). After thawing, the pellet is incubated for 45 min at 25°C in Krebs-Henseleit solution or Tris-HCl buffer (50 n*M*, pH 7.4) containing 2.5 m*M* $CaCl_2$. The membranes are then centrifuged and washed four times in buffer with a resuspension and 15-min incubation between each wash. After the final wash, the pellet is resuspended in sufficient buffer to yield 0.6–1.25 mg protein/mL.

The exception to this procedure for membrane preparation is that described by Rago and Zarkovsky (1983). In this procedure, the washing and preparation of membranes was accomplished over 4 d using Tris-HCl buffer (50 m*M*,pH 7.4).

3.2. $GABA_B$ Binding Assay

Binding to the $GABA_B$ recognition site has been performed using [^{3}H]-GABA (in the presence of isoguvacine to saturate $GABA_A$ sites), [^{3}H]-(±)-baclofen, and [^{3}H]-(–)-baclofen. As will be seen below, the preferable ligand is probably [^{3}H]-GABA (with suitable precaution to block $GABA_A$ receptors), although not enough data with [^{3}H]-(–)-baclofen is yet available to be able to properly assess this ligand.

In their initial description of $GABA_B$ binding, Hill and Bowery (1981) described the use of two buffer systems, either Krebs-Henseleit solution or Tris-HCl (50 m*M*,pH 7.4, containing 2.5 m*M* $CaCl_2$). Most others have subsequently used the Tris-HCl-$CaCl_2$ buffer.

With the use of [^{3}H]-GABA (10 n*M*) as radioligand, the incubation mixture contains 40 μ*M* isoguvacine, which is inactive at $GABA_B$ receptors (see below), and at this concentration fully occupies $GABA_A$ receptors (see above). Nonspecific binding is estimated by the use of 100–1000 μ*M* GABA or (±)-baclofen in parallel incubation tubes. The membrane protein concentration is normally within 0.5–1.0 mg protein/mL. A lesser amount may not accumulate properly on centrifugation. The incubation normally lasts 10 min at 20°C (although 2–20 min incubation gives similar results) (Bowery et al., 1983). If a lower temperature is used, a longer incubation time is needed (e.g., 30 min at 4°C), whereas at 35°C, there is a marked decrease in binding (Bowery et al., 1983). Termination of the assay is by centrifugation (e.g., 8000–12,000*g* for 10 min). After a superficial rinse, the pellet is solubilized and assayed for radioactivity.

The only marked deviation from this procedure has been described by Rago and Zarkovsky (1983) who used 75 μ*M* bicuculline to block $GABA_A$ receptor binding and terminated the assay by means of filtration. Since the results using this procedure differ considerably from the standardized assay (see below), it is not certain that the same GABA recognition sites are analyzed under the different conditions (cf. also differences in membrane preparation, above).

When using [^{3}H]-(±)-baclofen (20 n*M*) (Hill and Bowery, 1981; Bowery et al., 1983) or [^{3}H]-(–)-baclofen (8–10 n*M*) (Drew et al., 1984) as the radioligand, all steps are the same as when using [^{3}H]-GABA as the radioligand. Drew et al. (1984) include 40 μ*M* isoguvacine in the incubation mixture, whereas Bowery and colleagues (1983) omit this $GABA_A$ agonist when using [^{3}H]-(±)-baclofen as the ligand.

Under these standardized conditions, "specific" $GABA_B$ binding is 35–45% of total binding using [^{3}H]-GABA as the ligand (Hill and Bowery, 1981; Bowery et al., 1983), 20–25% using [^{3}H]-(±)-baclofen as the ligand (Bowery et al., 1983; Hill and Bowery, 1981; Drew et al., 1984; Kato et al., 1983), and 30% using [^{3}H]-(–)-baclofen as the radioligand (Drew et al., 1984). It should be noted that Kato et al. (1983) report that if 2.5 m*M* $MnCl_2$ is used instead of $CaCl_2$, the specific binding increases from 24 to 51%. Other cations such as Mg^{2+} or Ni^{2+} also substitute for Ca^{2+} in the incubation medium (Bowery et al., 1983; Kato et al., 1983).

In contrast to $GABA_A$ binding, the use of Triton-X-100 in the binding assay, or during membrane preparation, results in a marked decrease (50%) of $GABA_B$ binding (Bowery et al., 1983).

Using the standard assay conditions, a reproducible set of kinetic parameters are obtained for $GABA_B$ binding sites (Table 5). Using [^{3}H]-GABA or [^{3}H]-(–)-baclofen as the ligand in Tris-HCl buffer with 2.5 m*M* $CaCl_2$, a single binding site is obtained with a K_d of 40–80 n*M* and a B_{max} of 0.65–1.27 pmol/mg protein. However, when [^{3}H]-(±)-baclofen is used as the ligand, the kinetic parameters vary according to the buffer. In Krebs-Henseleit solution, two

TABLE 5
Kinetic Parameters of $GABA_B$ Binding

Radioligand	Buffer[f]	Kinetic parameters: K_d, nM High	K_d, nM Low	B_{max}, pmol/mg prot High	B_{max}, pmol/mg prot Low
[^{3}H]-GABA	Tris-HCl $CaCl_2$	44[a] 77[b,c]	— —	1.27[a] 1.22[b,c]	— —
[^{3}H]-(±)-Baclofen	Krebs Henseleit	82[b,c] 313[d]	5300[b,c]	0.68[b,c] 2.35[d]	6.9[b,c]
	Tris-HCl $CaCl_2$	110[e] 149[b,c] 337[d]	—	1.17[b,c] 1.3[e] 2.38[d]	—
	Tris-HCl $CaCl_2$ NaCl	68[c]	6800[c]	0.65[c]	5.7[c]
	Tris-HCl $MnCl_2$	185[d]	—	2.14[d]	—
[^{3}H]-(–)-Baclofen	Tris-HCl $CaCl_2$	47[e]	—	0.65[e]	—

[a]From Pilc and Lloyd, 1984.
[b]From Hill and Bowery, 1981.
[c]From Bowery et al., 1983.
[d]From Kato et al., 1983.
[e]From Drew et al., 1984.
[f]The standardized binding assay is used on frozen-thawed membranes. NaCl was added at 143 nM.

binding sites are obtained, the high-affinity site being similar to K_d and B_{max} to that described above; the low-affinity site has a μM affinity and a very high capacity. It should be noted that in a Krebs-Henseleit-Hepes solution, Kato et al. (1983) obtained a single site for [^{3}H]-(±)-baclofen with a K_d and B_{max} intermediary to the two described above.

Using [^{3}H]-(±)-baclofen in a Tris-HCl-$CaCl_2$ buffer, a single binding site is described with a lower affinity, but similar capacity to that described for [^{3}H]-GABA or [^{3}H]-(–)-baclofen. Since the K_d is approximately twice that found for [^{3}H]-GABA or [^{3}H]-(–)-baclofen as ligands, it may be assumed that the K_d for the active (–)-isomer is the same as for the other radioligands under the same experimental conditions. Substitution of $MnCl_2$ or $CaCl_2$ apparently increases the affinity, but does not alter the B_{max}; however, it should be noted that in this series of experiments the K_d values obtained are of considerably lower affinity than that found by other groups (*see* Table 5).

The addition of NaCl to the Tris-HCl-$CaCl_2$ mixture (thus approaching the conditions found in the Krebs-Henseleit solution) apparently induces a second binding site. However, this site has a similar K_d to that of Na^+-dependent GABA binding to the GABA transport system (cf., Enna and Snyder, 1975, 1977), and thus may not represent a GABA receptor recognition site.

In the particular membrane and binding conditions described by Rago and Zarkovsky (1983), a completely different set of kinetic parameters is observed: a high-affinity K_d of 7–8 n*M* and a low-affinity K_d of 400–600 n*M*; B_{max} values of 0.024 and 0.036–0.109 pmol/mg protein were obtained for these sites. Since muscimol displaces [^{3}H]-GABA binding in this system (although weakly), it is not certain that a $GABA_B$ receptor population is present.

3.3. Pharmacological Characterization of $GABA_B$ Binding

Fewer studies have been performed on the $GABA_B$ site than on the $GABA_A$ site; however, the data available are sufficient to clearly distinguish the two receptors. Thus, the pharmacological profiles are completely different for the two sites. In contrast to the $GABA_A$ sites (Table 4), the ligands for the $GABA_B$ site (Table 6, [^{3}H]-baclofen; [^{3}H]-GABA in the presence of isoguvacine), are not displaced by THIP, imidazoleacetic acid, isoguvacine, and bicuculline. Baclofen potently displaces $GABA_B$ ligands in a sterospecific manner, but is inactive on $GABA_A$ receptor binding.

Some compounds are active at both $GABA_A$ and $GABA_B$ sites, notably SL 75102 (agonist at both sites), 3-aminovaleric acid, homotaurine ($GABA_A$ agonists, $GABA_B$ antagonists), 3-aminopropanesulfonic acid, and muscimol. The latter two compounds are relatively weak at the $GABA_B$ sites, whereas they are potent $GABA_A$ agonists.

In contrast to the $GABA_A$ site, $GABA_B$ binding is sensitive to the presence of heavy metals, e.g., mercury, cadmium, lead, or zinc (93–62% inhibition at 100 μ*M*) (Drew et al., 1984), and also by guanyl nucleotides (Hill et al., 1984).

4. Differentiation of GABA Receptor Binding Sites

Although the quantity of data is limited, it is sufficient to differentiate $GABA_A$ from $GABA_B$ receptors, as well as from GABA

TABLE 6
Pharmacological Definition of the $GABA_B$ Receptor

Displacer	IC_{50}, n*M* [³H]-Baclofen[b]	[³H]-GABA[c]
GABA	40	80–90
(–)-Baclofen	40	130
(±)-Baclofen	180	130–900
SL 75102	400	500–2000
GTP	1000	15,700
β-OH GABA	1600	1100
β-Cl GABA	11,400	4600
3-Aminopropane sulfonic acid	11,000	10,000
Muscimol	12,000	5400
GDP	24,000	48,000
(+)-Baclofen	33,000	74,000
5-Aminovaleric Acid	50,000	30,000
THIP	$>10^5$	$>10^5$
Progabide	$>10^5$	$>10^5$
Isoguvacine	$>10^6$	$>10^5$
P4S	$>10^5$	$>10^5$
Imidazoacetic acid	$>10^6$	—
Bicuculline MeBr	$>10^6$	$>10^5$

[a]From Hill and Bowery, 1981; Bowery et al., 1982, 1983; Hill et al., 1984; Lloyd et al., 1982; Lloyd, unpublished.
[b]Krebs-Henseleit solution.
[c]Tris-HCl buffer with 40 μ*M* isoguvacine.

uptake sites or other neurotransmitter or hormonal receptors. Thus, $GABA_A$ receptors are sodium-independent, bicuculline-sensitive sites linked to chloride-ion channels, with an order of displacement with [³H]-GABA as ligand equal to muscimol > 3-aminopropane sulfonic acid > imidazoleacetic acid > isoguvacine > GABA > THIP > SL 75102 >> baclofen. $GABA_B$ receptors are bicuculline-insensitive, calcium-dependent sites, and the order of displacement of [³H]-GABA at the $GABA_B$ receptor is GABA > baclofen = SL 75102 >> 3-aminopropane sulfonic acid >> THIP, isoguvacine, imidazoleacetic acid. Inhibitors of GABA uptake, such as nipecotic acid, beta-alanine, β-aminobutyric acid, or chlorpromazine (Iversen, 1978) are inactive on both $GABA_A$ and $GABA_B$ binding sites.

Thus, in vitro it is relatively easy to differentiate between $GABA_A$ and $GABA_B$ binding sites. Either specific [^{3}H]-ligands can be used (such as [^{3}H]-muscimol for $GABA_A$ and (–)-^{3}H-baclofen for $GABA_B$) or [^{3}H]-GABA itself can be used with appropriate incubation conditions. The latter approach is preferred by the present author since the sites identified are, by definition, GABA recognition sites, whereas with the other ligands the doubt remains that only certain subsites are identified and/or nonGABAergic sites contribute to the binding.

Furthermore, using these binding techniques it has been possible to identify not only specific ligands for each site, but also compounds that have an activity at both $GABA_A$ and $GABA_B$ sites, in effect mimicking the receptor activation of the natural neurotransmitter.

5. Conclusions

The [^{3}H]-ligand binding technique for GABA receptors has proved most effective for identifying compounds that have an affinity for $GABA_A$ and $GABA_B$ receptors without an effect at other recognition sites in the GABA synapse (e.g., uptake, synthesis, metabolism) or at other neurotransmitter receptors. Because the natural neurotransmitter can be used as a ligand, it is highly probable that the sites identified have a physiological, as well as pharmacological, significance. In this manner, not only can new GABA agonists and antagonists be identified, but up-regulation, down-regulation, and interactions between different membrane components can be identified.

References

Andersen P., Dingledine R., Gjerstad L., Langmoen J. A., and Mosfeldt Laursen A. M. (1980) Two different responses of hippocampal pyramidal cells to application of gamma-aminobutyric acid. *J. Physiol.* (Lond.) **305,** 279–296.

Bartholini G., Scatton B., Zivkovic B., and Lloyd K. G. (1979) Pharmacological Profile of SL 76002, A New GABAmimetic Drug, in *GABA-Neurotransmitters* (Krogsgaard-Larsen P., Scheel-Kruger J., and Kofod H., eds.), Munksgaard, Copenhagen.

Beaumont K., Chilton W. S., Yamamura H. I., and Enna S. J. (1978) Muscimol binding in rat brain: Association with synaptic GABA receptors. *Brain Res.* **148,** 153–162.

Bowery N. G., Collins J. F., Cryer G., Inch T. D., and McLaughlin N. J. (1979) The GABA Receptor: Stereospecificity and Structure–Activity Studies, in *GABA-Biochemistry and CNS Functions* (Mandel P. and DeFeudis F. V., eds.), Plenum, New York.

Bowery N., Hill D. R., and Hudson A. L. (1982) Evidence that SL 75102 is an agonist at $GABA_B$ as well as $GABA_A$ receptors. *Neuropharmacology* **21,** 391–395.

Bowery N., Hill D. R., and Hudson A. L. (1983) Characteristics of $GABA_B$ receptor binding sites on rat whole brain synaptic membranes. *Brit. J. Pharmacol.* **78,** 191–206.

Bowery N. G., Price G. W., Hudson A. L., Hill D. R., Wilkin G. P., and Turnbull M. J. (1984) Visualization of different receptor types in the mammalian CNS. *Neuropharmacology* **23-2B,** 219–331.

Brehm L., Krogsgaard-Larsen P., and Jacobsen P. (1979) GABA Uptake Inhibitors and Structurally Related "Pro-Drugs," in *GABA-Neurotransmitters* (Krogsgaard-Larsen P., Scheel-Kruger J., and Kofod H., eds.), Munksgaard, Copenhagen.

Collins J. F. and Cryer G. (1978) A Study of the GABA Receptor Using [^{3}H]-Bicuculline Methobromide, in *Amino Acids as Chemical Transmitters* (Fonnum F., ed.), Plenum, New York.

Collins J. F., McDonald J. A., and Newton R. F. (1980) Characterisation of the binding of [^{3}H]-isoguvacine, a γ-aminobutyric acid agonist, to brain synaptosomal membranes. *Brain Res. Bull.* **5,** Suppl. 2, 141–144.

Curtis D. R. (1979) GABAergic Transmission in the Mammalian Central Nervous System, in *GABA-Neurotransmitters* (Krogsgaard-Larsen P., Scheel-Kruger J., and Kofod H., eds.), Munksgaard, Copenhagen.

DeFeudis F. (1973) Sodium dependency of γ-aminobutyric acid binding to particulate fractions of mouse brain. *Exp. Neurol.* **41,** 54–62.

DeFeudis B. (1981) Muscimol binding and GABA receptors. *Drug Dev. Res.* **1,** 93–105.

Desarmenien M., Headley P. M., Santangelo F., and Feltz P. (1980) The effects of various GABA-mimetics on dorsal root ganglion neurons in vitro: A physiological analysis. *Brain Res. Bull.* **5,** Suppl. 2, 471–475.

Di Perri B., Calderini G., Battistella A., Raciti R., and Toffano G. (1983) Phospholipid methylation increases [^{3}H] diazepam and [^{3}H] GABA binding in membrane preparation of rat cerebellum. *J. Neurochem.* **41,** 302–308.

Drew C. A., Johnston G. A. R., and Weatherby R. P. (1984) Bicuculline-insensitive GABA receptors: Studies on the binding of (–) baclofen to rat cerebellar membranes. *Neurosci. Lett.* **52,** 317–321.

Enna S. J. and Karbon E. W. (1984) $GABA_B$ receptors and transmitter-stimulated cAMP accumulation in rat brain. *Neuropharmacology* **23-7B,** 821–822.

Enna S. J. and Snyder S. H. (1975) Properties of γ-aminobutyric acid (GABA) receptor binding in rat brain synaptic membrane fractions. *Brain Res.* **100,** 81–97.

Enna S. J. and Snyder S. H. (1977) Influences of ions, enzymes, and detergents on γ-aminobutyric acid-receptor binding in synaptic membranes of rat brain. *Mol. Pharmacol.* **13,** 442–453.

Enna S. J., Beaumont K., and Yamamura H. J. (1978) Comparison of [^{3}H]-Muscimol and [^{3}H]-GABA Binding in Rat Brain, in *Amino Acids as Chemical Transmitters* (Fonnum F., ed.), Plenum, New York.

Enna S. J., Ferkany J. W., and Krogsgaard-Larsen P. (1979) Pharmacological Characteristics of GABA Receptors in Different Brain Regions, in *GABA-Neurotransmitters* (Krogsgaard-Larsen P., Scheel-Kruger J., and Kofod H., eds.), Munksgaard, Copenhagen.

Falch E. and Krogsgaard-Larsen P. (1982) The binding of the GABA agonist [^{3}H] THIP to rat brain synaptic membranes. *J. Neurochem.* **38,** 1123–1129.

Gallagher J. P., Nakamura J., and Shinnick-Gallagher P. (1983) Effects of glial uptake and desensitization on the activity of γ-aminobutyric acid (GABA) and its analogs at the cat dorsal root ganglion. *J. Pharmacol. Exp. Ther.* **226,** 876–884.

Giambalvo C. T. and Rosenberg P. (1976) The effect of phospholipases and proteases on the binding of γ-aminobutyric acid to functional complexes of rat cerebellum. *Biochem. Biophys. Acta* **436,** 741–756.

Giotti A., Luzzi S., Spagnesi S., and Zilletti L. (1983) Homotaurine: A $GABA_B$ antagonist in the guinea pig ileum. *Brit. J. Pharmacol.* **79,** 855–862.

Gottesfeld Z. and Elliott K. A. C. (1971) Factors that affect the binding and uptake of GABA by brain tissue. *J. Neurochem.* **18,** 683–690.

Greenlee D. V., Van Ness P. C., and Olsen R. W. (1978) Endogenous inhibitor of GABA binding in mammalian brain. *Life Sci.* **22,** 1653–1662.

Guidotti A. and Ferrero P. (1985) Ex Vivo Binding of ^{3}H-Muscimol to $GABA_A$ Recognition Sites: A Tool to Characterize GABA Receptor Agonists, in *Epilepsy and GABA Receptor Agonists* (Bartholini G., Bossi L., Lloyd K. G., and Morselli P. L., eds.), Raven, New York.

Guidotti A., Toffano G., Grandison L., and Costa E. (1978) Second Messenger Responses and Regulation of High Affinity Receptor Binding to Study Pharmacological Modifications of GABAergic Transmission, in *Amino Acids as Chemical Transmitters* (Fonnum F., ed.), Plenum, New York.

Guidotti A., Corda M. G., Vaccarino F. M., and Wise B. C. (1984) Role of GABA-Modulin and of an Endogenous Effect or of Beta-Carboline Binding Sites in the GABA-Benzodiazepine Receptor Interaction, in *Actions and Interactions of GABA and Benzodiazepines* (Bowery N. G., ed.), Raven, New York.

Haefely W., Pieri L., Polc P., and Schaffner R. (1981) General Pharmacology and Neuropharmacology of Benzodiazepine Derivatives, in *Handbook of Experimental Pharmacology 55/II* (Hoffmeister F. and Stille G., eds.), Springer Verlag, Berlin.

Hill D. R. (1985) $GABA_B$ receptor modulation of adenylate cyclase activity in rat brain slices. *Brit. J. Pharmacol.* **84,** 249–257.

Hill D. R. and Bowery N. G. (1981) ^{3}H-Baclofen and ^{3}H-GABA bind to bicuculline—insensitive $GABA_B$ sites in rat brain. *Nature* (Lond.) **290,** 149–152.

Hill D. R., Bowery N. G., and Hudson A. L. (1984) Inhibition of $GABA_B$ receptor binding by guanyl nucleotides. *J. Neurochem.* **42,** 652–657.

Horng J. S. and Wong D. T. (1979) γ-Aminobutyric acid receptors in cerebellar membranes of rat brain after a treatment with Triton-X-100. *J. Neurochem.* **32,** 1379–1386.

Iversen L. L. (1978) Biochemical Psychopharmacology of GABA, in *Psychopharmacology: A Generation of Progress* (Lipton M. A., DiMascio A., and Killam K. F., eds.), Raven, New York.

Iversen L. L., Bird E., Spokes E., Nicholson S. H., and Suckling C. J. (1979) Agonist Specificity of GABA Binding Sites in Human Brain and GABA in Huntington's Disease and Schizophrenia, in *GABA Neurotransmitters* (Krogsgaard-Larsen P., Scheel-Kruger J., and Kofod H., eds.), Munksgaard, Copenhagen.

Johnston G. A. R. (1976) Physiologic Pharmacology of GABA and Its Antagonists in the Vertebrate Nervous System, in *GABA in Nervous System Function* (Roberts E., Chase T. N., and Tower D. B., eds.), Raven, New York.

Johnston G. A. R. and Allan R. D. (1984) GABA agonists. *Neuropharmacology* **23-7B,** 831–832.

Johnston G. A. R. and Kennedy S. M. (1978) GABA Receptors and Phospholipids, in *Amino Acids as Chemical Transmitters* (Fonnum F., ed.), Plenum, New York.

Johnston G. A. R. and Skerritt J. H. (1984) GABArins and the Nexus Between GABA and Benzodiazepine Receptors, in *Actions and Interactions of GABA and Benzodiazepines* (Bowery N. G., ed.), Raven, New York.

Johnston G. A. R., Allan R. D., Kennedy S. M., and Twitchin B. (1979) Systematic Study of GABA Analogues of Restricted Conformation, in *GABA-Neurotransmitters* (Krogsgaard-Larsen P., Scheel-Kruger J., and Kofod H., eds.), Munksgaard, Copenhagen.

Karbon E. W., Duman R. S., and Enna S. J. (1984) $GABA_B$ receptors and norepinephrine-stimulated cAMP production in rat brain cortex. *Brain Res.* **306,** 327–332.

Kato K., Goto M., and Fukuda H. (1983) Regulation by divalent cations of ^{3}H-baclofen binding to $GABA_B$ sites in rat cerebellar membranes. *Life Sci.* **32,** 879–887.

Krogsgaard-Larsen P. and Arnt J. (1980) Pharmacological studies of interactions between benzodiazepines and GABA receptors. *Brain Res. Bull.* **5,** Suppl. 2, 867–872.

Krogsgaard-Larsen P., Johnston G. A. R., Curtis D. R., Game C. J. A., and

McCulloch R. M. (1975) Structure and biological activity of a series of conformationally restricted analogues of GABA. *J. Neurochem.* **25,** 797–802.

Krogsgaard-Larsen P., Honore T., and Thyssen K. (1979) GABA Receptor Agonists: Design and Structure–activity Studies, in *GABA-Neurotransmitters* (Krogsgaard-Larsen P., Scheel-Kruger J., and Kofod H., eds.), Munksgaard, Copenhagen.

Krogsgaard-Larsen P., Snowman A., Leemis S. C., and Olsen R. W. (1981) Characterization of the binding of the GABA agonist [^{3}H]-piperidine-4-sulphonic acid to bovine brain synaptic membranes. *J. Neurochem.* **37,** 401–409.

Krogsgaard-Larsen P., Falch E., Peet M. J., Leach J. D., and Curtis D. R. (1983) Molecular Pharmacology of the GABA Receptors and GABA Agonists, in *CNS Receptors—From Molecular Pharmacology to Behaviour* (Mandel P. and De Feudis F. V., eds.), Raven, New York.

Krogsgaard-Larsen P., Falch E., and Jacobsen P. (1984) GABA Agonists: Structural Requirements for Interaction With the GABA-Benzodiazepine Receptor Complex, in *Actions and Interactions of GABA and Benzodiazepines* (Bowery N. G., ed.), Raven, New York.

Leach M. J. and Wilson J. A. (1978) GABA receptor binding with [^{3}H]-muscimol in calf cerebellum. *Eur. J. Pharmacol.* **48,** 329–330.

Lester B. R. and Peck E. J. (1979) Kinetic and pharmacologic characterization of gamma-aminobutyric acid receptive sites from mammalian brain. *Brain Res.* **161,** 79–97.

Lloyd K. G. and Beaumont K. (1980) Possible role of phospholipids in GABA receptor function in human and rat brain. *Brain Res. Bull.* **5,** Suppl. 2, 285–290.

Lloyd K. G. and Davidson L. (1979) ^{3}H-GABA binding in brains from Huntington's chorea patients: Altered regulation by phospholipids? *Science* **205,** 1147–1149.

Lloyd K. G. and Dreksler S. (1979) An analysis of ^{3}H-gamma-aminobutyric acid (GABA) binding in the human brain. *Brain Res.* **163,** 77–87.

Lloyd K. G., Dreksler S., and Bird E. (1977a) Alterations in ^{3}H-GABA binding in Huntington's chorea. *Life Sci.* **21,** 747–754.

Lloyd K. G., Shemen L., and Hornykiewicz O. (1977b) Distribution of high affinity sodium-independent (^{3}H)gamma-aminobutyric acid [^{3}(H)-GABA] binding in the human brain: Alterations in Parkinson's disease. *Brain Res.* **127,** 269–278.

Lloyd K. G., Arbilla S., Beaumont K., Briley M., de Montis G., Scatton B., Langer S. Z., and Bartholini G. (1982) γ-Aminobutyric acid (GABA) receptor stimulation. II. Specificity of progabide (SL 76002) and SL 75102 for the GABA receptor. *J. Pharmacol. Exp. Ther.* **220,** 672–677.

Lloyd K. G., Munari C., Worms P., Bossi L., and Morselli P. L. (1983) Indications for the Use of Gamma-Aminobutyric Acid (GABA) Agonists in Convulsant Disorders, in *Epilepsy: An Update on Research and Therapy* (Nistico G., Di Perri R., and Meinardi H., eds.), Alan Liss, New York.

Luzzi S., Maggi C. A., Spagnesi S., Santicoli P., and Zilletti L. (1985) 5-Aminovaleric acid interactions with $GABA_A$ and $GABA_B$ receptors in guinea pig ileum. *J. Auton. Pharmacol.* **5,** 65–69.

Madtes P. (1984) Chloride ions preferentially mask high-affinity GABA binding sites. *J. Neurochem.* **43,** 1434–1437.

Mao C. C., Guidotti A., and Costa E. (1975) Interactions between γ-aminobutyric acid and guanosine cyclic 3',5'-monophosphate in rat cerebellum. *Mol. Pharmacol.* **10,** 736–745.

Maurer R. (1979) [^{3}H]-Muscimol Binding to Rat Cerebellar Membranes, in *GABA-Biochemistry and CNS Functions* (Mandel P. and De Feudis F. V., eds.), Plenum, New York.

Möhler H. (1979) GABA Receptor Binding With [^{3}H]-GABA and [^{3}H]-(+)-Bicuculline Methiodide: An Improved Method, in *GABA-Biochemistry and CNS Functions* (Mandel P. and De Feudis F. V., eds.), Plenum, New York.

Möhler H. and Okada T. (1978) GABA Receptor in Rat Brain: Demonstration of an Antagonist Binding Site, in *Amino Acids as Chemical Transmitters* (Fonnum F., ed.), Plenum, New York.

Möhler H., Richards J. G., and Wu J. Y. (1981) Autoradiographic localization of benzodiazepine receptors in immunocytochemically identified γ-aminobutyrergic synapses. *Proc. Natl. Acad. Sci. USA* **78,** 1935–1938.

Morin A. M. and Wasterlain C. G. (1980) The binding of ^{3}H-isoguvacine to mouse brain synaptic membranes. *Life Sci.* **26,** 1239–1245.

Muhyaddin M., Roberts P. J., and Woodruff G. N. (1982) Presynaptic GABA receptors in the rat anococygelus muscle and their antagonism by 5-aminovaleric acid. *Brit. J. Pharmacol.* **77,** 163–168.

Olsen R. W. and Snowman A. M. (1982) Chloride-dependent enhancement by barbiturates of γ-aminobutyric acid receptor binding. *J. Neurosci.* **2,** 1812–1823.

Olsen R. W. and Snowman A. M. (1983) [^{3}H]-Bicuculline methochloride binding to low-affinity γ-aminobutyric acid receptor sites. *J. Neurochem.* **41,** 1653–1663.

Olsen R. W., Ticku M. K., Van Ness P. C., and Greenlee D. (1978) Effects of drugs on γ-aminobutyric acid receptors, uptake, release and synthesis. *Brain. Res.* **139,** 277–294.

Olsen R. W., Ticku M. K., Greenlee D., and Van Ness P. (1979) GABA Receptor and Ionophore Binding Sites: Interaction With Various Drugs, in *GABA-Neurotransmitters* (Krogsgaard-Larsen P., Scheel-Kruger S., and Kofod H., eds.), Munksgaard, Copenhagen.

Pilc A. and Lloyd K. G. (1984) Chronic antidepressants and GABA "B" receptors: A GABA hypothesis of antidepressant drug action. *Life Sci.* **35,** 2149–2154.

Rago L. K. and Zarkovsky A. M. (1983) Bicuculline-insensitive effects of baclofen. *Naunyn Schmiedeberg's Arch. Pharmacol.* **322,** 166–169.

Sano K. and Roberts E. (1963) Binding of γ-aminobutyric acid by brain preparations. *Biochem. Pharmacol.* **12,** 489–502.

Snodgrass S. R. (1978) Use of ^{3}H-muscimol for GABA receptor studies. *Nature* (Lond.) **273,** 392–394.

Svenneby G. and Roberts E. (1973) Bicuculline and *N*-methylbicuculline—competitive inhibitors of brain acetylcholinesterase. *J. Neurochem.* **21,** 1025–1026.

Toffano G. (1983) Endogenous Modulators of the GABA Receptor, in *The GABA Receptors* (Enna S. J., ed.), Humana, Clifton, New Jersey.

Toffano G., Guidotti A., and Costa E. (1978) Purification of endogenous protein inhibitor of the high affinity binding of aminobutyric acid to synaptic membranes of rat brain. *Proc. Natl. Acad. Sci. USA* **75,** 4024–4028.

Van Ness P. C. and Olsen R. W. (1979) Gamma-aminobutyric acid receptor binding in human brain regions. *J. Neurochem.* **33,** 593–596.

Wang Y. J., Salvaterra P., and Roberts E. (1979) Characterization of [^{3}H] muscimol binding to mouse brain membranes. *Biochem. Pharmacol.* **28,** 1123–1128.

White W. F. and Snodgrass S. R. (1983) Isoguvacine binding, uptake and release: Relation to the GABA system. *J. Neurochem.* **40,** 1701–1708.

Williams M. and Risley E. A. (1979) Characterization of the binding of [^{3}H] muscimol, a potent γ-aminobutyric acid agonist, to rat brain synaptosonal membranes using a filtration assay. *J. Neurochem.* **32,** 713–718.

Wojcik W. J. and Neff N. H. (1984) Aminobutyric acid β receptors are negatively coupled to adenylate cyclase in brain and in the cerebellum: These receptors may be associated with granule cells. *Mol. Pharmacol.* **25,** 24–28.

Wong D. T. and Horng J. S. (1977) Na^+-independent binding of GABA to the Triton-X-100-treated synaptic membranes from cerebellum of rat brain. *Life Sci.* **20,** 445–452.

Worms P., Depoortere H., Durand A., Morselli P. L., Lloyd K. G., and Bartholini G. (1982) γ-Aminobutyric acid (GABA) receptor stimulation. I. Neuropharmacological profiles of progabide (SL 76002) and SL 75102, with emphasis on their anticonvulsant spectra. *J. Pharmacol. Exp. Ther.* **220,** 660–671.

Zukin S. R., Young A. B., and Snyder S. H. (1974) Gamma aminobutyric acid binding to receptor sites in the rat central nervous system. *Proc. Natl. Acad. Sci. USA* **71,** 4802–4807.

Chapter 9
Receptors for Nonopioid Neuropeptides

Michel Goedert

1. Introduction

Until about 10 yr ago, it was generally assumed that the nervous system, the endocrine system, and the neuroendocrine system are not only anatomically distinct, but that they differ also with respect to the chemical messengers they use. There existed, however, a few exceptions, such as the presence of noradrenaline (NA) in the adrenal medulla or of dopamine (DA) in the external layer of the median eminence. It has since become apparent, however, that these exceptions reflect the general rule in the sense that many hormones are now known to occur in nerve cells—neuroendocrine hormones occur in both nerve cells and endocrine cells, and, more rarely, neurotransmitters occur in endocrine and neuroendocrine cells. In addition, a large number of previously unknown peptides has been found to occur at practically all levels of the nervous, endocrine, and neuroendocrine systems, thereby leading to a major conceptual change in the field of cell-to-cell communication through chemical signaling. Thus, although the anatomical distinctions between the three systems are still valid, the chemical specificity no longer holds. It appears that, depending on their location, several of the newly discovered peptides can function as either neurotransmitters, hormones, or neuroendocrine regulators.

The chemical structures of these messengers do not lend themselves to a useful classification, and their ubiquity makes it difficult to group them under one general heading according to their location. Moreover, it is at present impossible even to attempt a functional classification. Several names have been proposed, such as brain–gut peptides, regulatory peptides, or neuropeptides. The term "neuropeptides" will be used here, thereby taking the occurrence of these peptides in the mammalian nervous system as their common denominator. However, this by no means implies that their distribution is limited to nervous tissues.

In the following, I attempt to review what is known about the receptors for nonopioid neuropeptides. However, in view of the fact that an exhaustive review of the literature on neuropeptide receptors in general represents an impossible task in the space of one article, I will concentrate on a selection of neuropeptide receptors, thereby taking them as representative examples of the properties of neuropeptide receptors in general.

2. Nonopioid Neuropeptides: Functional Aspects

Over the last years, nonopioid neuropeptides (Table 1) have been implicated in almost every body manifestation, ranging from pain and satiety to sexual behavior and memory consolidation. The effects observed following the injection of totally nonphysiological peptide concentrations (up to 10,000 times their total brain content) into small brain areas have been taken as evidence of a physiological function of a given peptide. Not surprisingly, the main existing evidence for a physiological function of neuropeptides has been derived from studies performed on peripheral tissues. As for classical neurotransmitters, the relative anatomical simplicity, as well as the experimental accessibility of the autonomic nervous system, have greatly contributed to an understanding of some of the basic properties of neuropeptides.

The 11-amino acid neuropeptide substance P has been found in nerve fibers in all peripheral tissues investigated, and it always appears to be present in unmyelinated nerve fibers belonging to the peripheral branch of sensory neurons (Jessell, 1983). Substance P-positive cell bodies are present in sensory ganglia of the trigeminal, glossopharyngeal, and vagal nerves, as well as in dorsal root ganglia at all levels of the spinal cord. Substance P-positive sensory axon collaterals have been found in all pre- and paravertebral sympathetic ganglia investigated, thereby raising the possibility of a direct link between the sensory nervous system and the sympathetic division of the autonomic nervous system (Hökfelt et al., 1977). Ultrastructurally, these axon collaterals make direct contact with the sympathetic ganglion cell dendrites (Matthews and Cuello, 1982). Electrophysiologically, substance P depolarizes sympathetic ganglion cells, and the time-course of its action, as well as the conductance changes it produces, are strikingly similar to a slow noncholinergic excitatory postsynaptic

TABLE 1
Nonopioid Neuropeptides Present in the Mammalian Central Nervous System[a]

Peptide	Peptide (continued)
Adrenocorticotropin	Melanocyte-stimulating hormones
Angiotensins	Motilin
Atrial natriuretic peptides	Neurokinins
Bradykinin	Neuromedins
Calcitonin gene-related peptide	Neuropeptide Y
Carnosine	Neurotensin
Cholecystokinin	Oxytocin
Corticotropin-releasing factor	Peptide histidine-isoleucine
Delta sleep-inducing peptide	Proctolin
Epidermal growth factor	Prolactin
Galanin	Secretin
Gastrin-releasing peptide	Somatostatins
Glucagon	Substance P
Growth hormone	Thyroid-stimulating hormone
Growth hormone-releasing hormone	Thyrotropin-releasing hormone
Hydra head activator peptide	Vasoactive intestinal polypeptide
Insulin	Vasopressin
Luteinizing hormone-releasing hormone	Vasotocin

[a]Material immunoreactive with antibodies raised against these substances has been found in neurons and nerve terminals different from those mediating neuroendocrine functions. In some cases, multiple molecular forms of one substance have been described; these have not been listed individually. However, this fact is indicated by the use of the plural.

potential observed following preganglionic nerve stimulation. Moreover, a substance P receptor antagonist has been shown to block selectively, not only the action of substance P, but also the noncholinergic slow excitatory postsynaptic potential (Jiang et al., 1982). Thus, there can be little doubt that the substance P released from sensory axon collaterals is responsible for the noncholinergic excitatory postsynaptic potential in sympathetic ganglia.

In a similar fashion, a peptide closely related to luteinizing hormone-releasing hormone (LHRH) has been demonstrated to mediate a noncholinergic, late, slow excitatory postsynaptic potential in paravertebral sympathetic ganglia of the bullfrog (Jan and Jan, 1982; Jones et al., 1984). Luteinizing hormone-releasing hormone-like immunoreactivity is present in the same preganglionic nerve fibers as acetylcholine, and both substances are released upon electrical stimulation of the preganglionic nerve. The effects of LHRH last for several minutes and the late, slow excitatory postsynaptic potential can be truncated by a specific LHRH antagonist as late as 30 s after the application of the peptide, indicating an unusually long half-life for the LHRH-like peptide consequent to its release. Morphological studies have indicated that the presynaptic terminals do not establish synaptic contacts with the LHRH-responsive postganglionic cells, and that the peptide has to diffuse over a distance of several tens of micrometers before it can act on its receptor on the postganglionic cell. Therefore, the bullfrog sympathetic ganglion represents an example of a novel type of chemical cell-to-cell communication that one can describe as "neuroparacrine." It will be interesting to see whether this form of communication represents a more general feature of peptidergic neurotransmission. It appears likely that slow synaptic potentials, such as those described above, are able to modify ganglionic transmission over quite long time periods.

Neuropeptides are found not only in preganglionic autonomic nerve fibers, but also in postganglionic cell bodies. For instance, vasoactive intestinal polypeptide (VIP) is present in a large proportion of the acetylcholine-containing cells in the cat submandibular and otic ganglia and in apparently all cells of the sphenopalatine ganglion (Lundberg et al., 1981a). Ultrastructurally, it is found in large dense-core vesicles; it remains to be established whether both substances are localized in the same granules. High-frequency electrical stimulation of the preganglionic parasympathetic nerve fibers to the submandibular gland leads to a large atropine-sensitive increase in salivary secretion and to an atropine-resistant vasodilation. The intraarterial injection of VIP leads to an atropine-resistant vasodilation and VIP is released upon preganglionic nerve stimulation (Lundberg et al., 1981b). There also appears to exist a synergistic interaction between the effects of acetylcholine and VIP on salivary secretion. When infused together, they produce more than an additive effect on salivary secretion, whereas VIP has no action on its own. In addition, the intraarterial injection of antibodies raised against VIP blocks the vasodilation consequent to nerve stimulation (Lundberg et al., 1981b). These results therefore strongly suggest that the acetylcholine and the VIP that are costored and coreleased from postganglionic parasympathetic nerve terminals exert complementary actions on salivary gland secretion.

Neuropeptides are also found in various neuroendocrine and endocrine tissues, where they may serve a physiological function.

Vasoactive intestinal polypeptide or the structurally related peptide histidine-isoleucine may be a regulator of prolactin secretion. Vasoactive intestinal polypeptide is present in the external layer of the median eminence and its concentration in the hypophyseal portal blood is about 20 times higher than in the peripheral circulation (Leong et al., 1983). When added to pituitary cells in culture, VIP exerts a potent effect on prolactin secretion through a cyclic AMP-dependent mechanism. Dopamine and glucocorticoids inhibit the VIP-induced increase in prolactin secretion at the level of the cyclic AMP response (Rotsztejn et al., 1981; Onali et al., 1981). Although these results are compatible with a role for VIP as a hypothalamic hormone, there is at present no evidence available concerning its possible physiological function. Luteinising hormone-releasing hormone constitutes the best understood peptidergic neuroendocrine regulator. This 10-amino acid peptide is not released continuously, but in a pulsatile manner, leading in turn to an intermittent secretion of the gonadotropins from the anterior pituitary gland (Knobil, 1980). The physiological importance of this phenomenon became apparent when it was found that the continuous administration of LHRH failed to restore gonadotropin secretion in monkeys with an abolished endogenous LHRH production consequent to a specific hypothalamic lesion. However, luteinizing hormone and follicle-stimulating hormone secretions did normalize when LHRH was administered in an intermittent manner with one 6-min pulse every hour. Moreover, the same regimen initiated puberty and ovulatory menstrual cycles in immature prepubertal rhesus monkeys (Knobil, 1980). These findings led to the prediction that a pulsatile administration of LHRH may have beneficial effects in the treatment of hypothalamic amenorrhea and that a LHRH antagonist may prove useful in the treatment of precocious puberty. The pulsatile administration of LHRH to women with hypothalamic amenorrhea has led to ovarian development, ovulation, and uneventful pregnancy in a large percentage of cases (Leyendecker et al., 1980). The long-term treatment of children with precocious puberty with a stable analog of LHRH that behaves like a functional antagonist resulted in the cessation of menstruation, a regression of secondary sexual characteristics, a slowing of bone maturation, and a decrease in growth velocity (Mansfield et al., 1983).

3. Substance P Receptors

Substance P is an 11-amino acid peptide (Table 2), first identified in equine intestinal extracts (Von Euler and Gaddum, 1931) and later isolated and sequenced from bovine hypothalamus (Chang and Leeman, 1970; Chang et al., 1971) and horse intestine (Studer et al., 1973). It belongs to a family of structurally related peptides with rapid stimulant actions on vascular and avascular smooth muscle, known as the tachykinins (Erspamer and Melchiorri, 1983). They share the common carboxy-terminus sequence Phe-X-Gly-Leu-Met-NH_2 (X = Phe for substance P). Until recently, substance P was the only known mammalian tachykinin, the others being of amphibian origin, with the exception of eledoisin, which is found in molluscan salivary glands. More recently, however, neurokinin A (substance K or neuromedin L) and neurokinin B (neuromedin K), two peptides belonging to the tachykinin family, have been isolated and sequenced from porcine spinal cord (Table 2) (Kimura et al., 1983; Kangawa et al., 1983; Minamino et al., 1984a). In addition, a sequence identical to that of neurokinin A has been found in one of the two types of bovine substance P precursors that are generated from a single gene through alternative splicing (Nawa et al., 1983, 1984). Whereas little is known about the distribution and possible functions of the two neurokinins, a considerable amount of information is available on the distribution of substance P in both central and peripheral mammalian tissues. The evidence for a possible neurotransmitter role for substance P has been recently reviewed (Nicoll et al., 1980; Jessell, 1983).

Receptors for substance P have been investigated mainly using biological assays, radioligand binding techniques, and inositol phospholipid hydrolysis, the latter representing a possible second messenger linked to substance P receptor occupancy.

3.1. Biological Assays

The atropine-resistant contraction of the isolated rabbit jejunum by substance P was the preparation used by Von Euler and Gaddum (1931) in their initial description of substance P. A wide variety of vascular and nonvascular smooth muscle preparations has since been shown to respond to substance P (for a review, *see* Hanley and Iversen, 1980). At present, the most detailed information is available on the guinea pig ileum. Substance P exerts a potent contractile action on the longitudinal smooth muscle layer with an EC_{50} value on the order of 2–5 nM (Bury and Mashford, 1977). On a molar basis, it has been found to be 40 times more potent than acetylcholine and 160 times more potent than histamine. Structure–activity studies have indicated that the biological activity resides in the carboxy-terminal end of the molecule, with substance P (6–11) representing the shortest sequence with full

TABLE 2
Amino Acid Sequences of Mammalian Tachykinins

Tachykinin	Amino acid sequence
Substance P	H – Arg – Pro – Lys – Pro – Gln – Gln – Phe – Phe – Gly – Leu – Met – NH_2
Neurokinin A (Substance K) Neuromedin L	H – His – Lys – Thr – Asp – Ser – Phe – Val – Gly – Leu – Met – NH_2
Neurokinin B (Neuromedin K)	H – Asp – Met – His – Asp – Phe – Phe – Val – Gly – Leu – Met – NH_2

biological activity (Table 3). On the guinea pig ileum, the nonmammalian tachykinins physalemin, eledoisin, and kassinin and the mammalian tachykinins neurokinin A and neurokinin B are all roughly equipotent (Lee et al., 1982; Hunter and Maggio, 1984a; Minamino et al., 1984a). The study of the structure–activity requirements of the carboxy-terminal end of the substance P molecule has led to the formulation of a tentative model of substance P–receptor interactions (Sandberg and Iversen, 1982). This model involves a hydrophobic interaction between the aromatic sidechain of the Phe residue in position 7 and the substance P receptor, leading to a sharp bend in the polypeptide backbone; the molecule assumes a U shape as a consequence of another bend in the vicinity of Gly^9 and Leu^{10}. Substitutions by D-amino acids in any part of the carboxy-terminal hexapeptide of substance P result in a loss of biological activity; however, a systematic investigation of the biological activity of substance P analogs containing D-amino acid substitutions has led to the development of several substance P antagonists with little agonist activity, the prototype being represented by [D-Pro^2,D-$Trp^{7,9}$]-substance P (Folkers et al., 1981; Rosell and Folkers, 1982). It inhibits the substance P-induced contraction of the isolated guinea pig ileum, resulting in a parallel shift to the right of the substance P dose–response curve, indicative of a blockade of the competitive type.

With radioimmunoassay, substance P-like immunoreactivity is found in high concentrations in the muscle layers of the guinea pig ileum (Holzer et al., 1982). Immunohistochemical studies have indicated that it is present in cell bodies of the myenteric and submucous plexi and in nerve fibers throughout the muscle layers (Nilsson et al., 1975; Pearse and Polak, 1975; Schultzberg et al., 1980). Interestingly, the application of a substance P antagonist leads to a blockade of the noncholinergic contractile action of nerve stimulation in the guinea pig taenia coli, indicating that substance P may function as an excitatory neurotransmitter in the enteric nervous system (Leander et al., 1981).

Although the nonmammalian tachykinins physalemin, eledoisin, and kassinin and the mammalian tachykinin substance P share a similar qualitative pharmacological profile, they have different quantitative effects on various biological test systems (Erspamer, 1981). This was used originally as a means of distinguishing two subtypes of substance P receptors (Lee et al., 1982); the "SP-P" type, where physalemin $\simeq$ substance P $\simeq$ eledoisin $\simeq$ kassinin, and exemplified by the guinea pig ileum longitudinal smooth muscle layer; and the "SP-E" type, where eledoisin $\simeq$ kassinin $>>$ substance P $\simeq$ physalemin, and exemplified by the rat

TABLE 3
Relative Potencies of Tachykinins and Substance P Fragments in Peripheral and Central Biological Assays

Peptide	Guinea pig ileum[a,b,c]	Rat salivation[d]	Rat spinal cord depolarization[e]
Substance P	1.0	1.0	1.0
Neurokinin A	0.50	ND	ND
Neurokinin B	0.39	ND	ND
Eledoisin	0.90	1.30	ND
Physalemin	1.60	2.70	ND
Kassinin	ND[f]	0.10	ND
Substance P (2–11)	1.0	0.26	0.6–0.9
Substance P (3–11)	1.90	0.16	0.4–1.0
Substance P (4–11)	0.90	0.37	0.8–1.0
Substance P (5–11)	0.30	0.28	2–12
Substance P (6–11)	0.40	0.40	5–12
Substance P (7–11)	0.0005	0.003	<0.02
Substance P free acid	0.0004	0.0003	ND
Substance P (1–9)	0.00005	ND	ND

[a]Lee et al., 1982.
[b]Hunter and Maggio, 1984a.
[c]Minamino et al., 1984a.
[d]Hanley et al., 1980a.
[e]Otsuka and Konishi, 1976.
[f]ND, not determined.

vas deferens. It has recently been shown that the mammalian tachykinin neurokinin A exhibits similar biological activities to the amphibian peptide kassinin (Hunter and Maggio, 1984a). This, together with the fact that all three mammalian tachykinins are widely distributed throughout the CNS and gastrointestinal tract (Maggio and Hunter, 1984 Minamino et al., 1984c) would tend to suggest that neurokinin A and B are the endogenous ligands for the "SP-E" receptor. Moreover, using a substance P receptor antagonist, evidence has been obtained for the presence of two distinct tachykinin receptors in the guinea pig urinary bladder, but not in the rat vas deferens (Hunter and Maggio, 1984b). Studies of this type will allow unambiguous establishment of the presence and distribution of multiple tachykinin receptors.

The pronounced stimulation of salivation was the biological assay that permitted the isolation of substance P (Leeman and Hammerschlag, 1967). Upon intravenous injection, it causes up to a 50-fold increase in amylase release per unit volume of saliva (Liang and Cascieri, 1979). This effect appears to be direct, since it is not affected by either adrenergic or cholinergic receptor blockade. However, structure–activity studies in vitro have indicated a different profile from that of the intestinal smooth muscle in the sense that shortening of the carboxy-terminal end of substance P leads to a progressive reduction in its effects on amylase release and salivation (Brown and Hanley, 1981). Substance P-like immunoreactivity has been found in salivary glands by both radioimmunoassay and immunohistochemistry, where at least part of it appears to be present in primary sensory nerve fibers (Goedert et al., 1982; Sharkey and Templeton, 1984). The effects of substance P on salivary secretion can be mimicked by field stimulation of the parotid gland in a noncholinergic, nonadrenergic manner. Moreover, this effect of nerve stimulation is abolished in the presence of a substance P receptor antagonist (Gallacher, 1983). It therefore appears that in the rat parotid gland substance P acts as a nonadrenergic, noncholinergic neurotransmitter.

As for most neuropeptides, there is more information available on the possible roles of substance P in peripheral tissues than on its function in the CNS. The most extensive study on the structure–activity requirements of CNS neurons to substance P has been performed on frog and rat spinal motoneurons using electrophysiological techniques (Otsuka and Konishi, 1976). With the guinea pig ileum longitudinal smooth muscle layer, the biological activity was found to reside in the carboxy-terminal hexapeptide (Table 3). Similar results were obtained in more limited studies performed on rat cerebral cortex and rat locus ceruleus (Jones and Olpe, 1982; Cheeseman et al., 1983). In the latter preparation, eledoisin was found to be more potent than substance P. In contrast to peripheral tissues, the currently available substance P receptor antagonists do not appear to antagonize the actions of substance P in the CNS, where they may even have neurotoxic effects (Hökfelt et al., 1981; Jones et al., 1983). A recent study has indicated that a monoclonal antibody directed against substance P can inhibit the stress-induced activation of the mesocortical dopaminergic system (Bannon et al., 1983), indicating that such an immunopharmacological approach may provide an alternative way for studying the physiological importance of endogenous substance P in the CNS.

3.2. Radioligand Binding Studies

Several studies have investigated the properties of specific substance P binding sites in central and peripheral tissues by using radioligand binding techniques. They have all made use of radiolabeled agonists, such as [^{3}H]-substance P, [^{125}I]-Bolton-Hunter substance P, [^{125}I]-Bolton-Hunter physalemin, [^{125}I]-Bolton-Hunter eledoisin, or [^{125}I]-Bolton-Hunter kassinin. Most studies have investigated the binding of radiolabeled tachykinins to membrane preparations, the rat brain representing the most widely used tissue (Hanley et al., 1980b; Cascieri and Liang, 1983; Perrone et al., 1983; Viger et al., 1983; Beaujouan et al., 1984; Cascieri et al., 1985; Torrens et al., 1985). Other studies have made use of retina (Osborne, 1984), salivary gland tissue (Putney et al., 1980; Liang and Cascieri, 1981; Lee et al., 1983), pancreatic acinar cells (Jensen and Gardner, 1979; Hanley and Jessell, 1984), and ileum longitudinal smooth muscle (Buck et al., 1984; Watson and Iversen, 1984). All studies find the binding to be specific, saturable, and reversible and all report a single population of binding sites. Although the specific substance P binding sites in the CNS appear to be of high affinity and low density, there exists less agreement concerning the equilibrium dissociation constant. Four studies have reported an equilibrium dissociation constant of less than 1 n*M* (Hanley et al., 1980b; Perrone et al., 1983; Viger et al., 1983; Osborne, 1984; Beaujouan et al., 1984; Torrens et al., 1985), with two studies reporting a value of 12 n*M* (Cascieri and Liang, 1983; Cascieri et al., 1985). The latter study described a maximum number of binding sites of 500 fmol/mg protein, whereas the number of binding sites ranged between 13 and 87 fmol/mg protein in the other three studies. Similarly, in peripheral tissues, the specific binding of either [^{3}H]-substance P, [^{125}I]-Bolton-Hunter substance

P, or [^{125}I]-Bolton-Hunter physalemin was found to be specific, saturable, and reversible and only one population of binding sites was detected. The equilibrium dissociation constant was 0.5 and 1.8 n*M* for the guinea pig ileum longitudinal smooth muscle (Buck et al., 1984; Watson and Iversen, 1984), 1.4 n*M* for the rat parotid gland (Putney et al., 1980), 1 n*M* for the rat submandibular gland (Lee et al., 1983), 5 n*M* for acini of the guinea pig pancreas (Jensen and Gardner, 1979), and 40 p*M* for a cell line derived from a rat exocrine pancreatic tumour (Hanley and Jessell, 1984). As for the rat brain, the maximum number of specific binding sites was low, with the rat submandibular gland representing an exception at 60 fmol/mg protein.

The potencies of various substance P fragments in competing with labeled substance P for its specific binding site in central and peripheral tissues indicate that almost the entire substance P sequence is required for full binding activity (Table 4). Whereas substance P (2–11) and substance P (3–11) retain substantial binding activity, all other substance P fragments investigated are either weakly active or completely inactive in competing with labeled substance P for its specific binding site. Of the nonmammalian tachykinins investigated, physalemin was the most active in rat brain, guinea pig ileum, and rat submandibular gland, whereas eledoisin and kassinin showed little binding activity (Perrone et al., 1983; Viger et al., 1983). The mammalian tachykinins neurokinin A and neurokinin B were only weakly active in competing with labeled substance P for its specific binding site in rat brain (Quirion and Pilapil, 1984; Cascieri et al., 1985). In a tumor cell line derived from the exocrine rat pancreas, neurokinin A was also found to be weakly active in competing with [^{125}I]-Bolton-Hunter substance P for its specific binding site (Hanley and Jessell, 1984). In addition, it has been demonstrated that in rat brain [^{125}I] Bolton-Hunter eledoisin binds to specific binding sites distinct from the substance P binding sites (Cascieri and Liang, 1984; Beaujouan et al., 1984; Rothman et al., 1984; Buck et al., 1984; Cascieri et al., 1985; Torrens et al., 1984, 1985). Substance P was much more potent in competing with [^{125}I] Bolton-Hunter substance P than with [^{125}I] Bolton-Hunter eledoisin for their respective binding sites, whereas the reverse was true for eledoisin. Kassinin and neurokinin B were roughly equipotent with eledoisin in competing for its binding site, whereas neurokinin A and physalemin were approximately 20% as potent as eledoisin (Torrens et al., 1984; Cascieri et al., 1985). The two binding sites could be further differentiated pharmacologically (Torrens et al., 1985). Substance P methyl ester was found to only recognize the [^{125}I] Bolton-Hunter substance P binding site, whereas the metabolically stable substance P analog [pGlu^5 Me Phe^8 Sar^9]-substance P (5–11) was a potent inhibitor of [^{125}I] Bolton-Hunter eledoisin binding. In addition, several substance P antagonists inhibited [^{125}I] Bolton-Hunter substance P, but not [^{125}I] Bolton-Hunter eledoisin binding. These results thus strongly suggest the presence of two different tachykinin binding sites in the CNS. Substance P is the best candidate for interacting with the site labeled by [^{125}I] Bolton-Hunter substance P, whereas neurokinin B is the best mammalian tachykinin candidate for interacting with the [^{125}I] Bolton-Hunter eledoisin binding site (Torrens et al., 1984; Cascieri et al., 1985). In addition, a third type of tachykinin receptor has been described in rat duodenum and mouse bladder (Buck et al., 1984), with neurokinin A being the probable endogenous ligand. It thus appears that the three endogenous mammalian tachykinins are matched by the presence of three different receptor subtypes.

A comparison between the relative potencies of substance P fragments and other tachykinins in biological assay systems and in radioligand binding studies discloses some surprising discrepancies (Tables 3 and 4). For instance, eledoisin is approximately equipotent or more potent than substance P in almost all biological assay systems (Hanley and Iversen, 1980), whereas it is only very weakly active in competing with labeled substance P for its specific binding site in both central and peripheral tissues. These discrepancies are probably caused by the different conditions under which biological assays and radioligand binding studies are conducted (intact tissue slices in Krebs-saline vs membrane preparations in less physiological media; no control over peptide metabolism vs stabilization of substance P against enzymatic degradation). Nonetheless, it remains an important task for future studies to provide a rational explanation for these discrepancies.

One study has investigated the distribution of specific substance P binding sites using membranes prepared from different regions of the rat brain (Torrens et al., 1983). The highest number of binding sites was present in the septum, followed by the striatum, the hippocampus, the hypothalamus, the mesencephalon, the cerebral cortex, the cerebellum, and the substantia nigra. Neither the equilibrium dissociation constant, nor the substance P fragment potencies, varied significantly among the different brain regions investigated. More extensive studies on the distribution of specific substance P binding sites have been conducted using the autoradiographic technique for diffusible ligands first described for opiate receptors (Young and Kuhar, 1979). In the rat CNS (Fig. 1), substance P binding sites presented a wide distribution (Shults

TABLE 4
Relative Potencies of Tachykinins and Substance P Fragments in Radioligand Binding Assays for Substance P

Peptide	Guinea pig ileum longitudinal muscle[a]	Rat submandibular gland[b]	Rat brain[c,d]
Substance P	1.0	1.0	1.0
Neurokinin A	ND[a]	ND	0.0005
Neurokinin B	ND	ND	<0.0001
Eledoisin	0.003	0.03	0.0058
Physalemin	0.10	1.5	0.26
Kassinin	ND	ND	0.0006
Substance P (2–11)	0.30	ND	0.16
Substance P (3–11)	0.30	ND	0.081
Substance P (4–11)	0.03	0.21	0.013
Substance P (5–11)	0.005	ND	0.0067
Substance P (6–11)	ND	0.05	0.0009
Substance P (7–11)	0.00015	<0.001	<0.0001
Substance P free acid	ND	ND	<0.0001
Substance P (1–9)	ND	ND	<0.0001

[a]Buck et al., 1984.
[b]Lee et al., 1983.
[c]Viger et al., 1983.
[d]Quirion and Pilapil, 1984
[e]ND, not determined.

et al., 1982; Quirion et al., 1983; Mantyh et al., 1984c; Maurin et al., 1984; Rothman et al., 1984). A high concentration was found in the external layers of the olfactory bulb, the olfactory tubercle, the septal nucleus, the medial nucleus of the amygdala, the medial habenula, the dentate gyrus, the superior colliculus, the locus ceruleus, and the dorsal parabrachial nucleus. Moderate levels of binding sites were present in the nucleus accumbens, striatum, piriform cortex, central nucleus of the amygdala, paraventricular nucleus of the hypothalamus, periaqueductal grey, subiculum, entorhinal cortex, inferior colliculus, and dorsal horn of the spinal cord. Low to moderate levels were present in the globus pallidus, most cortical and hypothalamic areas, and the dorsal raphe nucleus. The cerebellum, substantia nigra, and most thalamic regions contained only low concentrations of substance P binding sites. In the rat spinal cord, the highest concentration of substance P receptors is found in lamina X with lower amounts in laminae I, II, and V–IX. Very low concentrations were found in laminae III and IV (Mantyh and Hunt, 1985). There existed a good agreement between the distribution of substance P receptors and substance P-like immunoreactivity in several brain areas, such as the striatum, central grey, hypothalamus, and amygdala. However, clear examples of mismatch were also found. The two most striking examples are represented by the external plexiform layer of the olfactory bulb and the pars reticulata of the substantia nigra. When examined for substance P-like immunoreactivity, the olfactory bulb contains very low levels, whereas the substantia nigra contains the highest concentration of substance P-like immunoreactivity in the central nervous system (Brownstein et al., 1976). However, when investigated for the presence of substance P binding sites, the external plexiform layer of the olfactory bulb contains one of the highest levels, whereas the substantia nigra has one of the lowest (Fig. 2). A similar, although less dramatic mismatch was observed in the cerebral cortex, hippocampus, locus ceruleus, and the transition zone between the amygdala and the cerebral cortex. Interestingly, [^{3}H]-substance P binding sites and [^{125}I]-Bolton-Hunter kassinin or eledoisin binding sites were found to be differentially distributed in at least some rat CNS regions (Mantyh et al., 1984d; Ninkovic et al., 1985), again suggesting the existence of two different binding sites.

With the exception of the gastrointestinal tract virtually nothing is known about the autoradiographic distribution of substance P binding in peripheral tissues (Burchen et al., 1984; Mantyh et al., 1984b). In the duodenum, a high density of binding sites was observed in the lamina muscularis mucosae, with lower densities in the longitudinal and circular muscle layers (Fig. 3). In the ileum

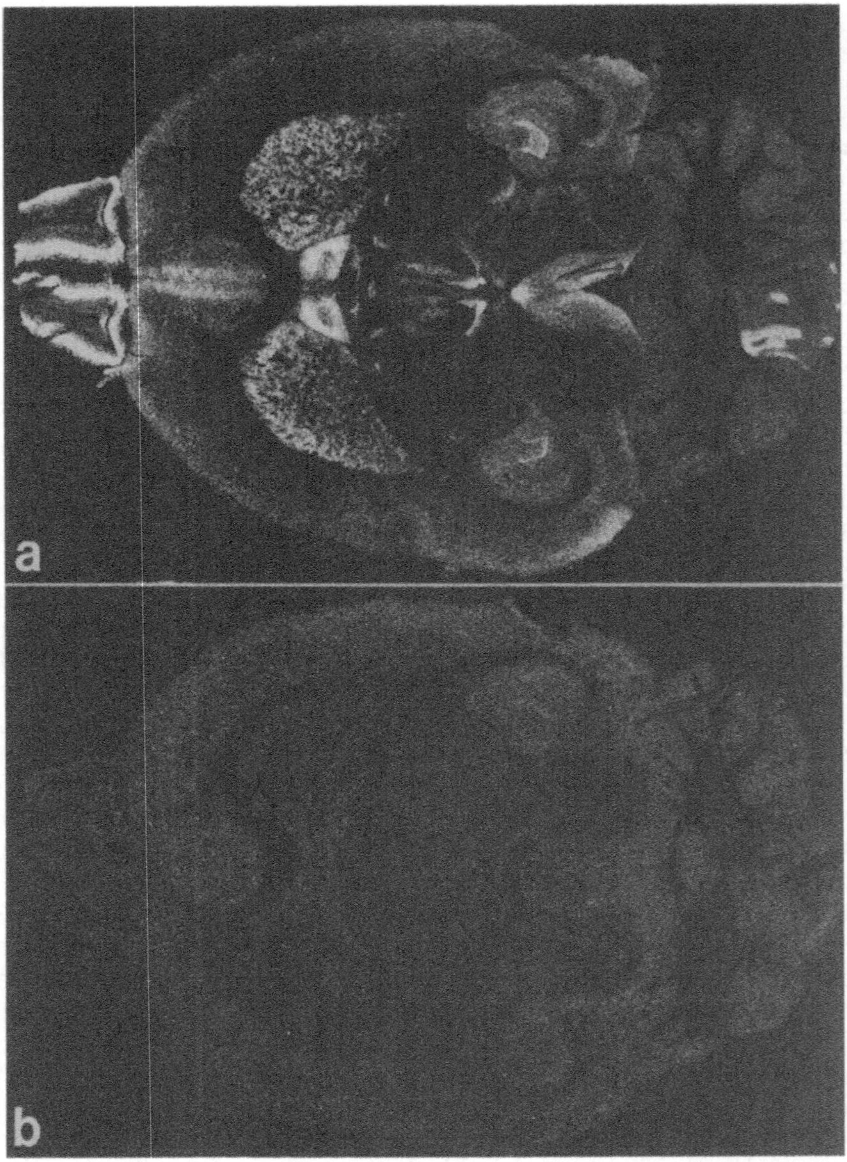

Fig. 1. Autoradiographic localization of [^{3}H]-substance P binding sites in the rat brain. (a) Darkfield photomicrograph of [^{3}H]-substance P (2 n*M*) binding in a horizontal brain section. (b) Darkfield photomicrograph of an adjacent section treated identically to (a), except that 1 μ*M* substance P had been added to the incubation solution. Note the heavy labeling in the external plexiform layer of the olfactory bulb, the striatum, the septum, the central grey of the midbrain, and folia IX and X of the cerebellum, and moderate labeling in the cerebral cortex and hippocampus, and absence of label in the thalamus (reprinted with permission from Mantyh et al., 1984a).

and colon, substance P binding sites were most heavily concentrated in the circular smooth muscle layer (Fig. 4). In all three regions, the epithelium, lamina propria, and serosa were only weakly labeled.

3.3. Second Messengers

Substance P represents one of the neuropeptides for which the most information is available with respect to the events consequent to receptor occupancy. Whereas it has no effect on cyclic AMP formation (Quik et al., 1978), substance P produces an increase in inositol phospholipid hydrolysis in all central and peripheral tissues investigated to date (Hanley et al., 1980a; Watson and Downes, 1983; Mantyh et al., 1984a). The hydrolysis of inositol phospholipids generates two products, diacylglycerol and inositol 1,4,5-trisphosphate, which have independent roles as second messengers. Diacylglycerol activates protein kinase C (Nishizuka, 1984), whereas inositol 1,4,5,-trisphosphate stimulates the release of calcium from nonmitochondrial intracellular membrane stores that are probably microsomal in nature (Streb et al., 1983; Burgess et al., 1984; Joseph et al., 1984; Prentki et al., 1984). In peripheral tissues the magnitude of inositol phospholipid hydrolysis effected by substance P, substance P fragments, and other tachykinins appears to be closely associated with the biological potencies of these peptides on amylase release from the rat parotid gland and on contraction of the guinea pig ileum longitudinal smooth muscle layer (Hanley et al., 1980a; Watson and Downes, 1983). In the rat CNS, a good correlation was obtained between the autoradiographic distribution of substance P binding sites and the magnitude of inositol phospholipid hydrolysis (Mantyh et al., 1984a), supporting the view that the mismatches observed between peptide levels and binding site density are real and not caused by an artifact of the receptor binding technique.

The mammalian tachykinins neurokinin A and neurokinin B lead also to an increase in inositol phospholipid hydrolysis in the rat CNS (Hunter et al., 1985), indicating that, although they bind to a different receptor subtype, they use the same second messenger system as substance P.

4. Vasoactive Intestinal Polypeptide Receptors

Vasoactive intestinal polypeptide was first isolated and sequenced from porcine duodenum and later also from bovine small intestine (Mutt and Said, 1974; Carlquist et al., 1979). Porcine and

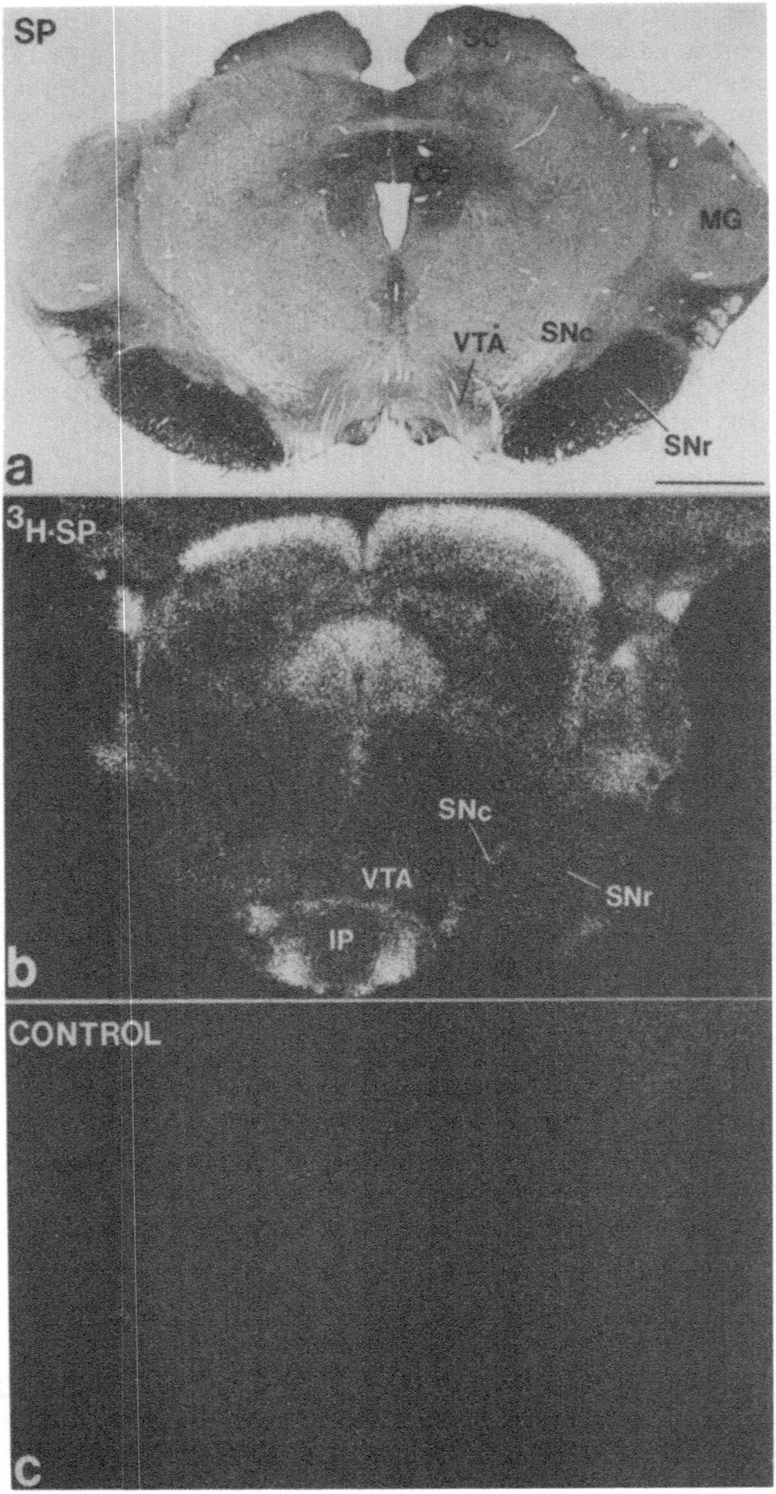

SP
SC
CG
MG
VTA
SNc
SNr
a
³H·SP
SNc
VTA
SNr
IP
b
CONTROL
c

bovine VIP is a 28-amino acid peptide (Table 5) that shows extensive sequence homologies with secretin and to a lesser extent with glucagon, gastric inhibitory peptide, and growth hormone-releasing hormone (GHRH). It has been isolated and sequenced from chicken intestine and, like porcine and bovine VIP, chicken VIP has 28 amino acids and the sequence differs only in four positions with conservative substitutions (Nilsson, 1975). More recently, a 27-amino acid peptide very closely related structurally to mammalian VIP and named peptide histidine-isoleucine (PHI) was isolated from porcine small intestine (Tatemoto and Mutt, 1981) (Table 5). The cloning of the messenger RNA sequence encoding human VIP has disclosed that it shows the same amino acid sequence as the porcine and bovine peptide and that a peptide designated peptide histidine-methionine (PHM) that differs only in two amino acids from PHI is present on the same peptide precursor (Itoh et al., 1983; Bloom et al., 1983) (Table 5). Although it was first isolated from small intestine, VIP-like immunoreactivity has since been found to be widely distributed throughout the central and peripheral nervous systems of various mammalian species, including man. It produces a wide variety of biological activities, such as relaxation of smooth muscle, vasodilation, electrolyte secretion, enhanced secretion from exocrine glands, and hormone release from the anterior pituitary gland (for a recent review, *see* Emson and Marley, 1983).

Receptors for VIP have been investigated using mainly radioligand binding assays and the peptide's effects on cyclic AMP formation.

4.1. Radioligand Binding Studies

Several studies have investigated the specific binding of [^{125}I]-labeled VIP to membranes prepared from central and peripheral

←

Fig. 2. Immunohistochemical localization of substance P-like immunoreactivity (a) and autoradiographic localization of [^{3}H]-substance P binding sites (b,c) in the midbrain of the rat. (c) was treated identically to (b), except that 1 μM substance P was added to the incubation medium. Note the dense accumulation of substance P-positive immunoreactive nerve fibers in the pars reticulata of the substantia nigra in (a) and the virtual absence of [^{3}H]-substance P binding sites in the substantia nigra pars reticulata in (b). Abbreviations: SC, superior colliculus; CG, central grey; MG, medial geniculate; SNc, substantia nigra, pars compacta; SNr, substantia nigra, pars reticulata; IP, interpeduncular nucleus. Scale bar, 800 μm.

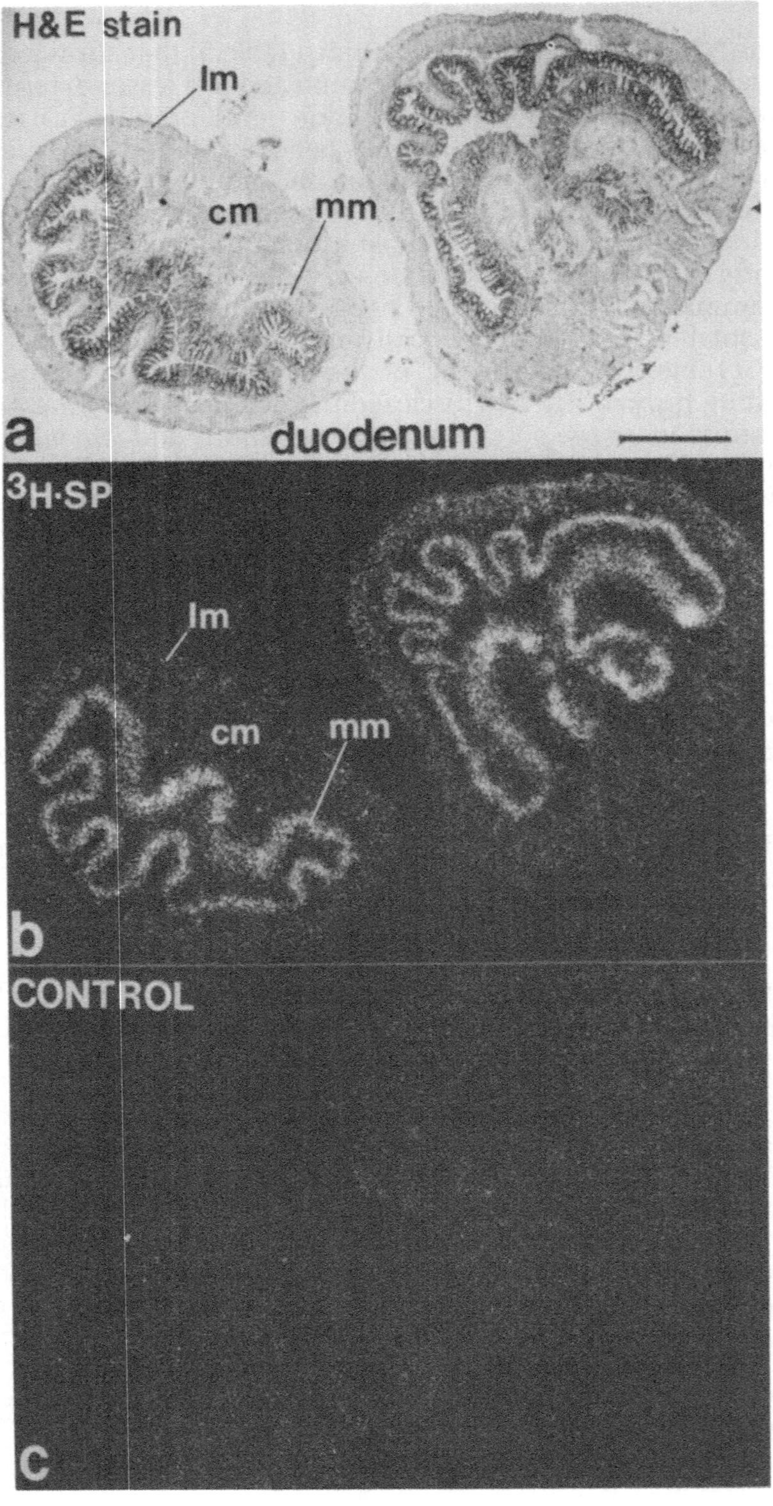
H&E stain
lm
cm
mm
a
duodenum
^{3}H·SP
lm
cm
mm
b
CONTROL
c

mammalian tissues; some have provided evidence that the iodinated peptide exhibits the same biological activity as the unlabeled peptide.

In the CNS, the specific binding of [^{125}I]-VIP was found to be saturable, specific, and reversible (Robberecht et al., 1978; Taylor and Pert, 1979; Staun-Olsen et al., 1982, 1985). One study on rat brain membranes (Taylor and Pert, 1979) reported the presence of one class of noninteracting high-affinity binding sites with an equilibrium dissociation constant of 1 n*M* and a maximum number of binding sites of 220 fmol/mg protein. In guinea pig brain synaptosomes (Robberecht et al., 1978), two noninteracting binding sites with equilibrium dissociation constants of 36 and 295 n*M* were described; the maximum number of binding sites amounted to 4000 and 20,000 fmol/mg protein for the high- and low-affinity binding sites, respectively. Similarly, two binding sites for [^{125}I]-VIP with equilibrium dissociation constants of 2.5 and 295 n*M* were found in synaptosomes prepared from rat cerebral cortex (Staun-Olsen et al., 1982). All three studies found secretin to be weakly active in competing with [^{125}I]-VIP for its specific binding site in the central nervous system, whereas glucagon was inactive. With the exception of VIP (18–28), which was inactive in guinea pig brain synaptosomes (Robberecht et al., 1979a,b), nothing is known at present about the relative potencies, either of other VIP fragments, or of PHI and GHRH. In the rat CNS the highest number of specific binding sites was found in cerebral cortex, limbic forebrain, and cerebellum (Staun-Olsen et al., 1985). Intermediate levels were present in hypothalamus, thalamus, and brainstem, and low levels in pons and spinal cord.

Specific binding sites for [^{125}I]-VIP have been described in a variety of peripheral tissues, such as mouse adrenal cortex, rat liver, rat lung, rat adipocytes, guinea pig pancreatic acini, and pig myometrium (Desbuquois, 1974; Bataille et al., 1974; Christophe et al., 1976; Morera et al., 1979; Ottesen et al., 1982; Leroux et al.,

←

Fig. 3. Autoradiographic localization of [^{3}H]-substance P binding sites in the guinea pig duodenum. (a) Lightfield photomicrograph of a transverse section through the duodenum stained with hematoxylin/eosin. (b) Darkfield photomicrograph of the same section as in a following incubation with 2 n*M* [^{3}H]-substance P. (c) Darkfield photomicrograph of an adjacent section to (b) incubated in the presence of 2 n*M* [^{3}H]-substance P and 1 μ*M* substance P. Note the heavy labeling of the lamina muscularis mucosae (mm) and the labeling of the longitudinal (lm) and circular (cm) smooth muscle layers in (b). Scale bar, 1 mm (reprinted by permission from Mantyh et al., 1984b).

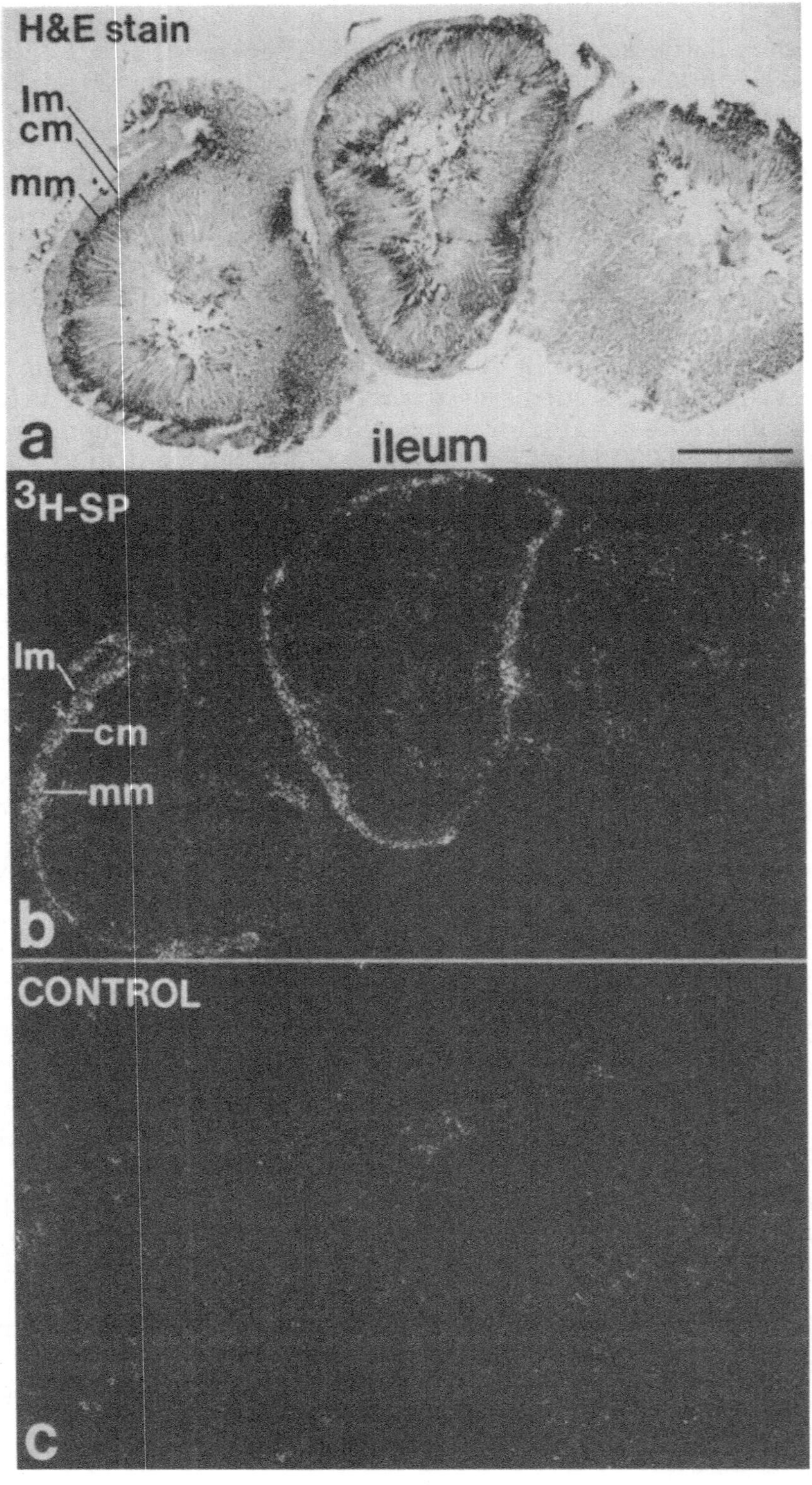

H&E stain
lm
cm
mm
a
ileum
^{3}H-SP
lm
cm
mm
b
CONTROL
c

1984). In general, these studies are in agreement with respect to the presence of two classes of noninteracting binding sites for [^{125}I]-VIP. For instance, in rat lung, the equilibrium dissociation constants were 0.32 and 23 n*M*, with a maximum number of binding sites of 89 and 584 fmol/mg protein. Autoradiographically, the binding sites were localized over alveoli and bronchi (Leroux et al., 1984). Similarly, in pig myometrium, the equilibrium dissociation constants were 0.23 and 9.5 n*M* (Ottesen et al., 1982). As in the CNS, secretin was only weakly active in competing with [^{125}I]-VIP for its specific binding site, and glucagon had either no or very little activity. One study on guinea pig pancreatic acinar cells investigated the relative potencies of VIP (1–8), VIP (14–28), VIP (15–28), and VIP (18–28) in competing with [^{125}I]-VIP for its specific binding site (Christophe et al., 1976). Neither the amino-terminal fragment, nor the carboxy-terminal fragments investigated exhibited any significant activity (relative potencies of less than 0.001).

To date, most information is available on the specific [^{125}I]-VIP binding sites present on rat small intestinal epithelial cells (Prieto et al., 1979). Scatchard transformation of the specific binding data indicated the presence of two classes of noninteracting binding sites, with a high-affinity, low-capacity binding site (equilibrium dissociation constant 1.6 n*M* and 1.4×10^5 binding sites/cell) and a low-affinity, high-capacity binding site (equilibrium dissociation constant 74 n*M* and 1.1×10^6 binding sites/cell). Detailed structure–activity relationships have been carried out on this tissue (Table 6). Surprisingly, chicken VIP was found to be more potent than the porcine peptide in competing with [^{125}I]-VIP for its specific binding sites. Of the nine partial VIP sequences investigated, VIP (2–28) was the only one to show some potency—1% of VIP (1–28), taken as 100% (Couvineau et al., 1984). Peptide histidine-isoleucine had a relative potency of 20%, whereas the potency of secretin was of the order of 1% (Bataille et al., 1980). Growth hormone-releasing hormone had a relative potency of 0.1%, whereas glucagon and gastric inhibitory peptide were inactive

←

Fig. 4. Autoradiographic localization of [^{3}H]-substance P binding sites in the guinea pig ileum. (a) Lightfield photomicrograph of a transverse section through the ileum stained with hematoxylin/eosin. (b) Darkfield photomicrograph of the same section as in (a) following incubation with 2 n*M* [^{3}H]-substance P. (c) Darkfield photomicrograph of an adjacent section to (b) incubated in the presence of 2 n*M* [^{3}H]-substance P and 1 μ*M* substance P. Note the labeling of the longitudinal (lm) and circular (cm) smooth muscle layers, as well as of the lamina muscularis mucosae (mm) in (b). Scale bar, 1 mm.

TABLE 5
Amino Acid Sequences of Peptides Belonging to the Vasoactive Intestinal Polypeptide Family

Peptide	Amino acid sequence
Vasoactive intestinal polypeptide	H – His – Ser – Asp – Ala – Val – Phe – Thr – Asp – Asn – Tyr – Thr – Arg – Leu – Arg – Lys – Gln – Met – Ala – Val – Lys – Lys – Tyr – Leu – Asn – Ser – Ile – Leu – Asn – NH_2
Peptide histidine-isoleucine	H – His – Ala – Asp – Gly – Val – Phe – Thr – Ser – Asp – Phe – Ser – Arg – Leu – Leu – Gly – Gln – Leu – Ser – Ala – Lys – Lys – Tyr – Leu – Glu – Ser – Leu – Ile – NH_2
Peptide histidine-methionine	H – His – Ala – Asp – Gly – Val – Phe – Thr – Ser – Asp – Phe – Ser – Lys – Leu – Leu – Gly – Gln – Leu – Ser – Ala – Lys – Lys – Tyr – Leu – Glu – Ser – Leu – Met – NH_2

TABLE 6
Relative Potencies of VIP Fragments and Related Peptides in Binding to Rat Small Intestinal Epithelial Cells and in Stimulating Cyclic AMP Formation in Intestinal Epithelial Cells

Peptide	Radioligand binding assay[a,b,c]	Cyclic AMP formation[a,b,c]
VIP (1–28)	1.00	1.00
VIP (1–14)	<0.0001	<0.0001
VIP (2–14)	<0.0001	<0.0001
VIP (2–28)	0.01	<0.0001
VIP (14–28)	<0.0001	<0.0001
VIP (15–28)	<0.0001	<0.0001
VIP (20–28)	<0.0001	<0.0001
VIP (21–28)	<0.0001	<0.0001
Peptide histidine isoleucine	0.20	0.30
Secretin	0.01	0.001
Growth hormone-releasing hormone	0.001	<0.001
Glucagon	<0.0001	<0.0001

[a]Bataille et al., 1980.
[b]Laburthe et al., 1983.
[c]Couvineau et al., 1984.

(Laburthe et al., 1983). Despite its relatively high potency on rat intestinal cells, peptide histidine-isoleucine was three orders of magnitude less active than VIP in competing with [^{125}I]-VIP for its specific binding site in human colonic epithelium (Laburthe et al., 1985). The same was true for peptide histidine-methionine, implying the existence of two separate receptors in human tissues. In contrast to other neuropeptides, the whole amino acid sequence of VIP thus seems to be required for full binding activity. Guanosine triphosphate and its stable analog 5'-guanylimidophosphate markedly inhibited the interaction of [^{125}I]-VIP with its specific, high-affinity binding site, probably by increasing the rate of dissociation (Amiranoff et al., 1980). Other nucleotides, such as adenosine triphosphate, required 1000-fold higher concentrations in order to inhibit the specific binding of [^{125}I]-VIP.

The receptor for [^{125}I]-VIP in rat small intestine has been characterized at the molecular level using covalent cross-linking of the ligand to a membrane preparation (Laburthe et al., 1984). Polyacrylamide gel electrophoresis of the membrane proteins followed by autoradiography revealed three labeled bands with molecular weights of 73,000, 33,000, and 14,000. Labeling of the 73,000 and 33,000 molecular weight components could be abolished by VIP, whereas the labeling of the 14,000 molecular weight component was probably nonspecific in nature. The labeling of the 73,000 molecular weight component was inhibited by concentrations of VIP in the nanomolar range, as well as by PHI and secretin. The labeling of the 33,000 molecular weight band was inhibited by only high concentrations of VIP. Guanosine triphosphate inhibited the binding to the 73,000 molecular weight band, without affecting the 33,000 molecular weight component. These results, which agree with most radioligand binding studies, thus provide direct evidence for the existence of two distinct macromolecules with the ability to bind $[^{125}I]$-VIP in rat intestinal membranes. Future experiments will have to determine whether the low-affinity, high-capacity VIP binding site represents a true receptor.

4.2. Second Messengers

As first demonstrated in liver and adipose tissue (Desbuquois et al., 1973), VIP is a potent stimulator of cyclic AMP formation in all central and peripheral tissues investigated.

In rat hippocampal slices, concentrations as low as 20 n*M* produced a doubling of the cyclic AMP content, with a maximal 90-fold increase at 10 μ*M* (Etgen and Browning, 1983). VIP stimulated cyclic AMP formation in all rat brain regions studied (Quik et al., 1978; Borghi et al., 1979; Etgen and Browning, 1983), and a good correlation was obtained between the levels of VIP-like immunoreactivity and the magnitude of VIP-induced cyclic AMP formation. At a concentration of 1 μ*M*, VIP produced a 30–50-fold increase in cyclic AMP formation in cerebral cortex, a 10-fold increase in midbrain and hypothalamus, and no significant change in the cerebellum. These effects were probably direct, as tetrodotoxin did not alter the cyclic AMP response to VIP. In rat cerebral cortex, the effect of VIP on cyclic AMP formation was potentiated in the presence of NA through an alpha-adrenergic receptor-mediated mechanism (Magistretti and Schorderet, 1984). Vasoactive intestinal polypeptide has also been shown to produce an increase in cyclic AMP formation in the retina (Longshore and Makman, 1981; Schorderet et al., 1981; Watling and Dowling, 1983), where VIP-like immunoreactivity is present in a subpopulation of amacrine cells (Loren et al., 1980).

Vasoactive intestinal polypeptide produces a marked glycogenolytic effect (EC_{50}, 26 n*M*) in slices of mouse cerebral cortex (Magistretti et al., 1981); this effect was potentiated in the presence of a phosphodiesterase inhibitor, suggesting that it was probably mediated through the activation of a cyclic AMP-dependent protein kinase. Glycogenolysis in the CNS appears to be mediated mainly through glial cells that are probably astrocytic in nature. Interestingly, VIP has been shown to increase cyclic AMP levels in primary cultures of both rat brain astrocytes and chicken retinal Müller cells (Rougon et al., 1983; Koh et al., 1984), indicating the presence of VIP receptors on glial cells. These results suggest that VIP may play a physiological role in CNS energy metabolism and that neuropeptides in general may exert effects other than those on neuronal excitability.

Relatively little is known about the structure–activity requirements of the VIP-induced increase in cyclic AMP formation. Vasoactive intestinal polypeptide (10–28) produced no significant increase in cyclic AMP formation in rat hippocampus and carp retina (Etgen and Browning, 1983; Watling and Dowling, 1983). In the latter tissue the effect of 10 μ*M* PHI amounted to approximately 40% of that of 10 μ*M* VIP (1–28), whereas secretin had 10% of the activity of VIP and glucagon was inactive. In peripheral tissues VIP has been shown to stimulate cyclic AMP formation in tissues as diverse as liver, adipocytes, submandibular gland, pituitary gland, adrenal cortex, gall bladder, pancreatic acini, and small intestinal epithelium (Desbuquois et al., 1973; Fredholm and Lundberg, 1982; Robberecht et al., 1979a,b; Morera et al., 1979; Robberecht et al., 1978; Laburthe et al., 1979a).

In the anterior pituitary gland, VIP has been shown to stimulate cyclic AMP formation and to release prolactin (Gourdji et al., 1979). Interestingly, although having no effect on basal adenylate cyclase activity, DA has been found to inhibit the stimulation of adenylate cyclase effected by VIP (Onali et al., 1981). In the submandibular gland of the cat, the VIP-stimulated increase in cyclic AMP formation was markedly potentiated by carbachol (Fredholm and Lundberg, 1982). The two substances are present in the same cells in the submandibular ganglion and they exert complementary effects on salivary gland secretion and blood flow (Lundberg et al., 1981b).

A direct relationship between the stimulatory effect of VIP on cyclic AMP formation and a biological effect has been most clearly demonstrated in the case of the gastrointestinal tract. Vasoactive intestinal polypeptide is a potent stimulator of cyclic AMP formation in rat small intestinal epithelial cells with an EC_{50} value of 2 n*M*, a threshold at 0.1 n*M*, and a maximal 11-fold increase over basal, at 30 n*M* (Laburthe et al., 1979a). Cell fusion experiments have indicated that the VIP receptor and adenylate cyclase constitute separate physical entities (Laburthe et al., 1979b). Structure–activity studies indicate that, as for radioligand binding, the whole VIP sequence is required for the stimulation of cyclic AMP formation (Couvineau et al., 1984) (Table 6). Peptide histidine isoleucine had 30% of the activity of VIP (Bataille et al., 1980), and secretin had 0.1%, whereas GHRH and glucagon were inactive (Laburthe et al., 1983). It has been demonstrated that VIP activates cyclic AMP-dependent protein kinases in rat small intestinal epithelial cells, with an EC_{50} value of 0.7 n*M* (Laburthe et al., 1979c); this leads to the phosphorylation of microvillar and basolateral membrane proteins (Shlatz et al., 1979). It is well established that cyclic AMP is causally involved in the secretory diarrhea of cholera, and VIP has also been shown to cause net secretion of Na^+ and Cl^- ions in the gastrointestinal tract (Schwartz et al., 1974). The finding that VIP-like immunoreactivity is found in pancreatic tumors from patients with the watery-diarrhea syndrome (Verner and Morrison, 1958; Bloom et al., 1973) has led to the suggestion that VIP might be responsible for the symptomatology of pancreatic cholera. This proposal has received considerable support from the demonstration that chronic infusion of VIP into healthy subjects invariably leads to profuse watery diarrhea (Kane et al., 1983). Thus, there can be little doubt that VIP represents a physiologically important regulator of gastrointestinal secretion.

5. Cholecystokinin Receptors

Cholecystokinin (CCK), one of the first hormones known, was discovered through its contractile actions on the gall bladder (Ivy and Oldberg, 1928). Later experiments demonstrated the existence in intestinal extracts of a substance named pancreozymin, which, on intravenous infusion, stimulated the secretion of pancreatic enzymes (Harper and Raper, 1943). The elucidation of the amino acid sequence of CCK indicated that the CCK and pancreozymin activities reside in one molecule (Jorpes and Mutt, 1966). The material purified and sequenced from porcine intestine was 33 amino acids long (CCK 33) (Table 7); in addition, a variant with an amino-terminal hexapeptide extension (CCK 39) was isolated. Subsequently, it was shown that the CNS contains CCK-like immunoreactivity, the vast majority of which corresponds to CCK 8, the carboxy-terminal octapeptide of CCK 33 (Dockray et al., 1978) (Table 7). The carboxy-terminal pentapeptide sequence of CCK is

TABLE 7
Amino Acid Sequences of Peptides Belonging to the Cholecystokinin Family

Peptide	Amino acid sequence
Cholecystokinin 33	H – Lys – Ala – Pro – Ser – Gly – Arg – Val – Ser – Met – Ile – Lys – Asn – Leu – Gln – Ser – Leu – Asp – Pro – Ser – His – Arg – Ile – Ser – Asp – Arg – Asp – Tyr – Met – Gly – Trp – Met – Asp – Phe – ONH_2 (Tyr: | SO_3)
Cholecystokinin 8	H – Asp – Tyr – Met – Gly – Trp – Met – Asp – Phe – ONH_2 (Tyr: | SO_3)
Cerulein	Pyr– Gln – Asp– Tyr – Thr – Gly – Trp – Met – Asp – Phe – ONH_2 (Tyr: | SO_3)

identical to the corresponding sequence in gastrin and in the amphibian skin peptide cerulein. Peptides of the gastrin/CCK family contain a sulfated tyrosine residue. In the gastrin family, the sulfated residue is at the sixth amino acid from the carboxy-terminus.

Despite their structural similarities, CCK, gastrin, and cerulein are encoded by three distinct genes (Yoo et al., 1982; Boel et al., 1983; Deschenes et al., 1984; Gubler et al., 1984; Wakabayashi et al., 1985). Rat procholecystokinin is a 95-amino-acid protein that contains all the known CCK sequences (Deschenes et al., 1984). Similarly, pig procholecystokinin contains all the known CCK sequences, and the same precursor molecule of 114 amino acids is present in the brain and the gastrointestinal tract (Gubler et al., 1984). Therefore, it appears probable that the tissue-specific expression of CCK-like peptides is caused by post-translational processing, especially in view of the fact that the smaller molecular weight forms can all be generated from the larger peptides through the action of trypsin-like enzymes.

Besides their well-established role as gastrointestinal hormones, CCK peptides have been implicated in a variety of physiological processes, such as the control of satiety, neuroendocrine regulation, and neurotransmission in the central and peripheral nervous systems (for a recent review, *see* Emson and Marley, 1983).

Central and peripheral CCK receptors have been investigated mainly through radioligand binding assays and, to a more limited extent, through second messenger systems.

5.1. Radioligand Binding Studies

The presence of CCK receptors in the CNS has been investigated using either radiolabeled CCK 33 or radiolabeled CCK 8 (Hays et al., 1980; Innis and Snyder, 1980; Saito et al., 1980, 1981; Praissman et al., 1983; Steigerwalt and Williams, 1984; Van Dijk et al., 1984). With the exception of one study that made use of [^{3}H]-CCK 8 (Van Dijk et al., 1984) all studies have used [^{125}I]-Bolton-Hunter CCK as the radioligand; some of these studies have demonstrated full biological activity of the radioligand. The specific binding in rat (Hays et al., 1980; Saito et al., 1980; Praissman et al., 1983; Van Dijk et al., 1984), mouse (Saito et al., 1981), and guinea pig (Innis and Snyder, 1980) brain was saturable, specific, and reversible, and one class of noninteracting, high-affinity binding sites with an equilibrium dissociation constant of 1–2 n*M* was detected.

In view of the fact that CCK 8 is the natural ligand for CCK receptors in the CNS, radiolabeled CCK 8 is likely to represent the

TABLE 8
Relative Potencies of CCK Fragments and Related Peptides in Binding to Rat Brain and Pancreatic Membranes

Peptide	Brain membranes[a]	Pancreatic membranes[a]
CCK 8	1.00	1.00
Cerulein	1.12	2.26
CCK 33	0.18	0.21
Desulfated CCK 8	0.06	<0.0001
CCK 4	0.05	<0.0001
CCK 3	<0.0001	<0.0001
CCK 2	<0.0001	<0.0001
CCK 8 free acid	<0.0001	<0.0001
CCK (26–29)	<0.0001	<0.0001
CCK (30–32)	<0.0001	<0.0001
Pentagastrin	0.22	0.0009
Gastrin (1–17)	0.07	<0.0001
Gastrin (4–17)	0.11	<0.0001
Gastrin (5–17)	0.08	<0.0001

[a]Van Dijk et al., 1984.

best ligand for central CCK receptors (Praissman et al., 1983; Van Dijk et al., 1984), especially with tritium as the radiolabel (Van Dijk et al., 1984). Structure–activity studies on rat brain membranes (Table 8) have indicated that CCK 8 and cerulein are equipotent in competing with [^{3}H]-CCK 8 for its specific binding site. A sulfated tyrosine residue and an amide function at the carboxy-terminus were essential for full binding activity. The binding activity of CCK 8 was located in the carboxy-terminal end of the molecule, with CCK 4 representing the shortest peptide with binding activity. Cholecystokinin 33 and pentagastrin showed only about 20% of the binding activity of CCK 8.

The equilibrium dissociation constants between different brain regions were very similar, and the number of specific [^{3}H]-CCK 8 binding sites was highest in the olfactory bulb (34 fmol/mg protein), followed by the striatum, cerebral cortex, hypothalamus, hippocampus, midbrain, and pons-medulla. No specific binding sites were found in the cerebellum. These results are in reasonable, although not strict, agreement with the distribution of CCK-like immunoreactivity in the rat brain. The autoradiographic distribution of CCK binding sites has been investigated in the guinea pig brain using [^{125}I]-Bolton-Hunter CCK 33 as the radioligand (Zarbin et al., 1983), and in the rat brain using [^{3}H]-CCK 8 (Van Dijk et al., 1984). In the latter species, a high density of CCK binding sites was observed in layers III and IV of the cerebral cortex, mainly in cingular and occipital areas. Dense labeling was found in the external plexiform layer of the olfactory bulb and the olfactory tubercle, with only background levels in the olfactory tract. High levels of binding sites were also found in the nucleus accumbens, sulcus rhinalis, and amygdala, with no appreciable binding in the cerebellum.

Cholecystokinin receptors in peripheral tissues have been investigated, mainly in the exocrine pancreas of various mammalian species (Jensen et al., 1980; Innis and Snyder, 1980; Sankaran et al., 1980; Steigerwalt and Williams, 1981; Van Dijk et al., 1984). All studies have made use of [^{125}I]-Bolton-Hunter CCK 33, with the exception of one study (Van Dijk et al., 1984), which used [^{3}H]-CCK 8 as the radioligand. As for the CNS, the binding was specific, saturable, and reversible and, with the exception of one study (Steigerwalt and Williams, 1981), one population of high-affinity binding sites was detected. The equilibrium dissociation constant was typically on the order of 0.5 n*M*, and the maximum number of binding sites was about 30 times higher than in the brain (1140 fmol/mg protein in the rat pancreas, as opposed to 34 fmol/mg protein in the olfactory bulb with [^{3}H]-CCK 8 as the radioligand). Autoradiographically, the rat pancreas was uniformly labeled, with no apparent difference between exocrine and endocrine parts (Van Dijk et al., 1984). A quantitative autoradiographic analysis of [^{125}I]-Bolton-Hunter CCK 33 binding to rat and mouse pancreatic acini (Williams et al., 1982; Rosenzweig et al., 1983) has shown the majority of silver grains to be localized over the basolateral portion of the acinus with no labeling over the apical region of the cell. Similar to other pancreatic secretagogs, it thus appears that circulating CCK interacts with its receptor at the basal pole of the acinus, thereby triggering the events leading to exocytosis at the apex.

Structure–activity studies using either [^{125}I]-Bolton-Hunter CCK 33 or [^{3}H]-CCK 8 revealed some differences between central and peripheral CCK receptors (Table 8). As for the CNS, cerulein, CCK 8, and CCK 33 were the most potent substances in competing with radiolabeled CCK for its specific binding site in the pancreas. However, in contrast to the central receptor, it was found that all gastrin peptides, as well as CCK 4, were inactive on the pancreatic receptor; it therefore appears that the pancreatic receptor requires the full CCK 8 sequence. This is in good agreement with structure–activity studies on CCK-induced amylase release from isolated pancreatic acini (Jensen et al., 1980; Sankaran et al., 1980).

The receptor for [^{125}I]-CCK 33 in rat and mouse pancreatic acini has been characterized at the molecular level by cross-linking of the radioligand (Svoboda et al., 1982; Rosenzweig et al., 1983; Sakamoto et al., 1983). Polyacrylamide gel electrophoresis of rat pancreatic membranes prepared in the presence of a reducing agent revealed predominant labeling of a protein band with a molecular weight of 85,000 (Rosenzweig et al., 1983) or 76,000 (Svoboda et al., 1982); the labeled proteins are sialoglycoproteins (Rosenzweig et al., 1984). In the mouse pancreas, the major labeled band had a molecular weight of 120,000 in the absence of a reducing agent and of 80,000 in the presence of mercaptoethanol (Sakamoto et al., 1983). Cholecystokinin 8 and the competitive receptor antagonist dibutyryl cyclic GMP inhibited the binding of [^{125}I]-CCK 33 to either the 120,000 or the 80,000 molecular weight component. These results thus indicate that the CCK receptor in the mouse pancreas is composed of two subunits—an 80,000 molecular weight component that binds the ligand and a 40,000 molecular weight subunit that is linked to the 80,000 molecular weight component through disulfide bonds. Moreover, functional CCK receptors have been solubilized from mouse pancreatic membranes (Szecòwka et al., 1985), thereby allowing the large-scale purification of the receptor molecule in view of the determination of its amino acid sequence.

In the mouse brain cross-linking studies revealed the labeling of a single band of 55,000 molecular weight in the presence or absence of reducing agents, indicating that mouse pancreatic and brain CCK receptors are structurally dissimilar (Sakamoto et al., 1984).

Over the last 5 yr, three classes of substances have been described that act as competitive antagonists of CCK at its receptor site. They consist of cyclic nucleotide and amino acid derivatives, as well as partial CCK sequences. Interestingly, some of these receptor antagonists are nonpeptidergic in nature. They have been used mostly in the exocrine pancreas, and in each case the antagonists were shown to inhibit CCK-stimulated amylase secretion and the binding of radiolabeled CCK to its receptor. The antagonism was reversible, competitive, and specific in the sense that the action of other pancreatic secretagogs was not inhibited. Dibutyryl cyclic GMP was the first substance shown to act as a competitive antagonist on pancreatic CCK receptors (Peikin et al., 1979). It has since been shown to antagonize the actions of CCK in a variety of tissues, such as the gallbladder (Poitras et al., 1980), gastric and ileal smooth muscle (Bitar and Makhlouf, 1982; Davison and Najafi-Farashah, 1981), and the myenteric plexus (Hutchinson and Dockray, 1980). The second class of antagonists consists of the amino acid derivatives proglumide (D,L-4-benzoamido-*N*,*N*-dipropyl-glutaramic acid) and benzotript (*N*-p-chlorobenzoyl-L-tryptophan), substances that had previously been shown to inhibit some of the actions of gastrin and cerulein (Hahne et al., 1981). Proglumide has in fact been used therapeutically for the treatment of patients with gastric ulcers. Besides its action on CCK receptors in the exocrine pancreas, proglumide has also been shown to antagonize the action of CCK on tissues, such as the gallbladder, ileal smooth muscle (Davison and Najafi-Farashah, 1981), and gastric smooth muscle (Bitar and Makhlouf, 1982). Interestingly, it also selectively antagonized the electrophysiological excitatory action of CCK on dopaminergic neurons in the ventral tegmental area and on nondopaminergic neurons in the prefrontal cortex of the rat (Chiodo and Bunney, 1983). Benzotript is slightly more potent than proglumide in competing with [^{125}I]-Bolton-Hunter CCK 33 for its specific binding site in the rat pancreas, and both substances are about five times less potent than dibutyryl cyclic GMP (Hahne et al., 1981). To date, the most potent CCK receptor antagonist known is CCK (27–32) amide, which is structurally equivalent to the carboxy-terminal heptapeptide of CCK 8 minus the carboxy-terminal phenylalanine (Spanarkel et al., 1983). It does not stimulate amylase secretion from pancreatic acini, but does function as a competitive CCK antagonist. Although CCK (27–32) amide is 30 times more potent than dibutyryl cyclic GMP, it causes half-maximal inhibition of either CCK 8-stimulated amylase release or [^{125}I]-Bolton-Hunter CCK 33 binding at a concentration of 3 μM. This relatively low potency limits somewhat the usefulness of the currently available CCK antagonists for in vivo studies and as ligands for the in vitro study of CCK receptors.

5.2. Second Messengers

Relatively little is known about the biochemical events consequent to CCK receptor occupancy. In the exocrine pancreas, the CCK-induced amylase secretion is preceded by the mobilization of intracellular calcium and an increase in cyclic GMP formation. Both of these events are thought to be consequent to the breakdown of inositol phospholipids (Berridge, 1984) and CCK 8 has indeed been shown to stimulate inositol phospholipid hydrolysis in mouse pancreatic acini (Hokin-Neaverson and Sadeghian, 1984). The

CCK-induced increase in the intracellular calcium concentration results in exocytosis at the luminal membrane and in the opening of voltage-sensitive potassium channels in the basolateral membrane, thereby resulting in membrane hyperpolarization (Maruyama et al., 1983). Similarly, CCK 8 stimulates inositol phospholipid hydrolysis in guinea pig gastric glands (Chang et al., 1985). This effect appears to be direct and may be the second messenger leading to pepsinogen secretion. It remains to be shown whether central CCK receptors are linked to the same transmembrane events.

6. Neurotensin Receptors

Neurotensin is a 13-amino acid peptide (Table 9) first isolated and sequenced from bovine hypothalamus (Carraway and Leeman, 1973, 1975) and later from both bovine and human small intestine (Carraway et al., 1978; Hammer et al., 1980). Xenopsin, a peptide isolated from the skin of *Xenopus laevis* (Araki et al., 1973), and two peptides isolated from chicken intestine (Carraway and Bhatnagar, 1980; Carraway and Ferris, 1983) share marked sequence homologies with the carboxy-terminal end of mammalian neurotensin (Table 9). However, neither xenopsin, nor [Lys^8Asn^9] neurotensin (8–13) are present in central and peripheral tissues of various mammalian species (Goedert et al., 1984a; Goedert et al., 1985). Recently, a six-amino-acid peptide called neuromedin N has been isolated and sequenced from porcine spinal cord (Minamino et al., 1984b). It differs only in two amino acids from the carboxy-terminal end of mammalian neurotensin (1–13) (Table 9). At present, its distribution and possible physiological functions are not known. It is also unknown whether neurotensin and neuromedin N are present on the same protein precursor molecule.

As is the case for most peptides whose amino acid sequence is known and for which antibodies are available, far more is known about the distribution of neurotensin than about its possible functions. The existing evidence for a neurotransmitter role of neurotensin in the central and peripheral nervous systems and for an endocrine role in peripheral tissues has been reviewed recently (Nemeroff et al., 1983; Emson et al., 1985).

Receptors for neurotensin have been investigated using biological assays, radioligand binding studies, and, to a more limited extent, transmembrane events consequent to neurotensin receptor occupancy.

TABLE 9
Amino Acid Sequences of Peptides of the Neurotensin Family

Peptide	Amino acid sequence
Mammalian neurotensin	pGlu – Leu – Tyr – Glu – Asn – Lys – Pro – Arg – Arg – Pro – Tyr – Ile – Leu – OH
Neuromedin N	H – Lys – Ile – Pro – Tyr – Ile – Leu – OH
Chicken neurotensin	pGlu – Leu – His – Val – Asn – Lys – Ala – Arg – Arg – Pro – Tyr – Ile – Leu – OH
[Lys^8Asn^9]-neurotensin (8–13)	H – Lys – Asn – Pro – Tyr – Ile – Leu – OH
Xenopsin	pGlu – Gly – Lys – Arg – Pro – Trp – Ile – Leu – OH

6.1. Biological Assays

Neurotensin was originally discovered through its vasoactive properties (Carraway and Leeman, 1973). It has since been shown to produce a whole variety of smooth muscle effects, some of which occur only in certain species. At present, the most information is available on the actions of neurotensin on the guinea pig ileum smooth muscle, in which it exerts both contractile and relaxant effects on the longitudinal smooth muscle layer (Segawa et al., 1977; Kitabgi and Freychet, 1979a). The relaxant effect appears to be direct, since it was not blocked in the presence of tetrodotoxin. In contrast, the contratile effect of neurotensin was inhibited by tetrodotoxin and markedly reduced in the presence of atropine and following desensitization to substance P, indicating that it is probably mediated through the release of acetylcholine and substance P from enteric nerves (Kitabgi and Freychet, 1979b; Monier and Kitabgi, 1980). Moreover, electrophysiological studies have indicated that neurotensin produces an excitatory effect on neurons in the guinea pig ileum myenteric plexus (Kitabgi et al., 1979; Williams et al., 1979). Structure–activity studies have indicated that the carboxy-terminal neurotensin fragment (8–13) shows 46% of the biological activity of neurotensin (1–13) and neurotensin (9–13) only 3% (Kitabgi and Freychet, 1979a). All other amino- and carboxy-terminal fragments tested were inactive. Xenopsin had 21% of the potency of neurotensin (1–13) (Kitabgi and Freychet, 1979a) and the chicken peptide [Lys^8Asn^9] neurotensin (8–13) 2% (Kitabgi et al., 1984). The mammalian peptide neuromedin N had 15% of the biological activity of neurotensin (1–13) on the contraction of the guinea pig ileum longitudinal smooth muscle layer (Minamino et al., 1984b) (Table 10).

In contrast to its effects on the longitudinal smooth muscle layer, neurotensin exerts only a direct relaxant action on the guinea pig ileum circular smooth muscle layer (Goedert et al., 1984b). As for its relaxant action on the guinea pig colon and the guinea pig and rabbit ileum longitudinal smooth muscle layers (Kitabgi and Vincent, 1981; Huidobro-Toro and Yoshimura, 1983; Huidobro-Toro and Zhu, 1984), the effect of neurotensin was inhibited by the bee venom apamin, suggesting its mediation through calcium-dependent potassium channels. Electrophysiological studies have indicated that the circular smooth muscle layer of the guinea pig ileum is innervated by nonadrenergic, noncholinergic inhibitory nerves, the action of which is inhibited by apamin (Bauer and Kuriyama, 1982a,b). These observations suggested that neurotensin might be released from nonadrenergic inhibitory

TABLE 10
Relative Potencies of Neurotensin Fragments and Peptides of the Neurotensin Family in Peripheral Biological Assays

Peptide	Guinea pig ileum longitudinal muscle[a,b,c]	Guinea pig ileum circular muscle[d]	Guinea pig auricle[e]	Rat stomach[e]
Neurotensin (1–13)	1.00	1.00	1.00	1.00
Neurotensin (8–13)	0.20	0.29	0.52	1.02
Neurotensin (9–13)	0.03	0.01	0.01	0.01
Neurotensin (10–13)	<0.0001	<0.0001	<0.0001	<0.0001
Neurotensin (1–6)	ND[f]	<0.0001	ND	ND
Neurotensin (1–12)	<0.001	ND	<0.0001	<0.0001
Neuromedin N	0.15	ND	ND	ND
[Lys^{8}Asn^9]-neurotensin (8–13)	0.02	0.01	ND	ND
Xenopsin	0.55	ND	ND	ND

[a]Granier et al., 1982.
[b]Kitabgi et al, 1984.
[c]Minamino et al., 1984b.
[d]Goedert et al., 1984b.
[e]St Pierre et al., 1981.
[f]ND, not determined.

nerves consequent to nerve stimulation. Although the majority of ileal neurotensin-like immunoreactivity is found in endocrine-like cells, the presence of neurotensin-positive nerve fibers has been demonstrated in the guinea pig ileum (Reinecke et al., 1983). The hypothesis that neurotensin may function as a nonadrenergic, noncholinergic neurotransmitter in the guinea pig ileum received support from the finding that the relaxant effect of nerve stimulation was substantially reduced in the presence of an antiserum raised against synthetic neurotensin (1–13) (Goedert et al., 1984b). Both preimmune serum and an antiserum raised against [Lys^8Asn^9]-neurotensin (8–13) were without an effect. Structure–activity studies on the guinea pig ileum circular smooth muscle layer (Table 10) showed that neurotensin (8–13) possesses 29% of the activity of neurotensin (1–13), and neurotensin (9–13) possesses 1%, whereas neurotensin (1–6) and neurotensin (10–13) are inactive. It thus appears that in the guinea pig ileum the biological activity of neurotensin resides entirely in the carboxy-terminus and that the presence of basic amino acid residues in positions 8 and 9 represents an important prerequisite for full biological activity.

Neurotensin possesses a variety of cardiovascular actions, of which the positive inotropic and chronotropic effects on the spontaneously beating guinea pig auricle represent the best characterized (Quirion et al., 1978, 1980b). When expressed on a molar basis, neurotensin was more potent than either adrenaline or NA. As for the guinea pig ileum smooth muscle, the biological activity was found to reside in the carboxy-terminal end of neurotensin (1–13) (Table 10). Neurotensin-like immunoreactivity has been found in the atrium of various mammalian species, where it is present in nerve fibers that are probably preganglionic parasympathetic in nature (Weihe and Reinecke, 1981; Reinecke et al., 1982; Goedert et al., 1984a). It therefore is tempting to speculate that neurotensin represents a physiological regulator of the mammalian myocardium.

The rat stomach fundus strip represents a third peripheral assay system used for structure–activity studies on neurotensin and neurotensin fragments. Neurotensin exerts a potent contractile action on the rat stomach fundus (EC_{50} 3–11 n*M*) through what appears to be a direct effect (Quirion et al., 1980b). As for the guinea pig ileum and the guinea pig atrium, the biological activity resides entirely in the carboxy-terminus of neurotensin (1–13), with neurotensin (8–13) possessing 69% and neurotensin (9–13) 11% of the biological activity of neurotensin (1–13) (Table 10).

Neurotensin exerts a variety of central effects, such as hypothermia, antinociception, release of anterior pituitary hormones, and changes in DA turnover (for a review, *see* Nemeroff et al., 1983). Relatively little is known about the structure–activity requirements of the neurotensin receptors mediating these central effects. This is mainly because they require in general the in vivo administration of large concentrations of neurotensin, and factors such as distribution in the extracellular space, access to receptor sites, and inactivation by peptidases are likely to vary between neurotensin and its partial sequences. Moreover, to date no detailed structure–activity studies have been performed using electrophysiological techniques. However, the limited data available, obtained mainly using the hypothermic action of neurotensin in vivo (Loosen et al., 1978) and its effects on DA release in vitro (De Quidt and Emson, 1983), are consistent with the view that the biological activity resides in the carboxy-terminus of the neurotensin molecule.

These studies indicate that the biological activity of neurotensin resides in the carboxy-terminal end of the molecule, with neurotensin (9–13) representing the shortest sequence with full intrinsic activity, albeit low potency. Detailed structure–activity studies performed on the guinea pig ileum longitudinal smooth muscle (Granier et al., 1982), the spontaneously beating guinea pig auricles, and the rat stomach fundus (St Pierre et al., 1981) have demonstrated an absolute requirement for two basic amino acids in positions 8 and 9. Thus, substituting these residues with citrulline or replacing the arginine residue in position 9 by alanine leads to a large decrease in potency. Accordingly, xenopsin with two basic amino acids in positions 8 and 9 exhibits substantial biological activity, whereas [Lys^8Asn^9]-neurotensin (8–13) and neuromedin N with only one basic amino acid residue show little biological activity. An aromatic amino acid in position 11 in neurotensin (1–13) appears to be essential, since its substitution with phenylalanine or tryptophan (as in xenopsin) yields analogs with substantial biological activity, whereas the substitution with nonaromatic amino acids results in a large reduction in biological activity. In addition, the methyl groups on the side-chains of isoleucine and leucine are indispensible, since their removal results in the loss of biological activity.

On the basis of these findings, it has been proposed (Leeman and Carraway, 1982) that the interaction of neurotensin with its receptor involves strong ionic interactions between the receptor and the positive charges of the two arginine residues in positions 8 and 9, as well as hydrophobic interactions of the receptor with the tyrosine, isoleucine, and leucine residues in positions 11, 12, and 13.

6.2. Radioligand Binding Studies

The presence of specific neurotensin binding sites in rat brain has been demonstrated by using either iodinated (Lazarus et al., 1977a; Uhl et al., 1977) or tritiated (Kitabgi et al., 1977; Granier et al., 1982; Goedert et al., 1984c) neurotensin (1–13) as the radioligand. The iodination of neurotensin results in a mixture of peptides in which one or both tyrosine molecules have incorporated iodine atoms, thereby rendering an unambiguous interpretation of binding data very difficult. In addition, it has been shown that the iodination of the tyrosine residue in position 11 reduces the biological activity of the neurotensin molecule (Mazella et al., 1983). Tritiated neurotensin has been shown to bind to either rat brain synaptic membranes (Kitabgi et al., 1977) or washed rat brain membranes (Goedert et al., 1984c) in a specific, saturable, and reversible manner. Scatchard transformation of the saturation data indicated the presence of a single population of binding sites with an equilibrium dissociation constant of 0.9 n*M* (Kitabgi et al., 1977) or 2.8 n*M* (Goedert et al., 1984c) and a maximum number of binding sites of 135 fmol/mg protein (Kitabgi et al., 1977) or 432 fmol/mg protein (Goedert et al., 1984c). Besides neurotensin (1–13) and neurotensin fragments and analogs, dynorphin (1–13) competed with [^{3}H]-neurotensin for its specific binding site. The sequence similarity between the carboxy-terminal ends of neurotensin (1–13) and dynorphin (1–13) probably accounts for these results (Goedert et al., 1984c). However, if phenomena such as these apply to other peptides and their receptors, results from studies in which intraventricular or local injections of several thousand times the total brain content of a given peptide are applied should be interpreted with caution.

Structure–activity studies (Table 11) on [^{3}H]-neurotensin binding to rat brain have indicated that amino-terminal fragments and carboxy-terminal fragments shorter than the pentapeptide are unable to compete with labeled neurotensin (Kitabgi et al., 1977; Goedert et al., 1984c). Divergent results were obtained for the relative activities of neurotensin (8–13) and neurotensin (9–13), which were 700% and 1.5%, respectively (Kitabgi et al., 1977). In a subsequent investigation, the same authors found a value of 15% for neurotensin (8–13) and 1% for neurotensin (9–13), and they concluded that acetyl-neurotensin (8–13) represents the shortest neurotensin analog with full binding activity (Granier et al., 1982). In our study on washed rat brain membranes, we have found relative activities of 116% for neurotensin (8–13) and 0.51% for neurotensin (9–13) (Goedert et al., 1984c). This discrepancy may be explained by our use of EDTA and bacitracin to stabilize neurotensin (8–13) against aminopeptidase activity. It therefore appears that in the rat brain, neurotensin (8–13) is the shortest fragment with full binding activity. Xenopsin was equipotent with neurotensin (1–13) in competing with [^{3}H]-neurotensin for its specific binding site (Checler et al., 1982; Goedert et al., 1984c). This is probably of no physiological significance, since no xenopsin-like immunoreactivity could be detected in the rat brain (Goedert et al., 1984a). The same applies to [Lys^{8}Asn^9]-neurotensin (8–13), which was only weakly active in competing with [^{3}H]-neurotensin (Goedert et al., 1984c; Kitabgi et al., 1984; Goedert et al., 1985), being consistent with the requirement of two basic residues in positions 8 and 9 for full binding activity; it also agrees with the relatively low potency of the mammalian peptide neuromedin N.

Similar structure–activity relationships were found in rat brain and *postmortem* human brain using monoiodo [^{125}I-Tyr^{3}Trp^{11}]-neurotensin (1–13) as the radioligand (Mazella et al., 1983; Sadoul et al., 1984). The latter retains full biological activity and can be labeled to a high specific activity. Saturation experiments performed with this ligand indicate the presence of two populations of neurotensin binding sites with equilibrium dissociation constants of 0.1 and 4.7 n*M* for rat brain and of 0.26 and 4.3 n*M* for *postmortem* human brain. The two types of binding sites very probably represent two different affinity states of the neurotensin receptor rather than two different receptors. Monoiodo [^{125}I-Tyr^{3}Trp^{11}]-neurotensin (1–13) has been linked covalently to the neurotensin receptor in rat brain membranes, resulting in the labeling of two protein bands with molecular weights of 49,000 and 51,000 (Mazella et al., 1985). These results indicate that the neurotensin receptor may be composed of two protein subunits with a molecular weight of 50,000 held together by noncovalent bonds.

Radioligand binding studies performed on the mammalian CNS are thus in full agreement with the structure–activity requirements observed in peripheral biological assays, suggesting the presence of one population of non-interacting specific neurotensin binding sites in central and peripheral tissues.

The distribution of specific [^{3}H]-neurotensin binding sites in the rat brain has been investigated by both tissue homogenate binding and autoradiographic techniques. In tissue homogenates (Goedert et al., 1984c), the total number of binding sites varied among different brain regions, whereas the equilibrium dissociation constant did not differ significantly among the nine regions investigated (Table 12). The density of binding sites was highest in

TABLE 11

Relative Potencies of Neurotensin Fragments and Related Peptides of the Neurotensin Family in Binding to Central and Peripheral Tissues

Peptide	Rat brain membranes[a]	Rat adrenal gland membranes[b]	Guinea pig ileum membranes[c]
Neurotensin (1–13)	1.00	1.00	1.00
Neurotensin (8–13)	1.16	1.08	0.43
Neurotensin (9–13)	0.51	0.90	0.50
Neurotensin (10–13)	<0.0004	<0.0004	<0.0004
Neurotensin (1–6)	<0.0004	<0.0004	<0.0004
Neurotensin (1–11)	<0.0004	<0.0004	<0.0004
Neuromedin N	0.14	ND[d]	ND
[Lys^8Asn^9]-neurotensin (8–13)	0.02	ND	ND
Chicken neurotensin (1–13)	0.96	ND	ND
Xenopsin	0.89	ND	ND

[a]Goedert et al., 1984c.
[b]Goedert et al., 1984f.
[c]Goedert et al., 1984b.
[d]ND, not determined.

the hypothalamus and lowest in the cerebellum and pons-medulla; this distribution agreed reasonably well with the levels of neurotensin-like immunoreactivity in these tissues. However, the frontal cortex contained a high density of neurotensin binding sites, whereas little neurotensin-like immunoreactivity was detected. Similar contrasts were found for the striatum, the thalamus, and the olfactory bulb. In general, the variation among different brain regions was much less for [^{3}H]-neurotensin binding than for neurotensin-like immunoreactivity.

The distribution of [^{3}H]-neurotensin binding sites in the rat CNS was also investigated by in vitro receptor autoradiography (Figure 5) (Young and Kuhar, 1981; Quirion et al., 1982). Saturation analysis on brain slices revealed the presence of one population of binding sites with an equilibrium dissociation constant of 6 n*M* and a maximum number of binding sites of 20 fmol/mg tissue. Moreover, the carboxy-terminal fragment neurotensin (4–13) was found to be 140 times more potent than neurotensin (9–13) in competing with tritiated neurotensin for its binding site (Young and Kuhar, 1981). The highest concentration of binding sites was found in the external layer of the olfactory bulb. High levels were also present in the anterior cingulate cortex, the rhinal sulcus, the habenula, the zona incerta, the pars compacta of the substantia nigra, and the ventral tegmental area. Medium levels were present in the striatum, the amygdala, the superior colliculus, the periaqueductal grey, and the solitary tract nucleus. Low levels were observed in the hypothalamus and most cortical areas. A high concentration of neurotensin binding sites was found in the substantia gelatinosa of the dorsal horn of the spinal cord (Ninkovic et al., 1981). The injection of 6-hydroxydopamine into the rat substantia nigra resulted in a marked reduction in [^{3}H]-neurotensin binding sites, indicating that a large proportion of these binding sites is located on dopaminergic cell bodies (Palacios and Kuhar, 1981). Similarly, the number of [^{3}H]-neurotensin binding sites appeared to be reduced in the substantia nigra of patients with Parkinson's disease (Uhl et al., 1984). Interestingly, the chronic administration of neuroleptics has been shown to result in an apparent increase in [^{3}H]-neurotensin binding sites in the rat substantia nigra (Uhl and Kuhar, 1984). [^{3}H]-neurotensin binding sites were also found in the internal plexiform layer of both chicken and rat retinae (manuscript in preparation), in which neurotensin-like immunoreactivity is present in a population of amacrine cells (Brecha et al., 1981; Fukuda et al., 1981). As was found by using the tissue homogenate binding assay for [^{3}H]-neurotensin, the autoradiographic technique revealed examples of mismatch between levels of neu-

TABLE 12

Distribution of [^{3}H]-Neurotensin Binding Sites and Neurotensin-Like Immunoreactivity in the Rat Brain

Brain region	[^{3}H]-neurotensin binding sites, fmol/mg protein	K_d, nM	Neurotensin-like immunoreactivity, pmol/g tissue
Hypothalamus	565 ± 62	2.43 ± 0.10	76.15 ± 8.30
Frontal cortex	524 ± 56	3.09 ± 0.39	1.48 ± 0.25
Midbrain	448 ± 25	3.72 ± 0.63	45.84 ± 6.44
Striatum	418 ± 43	2.64 ± 0.40	19.02 ± 1.94
Thalamus	392 ± 35	3.08 ± 0.63	24.94 ± 2.70
Hippocampus	374 ± 22	2.37 ± 0.20	8.16 ± 0.92
Olfactory bulb	366 ± 18	3.64 ± 0.21	3.56 ± 0.28
Cerebellum	212 ± 18	3.71 ± 0.39	0.85 ± 0.18
Pons-medulla	202 ± 14	2.39 ± 0.25	22.75 ± 1.06

rotensin-like immunoreactivity and density of [^{3}H]-neurotensin binding sites, particularly in the entorhinal cortex and the cerebral cortex around the rhinal fissure.

A strikingly different topographical localization of neurotensin-like immunoreactivity (Goedert et al., 1983) and [^{3}H]-neurotensin binding sites (Goedert et al., 1984d) within a structure characterized by distinct histochemical compartments was found in the case of the cat corpus striatum (Figs. 6 and 7). In the caudate nucleus, neurotensin-like immunoreactivity presents a mosaic distribution that coincides with the distribution of striosomes defined by their high content of [Met]-enkephalin-like immunoreactivity and opiate receptor binding, as well as by their low acetylcholinesterase activity (Graybiel and Ragsdale, 1983). In contrast to [Met]-enkephalin-like immunoreactivity and opiate receptors, neurotensin-like immunoreactivity and [^{3}H]-neurotensin binding sites were inversely related (Figs. 6 and 7). The localization of striatal neurotensin receptors in the rat has been investigated by using kainic acid and 6-hydroxydopamine injections, as well as frontal cortex ablation (Goedert et al., 1984e). More than 50% of the specific [^{3}H]-neurotensin binding sites was localized on intrinsic striatal neurons, approximately 35% on dopaminergic nerve terminals, and 20% on corticostriatal nerve fibers. This indicates that specific [^{3}H]-neurotensin binding sites in the rat striatum are localized on neurons and their processes, rather than on glial cells.

The reason for the inverse relationship between peptide-like immunoreactivity and peptide receptors is unknown at present. It could be caused by different turnover rates for a given peptide in different brain regions. An excess of peptide levels over binding density could also be caused by the presence of low-affinity bind-

→

Fig. 5. Autoradiographic localization of [^{3}H]-neurotensin binding sites in the rat brain. (a) Adjacent section to (b) incubated in the presence of 2 n*M* [^{3}H]-neurotensin and 1 μ*M* neurotensin (1–13). (b–f) Horizontal sections taken at different levels of the brain, with (b) representing the most dorsal and (f) the most ventral sections. All the sections were incubated in the presence of 2 n*M* [^{3}H]-neurotensin. Scale bars, 2.5 mm. Abbreviations: Fr, frontal cortex; DG, dentate gyrus; SC, superior colliculus; IC, inferior colliculus; Ent, entorhinal cortex; Cb, cerebellum; ACg, anterior cingulate cortex; CPu, caudate-putamen; sm, stria medullaris; CG, central grey; OlfT, olfactory tubercle; DR, dorsal raphe; Tg, tegmentum; Sol, solitary tract nucleus; RF, rhinal fissure; LS, lateral septum; ZI, zona incerta; PH, posterior hypothalamus; SNc, substantia nigra, pars compacta; Pa, paraventricular nucleus; VTA, ventral tegmental area; SNr, substantia nigra, pars reticulata.

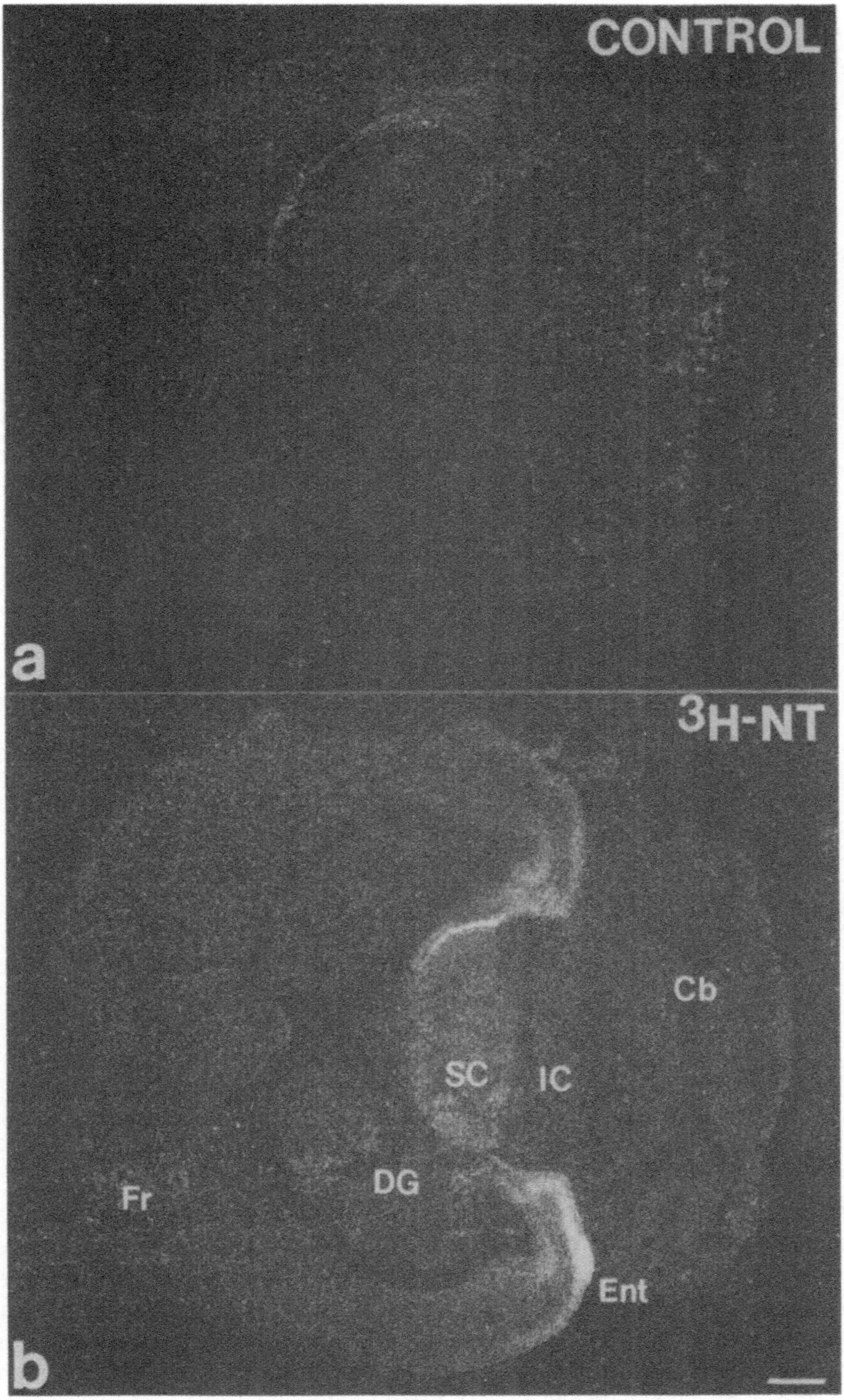
CONTROL
a
3H-NT
Cb
SC
IC
DG
Fr
Ent
b

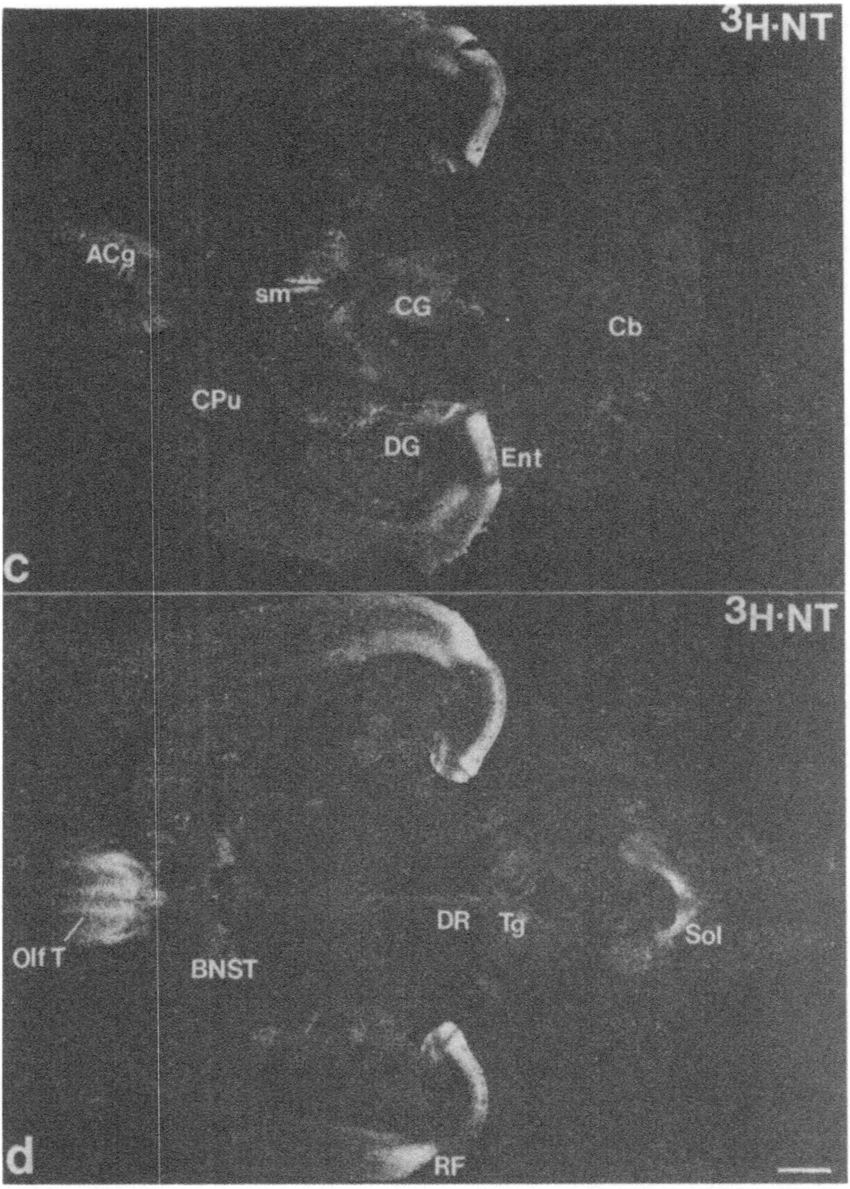

ing sites not detected in binding studies operating at nanomolar range. An excess of binding sites over peptide levels could occur because the density of receptors is determined by the presence or absence of peptide-containing nerve fibers rather than by the number of these fibers, as suggested previously for classical neurotransmitters (Snyder, 1975). Alternatively, it could be because the endogenous ligands are able to bind to the binding sites but are not

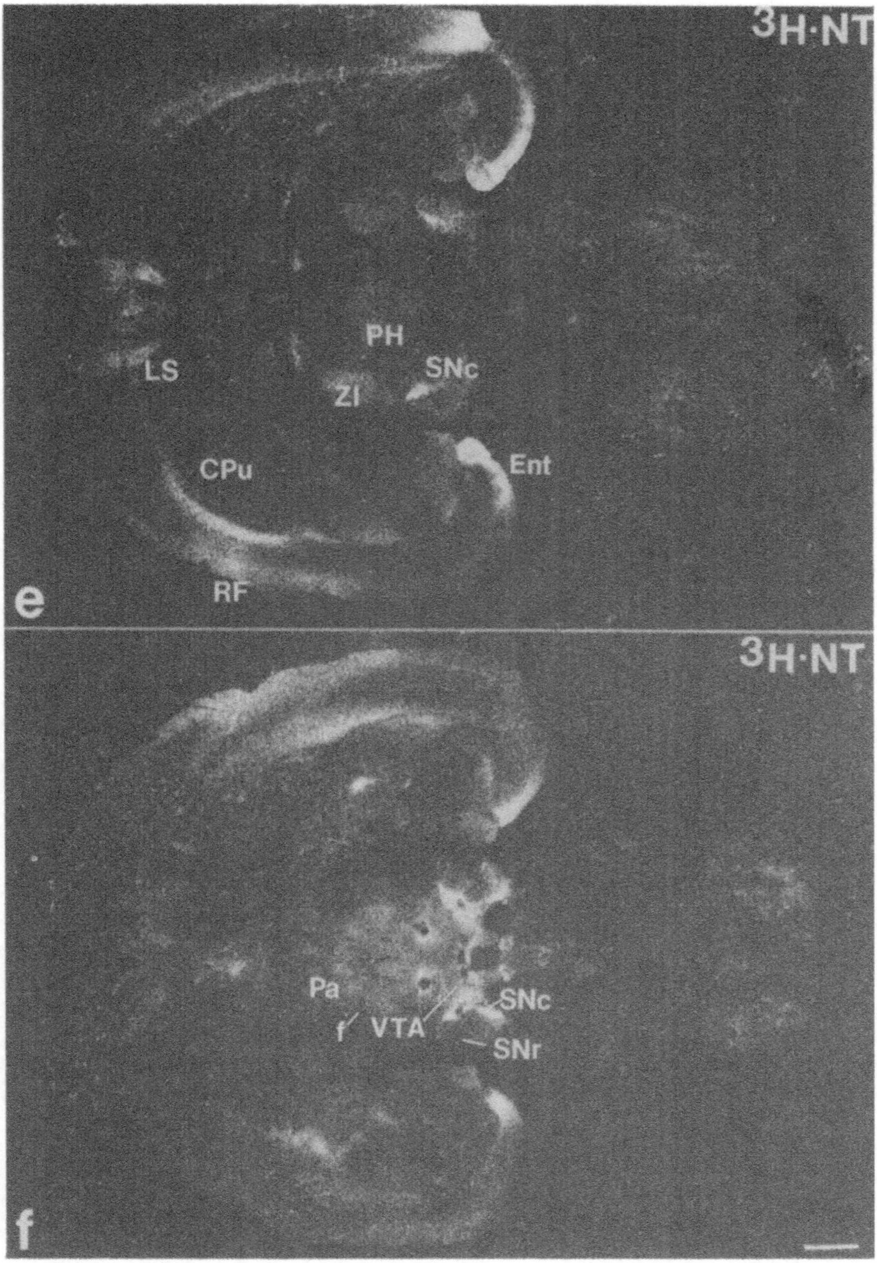

recognized by radioimmunoassay or immunohistochemistry. In the future it will be important to find a rational explanation for these discrepancies.

Comparatively little is known about neurotensin receptors in peripheral tissues. The binding sites on rat mast cells are clearly

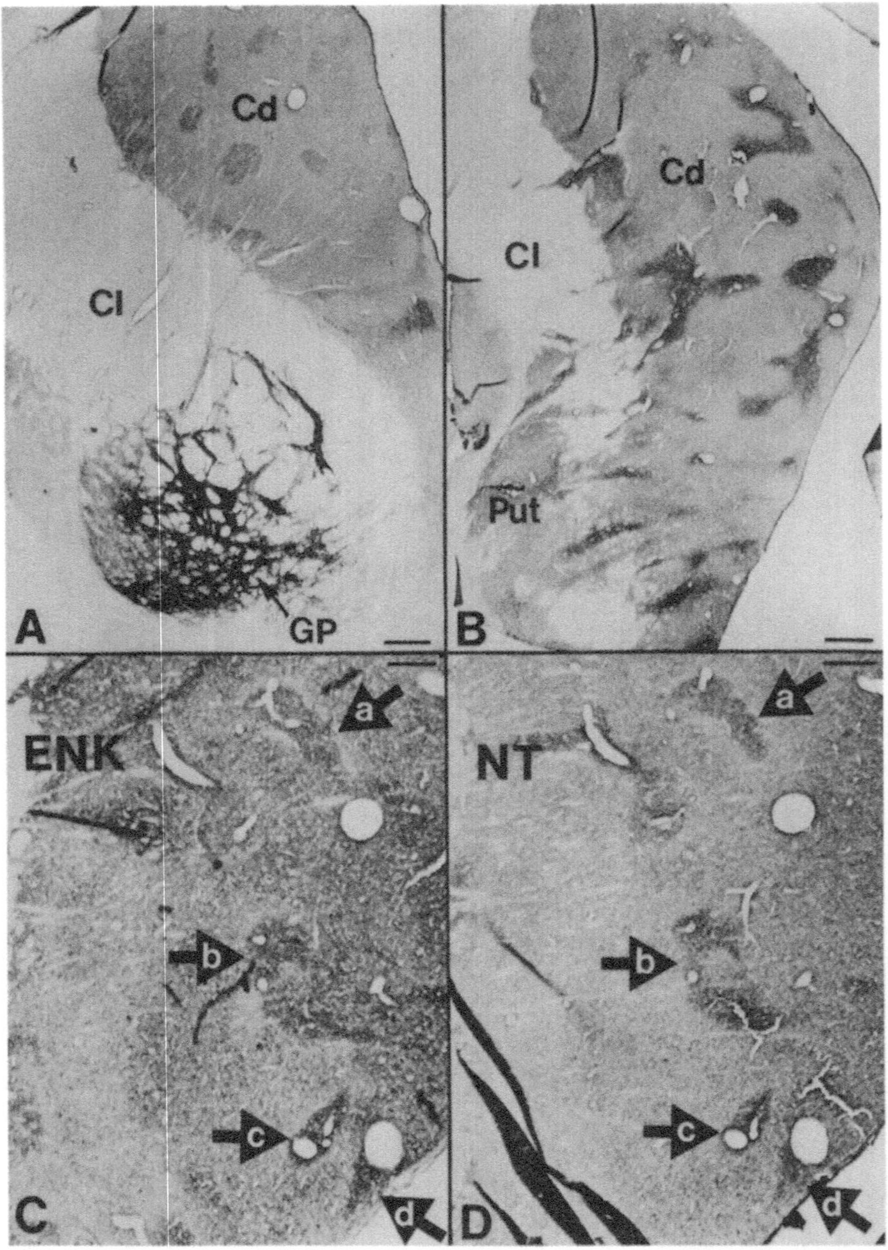

Fig. 6. Immunohistochemical localization of neurotensin-like immunoreactivity in the striatum of the cat (A,B). Note that in the caudate nucleus (Cd) there is a patchy distribution, whereas the globus pallidus (GP) has a homogeneous distribution broken up by myelinated fiber bundles. Scale bars, 450 μm. Photomicrographs (C) and (D) show

distinct from those in the CNS, with an equilibrium dissociation constant of 154 n*M* (Lazarus et al., 1977b). Autoradiographically, the majority of [^{3}H]-neurotensin binding sites in the guinea pig ileum was found to be localized in the circular smooth muscle layer, with lower levels in the longitudinal muscle (Fig. 8) (Goedert et al., 1984b). The characteristics of the specific [^{3}H]-neurotensin binding sites in ileal homogenates were consistent with their binding to a physiological neurotensin receptor. The potencies of the various neurotensin fragments (Table 11) were in good agreement with their effects on the relaxation of the circular smooth muscle preparation and with their actions on the longitudinal smooth muscle layer (Kitabgi, 1982; Goedert et al., 1984b). In endocrine tissues, [^{3}H]-neurotensin binding sites have been found in the pituitary gland and the adrenal gland. [^{3}H]-neurotensin binding sites were concentrated in the intermediate lobe of the rat (Fig. 9) and cat pituitary glands by autoradiography, with smaller amounts in both anterior and neural lobes. In the rat, the [^{3}H]-neurotensin binding sites did not appear to be affected by pituitary stalk transection, indicating that they are localized on cells of the intermediate lobe. Using the test-tube binding assay for [^{3}H]-neurotensin, low levels of specific binding sites were found in whole rat pituitary glands. However, considering that the intermediate lobe comprises less than 5% of the total pituitary gland, it appears that the local concentration of [^{3}H]-neurotensin binding sites is comparable with that found in the CNS. Hopefully, future experiments will investigate the effects of neurotensin on the synthesis and release of intermediate lobe hormones. In the rat adrenal gland, high levels of [^{3}H]-neurotensin binding sites were found in the inner layer of the rat adrenal cortex and lower amounts in the adrenal medulla (Fig. 10) (Goedert et al., 1984f). The structure–activity profile of these specific binding sites in whole rat adrenal gland homogenates (Table 11) closely resembled the values obtained in peripheral bioassays and in binding assays in the CNS, therefore being consistent with binding to a physiological neurotensin receptor. As in the pituitary gland, the number of specific [^{3}H]-neurotensin binding sites was low; however, the fact

---

[Met]enkephalin-like immunoreactivity (ENK) and neurotensin-like immunoreactivity (NT) in adjacent sections of the caudate nucleus. The arrows indicate patches of immunoreactivity that appear to overlap in the two adjacent 30 μm brain sections. Scale bars, 250 μm. Abbreviations: Cd, caudate nucleus; CI, internal capsule; GP, globus pallidus; Put, putamen; ENK, enkephalin; NT, neurotensin (reprinted by permission from Goedert et al., 1985).

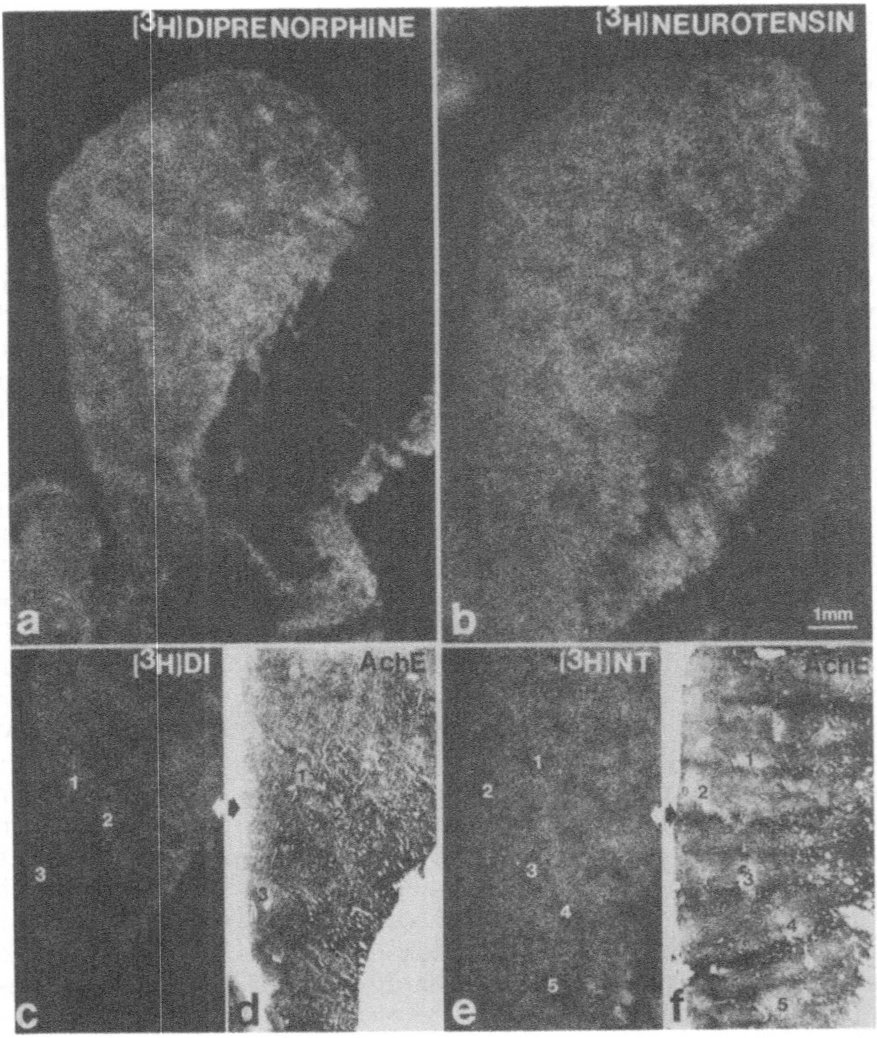

Fig. 7. Autoradiographic localization of [^{3}H]-neurotensin and [^{3}H]-diprenorphine binding sites in the cat caudate nucleus. The tissue sections were incubated in the presence of either 2 n*M* [^{3}H]-neurotensin or 4 n*M* [^{3}H]-diprenorphine. (a,b) The cat caudate nucleus labeled with [^{3}H]-diprenorphine and [^{3}H]-neurotensin, respectively. (c,d) Labeling of the sections with [^{3}H]-diprenorphine ([^{3}H]-DI) and visualization of acetylcholinesterase activity (AchE). (e,f) Labeling of the sections with [^{3}H]-neurotensin ([^{3}H]-NT) and visualization of acetylcholinesterase activity (AchE). The numbers in (c–f) refer to the acetylcholinesterase-poor striosomes. Note the correspondence between striosomes and dense [^{3}H]-diprenorphine labeling and the inverse relationship between striosomes and [^{3}H]-neurotensin binding sites. Scale bar, 1 mm (reprinted by permission from Goedert et al., 1984e).

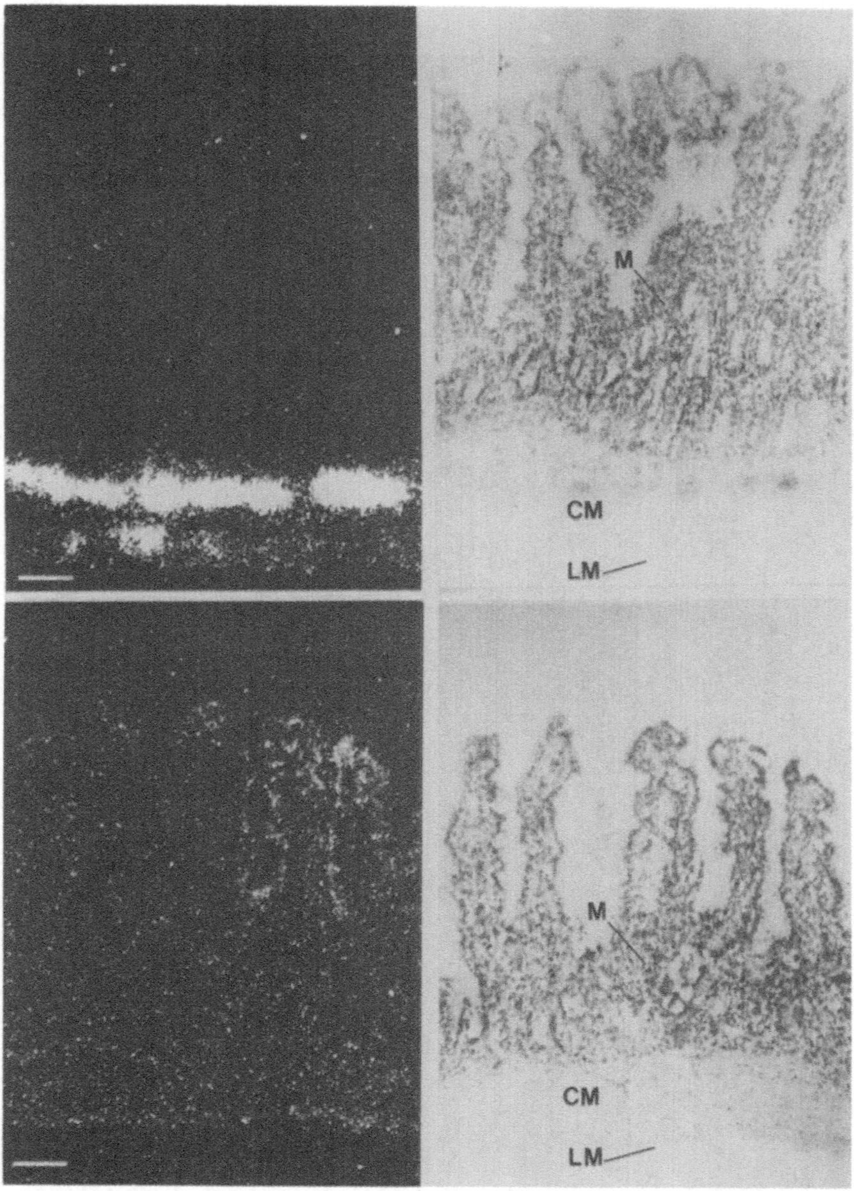

Fig. 8. Autoradiographic localization of [^{3}H]-neurotensin binding sites in the guinea pig ileum. (upper left) Darkfield photomicrograph of the autoradiogram of a transverse section after incubation in the presence of 2 n*M* [^{3}H]-neurotensin. (upper right) Lightfield photomicrograph of the Nissl stain of the same section as in upper left panel. (Lower left) Darkfield photomicrograph of the autoradiogram of a section adjacent to upper left panel after incubation in the presence of 2 n*M* [^{3}H]-neurotensin and 1 μ*M* neurotensin (1–13). (Lower right) Lightfield photomicrograph of the Nissl stain of the same section as in lower left panel. Scale bar, 100 μm. Note the heavy labeling of the circular and longitudinal muscle layers in upper left. Abbreviations: M, mucosa; CM, circular muscle; LM, longitudinal muscle (reprinted by permission from Goedert et al., 1984c).

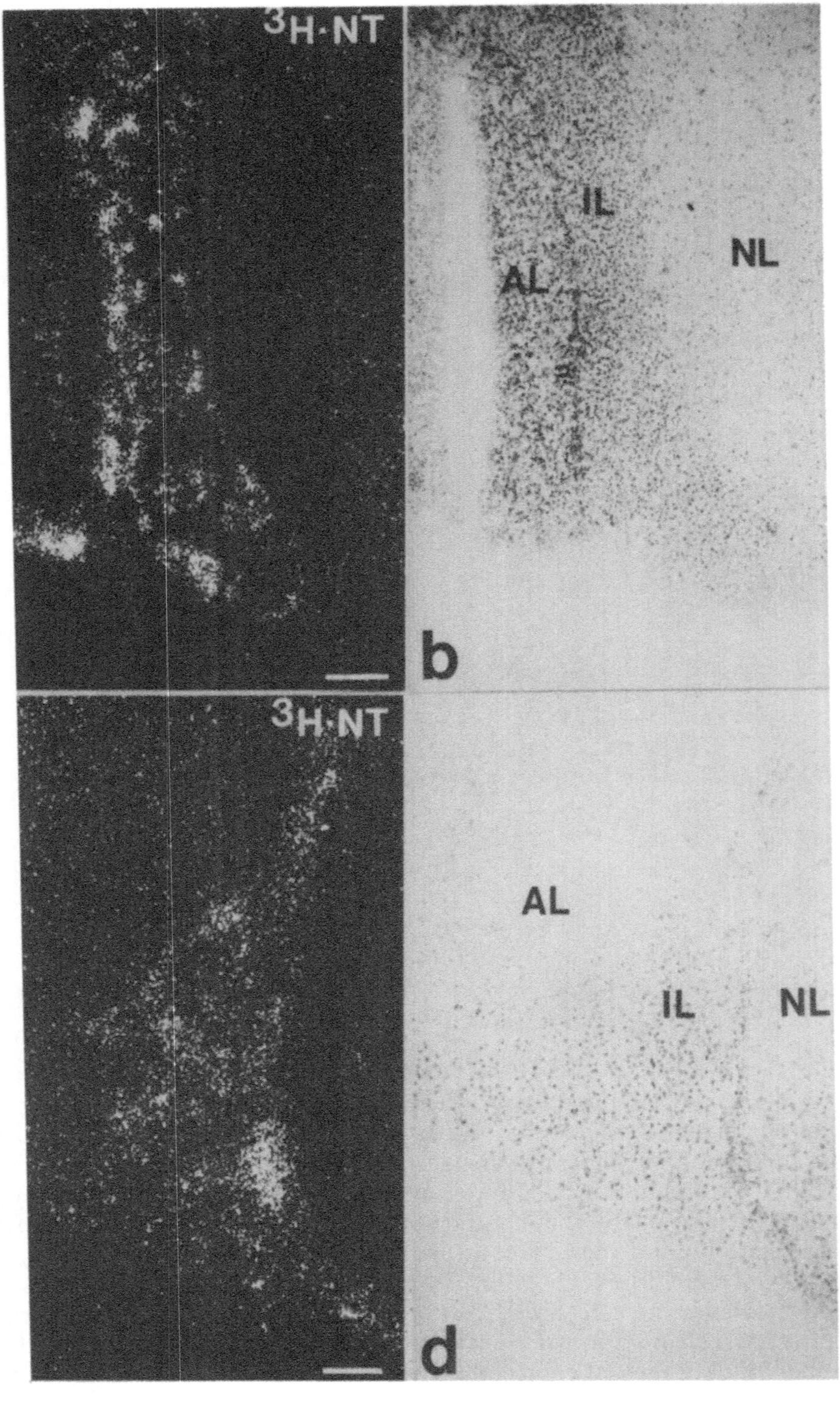

³H·NT
IL
NL
AL
b
³H·NT
AL
IL
NL
d

that 75% of the binding was present in one layer of the adrenal cortex implies that the local concentration of neurotensin binding sites may be comparable with the levels found in the CNS. The physiological significance of the specific [^{3}H]-neurotensin binding sites in the zona reticularis of the adrenal cortex is unknown. It is possible that circulating neurotensin levels could influence steroid secretion from the adrenal cortex; alternatively, the neurotensin from the adrenal medulla could influence the adrenal cortex via retrograde blood flow. Approximately 25% of the total number of adrenal gland [^{3}H]-neurotensin binding sites were found in the adrenal medulla. It has been shown that both the increase in blood pressure and the tachycardia following the intravenous injection of neurotensin are markedly reduced in adrenalectomized rats (Rioux et al., 1982). It seems likely that this effect is mediated through the [^{3}H]-neurotensin binding sites present in the adrenal medulla. In the cat, neurotensin-like immunoreactivity was also found in the splanchnic nerve (Goedert and Emson, 1983). If this also occurs in the rat, immunoreactive material released from the splanchnic nerve could influence adrenal medullary function through the [^{3}H]-neurotensin binding sites.

6.3. Second Messengers

Neurotensin has been shown to have no significant effect on cyclic AMP formation in guinea pig ileum longitudinal smooth muscle and rat hypothalamus (Quik et al., 1978; Kitabgi and Freychet, 1979a). In addition, it does not affect the stimulation of cyclic AMP formation effected by either forskolin or VIP (Goedert et al., 1984g). These results are in agreement with the finding that [^{3}H]-neurotensin binding was not influenced by either divalent cations or guanyl nucleotides (Goedert et al., 1984c).

Some experiments have indicated that neurotensin may produce an increase in intracellular calcium concentrations (Goldman et al., 1983; Kitabgi and Vincent, 1981). Not surprisingly, therefore, neurotensin was found to stimulate inositol phospholipid hydroly-

←

Fig. 9. Autoradiographic localization of [^{3}H]-neurotensin binding sites in the rat pituitary gland. (a,c) Darkfield photomicrographs of the autoradiogram of transverse sections through the rat pituitary gland after incubation in the presence of 2 n*M* [^{3}H]-neurotensin. (b,d) Lightfield photomicrographs of the Nissl stains of the same sections as in (a) and (c). Scale bars, 100 μm. Note the patchy accumulation of silver grains over the intermediate lobe in (a) and (c). Abbreviations: AL, anterior lobe; IL, intermediate lobe; NL, neural lobe.

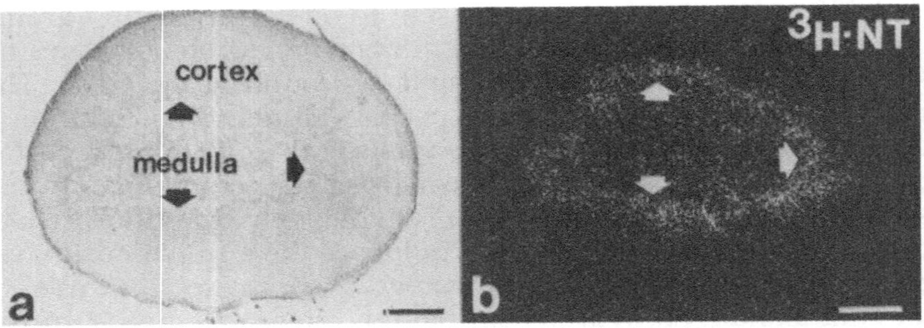

Fig. 10. Autoradiographic localization of [^{3}H]-neurotensin binding sites in the rat adrenal gland. (a,c) Lightfield photomicrograph of the Nissl strain of a transverse section through the adrenal gland. (b) Dark-field photomicrograph of the autoradiogram of the same section as in (a) after incubation in the presence of 2 n*M* [^{3}H]-neurotensin. Note the dense labeling in the inner cortical layer (arrows) and the labeling in the adrenal medulla. Scale bar, 540 μm (reprinted by permission from Goedert et al., 1984g).

sis in rat brain slices (Goedert et al., 1984g). This effect appeared to be direct, since it was not significantly influenced by tetrodotoxin. Moreover, there existed a good correlation between the magnitude of inositol phospholipid hydrolysis and the number of specific [^{3}H]-neurotensin binding sites in various regions of the rat brain (Table 13). Recently, neurotensin has also been shown to increase cyclic GMP levels in neuroblastoma cells maintained in tissue culture (Gilbert and Richelson, 1984). It has been repeatedly shown that the occupancy of receptors linked to inositol phospholipid hydrolysis leads also to increases in arachidonic acid formation and cyclic GMP production (Schultz et al., 1977; Bell et al., 1979). Neither the exact mechanism of cyclic GMP formation nor its functional importance are clearly understood (Nishizuka, 1984). However, it is well established that the increase in cyclic GMP formation is consequent to inositol phospholipid hydrolysis. Inositol phospholipid hydrolysis may thus constitute a valuable biochemical index for neurotensin receptors in the CNS.

7. Conclusion

The study of receptors for nonopioid neuropeptides is still in its infancy. This is hardly surprising, since the amino acid sequence of a substantial number of these neuropeptides has only been elucidated over the last few years. Nevertheless, a remarkable

TABLE 13

Comparison Between the Relative Number of Specific Neurotensin Binding Sites and the Relative Magnitude of Neurotensin-Stimulated Inositol Phospholipid Hydrolysis in Various Regions of the Rat Brain

Brain area	Number of binding sites[a]	Magnitude of inositol phospholipid hydrolysis[a]
Hypothalamus	1.00 ± 0.11	1.00 ± 0.06
Frontal cortex	0.93 ± 0.10	0.81 ± 0.03
Midbrain	0.79 ± 0.04	0.88 ± 0.06
Striatum	0.74 ± 0.01	0.50 ± 0.01
Hippocampus	0.66 ± 0.04	0.67 ± 0.01
Cerebellum	0.38 ± 0.03	0.33 ± 0.02
Pons-medulla	0.36 ± 0.02	0.27 ± 0.01

[a]The results are expressed relative to the hypothalamus (taken as 1.00).

amount of information has already been obtained about the characteristics of some of these receptors and the biochemical events consequent to receptor occupancy. However, clearly much more remains to be learned.

To this end, the single most important development will be the availability of potent and specific receptor antagonists. They constitute the necessary tools for establishing the physiological function of nonopioid neuropeptides. They also represent a necessary prerequisite for the classification of receptor subtypes; this is particularly important for neuropeptides, since the latter exert such a bewildering variety of biological actions that is very likely to be mediated through multiple receptor subtypes.

The development of autoradiographic techniques for the study of neuropeptide receptors has rendered possible a more detailed investigation of their cellular localization. At present, the autoradiographic technique cannot provide any information about the ultrastructural localization of neuropeptide receptors. Thus, an important area for future research will consist of the development of methods that allow the localization of neuropeptide receptors at the electron microscopic level. This has already been achieved for the nicotinic acetylcholine receptor and the low-density lipoprotein receptor by using antibodies raised against the purified receptor molecule (Sealock et al., 1984; Schneider et al., 1983). An alternative way to obtain such antibodies would be to raise anti-idiotypic antibodies that have the availability to bind to the receptor molecule (Jerne, 1974; Sege and Peterson, 1978).

Radioligand binding studies are by their very nature indirect, deducing the existence of a receptor from the characteristics of membrane sites identified through radioligand binding. Therefore, the purification and structural characterization of receptors must constitute the ultimate aim of receptor research. The development of protein microsequencing techniques and the advent of recombinant DNA technology, coupled to techniques such as *Xenopus laevis* oocyte injections of messenger RNA, have already led to the elucidation of the primary structure of several receptors, such as the nicotinic acetylcholine and the insulin receptor molecules (Noda et al., 1983; Ullrich et al., 1985). The nicotinic acetylcholine receptor has been functionally reconstituted through the introduction of the cloned genes into mammalian cells and some of the sequences involved in ligand binding and signal transduction have been determined by site-directed mutagenesis (Hishina et al., 1984, 1985).

If the last decade was marked by the advent of radioligand binding techniques, the next decade will undoubtedly belong to the structural characterization of receptor molecules. Hopefully, this will also be true for nonopioid peptide receptors.

Acknowledgment

I would like to thank Mrs. M. Wynn for her great help in the preparation of this manuscript.

References

Amiranoff B., Laburthe M., and Rosselin G. (1980) Characterisation of specific binding sites for VIP in rat intestinal epithelial cell membranes. *Biochim. Biophys. Acta* **3,** 215–224.

Araki K., Tachibana S., Uchiyama M., Nakajima T., and Yasuhara T. (1973) Isolation and structure of a new active peptide "xenopsin" on the smooth muscle, especially on a strip of fundus from a rat stomach, from the skin of *Xenopus laevis*. *Chem. Pharm. Bull.* **21,** 2801–2804.

Bannon M. J., Elliott P. J., Alpert J. E., Goedert M., Iversen S. D., and Iversen L. L. (1983) Role of endogenous substance P in stress-induced activation of mesocortical dopamine neurones. *Nature* (Lond.) **306,** 791–792.

Bataille D., Freychet P., and Rosselin G. (1974) Interactions of glucagon, vasoactive intestinal polypeptide and secretin with liver and fat cell plasma membranes: Binding to specific sites and stimulation of adenylate cyclase. *Endocrinology* **95,** 713–721.

Bataille D., Gespach C., Laburthe M., Amiranoff B., Tatemoto K., Vauclin N., Mutt V., and Rosselin G. (1980) Porcine peptide having *N*-terminal histidine and *C*-terminal isoleucine amide. Vasoactive intestinal polypeptide and secretin-like effects in different tissues from the rat. *FEBS Lett.* **114,** 240–242.

Bauer V. and Kuriyama H. (1982a) Evidence for non-cholinergic, non-adrenergic transmission in the guinea-pig ileum. *J. Physiol.* (Lond.) **330,** 95–110.

Bauer V. and Kuriyama H. (1982b) The nature of non-cholinergic, non-adrenergic transmission in longitudinal and circular muscles of the guinea pig ileum. *J. Physiol.* (Lond.) **332,** 375–391.

Beaujouan J. C., Torrens Y., Viger A., and Glowinski J. (1984) A new type of tachykinin binding site in the rat brain characterized by specific binding of a labeled eledoisin derivative. *Mol. Pharmacol.* **26,** 248–256.

Bell R. L., Kennerly D. A., Stanford N., and Majerus P. W. (1979) Diglyceride lipase: A pathway for arachidonate release from human platelets. *Proc. Natl. Acad. Sci. USA* **76,** 3238–3241.

Berridge M. J. (1984) Inositol trisphosphate and diacylglycerol as second messengers. *Biochem. J.* **220,** 345–355.

Bitar K. N. and Makhlouf G. M. (1982) Receptors on smooth muscle cells: Characterization by contraction and specific antagonists. *Am. J. Physiol.* **242,** G400–G407.

Bloom S. R., Polak J. M., and Pearse A. G. E. (1973) Vasoactive intestinal peptide and watery diarrhea syndrome. *Lancet* **ii,** 14–16.

Bloom S. R., Christofides N. D., Delamarter J., Buell G., Kawashima E., and Polak J. M. (1983) Diarrhoea in vipoma patients associated with cosecretion of a second active peptide (peptide histidine isoleucine) explained by single coding gene. *Lancet* **ii,** 1163–1165.

Boel E., Vuust J., Norris F., Norris K., Wind A., Rehfeld J. F., and Marcker K. A. (1983) Molecular cloning of human gastrin cDNA: Evidence for evolution of gastrin gene by duplication. *Proc. Natl. Acad. Sci. USA* **80,** 2866–2869.

Borghi C., Nicosia S., Giachetti A., and Said S. I. (1979) Vasoactive intestinal polypeptide-stimulated adenylate cyclase in selected areas of rat brain. *Life Sci.* **24,** 65–70.

Brecha N., Karten H. J., and Schenker C. (1981) Neurotensin-like and somatostatin-like immunoreactivity within amacrine cells of the retina. *Neuroscience* **6,** 1329–1340.

Brown C. L. and Hanley M. R. (1981) The effects of substance P and related peptides on α-amylase release from rat parotid gland slices. *Brit. J. Pharmacol.* **73,** 517–523.

Brownstein M. J., Mroz E. A., Kizer J. S., Palkovits M., and Leeman S. E. (1976) Regional distribution of substance P in the brain of the rat. *Brain Res.* **116,** 299–305.

Buck S. H., Burcher E., Shults C. W., Lovenberg W., and O'Donohue T. L. (1984) Novel pharmacology of substance K-binding sites: A third type of tachykinin receptor. *Science* **226,** 987–989.

Buck S. H., Maurin Y., Burks T. F., and Yamamura H. I. (1984) High-affinity ^{3}H-substance P binding to longitudinal muscle membranes of the guinea-pig small intestine. *Life Sci.* **34,** 497–507.

Burcher E. Shults C. W., Buck S. H., Chase T. N., and O'Donohue T. L. (1984) Autoradiographic distribution of substance K binding sites in rat gastrointestinal tract: A comparison with substance P. *Eur. J. Pharmacol.* **102,** 561–562.

Burgess G. M., Godfrey P. P., McKinney J. S., Berridge M. J., Irvine R. F., and Putney J. W. (1984) The second messenger linking receptor activation to internal Ca release in liver. *Nature* (Lond.) **309,** 63–66.

Bury R. W. and Mashford M. L. (1977) Substance P: Its pharmacology and physiological roles. *Aust. J. Exp. Biol. Med. Sci.* **55,** 671–735.

Carlquist M., Mutt V., and Jörnvall H. (1979) Isolation and characterisation of bovine vasoactive intestinal peptide. *FEBS Lett.* **108,** 457–460.

Carraway R. and Bhatnagar Y. M. (1980) Isolation, structure and biologic activity of chicken intestinal neurotensin. *Peptides* **1,** 167–174.

Carraway R. and Ferris C. F. (1983) Isolation, biological and chemical characterization, and synthesis of a neurotensin-related hexapeptide from chicken intestine. *J. Biol. Chem.* **258,** 2475–2479.

Carraway R. and Leeman S. E. (1973) The isolation of a new hypotensive peptide, neurotensin, from bovine hypothalami. *J. Biol. Chem.* **248,** 6854–6861.

Carraway R. and Leeman S. E. (1975) The amino acid sequence of a hypothalamic peptide, neurotensin. *J. Biol. Chem.* **250,** 1907–1911.

Carraway R., Kitabgi P., and Leeman S. E. (1978) The amino acid sequence of radioimmunoassayable neurotensin from bovine intestine. *J. Biol. Chem.* **253,** 7996–7998.

Cascieri M. A. and Liang T. (1983) Characterisation of the substance P receptor in brain cortex membranes and the inhibition of radioligand binding by guanine nucleotides. *J. Biol. Chem.* **258,** 5154–5161.

Cascieri M. A. and Liang T. (1984) Binding of [^{125}I]Bolton-Hunter conjugated eledoisin to rat brain cortex membranes: Evidence for two classes of tachykinin receptors in the mammalian central nervous system. *Life Sci.* **35,** 179–185.

Cascieri M. A., Chicchi G. G., and Liang T. (1985) Demonstration of two distinct tachykinin receptors in rat brain. *J. Biol. Chem.* **260,** 1501–1507.

Chang M. M. and Leeman S. E. (1970) Isolation of a sialogogic peptide from bovine hypothalamic tissue and its characterization as substance P. *J. Biol. Chem.* **245,** 4784–4790.

Chang M. M., Leeman S. E., and Niall H. D. (1971) Amino acid sequence of substance P. *Nature New Biol.* **232,** 86–87.

Chang R. S. L., Lotti V. J., and Chen T. B. (1985) CCK receptor mediated hydrolysis of inositol phospholipids in guinea pig gastric glands. *Life Sci.* **36,** 965–973.

Checler F., Labbé C., Granier G., Van Rietschoten J., Kitabgi P., and Vincent J. P. (1982) [Trp^{11}]-neurotensin and xenopsin discriminate between rat and guinea-pig neurotensin receptors. *Life Sci.* **31,** 1145–1150.

Cheeseman H. J., Pinnock R. D., and Henderson G. (1983) Substance P excitation of rat locus coeruleus neurones. *Eur. J. Pharmacol.* **94,** 93–99.

Chiodo L. A. and Bunney B. S. (1983) Proglumide: Selective antagonism of excitatory effects of cholecystokinin in central nervous system. *Science* **219,** 1449–1450.

Christophe J. P., Conlon T. P., and Gardner J. D. (1976) Interaction of porcine vasoactive intestinal peptide with dispersed pancreatic acinar cells from the guinea pig. *J. Biol. Chem.* **251,** 4629–4634.

Couvineau A., Royer-Fessard C., Fournier A., St Pierre S., Pipkorn R., and Laburthe M. (1984) Structural requirements for VIP interaction with specific receptors in human and rat intestinal membranes: Effect of nine partial sequences. *Biochem. Biophys. Res. Commun.* **121,** 493–498.

Davison J. S. and Najafi-Farashah A. (1981) Dibutyryl cyclic GMP, a competitive inhibitor of cholecystokinin/pancreozymin and related peptides in the gallbladder and ileum. *Can. J. Physiol. Pharmacol.* **59,** 1100–1104.

De Quidt M. E. and Emson P. C. (1983) Neurotensin facilitates dopamine release *in vitro* from rat striatal slices. *Brain Res.* **274,** 376–380.

Desbuquois B. (1974) The interaction of vasoactive intestinal polypeptide and secretin with liver cell membranes. *Eur. J. Biochem.* **46,** 439–450.

Desbuquois B., Laudat M. J., and Laudat P. (1973) Vasoactive intestinal polypeptide and glucagon: Stimulation of adenylate cyclase activity via distinct receptors in liver and fat cell membranes. *Biochem. Biophys. Res. Commun.* **53,** 1187–1194.

Deschenes R. J., Lorenz L. J., Haun R. S., Roos B. A., Collier K. J., and Dixon J. E. (1984) Cloning and sequence analysis of cDNA encoding rat preprocholecystokinin. *Proc. Natl. Acad. Sci. USA* **81,** 726–730.

Dockray G. J., Gregory R. A., Hutchinson J. B., Harris J. I., and Funswick M. J. (1978) Isolation, structure, and biological activity of two cholecystokinin octapeptides from sheep brain. *Nature* (Lond.) **274,** 711–713.

Emson P. C. and Marley P. D. (1983) Cholecystokinin and Vasoactive Intestinal Polypeptide, in *Handbook of Psychopharmacology* (Iversen L. L., Iversen S. D., and Snyder S. H., eds.), Vol. 16, pp. 255–306, Plenum, New York.

Emson P. C., Goedert M., and Mantyh P. W. (1985) Neurotensin, in *Handbook of Chemical Neuroanatomy* (Hökfelt T, and Björklund A., eds.), Vol. 4, Elsevier, Amsterdam, in press.

Erspamer V. (1981) The tachykinin peptide family: Present status and perspectives. *Trends Neurosci.* **4,** 267–269.

Erspamer V. and Melchiorri P. (1983) Actions of Amphibian Skin Peptides on the Central Nervous System and the Anterior Pituitary, in *Neuroendocrine Perspectives* (Müller E. E., and McLeod R. M., eds.), Vol. 2, pp. 37–106,. Elsevier, Amsterdam.

Etgen A. M. and Browning E. T. (1983) Activators of cyclic adenosine 3′:5′-monophosphate accumulation in rat hippocampal slices: Action of vasoactive intestinal peptide. *J. Neurosci.* **3,** 2487–2493.

Folkers K., Hong J., Rosell S., and Björkroth V. (1981) Chemical design of antagonists of substance P. *Acta Physiol. Scand* **111,** 505–506.

Fredholm B. B. and Lundberg J. M. (1982) VIP-induced cyclic AMP formation in the cat submandibular gland. *Acta Physiol. Scand.,* **114,** 157–159.

Fukuda M., Kuwayama Y., Shiosaka S., Inagaki S., Ishimoto I., Shimizu Y., Takagi H., Sakanaka M., Takatsuki K., Senba E., and Tokyama M. (1981) Localization of vasoactive intestinal peptide and neurotensin immunoreactivities in the avian retina. *Curr. Eye Res.* **1,** 115–118.

Gallacher D. V. (1983) Substance P is a functional neurotransmitter in the rat parotid gland. *J. Physiol.* (Lond.) **342,** 483–498.

Gilbert J. A. and Richelson E. (1984) Neurotensin stimulates formation of cyclic GMP in murine neuroblastoma clone N1E-115. *Eur. J. Pharmacol.* **99,** 245–246.

Goedert M. and Emson P. C. (1983) The regional distribution of neurotensin-like immunoreactivity in central and peripheral tissues of the cat. *Brain Res.* **272,** 291–297.

Goedert M., Nagy J. I., and Emson P. C. (1982) The origin of substance P in the rat submandibular gland and its major duct. *Brain Res.* **252,** 327–333.

Goedert M., Mantyh P. W., Hunt S. P., and Emson P. C. (1983) Mosaic distribution of neurotensin-like immunoreactivity in the cat striatum. *Brain Res.* **274,** 176–179.

Goedert M., Sturmey N., Williams B. J., and Emson P. C. (1984a) The comparative distribution of neurotensin- and xenopsin-like immunoreactivity in *Xenopus laevis* and rat tissues. *Brain Res.* **308,** 273–280.

Goedert M., Hunter J. C., and Ninkovic M. (1984b) Evidence for neurotensin as a non-adrenergic, non-cholinergic neurotransmitter in the guinea pig ileum. *Nature* (Lond.) **311,** 59–62.

Goedert M., Pittaway K., Williams B. J., and Emson P. C. (1984c) Specific binding of tritiated neurotensin to rat brain membranes: Characterisation and regional distribution. *Brain Res.* **304,** 71–81.

Goedert M., Mantyh P. W., Emson P. C., and Hunt S. P. (1984d) Inverse relationship between neurotensin receptors and neurotensin-like immunoreactivity in cat striatum. *Nature* (Lond.) **307,** 543–546.

Goedert M., Pittaway K., and Emson P. C. (1984e) Neurotensin receptors in the rat striatum: Lesion studies. *Brain Res.* **299,** 164–168.

Goedert M., Mantyh P. W., Hunt S. P., and Emson P. C. (1984f) Localisation of specific neurotensin binding sites in the rat adrenal gland. *Brain Res.* **299,** 389–392.

Goedert M., Pinnock R. D., Downes C. P., Mantyh P. W., and Emson P. C. (1984g) Neurotensin stimulates inositol phospholipid hydrolysis in rat brain slices. *Brain Res.* **323,** 193–197.

Goedert M., Schwartz W. N., and Williams B. J. (1985) The comparative distribution of neurotensin- and [Lys^8Asn^9]neurotensin (8–13)-like immunoreactivity in chicken and rat tissues. *Brain Res.* **342,** 259–265.

Goldman R., Bar-Shavit Z., and Romeo D. (1983) Neurotensin modulates human neutrophil locomotion and phagocytic capability. *FEBS Lett.* **159,** 63–67.

Gourdji D., Bataille D., Vauclin N., Grouselle D., Rosselin G., and Tixier-Vidal A. (1979) Vasoactive intestinal polypeptide stimulates prolactin release and cAMP production in a rat pituitary cell line. *FEBS Lett.* **104,** 165–168.

Granier J., Van Rietschoten J., Kitabgi P., Poustis C., and Freychet P. (1982) Synthesis and characterization of neurotensin analogues for structure/activity relationship studies. *Eur. J. Biochem.* **124,** 117–125.

Graybiel A. M. and Ragsdale C. W. (1983) Biochemical Anatomy of the Striatum, in *Chemical Neuroanatomy* (Emson P. C., ed.), pp. 427–504., Raven, New York.

Gubler U., Chua A. O., Hoffman B. J., Collier K. J., and Eng J. (1984) Cloned cDNA to cholecystokinin mRNA predicts an identical preprocholecystokinin in pig brain and gut. *Proc. Natl. Acad. Sci. USA* **81,** 4307–4310.

Hahne W. F., Jensen R. T., Lemp G. F., and Gardner J. D. (1981) Proglumide and benzotript: Members of a different class of cholecystokinin receptor antagonists. *Proc. Natl. Acad. Sci. USA* **78,** 6304–6308.

Hammer R. A., Leeman S. E., Carraway R., and Williams R. H. (1980) Isolation of human intestinal neurotensin. *J. Biol. Chem.* **255,** 2476–2480.

Hanley M. R. and Iversen L. L. (1980) Substance P Receptors in *Neurotransmitter Receptors,* Part 1 (Enna S. J. and Yamamura H. I., eds.), pp. 73–103,. Chapman and Hall, New York.

Hanley M. R. and Jessell T. M. (1984) Substance P receptors on a rat pancreatic acinar cell line. *J. Physiol.* (Lond.) **351,** 29P.

Hanley M. R., Lee C. M., Jones L. M., and Michell R. H. (1980a) Similar effects of substance P and related peptides on salivation and on phosphatidylinositol turnover in rat salivary glands. *Mol. Pharmacol.* **18,** 78–83.

Hanley M. R., Sandberg B. E. B., Lee C. M., Iversen L. L., Brundish D. E., and Wade R. (1980b) Specific binding of ^{3}H-substance P to rat brain membranes. *Nature* (Lond.) **286,** 810–812.

Harper A. A. and Raper H. S. (1943) Pancreozymin, a stimulant of secretion of the pancreas in cats. *J. Physiol.* **102,** 112–125.

Hays S. E., Beinfeld M. C., Jensen R. I., Goodwin F. D., and Paul S. N. (1980) Demonstration of a putative receptor site for cholecystokinin in rat brain. *Neuropeptides* **1,** 53–62.

Hökfelt T., Elfvin L. G., Schultzberg M., Goldstein M., and Nilsson G. (1977) On the occurrence of substance P-containing fibres in sympathetic ganglia: Immunohistochemical evidence. *Brain Res.* **132,** 29–41.

Hökfelt T., Vincent S., Hellsten L., Rosell S., Folkers K., Markey K.,

Goldstein M., and Cuello C. (1981) Immunohistochemical evidence for a "neurotoxic" action of [D-Pro^2,D-Trp^{7,9}]-substance P, an analogue with substance P antagonistic activity. *Acta Physiol. Scand.* **113,** 571–573.

Hokin-Neaverson M. and Sadeghian K. (1984) Lithium-induced accumulation of inositol 1-phosphate during cholecystokinin octapeptide- and acetylcholine-stimulated phosphatidylinositol breakdown in dispersed mouse pancreas acinar cells. *J. Biol. Chem.* **259,** 4346–4352.

Holzer P., Bucsics A., Saria A., and Lembeck F. (1982) A study of the concentrations of substance P and neurotensin in the gastrointestinal tract of various mammals. *Neuroscience* **7,** 2919–2924.

Huidobro-Toro J. P. and Yoshimura K. (1983) Pharmacological characterization of the inhibitory effects of neurotensin on the rabbit ileum myenteric plexus preparation. *Br. J. Pharmacol.* **80,** 645–653.

Huidobro-Toro J. P. and Zhu Y. X. (1984) Neurotensin receptors on the ileum of the guinea-pig: Evidence for the coexistence of inhibitory and excitatory properties. *Eur. J. Pharmacol.* **102,** 237–250.

Hunter J. C. and Maggio J. E. (1984a) Pharmacological characterisation of a novel tachykinin isolated from mammalian spinal cord. *Eur. J. Pharmacol.* **97,** 159–160.

Hunter J. C. and Maggio J. E. (1984b) A pharmacological study with substance K: Evidence for multiple tachykinin receptors. *Eur. J. Pharmacol.* **105,** 149–153.

Hunter J. C., Goedert M., and Pinnock R. D. (1985) Mammalian tachykinin-induced hydrolysis of inositol phospholipids in rat brain slices. *Biochem. Biophys. Res. Commun.*, **127,** 616–622.

Hutchinson J. B. and Dockray G. J. (1980) Inhibition of the action of cholecystokinin octapeptide on the guinea pig ileum myenteric plexus by dibutyryl cyclic guanosine monophosphate. *Brain Res.* **202,** 501–505.

Innis R. B. and Snyder S. H. (1980) Distinct cholecystokinin receptors in brain and pancreas. *Proc. Natl. Acad. Sci. USA* **77,** 6917–6921.

Itoh N., Obata K., Yanaihara N., and Okamoto H. (1983) Human preprovasoactive intestinal polypeptide contains a novel PHI-27-like peptide, PHM-27. *Nature* **304,** 547–549.

Ivy A. C. and Oldberg E. (1928) A hormone mechanism for gallbladder contraction and evacuation. *Am. J. Physiol.* **86,** 599–613.

Jan L. Y. and Jan Y. N. (1982) Peptidergic transmission in sympathetic ganglia of the frog. *J. Physiol.* **327,** 219–246.

Jensen R. T. and Gardner J. D. (1979) Interaction of physalaemin, substance P, and eledoisin with specific membrane receptors on pancreatic acinar cells. *Proc. Natl. Acad. Sci. USA* **76,** 5679–5683.

Jensen R. T., Lemp G. F., and Gardner J. D. (1980) Interaction of cholecystokinin with specific membrane receptors on pancreatic acinar cells. *Proc. Natl. Acad. Sci. USA* **77,** 2079–2083.

Jerne N. K. (1974) Towards a network theory of the immune system. *Annales d'Immunologie (Inst. Pasteur)* **125C,** 373–389.

Jessell T. M. (1983) Substance P in the Nervous System, in *Handbook of Psychopharmacology,* (Iversen L. L., Iversen S. D., and Snyder S. H., eds.), Volume 16, pp. 1–105, Plenum, New York.

Jiang Z. G., Simmons M. A., and Dunn N. J. (1982) Enkephalinergic modulation of non-cholinergic transmission in mammalian prevertebral ganglia. *Brain Res.* **125,** 185–191.

Jones R. S. G. and Olpe H. F. (1982) A structure–activity profile of substance P and some of its fragments on supraspinal neurones in the rat. *Neurosci. Lett.* **33,** 67–71.

Jones R. S. G., Wright D. M., and Olpe H. P. (1983) Testing of peripheral antagonists of substance P in the CNS. *Irish J. Med. Sci.* **152** (Suppl. 1), 36.

Jones S. W., Adams P. R., Brownstein M. J., and Rivier J. E. (1984) Teleost luteinising hormone-releasing hormone: Action on bullfrog sympathetic ganglia is consistent with role as neurotransmitter. *J. Neurosci.* **4,** 420–429.

Jorpes E. and Mutt V. (1966) Cholecystokinin and pancreozymin: One single hormone? *Acta Physiol. Scand.* **66,** 196–202.

Joseph S. K., Thomas A. P., Williams R. J., Irvine R. F., and Williamson J. R. (1984) Myo-Inositol 1,4,5-trisphosphate. A second messenger for the hormonal mobilisation of intracellular calcium in liver. *J. Biol. Chem.* **259,** 3077–3081.

Kane M. G., O'Dorisio T. M., and Krejs G. J. (1983) Production of secretory diarrhoea by intravenous infusion of vasoactive intestinal polypeptide. *N. Engl. J. Med.* **309,** 1482–1485.

Kangawa K., Minamino N., Fukuda A., and Matsuo H. (1983) Neuromedin K: A novel mammalian tachykinin in porcine spinal cord. *Biochem. Biophys. Res. Commun.* **114,** 533–540.

Kimura S., Okada M., Sugita Y., Kanazawa I., and Munekata E. (1983) Novel neuropeptides, neurokinin alpha and beta isolated from procine spinal cord. *Proc. Jpn. Acad. Ser. B.* **59,** 101–104.

Kitabgi P. (1982) Effects of neurotensin on intestinal smooth muscle: Application to the study of structure–activity relationships. *Ann. NY Acad. Sci.* **400,** 37–55.

Kitabgi P. and Freychet P. (1979a) Neurotensin: Contractile activity, specific binding, and lack of effect on cyclic nucleotides in intestinal smooth muscle. *Eur. J. Pharmacol.* **55,** 35–42.

Kitabgi P. and Freychet P. (1979b) Neurotensin contracts the guinea-pig longitudinal intestinal smooth muscle by inducing acetylcholine release. *Eur. J. Pharmacol.* **56,** 403–407.

Kitabgi P. and Vincent J. P. (1981) Neurotensin is a potent inhibitor of guinea-pig colon contractile activity. *Eur. J. Pharmacol.* **74,** 311–318.

Kitabgi P. and Vincent J. P. (1983) Effects of apamin and calcium antagonists on the neurotensin-induced myogenic relaxation in guinea-pig colon. *Regul. Peptides* **2** (Suppl.), 85.

Kitabgi B., Carraway R., van Rietschoten J., Granier C., Morgat J. L., Menez A., Leeman S. E., and Freychet P. (1977) Neurotensin: Specific binding to synaptic membranes from rat brain. *Proc. Natl. Acad. Sci. USA* **74,** 1846–1850.

Kitabgi P., Hamon G., and Worcel M. (1979) Electrophysiological study of the action of neurotensin on smooth muscle of the guinea-pig taenia coli. *Eur. J. Pharmacol.* **56,** 87–93.

Kitabgi P., Checler F., and Vincent J. P. (1984) Comparison of some biological properties of neurotensin and its natural analogue LANT-6. *Eur. J. Pharmacol.* **99,** 357–360.

Knobil E. (1980) The neuroendocrine control of the menstrual cycle. *Recent Prog. Horm. Res.* **36,** 53–88.

Koh S. W., Kyritsis A., and Chader G. J. (1984) Interaction of neuropeptides and cultured glial (Müller) cells of the chick retina: Elevation of intracellular cyclic AMP by vasoactive intestinal peptide and glucagon. *J. Neurochem.* **43,** 199–203.

Laburthe M., Prieto J. C., Amiranoff B., Dupont C., Hui Bon Hoa D., and Rosselin G. (1979a) Interaction of vasoactive intestinal peptide with isolated intestinal epithelial cells from rat. *Eur. J. Biochem.* **96,** 239–248.

Laburthe M., Rosselin G., Rousset M., Zweibaum A., Korner M., Selinger Z., and Schramm M. (1979b) Transfer of the hormone receptor for vaso-intestinal peptide to an adenylate cyclase system in another cell. *FEBS Lett.* **98,** 41–43.

Laburthe M., Mangeat P., Marchis-Mouren G., and Rosselin G., (1979c) Activation of cyclic AMP-dependent protein kinases by vasoactive intestinal polypeptide in isolated epithelial cells from rat. *Life Sci.* **25,** 1931–1938.

Laburthe M., Amiranoff B., Boige N., Rouyer-Fessard C., Tatemoto K., and Moroder L. (1983) Interaction of GRF with VIP receptors and stimulation of adenylate cyclase in rat and human intestinal epithelial membranes. *FEBS Lett.* **159,** 89–92.

Laburthe M., Bréant B., and Rouyer-Fessard C. (1984) Molecular identification of receptors for vasoactive intestinal polypeptide in rat intestinal epithelium by covalent cross-linking. *Eur. J. Biochem.* **139,** 181–187.

Laburthe M., Couvineau A., Rouyer-Fessard C., and Moroder L. (1985) Interaction of PHM, PHI, and 24-glutamine PHI with human VIP receptors from colonic epithelium: Comparison with rat intestinal receptors. *Life Sci.* **36,** 991–995.

Lazarus L. H., Brown M. R., and Perrin M. H. (1977a) Distribution, localization and characteristics of neurotensin binding sites in the rat brain. *Neuropharmacology* **16,** 625–629.

Lazarus L. H., Perrin M. H., and Brown M. R. (1977b) Mast cell binding of neurotensin. *J. Biol. Chem.* **252,** 7174–7179.

Leander S., Håkanson R., Rosell S., Folkers K., Sundler F., and Tornqvist K. (1981) A specific substance P antagonist blocks smooth muscle contractions induced by non-cholinergic, non-adrenergic nerve stimulation. *Nature* (Lond.) **294,** 467–469.

Lee C. M., Iversen L. L., Hanley M. R., and Sandberg B. E. B. (1982) The possible existence of multiple receptors for substance P. *Naunyn Schmiedeberg's Arch. Pharmacol.* **318,** 281–288.

Lee C. M., Javitch J. A., and Snyder S. H. (1983) [^{3}H]Substance P binding to salivary gland membranes; regulation by guanyl nucleotides and divalent cations. *Mol. Pharmacol.* **23,** 563–569.

Leeman S. E. and Carraway R. E. (1982) Neurotensin: Discovery, isolation, characterization, synthesis and possible physiological roles. *Ann. NY Acad. Sci.* **400,** 1–16.

Leeman S. E. and Hammerschlag R. (1967) Stimulation of salivary secretion by a factor extracted from hypothalamic tissue. *Endocrinology* **81,** 803–810.

Leong D. A., Frawley L. S., and Neill J. D. (1983) Neuroendocrine control of prolactin secretion. *Ann. Rev. Physiol.* **45,** 109–127.

Leroux P., Vaudry H., Fournier A., St Pierre S., and Pelletier G. (1984) Characterization and localization of vasoactive intestinal polypeptide receptors in the rat lung. *Endocrinology* **114,** 1506–1512.

Leyendecker G., Wildt L., and Hansmann M. (1980) Pregnancies following chronic pulsatile administration of LHRH by means of a portable pump—a new approach to the treatment of infertility in hypothalamic amenorrhea. *J. Clin. Endocrinol. Metabol.* **51,** 1214–1216.

Liang T. and Cascieri M. A. (1981) Substance P receptor on parotid cell membranes. *J. Neurosci.* **1,** 1131–1141.

Longshore M. and Makman M. H. (1981) Stimulation of retinal adenylate cyclase by vasoactive intestinal peptide. *Eur. J. Pharmacol.* **70,** 237–240.

Loosen P. T., Nemeroff C. B., Bissette G., Burnett G. B., Prange A. J., and Lipton M. A. (1978) Neurotensin-induced hypothermia in the rat: Structure–activity studies. *Neuropharmacology* **17,** 109–113.

Loren I., Tornqvist K., and Alumets J. (1980) VIP-immunoreactive neurons in the retina of the rat. *Cell Tissue Res.* **210,** 167–172.

Lundberg J. M., Änggård A., Emson P., Fahrenkrug J., and Hökfelt T. (1981a) Vasoactive intestinal polypeptide and cholinergic mechanisms in cat nasal mucosa: Studies on choline acetyltransferase and release of vasoactive intestinal polypeptide. *Proc. Natl. Acad. Sci. USA* **78,** 5255–5259.

Lundberg J. M., Änggård A., and Fahrenkrug J. (1981b) Complementary role of vasoactive intestinal polypeptide and acetylcholine for cat submandibular gland flow and secretion. *Acta Physiol. Scand.* **113,** 329–336.

Maggio J. E. and Hunter J. C. (1984) Regional distribution of kassinin-like immunoreactivity in rat central and peripheral tissues and the effect of capsaicin. *Brain Res.* **307,** 370–374.

Magistretti P. J. and Schorderet M. (1984) VIP and noradrenaline act synergistically to increase cyclic AMP in cerebral cortex. *Nature* (Lond.) **308,** 280–282.

Magistretti P. J., Morrison J. H., Shoemaker W. J., Sapin V., and Bloom F. E. (1981) Vasoactive intestinal polypeptide induces glycogenolysis in mouse cortical slices: A possible mechanism for the local control of energy metabolism. *Proc. Natl. Acad. Sci. USA* **78,** 6535–6539.

Mansfield M. J., Beardsworth D. E., Loughlin J. S., Crawford J. D., Bode H. H., Rivier J., Vale W., Kushner D. C., Crigler J. F., and Crowley W. F. (1983) Long-term treatment of central precocious puberty with a long-acting analogue of luteinizing hormone-releasing hormone. *New Engl. J. Med.* **309,** 1286–1290.

Mantyh P. W. and Hunt S. P. (1985) The autoradiographic localization of substance P receptors in the rat and bovine spinal cord and the rat and cat spinal trigeminal nucleus pars caudalis and the effects of neonatal capsaicin. *Brain Res.* **332,** 315–324.

Mantyh P. W., Pinnock R. D., Downes C. P., Goedert M., and Hunt S. P. (1984a) Correlation between inositol phospholipid hydrolysis and substance P receptors in rat CNS. *Nature* (Lond.) **309,** 795–797.

Mantyh P. W., Goedert M., and Hunt S. P. (1984b) Autoradiographic visualization of receptor binding sites for substance P in the gastrointestinal tract of the guinea-pig. *Eur. J. Pharmacol.* **100,** 133–134.

Mantyh P. W., Hunt S. P., and Maggio J. E. (1984c) Substance P receptors: Localisation by light microscopic autoradiography in rat brain using [^{3}H]SP as the radioligand. *Brain Res.* **307,** 147–167.

Mantyh P. W., Maggio J. E., and Hunt S. P. (1984d) The autoradiographic distribution of kassinin and substance K binding sites is different from the distribution of substance P binding sites in rat brain. *Eur. J. Pharmacol.* **102,** 361–364.

Maruyama Y., Petersen O. H., Flanagan P., and Pearson G. T. (1983) Quantification of Ca^{2+}-activated K^+ channels under hormonal control in pig pancreas acinar cells. *Nature* (Lond.) **305,** 228–232.

Matthews M. R. and Cuello A. C. (1982) Substance P-immunoreactive peripheral branches of sensory neurones innervate guinea-pig sympathetic neurones. *Proc. Natl. Acad. Sci. USA* **79,** 1668–1672.

Maurin Y., Buck S. H., Wamsley J. K., Burks T. F., and Yamamura H. I. (1984) Light microscopic autoradiographic localization of [^{3}H]-

substance P binding sites in rat thoracic spinal cord. *Life Sci.* **34,** 1713–1716.

Mazella J., Kitabgi P., and Vincent J. P. (1985) Molecular properties of neurotensin receptors in rat brain. Identification of subunits by covalent labeling. *J. Biol. Chem.* **260,** 508–514.

Mazella J., Poustis C., Labbé C., Checler F., Kitabgi P., Granier C., Van Rietschoten J., and Vincent J. P. (1983) Monoiodo-[Trp^{11}]neurotensin, a highly radioactive ligand for neurotensin receptors. *J. Biol. Chem.* **258,** 3476–3481.

Minamino N., Kangawa K., Fukuda A., and Matsuo H. (1984a) Neuromedin L: A novel mammalian tachykinin identified in porcine spinal cord. *Neuropeptides* **4,** 157–166.

Minamino N., Kangawa K., and Matsuo H. (1984b) Neuromedin N: A novel neurotensin-like peptide identified in porcine spinal cord. *Biochem. Biophys. Res. Commun.* **122,** 542–549.

Minamino N., Masuda H., Kangawa K., and Matsuo H. (1984) Regional distribution of neuromedin K and neuromedin L in rat brain and spinal cord. *Biochem. Biophys. Res. Commun.* **124,** 731–738.

Mishina M., Kurosaki T., Tobimatsu T., Morimoto Y., Noda M., Yamamoto T., Terao M., Lindstrom J., Takahashi T., Kuno M., and Numa S. (1984) Expression of functional acetylcholine receptor from cloned cDNAs. *Nature* (Lond.) **307,** 604–608.

Mishina M., Tobimatsu T., Imoto K., Tanaka K., Fujita Y., Fukuda K., Kurasaki M., Takahashi H., Murimoto Y., Hirose T., Inayama S., Takahashi T., Kuno M., and Numa S. (1985) Location of functional regions of acetylcholine receptor α-subunit by site-directed mutagenesis. *Nature* (Lond.) **313,** 364–369.

Monier S. and Kitabgi P. (1980) Substance P-induced autodesensitization inhibits atropine-resistant neurotensin-stimulated contractions in guinea-pig ileum. *Eur. J. Pharmacol.* **65,** 461–463.

Morera A. M., Cathiard A. M., Laburthe M., and Saez J. M. (1979) Interaction of vasoactive intestinal peptide with a mouse adrenal cell line (Y-I): Specific binding and biological effects. *Biochem. Biophys. Res. Commun.* **90,** 78–85.

Mutt V. and Said S. I. (1974) Structure of the porcine vasoactive intestinal octacosapeptide. The amino acid sequence. *Eur. J. Biochem.* **42,** 581–589.

Nawa H., Hirose T., Takashima H., Inayama S., and Nakanishi S. (1983) Nucleotide sequences of cloned cDNAs for two types of bovine brain substance P precursor. *Nature* (Lond.) **306,** 32–36.

Nawa H., Kotani H., and Nakanishi S. (1984) Tissue-specific generation of two preprotachykinin mRNAs from one gene by alternative RNA splicing. *Nature* (Lond.) **312,** 729–734.

Nemeroff C. B., Luttinger D., and Prange A. J. (1983) Neurotensin and Bombesin, in *Handbook of Psychopharmacology,* (Iversen L. L., Iversen

S. D., and Snyder S. H., eds.), Vol. 16, pp. 363–466., Plenum, New York.

Nicoll R. A., Schenker C., and Leeman S. E. (1980) Substance P as a transmitter candidate. *Ann. Rev. Neurosci.* **3,** 227–268.

Nilsson A. (1975) Structure of the vasoactive intestinal octacosapeptide from chicken intestine. The amino acid sequence. *FEBS Lett.* **60,** 322–326.

Nilsson G., Larsson L. I., Håkanson R., Brodin E., Pernow B., and Sundler F. (1975) Localization of substance P-like immunoreactivity in mouse gut. *Histochemistry* **43,** 97–99.

Ninkovic M., Hunt S. P., and Kelly J. S. (1981) Effect of dorsal rhizotomy on the autoradiographic distribution of opiate and neurotensin receptors and neurotensin-like immunoreactivity within the rat spinal cord. *Brain Res.* **230,** 111–119.

Ninkovic M., Beaujouan J. C., Torrens Y., Saffroy M., Hall M. D., and Glowinski J. (1985) Differential localization of tachykinin receptors in rat spinal cord. *Eur. J. Pharmacol.* **106,** 463–464.

Nishizuka Y. (1984) The role of protein kinase C in cell surface signal transduction and tumour promotion. *Nature* (Lond.) **308,** 693–698.

Noda M., Takahashi H., Tanabe T., Toyosato M., Kikyotani S., Furutani Y., Hirose T., Takashima H., Inayama S., Miyata T., and Numa S. (1983) Structural homology of *Torpedo californica* acetylcholine receptor subunits. *Nature* (Lond.) **302,** 528–532.

Onali P., Schwartz J. P., and Costa E. (1981) Dopaminergic modulation of adenylate cyclase stimulation by vasoactive intestinal peptide in anterior pituitary. *Proc. Natl. Acad. Sci. USA* **78,** 6351–6354.

Osborne N. (1984) Substance P in the bovine retina: Localization, identification, release, uptake and receptor analysis. *J. Physiol.* (Lond.) **349,** 83–93.

Otsuka M. and Konishi S. (1976) Substance P as an excitatory transmitter of primary sensory neurons. *Cold Spring Harbor Symp. Quant. Biol.* **40,** 135–143.

Ottesen B., Staun-Olsen P., Gammeltoft S., and Fahrenkrug J. (1982) Receptors for vasoactive intestinal polypeptide on crude smooth muscle membranes from porcine uterus. *Endocrinology* **110,** 2037–2043.

Palacios J. M. and Kuhar M. J. (1981) Neurotensin receptors are located on dopamine-containing neurons in rat midbrain. *Nature* (Lond.) **294,** 587–589.

Pearse A. G. E. and Polak J. (1975) Immunocytochemical localization of substance P in mammalian intestine. *Histochemistry* **41,** 373–375.

Peikin S. R., Costenbader C. L., and Gardner J. D. (1979) Actions of derivatives of cyclic neucleotides on dispersed acini from guinea pig pancreas. Discovery of a competitive antagonist of the action of cholecystokinin. *J. Biol. Chem.* **254,** 5321–5327.

Perrone M. H., Diehl R. E., and Haubrich D. R. (1983) Binding of [^{3}H]substance P to putative substance P receptors in rat brain membranes. *Eur. J. Pharmacol.* **95,** 131–133.

Poitras P., Iacino D., and Walsh J. H. (1980) Dibutyryl cGMP: Inhibitor of the effect of cholecystokinin and gastrin on the guinea pig gallbladder *in vitro. Biochem. Biophys. Res. Commun.* **96,** 476–482.

Praissman M., Martinez P. A., Saladino C. F., Berkowitz J. M., Steggles A. W., and Finkelstein J. A. (1983) Characterization of cholecystokinin binding sites in rat cerebral cortex using a ^{125}I-CCK-8 probe resistant to degradation. *J. Neurochem.* **40,** 1406–1413.

Prentki M., Bide T. J., Jancik D., Irvine R. F., Berridge M. J., and Wollheim C. B. (1984) Rapid mobilization of Ca^{2+} from rat insulinoma microsomes by inositol-1,4,5-trisphosphate. *Nature* (Lond.) **309,** 562–564.

Prieto J. C., Laburthe M., and Rosselin G. (1979) Interaction of vasoactive intestinal peptide with isolated intestinal epithelial cells. I. Characterization, quantitative aspects and structural requirements of binding sites. *Eur. J. Biochem.* **96,** 229–237.

Putney J. W., Van de Walle C. M., and Wheeler C. S. (1980) Binding of ^{125}I-physalaemin to rat parotid acinar cells. *J. Physiol.* (Lond.) **301,** 205–212.

Quik M., Iversen L. L., and Bloom S. L. (1978) Effect of vasoactive intestinal polypeptide and other peptides on cAMP accumulation in rat brain. *Biochem. Pharmacol.* **27,** 2209–2213.

Quirion R. and Pilapil C. (1984) Comparative potencies of substance P, substance K and neuromedin K on brain substance P receptors. *Neuropeptides* **4,** 325–329.

Quirion R., Rioux F., and Regoli D. (1978) Chronotropic and inotropic effects of neurotensin on spontaneously beating auricles. *Can. J. Physiol. Pharmacol.* **56,** 671–673.

Quirion R., Regoli D., Rioux F., and St Pierre S. (1980a) Structure–activity studies with neurotensin: Analysis of positions 9, 10 and 11. *Brit. J. Pharmacol.* **69,** 689–692.

Quirion R., Regoli D., Rioux F., and St Pierre S. (1980b) The stimulatory effects of neurotensin and related peptides in rat stomach strips and guinea pig auricles. *Brit. J. Pharmacol.* **68,** 83–91.

Quirion R., Gaudreau P., St Pierre S., Rioux F., and Pert C. B. (1982) Autoradiographic distribution of [^{3}H]neurotensin receptors in rat brain: Visualization by tritium-sensitive film. *Peptides* **3,** 757–763.

Quirion R., Shults C. W., Moody T. W., Pert C. B., Chase T. N., and O'Donohue T. L. (1983) Autoradiographic distribution of substance P receptors in the rat central nervous system. *Nature* (Lond.) **303,** 714–716.

Reinecke M., Weihe E., Carraway R., Leeman S. E., and Forssmann W. G. (1982) Localization of neurotensin immunoreactive nerve fibers in the guinea-pig heart: Evidence derived by immunohistochemistry, radioimmunoassay and chromatography. *Neuroscience* **7,** 1785–1795.

Reinecke M., Forssmann W. G., Thiekötter G., and Triepel J. (1983) Localization of neurotensin-immunoreactivity in the spinal cord and peripheral nervous system of the guinea-pig. *Neurosci. Lett.* **37,** 37–42.

Rioux F., Kérouac R., and St Pierre S. (1982) Analysis of the biphasic depressor–pressor effect and tachycardia caused by neurotensin in ganglion-blocked rats. *Neuropeptides* **3,** 113–127.

Robberecht P., De Neef P., Lammens M., Deschodt-Lanckman M., and Christophe J. (1978) Specific binding of vasoactive intestinal polypeptide to brain membranes from the guinea pig. *Eur. J. Biochem.* **90,** 147–154.

Robberecht P., Deschodt-Lanckman M., Camus J. C., De Neef P., Lambert M., and Christophe J. (1979a) VIP activation of rat anterior pituitary adenylate cyclase. *FEBS Lett.* **103,** 229–233.

Robberecht P., König W., Deschodt-Lanckman M., De Neef P., and Christophe J. (1979b) Specificity of receptors to vasoactive intestinal peptide in guinea-pig brain. *Life Sci.* **25,** 879–884.

Rosell S. and Folkers K. (1982) Substance P antagonists: A new type of pharmacological tool. *Trends Pharmacol. Sci.* **4,** 211–212.

Rosenzweig S. A., Miller L. J., and Jamieson J. D. (1983) Identification and localization of cholecystokinin-binding sites on rat pancreatic plasma membranes and acinar cells: A biochemical and autoradiographic study. *J. Cell Biol.* **96,** 1288–1297.

Rosenzweig S. A., Madison L. D., and Jamieson J. D. (1984) Analysis of cholecystokinin-binding proteins using endo-β-*N*-acetylglucosaminidase F. *J. Cell Biol.* **99,** 1110–1116.

Rothman R. B., Danks J. A., Herkenham M., Cascieri M. A., Chicchi G. G., Liang T., and Pert C. B. (1984) Autoradiographic localization of a novel peptide binding site in rat brain using the substance P analog, eledoisin. *Neuropeptides* **4,** 343–349.

Rothman R. B., Herkenham M., Pert C. B., Liang T., and Cascieri M. A. (1984) Visualization of rat brain receptors for the neuropeptide substance P. *Brain Res.* **309,** 47–54.

Rotsztejn W. H., Dussaillant M., Nobou F. and Rosselin G. (1981) Rapid glucocorticoid inhibition of vasoactive intestinal peptide-induced cyclic AMP accumulation and prolactin release in rat pituitary cells in culture. *Proc. Natl. Acad. Sci. USA* **78,** 7584–7588.

Rougon G., Noble M., and Mudge A. W. (1983) Neuropeptides modulate the β-adrenergic response of purified astrocytes *in vitro. Nature* (Lond.) **305,** 715–717.

Sadoul J. L., Kitabgi P., Rostene W., Javoy-Agid F., Agid Y., and Vincent J. P. (1984) Characterization and visualization of neurotensin binding to receptor sites in human brain. *Biochem. Biophys. Res. Commun.* **120,** 206–213.

Saito A., Sankaran H., Goldfine I. D., and Williams J. A. (1980) Cholecystokinin receptors in the brain. Characterisation and distribution. *Science* **208,** 1155–1156.

Saito A., Goldfine I. D., and Williams J. A. (1981) Characterization of receptors for cholecystokinin and related peptides in mouse cerebral cortex. *J. Neurochem.* **37,** 483–490.

Sakamoto C., Goldfine I. D., and Williams J. A. (1983) Characterization of cholecystokinin receptor subtypes on pancreatic plasma membranes. *J. Biol. Chem.* **258,** 12707–12712.

Sakamoto C., Williams J. A., and Goldfine I. D. (1984) Brain CCK receptors are structurally distinct from pancreas CCK receptors. *Biochem. Biophys. Res. Commun.* **124,** 497–502.

Sandberg B. E. B. and Iversen L. L. (1982) Substance P. *J. Med. Chem.* **25,** 1009–1015.

Sankaran H., Goldfine I. D., Deveney C. W., Wong K. Y., and Williams J. A. (1980) Binding of cholecystokinin to high affinity receptors on isolated rat pancreatic acini. *J. Biol. Chem.* **255,** 1849–1853.

Schneider W. J., Slaughter C. J., Goldstein J. L., Anderson G. W., Capra J. D., and Brown M. S. (1983) Use of antipeptide antibodies to demonstrate external orientation of the NH_2-terminus of the low density lipoprotein receptor in the plasma membrane of fibroblasts. *J. Cell Biol.* **97,** 1635–1640.

Schorderet M., Sovilla J. Y., and Magistretti P. J. (1981) VIP-and glucagon-induced formation of cyclic AMP in intact retinae *in vitro. Eur. J. Pharmacol.* **71,** 131–133.

Schultz K. D., Schultz K., and Schultz G. (1977) Sodium nitroprusside and other smooth muscle-relaxants increase cyclic GMP levels in rat ductus deferens. *Nature* (Lond.) **265,** 750–751.

Schultzberg M., Hökfelt T., Nilsson G., Terenius L., Rehfeld J. F., Brown M., Elde R., Goldstein M., and Said S. (1980) Distribution of peptide- and catecholamine-containing neurons in the gastrointestinal tract of rat and guinea-pig: Immunohistochemical studies with antisera to substance P, vasoactive intestinal polypeptide, enkephalins, somatostatin, gastrin/cholecystokinin, neurotensin and dopamine β-hydroxylase. *Neuroscience* **5,** 689–744.

Schwartz C. J., Kimberg D. V., Sheerin H. E., Fieldd M., and Said S. I. (1974) Vasoactive intestinal peptide stimulation of adenylate cyclase and active electrolyte secretion in intestinal mucosa. *J. Clin. Invest.* **54,** 536–544.

Sealock R., Wray B. E., and Froehner S. C. (1984) Ultrastructural localisation of the Mr 43,000 protein and the acetylcholine receptor in *Torpedo* postsynaptic membranes using monoclonal antibodies. *J. Cell Biol.* **98,** 2239–2244.

Segawa T., Hosokawa M., Kitagawa K., and Yajima H. (1977) Contractile activity of synthetic neurotensin and related polypeptides on guinea pig ileum. *J. Pharm. Pharmacol.* **29,** 57–58.

Sege K. and Peterson P. A. (1978) Use of anti-idiotypic antibodies as cell-surface receptor probes. *Proc. Natl. Acad. Sci. USA* **75,** 2443–2447.

Sharkey K. A. and Templeton D. (1984) Substance P in the rat parotid gland: Evidence for a dual origin from the otic and trigeminal ganglia. *Brain Res.* **304,** 392–396.

Shlatz L. J., Kimberg D. V., and Cattieu K. A. (1979) Phosphorylation of specific rat intestinal microvillus and basal-lateral membrane proteins by cyclic nucleotides. *Gastroenterology* **76,** 293–299.

Shults C. W., Quirion R., Jensen R. T., Moody T. W., O'Donohue T. L., and Chase T. N. (1982) Autoradiographic localization of substance P receptors using ^{125}I substance P. *Peptides* **3,** 1073–1075.

Snyder S. H. (1975) Neurotransmitter and drug receptors in the brain. *Biochem. Pharmacol.* **24,** 1371–1374.

Spanarkel M., Martinez J., Briet C., Jensen R. T., and Gardner J. D. (1983) Cholecystokinin-27-32-amide. A member of a new class of cholecystokinin receptor antagonists. *J. Biol. Chem.* **258,** 6746–6749.

Staun-Olsen P., Ottesen B., Bartels P. D., Nielsen M. H., Gammeltoft S., and Fahrenkrug J. (1982) Receptors for vasoactive intestinal polypeptide on isolated synaptosomes from rat cerebral cortex. *J. Neurochem.* **39,** 1242–1251.

Staun-Olsen P., Ottesen B., Gammeltoft S., and Fahrenkrug J. (1985) The regional distribution of receptors for vasoactive intestinal polypeptide (VIP) in the rat central nervous system. *Brain Res.* **330,** 317–321.

Steigerwalt R. W. and Williams J. A. (1981) Characterization of cholecystokinin receptors on rat pancreatic membranes. *Endocrinology* **109,** 1746–1753.

Steigerwalt R. W. and Williams J. A. (1984) Binding specificity of the mouse cerebral cortex receptor for small cholecystokinin peptides. *Regul. Pept.* **8,** 51–59.

St Pierre S., Lalonde J. M., Gendreau M., Quirion R., Regoli D., and Rioux F. (1981) Synthesis of peptides by the solid phase method. 6. Neurotensin, fragments and analogs. *J. Med. Chem.* **24,** 370–376.

Streb H., Irvine R. F., Berridge M. J., and Schulz I. (1983) Release of Ca^{2+} from a nonmitochondrial intracellular store in pancreatic acinar cells by inositol-1,4,5-trisphosphate. *Nature* (Lond.) **306,** 67–69.

Studer K. O., Trzeciak A., and Lergier W. (1973) Isolierung und Aminosäuresequenz von Substanz P aus Pferdedarm. *Helv. Chim. Acta* **56,** 860–866.

Svoboda M., Lambert M., Furnelle J., and Christophe J. (1982) Specific photoaffinity crosslinking of [^{125}I]cholecystokinin to pancreatic plasma membranes. Evidence for a disulfide-linked Mr 76,000 peptide in cholecystokinin receptors. *Regul. Pept.* **4,** 163–172.

Szecòwka J. Goldfine I. D., and Williams J. A. (1985) Solubilization and characterization of CCK receptors from mouse pancreas. *Regul. Pept.* **10,** 71–83.

Tatemoto K. and Mutt V. (1981) Isolation and characterization of the intestinal peptide procine PHI (PHI-27), a new member of the glucagon-secretin family. *Proc. Natl. Acad. Sci. USA* **78,** 6603–6607.

Taylor D. P. and Pert C. B. (1979) Vasoactive intestinal polypeptide: Specific binding to rat brain membranes. *Proc. Natl. Acad. Sci. USA* **76,** 660–664.

Torrens Y., Beaujouan J. C., and Glowinski J. (1985) Pharmacological characterisation of two tachykinin binding sites in the rat cerebral cortex. *Neuropeptides* **6,** 59–70.

Torrens Y., Beaujouan J. C., Viger A., and Glowinski J. (1983) Properties of a ^{125}I substance P derivative binding to synaptosomes from various brain structures and the spinal cord of the rat. *Naunyn Schmiedeberg's Arch. Pharmacol.* **324,** 134–139.

Torrens Y., Lavielle S., Chassaing G., Marquet A., Glowinski J., and Beaujouan J. C. (1984) Neuromedin K, a tool to further distinguish two central tachykinin binding sites. *Eur. J. Pharmacol.* **102,** 381–382.

Uhl G. R. and Kuhar M. J. (1984) Chronic neuroleptic treatment enhances neurotensin receptor binding in human and rat substantia nigra. *Nature* (Lond.) **309,** 350–352.

Uhl G. R., Bennett J. P., and Snyder S. H. (1977) Neurotensin, a central nervous system peptide: Apparent receptor binding in brain membranes. *Brain Res.* **130,** 299–313.

Uhl G. R., Whitehouse P. J., Price D. L., Tourtelotte W. W., and Kuhar M. J. (1984) Parkinson's disease: Depletion of substantia nigra neurotensin receptors. *Brain Res.* **308,** 186–190.

Ullrich A., Bell J. R., Chen E. Y., Herrera R., Petruzzelli L. M., Dull T. J., Gray A., Coussens L., Liao Y. C., Tsubokawa M., Mason A., Seeburg P. H., Grunfeld C., Rosen O. M., and Ramachandran J. (1985) Human insulin receptor and its relationship to the tyrosine kinase family of oncogenes. *Nature* (Lond.) **313,** 756–761.

Van Dijk A., Richards J. G., Trzeciak A., Gillessen D., and Möhler H. (1984) Cholecystokinin receptors: Biochemical demonstration and autoradiographical localization in rat brain and pancreas using [^{3}H]cholecystokinin 8 as radioligand. *J. Neurosci.* **4,** 1021–1033.

Verner J. V. and Morrison A. B. (1958) Islet cell tumor and a syndrome of refractory diarrhea and hypokalemia. *Am. J. Med.* **25,** 373–380.

Viger A., Beaujouan J. C., Torrens Y., and Glowinski J. (1983) Specific binding of a ^{125}I-substance P derivative to rat brain synaptosomes. *J. Neurochem.* **40,** 1030–1039.

Von Euler U. S. and Gaddum J. H. (1931) An unidentified depressor substance in certain tissue extracts. *J. Physiol.* **72,** 74–87.

Wakabayashi T., Kato H., and Tachibana S. (1985) Complete nucleotide sequence of mRNA for caerulein precursor from Xenopus skin: The mRNA contains an unusual repetitive structure. *Nucl. Ac. Res.* **13,** 1817–1828.

Watling K. J. and Dowling J. E. (1983) Effects of vasoactive intestinal peptide and other peptides on cyclic AMP accumulation in intact pieces and isolated horizontal cells of the teleost retina. *J. Neurochem.* **41,** 1205–1213.

Watson S. P. and Downes C. P. (1983) Substance P-induced hydrolysis of inositol phospolipids in guinea-pig ileum and rat hypothalamus. *Eur. J. Pharmacol.* **93,** 245–253.

Watson S. P. and Iversen L. L. (1984) ^{3}H-Substance P binding to guinea-pig ileum longitudinal smooth muscle membranes. *Regul. Pept.* **8,** 273–281.

Weihe E. and Reinecke M. (1981) Peptidergic innervation of the mammalian sinus nodes: Vasoactive intestinal polypeptide, neurotensin, substance P. *Neurosc. Lett.* **26,** 283–288.

Williams J. A., Sankaran H., Roach E., and Goldfine I. D. (1982) Quantitative electron microscope autoradiographs of ^{125}I cholecystokinin in pancreatic acini. *Am. J. Physiol.* **243,** G291–296.

Williams J. T., Katayama Y., and North R. A. (1979) The action of neurotensin on single myenteric neurones. *Eur. J. Pharmacol.* **59,** 181–186.

Yoo O. J., Powell C. T., and Agarwal K. L. (1982) Molecular cloning and nucleotide sequence of full-length cDNA coding for porcine gastrin. *Proc. Natl. Acad. Sci. USA* **79,** 1049–1053.

Young W. S. and Kuhar M. J. (1979) A new method for receptor autoradiography: [^{3}H]Opioid receptors in rat brain. *Brain Res.* **179,** 255–270.

Young W. S. and Kuhar M. J. (1981) Neurotensin receptor localization by light microscopic autoradiography in rat brain. *Brain Res.* **206,** 273–285.

Zarbin M. A., Innis R. B., Wamsley J. K., Snyder S. H., and Kuhar M. J. (1983) Autoradiographic localization of cholecystokinin receptors in rodent brain. *J. Neurosci.* **3,** 877–906.

Chapter 10
Multiple Opioid Receptors in the Central Nervous System

Paul L. Wood

1. Introduction

1.1. Multiple Opioid Receptors

The initial landmarks in opioid receptor research were the demonstration of stereospecificity (Becket and Casey, 1954) and the identification of receptor sites in central nervous system (CNS) membrane preparations (Pert and Snyder, 1973; Simon et al., 1973). Subsequent behavioral studies with a number of opiate agonists suggested that more than one opioid receptor population was present in the CNS (Martin et al., 1976; Gilbert and Martin, 1976). The concept of multiple opioid receptors has been extensively investigated, and present data indicate that three major opioid receptor populations are present in the brain (Iwamoto and Martin, 1981; Zukin and Zukin, 1981; Wood, 1982). These are the mu (μ), delta (δ), and kappa (κ) sites.

1.1.1. Behavior

The behavioral profiles of a number of narcotics have indicated that both μ and κ agonists are potent analgesic agents. However, in contrast to μ agonists, the κ agents: (a) are not self-injected by monkeys (Woods et al., 1978); (b) do not suppress abstinence in the morphine-dependent dog (Martin et al., 1976) or monkey (Woods et al., 1978); (c) do not express cross-tolerance with morphine (Martin et al., 1976; Gilbert and Martin, 1976; Tepper and Woods, 1978); and (d) do not elicit Straub tail or catalepsy in the mouse (Ward et al., 1978). The discriminative stimulus properties of κ agonists also differ from those of classical μ agonists (Herling and Woods, 1981). These behavioral data clearly indicate the unique pharmacological properties of μ and κ opiate receptors. The unavailability of a selective delta (δ) agonist has prohibited similar behavioral studies to date.

1.1.2. Receptor Binding

With the discovery of the enkephalins (Hughes, 1975), a new impetus to opiate receptor binding studies was in place. The preliminary studies comparing μ and δ binding sites in the CNS indicated regional variations in receptor numbers and differential ligand specificities (Wong and Horng, 1977; Simantov et al., 1978; Chang and Cuatrecasas, 1979; Childers et al., 1979). Subsequent unequivocal proof of separate μ and δ receptor populations was provided by the selective protection experiments of Robson and Kosterlitz (1979) and Smith and Simon (1980). In these studies, various cold ligands were used to protect opiate binding sites from inactivation by phenoxybenzamine or *N*-ethylmaleimide (NEM); μ and δ agonists were found to selectively protect their respective binding sites. Use of autoradiography has also supported regional variations in μ and δ receptors, with μ sites predominating in the thalamus, hippocampus, and periaqueductal gray and neocortex, whereas δ sites are concentrated in the amygdala, septum, nucleus accumbens and paraventricular hypothalamus (Goodman et al., 1980; Schubert et al., 1981). These receptor populations are further differentiated by the μ sites that appear to be coupled to adenyl cyclase, whereas no such coupling is observed with the δ sites (Pert, and Taylor, 1979).

In the case of presynaptic receptors located on sensory fibers entering the CNS, both μ and δ binding sites are numerous (Lamotte et al., 1976; Atweh and Kuhar, 1977; Gamse et al., 1979; Fields et al., 1980; Goodman et al., 1980). Within a given CNS region, these receptor sites may also be restricted to a defined cell population; for example, in the hippocampus, opiate receptors appear to be concentrated in area CA_2 (Meibach and Mayani, 1980; Duka et al., 1981), although in the neocortex, binding sites are localized to lamina IV (Goodman et al., 1980). These findings strongly support the concept that endogenous opiates are involved in the regulation of specific neuronal pathways. An excellent example is in the rat striatum, in which intrinsic neurons possess opiate receptors (Murrin et al., 1980), but of the afferent pathways (nigrostriatal, corticostriate, and thalamostriate), only the nigrostriatal system possesses significant numbers of presynaptic opiate receptors (Pollard et al., 1978; Trabucchi et al., 1978; Gardner et al., 1980; Murrin et al., 1980).

In contrast to the clear delineation of separate μ and δ receptor populations, initial binding studies with [^{3}H]ethylketazocine (EKC) were disappointing, and did not reveal a unique κ binding site in the CNS (Hiller and Simon, 1980, 1980; Harris and Sethy, 1980; Pasternak, 1980; Snyder and Goodman, 1980). However, subsequent studies (Kosterlitz and Paterson, 1980; Wood et al., 1981; Pfeiffer and Herz, 1981; Wolozin et al., 1982) reported a [^{3}H]EKC binding site that is insensitive to μ and δ receptor agonists, but sensitive to κ analgesics. Roemer et al. (1980) have also reported a distinctive [^{3}H]bremazocine binding site that may be a κ receptor. In addition, selective protection of the [^{3}H]EKC binding site from NEM inactivation has been observed with cold κ agonists, but not with the μ agonist morphine or the δ agonist D-Ala^2-D-Leu^5 enkephalin (DADLE) (Wood and Charleson, 1982). These experiments therefore present sufficient proof of a κ receptor within the CNS.

A problem inherent in all the above binding data is that the labeled ligands that have been utilized bind to more than one receptor site, so that total binding is not an absolute measure of a single receptor population. This problem can be partially circumvented by the displacement of bound label by various cold ligands of differing receptor specificities. Such studies (Kosterlitz and Paterson, 1980; Wood et al., 1981), as well as the more stringent selective receptor protection experiments (Robson and Kosterlitz, 1979; Smith and Simon, 1980; Wood and Charleson, 1982), clearly support the concept of separate μ, δ, and κ receptor sites in the rat CNS. However, species differences in the proportions of these opiate sites are evident (Kosterlitz and Paterson, 1980; Maurer, 1982), with δ sites representing about 25% of the opiate receptor population, and the remainder being 12–40% κ and 35–62% μ.

1.1.3. Neurochemistry of Defined CNS Pathways

1.1.3.1. CHOLINERGIC NEURONS. A large number of cholinergic systems appear to possess opiate receptor regulation. For example, morphine depresses acetylcholine turnover (TR_{ACh}) in the parietal and occipital cortices, hippocampus, and N. accumbens, but not in the frontal cortex, striatum, amygdala, substantia nigra, dorsal raphe, septum, or N. interpeduncularis (Cheney et al., 1974; Costa et al., 1975; Zsilla et al., 1976, 1977; Schmidt and Buxbaum, 1978; Wood and Stotland, 1980). It therefore appears that opiate receptors are not universal modulators of cholinergic neurons, but that specific pathways possess an inhibitory enkephalinergic regulation. Examination of the possible opiate receptor types involved in this regulation of neocortical and hippocampal cholinergic neurons has demonstrated that these pathways possess μ and δ receptor-mediated regulation (Moroni et al., 1977, 1978; Wood et al., 1979; Wood and Stotland, 1980). In contrast, these pathways appear to lack κ receptor regulation (Wood and Rackham, 1981).

In summary, several cholinergic systems possess μ and δ receptor regulation. However, the sites of these receptors remain to be determined. Preliminary studies of the septal–hippocampal pathway indicate that, in this case, the receptor sites are at the level of the cholinergic cell bodies or on afferent terminals or interneurons innervating these cells (Moroni et al., 1977, 1978; Wood et al., 1979).

1.1.3.2. Dopaminergic Neurons. The rat nigrostriatal dopaminergic pathway possesses presynaptic opiate receptors (Pollard et al., 1977, 1978; Trabucchi et al., 1978; Gardner et al., 1980; Murrin et al., 1980) that, when activated by μ and δ receptor agonists, elicit elevations in dopamine (DA) synthesis, but not in release (Ahtee and Kaarianinen, 1973; Gauchy et al., 1973; Sugrue, 1974; Costa et al., 1975; Guidotti et al., 1978; Moroni et al., 1979; Wood et al., 1980; Wood and Richard, 1982). This presynaptic clumping action results in depressed dopaminergic transmission (Algeri et al., 1978; Slater and Blundell, 1979; Moleman and Bruinvels, 1979; Wood et al., 1980; Wood and Richard, 1982) that, presumably, elicits changes in feedback activity responsible for the observed increases in substantia nigra cell firing after parenteral morphine (Iwatsuba and Clouet, 1977; Nowycky et al., 1978; Finnerty and Chan, 1979). However, in marked contrast to the μ and δ agonists, κ agonists do not alter DA metabolism in the nigrostriatal pathway (Wood et al., 1980), indicating that this pathway lacks κ receptors and that the κ agonists examined (EKC, tifluadom, U-50488H, and MR-2034) do not possess high agonist affinities for the μ or δ receptor sites on these neurons.

In summary, these data support the concept of multiple opiate receptors and further indicate that these receptors appear to possess unique topographies within the CNS, as indicated by autoradiographic and neurochemical studies.

1.1.4. Endogenous Ligands

This area of opiate research is fast evolving and not presently firmly established. However, it is becoming clear that there are multiple endogenous ligands. The enkephalins appear to be the ligands of the δ receptor, whereas the dynorphins are putative κ ligands (Chavkin et al., 1982). Unfortunately, no unique candidate for the μ receptor is presently indicated. Although β-endorphin has also been identified in the CNS, this opioid peptide is the ligand of the epsilon (ε) receptor, which is a minor population with a more restricted distribution in the brain (Wood, 1982).

2. Mu (μ) Receptor

The μ receptor can be labeled in rodent brain membrane preparations with [^{3}H]DHM or [^{3}H]DAGO (Wood, 1984a). Displacement of these tritiated ligands with cold displacers yields almost identical IC_{50} values (Table 1). [^{3}H]DAGO, however, is the ligand of choice because of its greater affinity for the μ receptor, lower nonspecific binding, and greater stability. The protocol used in our laboratories is as follows:

2.1. Membrane Preparation

All homogenizations are conducted in a Polytron homogenizer.

(a) Homogenize (30 s at setting 4) in 50 m*M* Tris·HCl (pH 7.7); 100 mg wet wt/mL.
(b) Centrifuge: 49,000*g* × 15 min at 4°C.
(c) Homogenize pellet in original volume of buffer.
(d) Centrifuge: 49,000*g* × 15 min at 4°C.
(e) Homogenize pellet in original volume of buffer and incubate at 37°C for 40 min.
(f) Centrifuge: 49,000*g* × 15 min at 4°C.
(g) Homogenize pellet in 22 mL of buffer per 250 mg original wet wt. This homogenate represents the working membrane solution.

2.2. Binding Assay

Solutions:

(a) [^{3}H]DAGO = 0.3 n*M* final concentration
= 10 μL of stock diluted to [1667/specific activity (Ci/mmol)] mL.
(b) Levallorphan = 10 μ*M* final concentration.
= 2 mg/11.53 mL buffer followed by a 1/4 dilution.
Use 20 μL of this dilution/2 mL assay.

(i) Incubate 1 mL of working membrane solution with 100 μL of [^{3}H]DAGO; make up to 2 mL with buffer (total binding), buffer + 20 μL levallorphan (non-

Table 1
IC_{50} Values of Cold Opiate Ligands for $[^3H]$DHM, $[^3H]$DAGO, $[^3H]$DADLE, $[^3H]$EKC, and $[^3H]$SKF 10047 Sites in Rat Whole Brain Membrane Preparations[a]

	IC_{50}, n*M*						
Drug	$[^3H]$Nal	(Na/R)	$[^3H]$DHM	$[^3H]$DAGO	$[^3H]$-DADLE	$[^3H]$EKC	$[^3H]$-SKF 10047
Mu agonists							
Morphine	4.3	(32)	2.0	3.4	37.1	1110	2.1
DHM	2.4	(49)	1.8	1.2	22.7	>1500	10.5
DAGO	0.76	(394)	3.6	1.5	16.0	139	2.2
Morphiceptin	343	(507)	28	11	10,554	>1500	194
Alletorphine	1.2	(62)	0.68	0.71	0.15	0.80	1.3
Partial mu agonists							
Buprenorphine	2.8	(1.6)	5.6	1.9	5.0	3.1	9.0
Zenazocine (WIN42964)	0.56	(4.1)	0.036	0.041	0.33	0.82	0.31
Delta agonists							
DADLE	10.1	(120)	13.4	19	1.8	>1500	23
DSLET	37.2	(46)	124	73	4.6	2752	26
Kappa agonists							
EKC	11.2	(8)	6.8	3.9	10.7	7.2	2.6
MR-2034	0.39	(17)	0.39	0.24	2.1	0.77	0.13
Ketazocine	21.0	(8.1)	2.6	9.0	9.1	11.6	21
Bremazocine	0.92	(1.3)	0.35	0.39	0.58	9.0	2.1
U-50488H	—	—	470	484	1625	91	220
Tifluadom	—	—	5.1	5.1	82	4.0	11.0

Sigma agonists							
SKF 10047	1.9	(3.0)	2.5	2.2	7.0	8.8	1.8
Ag/Ant							
Butorphanol	0.61	(5.0)	0.49	0.27	2.2	5	1.1
MRZ-2549	2.3	(9.2)	0.61	0.44	5.3	1.6	0.84
Nalorphine	6.3	(2.4)	1.1	1.9	19.1	87	2.3
Nalbuphine	4.9	(6.5)	4.1	1.7	30.3	60	3.3
Pentazocine	26.3	(11.8)	5.1	11.6	27.9	183	21
Cyclazocine	0.29	(2.3)	0.31	0.60	1.5	7.9	0.15
GPA-1657	10.4	(8.3)	2.9	1.3	3.6	508	7.8
GPA-1658	—		218	100	819	>2000	542
Tonazocine (WIN42156)	0.73	(2.4)	—	—	2.0	3.0	0.95
Ant							
Naloxone	2.0	(1.0)	2.2	4.9	33	12.7	0.5
MR-1464	14	(2.1)	3.5	2.2	25	6.3	21
MR-1466	15.5	(1.1)	8.8	3.2	55	4.5	47
WIN44,441-3	2.2	(1.0)	0.90	0.11	0.18	1.34	0.21
WIN44,441-2	1400	(—)	1800	>2000	>1500	>1200	1400
MR-2266	0.7	(1.6)	0.50	0.62	1.1	1.0	1.4
Peptides							
DYN (1-13)	2	(191)	1.5	3.4	4.7	0.52	22
$[Met^5]Enk \cdot Arg^6Phe^7$	17.8	(40)	8.4	12.4	6.1	> 500	1.2
β-Endorphin	1.3	(35)	0.51	0.60	0.45	>1500	1.1

[a]Unpublished data.

specific binding), or buffer plus cold displacer under study (competitive displacement).

(ii) Incubate at 25°C for 60 min.

(iii) Terminate by filtration (Brandel Cell Harvester) on GF/B filters.

(iv) Wash filters with 3 × 3 mL cold buffer.

Examination of the displacement of [^{3}H]DAGO by various cold ligands (Table 1) reveals that almost all opioids and synthetic opiates have high affinities for this receptor. Morphiceptin (Chang et al., 1981a) is, however, an extremely selective ligand for the μ receptor since its affinities for δ and κ receptors are 1000-fold less. The affinities of agonist/antagonist (Ag/Ant) analgesics and κ agonists for the μ receptor are complicated in that they can be agonist or antagonist (Wood et al., 1983a,b; Wood 1984a) (*see* also section 8 of this chapter).

The regional distribution of μ receptors has been studied in rat brain slices using autoradiography of [^{125}I]-[D-Ala^2, MePhe^4, Met(*o*)-ol^{5}enkephalin (Goodman et al., 1980). Using this approach, it was found that μ sites are most dense in laminae I and IV of the cerebral cortex, striatum, dorsomedial and ventral thalamus, hypothalamus, pyramidal layer of hippocampus, PAG, inferior colliculus, and interpeduncular nucleus.

It is also evident that there are μ isoreceptors present in the CNS. The first indications of this possibility were obtained by labeling a high affinity ($K_d <$ nM) binding site with [^{3}H]DHM (Pasternak and Snyder, 1975b). This site, which has been termed μ_1, can be selectively alkylated with naloxonazine (Hahn et al., 1982). In contrast, studies of the μ receptor affinities of κ analagesics have indicated that these compounds are selective μ_2 antagonists (Wood et al., 1980; Wood et al., 1982; Wood et al., 1983a; Wood and Pasternak, 1983; Wood, 1984a). The μ isoreceptor classification scheme is presented in Table 2. The list of μ_2-mediated actions is further complicated because these actions also generally possess a δ-mediated component (Pasternak, 1982) that is best determined by tolerance studies (Wood et al., 1983b).

3. Delta (δ) Receptor

[^{3}H]DADLE is a high affinity ligand for the δ receptor. As with the μ receptor, most opiates and synthetic opiates interact with this receptor (Table 1) both as agonists and antagonists. The membrane

Table 2
Mu Isoreceptor Classification

	μ_1	μ_2
Selective agonists	Meptazinol[a]	—
Selective antagonists	Naloxonazine[b]	κ agonists[c]
Actions	Analgesia[d]	GH release[f]
	Catalepsy[e]	Respiratory depression[i]
	Prolactin release[f]	Striatal DA neurons[j]
	Cortical ACh neurons[g]	Bradycardia[e]
	Hippocampal ACh neurons[g]	
	Hypothermia[e]	
	PAG analgesia[h]	

[a]Spiegel and Pasternak, 1984.
[b]Hahn et al., 1982.
[c]Wood et al., 1982; Wood et al., 1983b; Wood, 1984a.
[d]Pasternak, 1980.
[e]Pasternak, 1982.
[f]Spiegel et al., 1982.
[g]Wood and Stotland, 1980; Wood and Rackham, 1981; Wood et al., 1982.
[h]Wood et al., 1983a.
[i]Wood et al., 1981; Wood et al., 1982; Ling et al., 1983.
[j]Wood et al., 1982; Wood and Pasternak, 1983.

preparation and binding assay are identical to those described for the μ receptor, except that 1.0 n*M* [^{3}H]DADLE replaces [^{3}H]DAGO. The most selective δ ligands at the present time appear to be [D-Pen^2, L-Pen^5]enkephalin and [D-Pen^2, D-Pen^5]enkephalin (Mosberg et al., 1983).

The clear demonstration of a unique δ receptor, distinct from the μ receptor, was obtained with selective protection experiments. In these studies selective μ and δ agonists were shown to protect agonist alkylation of μ and δ receptors, respectively. The alkylating agents used were NEM (Smith and Simon, 1980) and phenoxybenzamine (Robson and Kosterlitz, 1979).

A unique CNS distribution of these receptors has also been demonstrated with autoradiography (Goodman et al., 1980). The δ sites predominate in the amygdala, N. accumbens, olfactory tubercle, and laminae II, III, and V of the cerebral cortex.

4. Kappa (κ) Receptor

Initial binding studies, as reviewed in the Introduction, did not demonstrate unique κ binding sites. This difficulty was mainly the result of the multiple binding affinities of κ drugs. To overcome this problem, three approaches have evolved, all of which support the concept of a unique κ receptor in the CNS.

4.1. Selective Suppression

In this approach, binding of κ radioligands to μ and δ sites is masked or suppressed by selective receptor agonists. The most popular conditions for suppression are 100 n*M* DADLE and 100 n*M* [D-Ala^2-MePhe^4, Gly-ol^5]enkephalin (Kosterlitz and Paterson, 1980). A variant of this approach is to selectively block with dynorphin the alkylation of κ receptors by β-chlornaltrexamine while μ and δ sites are inactivated. After washing, a relatively pure κ-rich membrane preparation can be obtained (James et al., 1982).

4.2. Computer Analysis of Multiphasic Displacement

Nonlinear computer analysis of multiphasic saturation or competitive displacement curves has also been used to demonstrate unique κ receptors (Pfeiffer and Herz, 1981; 1982).

4.3. Selective Incubation Conditions

By analogy with enzyme kinetics, the choice of buffer, of ionic composition, and of incubation temperature can be used to maximize binding to κ receptors. Using this approach, K_2HPO_4 buffer containing 100 m*M* NaCl and incubations at 0°C dramatically suppress μ and δ receptor binding while maximizing κ receptor binding (Wood et al., 1981; Quirion et al., 1982) (Table 3). The dramatic effects that the choice of buffer and ionic constituents can have on the evaluation of receptor binding at μ, δ, and κ sites cannot be emphasized enough at this point. This point is reviewed in detail by Barnard and Demoliou-Mason (1983). Unfortunately, this consideration has not been taken into account by laboratories that have attempted to determine the relative numbers of μ, δ, and κ sites in the CNS. It is unlikely, therefore, that the present estimates (Kosterlitz and Paterson, 1980; Gillan and Kosterlitz, 1982; Maurer, 1982) are in fact accurate.

The assay conditions we utilize for κ receptor binding are given below.

Table 3
Actions of NaCl on Binding Parameters for [^{3}H]DADLE, [^{3}H]DAGO, [^{3}H]EKC, and [^{3}H]Naloxone in Guinea Pig Brain Membranes[a]

[^{3}H]Ligand	NaCl, 100 m*M*	K_d n*M*	B_{max}, fmol/mg protein
DAGO	–	0.20	44
	+	0.50	4.3
DADLE	–	2.1	86
	+	4.3	18.8
EKC	–	2.7	207
	+	2.8	236
Naloxone	–	1.5	84
	+	1.0	182

[a]Incubations at 4°C, +/– NaCl for 2 h in 50 m*M* K_2HPO_4 buffer (*see* Section 4); unpublished data.

4.3.1. Membrane Preparation

This methodology is the same as for μ and δ receptors, except that the Tris buffer is replaced by 50 m*M* K_2HPO_4 (pH 7.4) containing 1 m*M* EDTA·K_2.

4.3.2. Binding Assay

Conditions are the same as described for μ binding, except that 2.0 n*M* [^{3}H]EKC is used as the ligand. The incubation buffer also contains 100 m*M* NaCl, with the 60 min incubation period conducted at 4°C rather than 25°C.

Other κ radioligands include [^{3}H]bremazocine (Roemer et al., 1980; Foote and Maurer, 1983), [^{3}H]dynorphin (1–9) (Robson et al., 1983), and [^{3}H]dynorphin (1–17) (Young et al., 1983). Bremazocine possesses higher affinity binding than EKC; however, no increase in receptor specificity is obtained with this ligand. The [^{3}H]-dynorphins also suffer from instability and high nonspecific binding.

The hallmark features of the κ receptor include:

(a) Low affinities (μ*M*) for μ, δ, and ϵ opioid agonists (Wood, 1982) (Table 1).
(b) Resistance to NEM inactivation (Wood and Charleson, 1982) (Table 4) with NEM reducing the affinity

Table 4
Physiochemical Properties of Opioid Receptor Populations

Parameter	Opioid receptor μ	δ	κ
Sensitivity to aliphatic alcohols[a]	—	↓ Affinity	—
Sensitivity to heat inactivation[b]	—	—	Most resistant
Sensitivity to NEM[c]	—	—	Most resistant
Sensitivity to tryptic agents[d]	—	—	Most resistant
Actions of NaCl[e]	↓ Binding	↓ Binding	↑ Binding
Sensitivity to freeze–thaw cycles	—	High	—

[a]Hiller et al., 1984.
[b]Wood and Pilapil, 1983.
[c]Wood and Charleson, 1982.
[d]Pasternak, 1980.
[e]Kelly et al., 1982; Wood et al., 1981; Quirion et al., 1982.

of [^{3}H]EKC for its binding sites (Table 5). This contrasts with μ sites in which NEM reduces the number and not the affinity of [^{3}H]DHM for its binding sites (Pasternak and Synder, 1975a).

(c) Resistance to heat inactivation (Wood and Pilapil, 1983) (Table 4). As with NEM treatment this process results in a decreased affinity for the [^{3}H]EKC binding sites (Table 5).

(d) Resistance to proteolytic attack with much slower rates of [^{3}H]EKC binding losses (Pasternak, 1980).

(e) Increased binding in the presence of Na^+ ions (Hiller and Simon, 1980; Wood et al., 1981; Kelly et al., 1982; Quirion et al., 1982).

Autoradiographic studies using [^{3}H]bremazocine (Foote and Maurer, 1982), [^{3}H]EKC (Goodman and Snyder, 1982), and [^{3}H]dynorphin (1–13) (Lewis et al., 1984) of κ receptor distribution in the CNS have also revealed a high concentration of κ sites in laminae V and VI of the cerebral cortex.

Table 5
Actions of Heat and NEM Treatments on κ Receptor Binding

	[^{3}H]EKC	
Membrane treatment	K_d, n*M*	B_{max}, fmol/mg protein
Control	1.9	166
Heat (60°C × 10 min)[1]	7.3	130
Control	1.8	150
NEM 0.25 m*M* (5 min)	4.1	155
(10 min)	3.6	144
(20 min)	5.7	129

[a]Wood and Pilapil (1983).

The recent solubilization and physical separation of κ receptors clearly indicate that these are separate molecular entities (Itzhak et al., 1984). Recent studies of κ receptor sites in the human amygdala (Pfeiffer et al., 1981), the guinea pig striatum (Audigier et al., 1982), the rat spinal cord (Attali et al., 1982), and rat forebrain (Pilapil and Wood, 1983) have all presented evidence for the existence of κ isoreceptor populations. These studies with [^{3}H]etorphine, [^{3}H]diprenorphine, and [^{3}H]SKF 10047 have described a tentative κ_1 site that possesses high affinities for dynorphin (1–17) and low affinities for β-endorphin and Met^5-enkephalin-Arg^6-Phe^7, with the reverse pharmacology for the κ_2 sites. More physiological data are required to determine if these binding criteria are of relevance in vivo.

5. Epsilon (ε) Receptor

The anatomically distinct neuronal system that contains β-endorphin originates in the arcuate hypothalamus and projects to a number of midbrain and limbic areas (Watson et al., 1978). This distinct anatomy, as well as the large size of β-endorphin, are both in themselves suggestive of unique opioid receptors for this peptide. Unfortunately, studies with β-endorphin are complicated by its high affinities for μ and δ receptors (Wood et al., 1982). This is reflected by binding of [^{3}H]β-endorphin in brain regions poor in β-endorphin levels (Law et al., 1979; Ferrara and Li, 1980; Akil et al., 1980). However, a unique ε receptor appears to be labeled in the

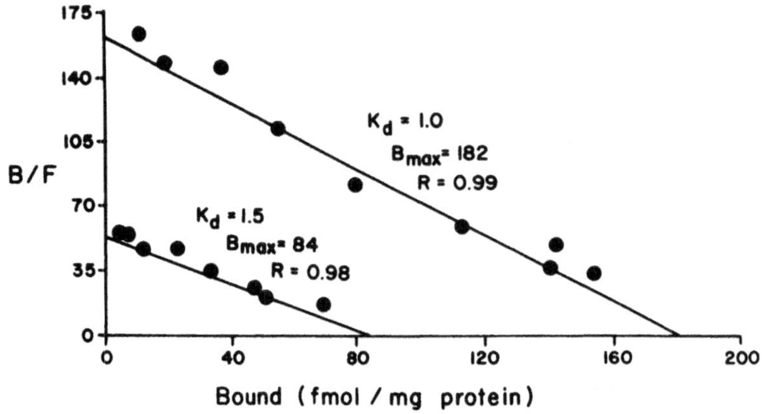

Fig. 1. Actions of 100 m*M* NaCl on [^{3}H]naloxone (K_d/B_{max}) binding in guinea pig whole brain minus cerebellum at 4°C.

septum (Law et al., 1979; Akil et al., 1980), which is an area rich in β-endorphin innervation (Watson et al., 1978). This septal receptor is of high density and extremely high affinity (Law et al., 1979). Pharmacological studies of intraseptal β-endorphin and the net actions on hippocampal acetylcholine turnover further support a physiological role for this receptor (Moroni et al., 1977; Wood et al., 1979), and suggest that this may be an ideal brain region in which to study the ε receptor.

6. Sigma (σ) Receptor

The dysphoric and hallucinogenic actions of opiates have been ascribed to actions on a unique σ receptor (Martin et al., 1976). However, the major limitation of laboratory studies based on this concept is that no clinical study has evaluated naloxone reversibility of these activities. Therefore, debates concerning whether the σ receptor is an opioid site accessible to [^{3}H]SKF 10047 (Pasternak et al., 1981, Quirion et al., 1981) or a non-opioid site labeled by [^{3}H]phencyclidine (Zukin and Zukin, 1979) are academic until this fundamental question of naloxone reversal is resolved.

7. [^{3}H]Naloxone Binding

The early binding studies with [^{3}H]naloxone clearly demonstrated that opiate antagonists displaced [^{3}H]naloxone equally well in the absence and presence of Na^+ ions, whereas opiate agonists were poor displacers of [^{3}H]naloxone in the presence of Na^+ ions (Pert and Snyder, 1974). This Na^+ ratio [IC_{50}(+NaCl)/IC_{50}(–NaCl)] was used as an index of opioid agonist activity and led to the concept of an agonist and antagonist form of the opioid receptor. Unfortunately, at that time the multiplicity of opioid receptors was not envisioned. With our new knowledge of μ, δ, and κ receptor populations, it has become clear that the Na^+ effect on [^{3}H]naloxone binding probably reflects a shift in the relative receptor affinities of [^{3}H]naloxone (Loew et al., 1983). Using computer-assisted analysis of [^{3}H]naloxone binding, these authors concluded that in the absence of Na^+ ions, [^{3}H]naloxone is 81% bound to μ sites, whereas in the presence of Na^+, 63% of the radioligand is bound to κ sites (Loew et al., 1983). These data clearly do not support the agonist–antagonist state hypothesis of opiate receptors (Pert and Snyder, 1974), and suggest that Na^+ suppresses binding of [^{3}H]naloxone to μ receptors and enhances binding to κ sites. Further evidence includes:

- (a) All agents with high Na^+ ratios possess low affinities for the κ receptor (Wood et al., 1981), which would be predicted by preferential binding of [^{3}H]naloxone to κ sites in the presence of Na^+ ions.
- (b) Pure opiate agonists like etorphine (Na^+ ratio = 7; Wood et al., 1981) and lofentanyl (Na^+ ratio = 1.3; Wood et al., 1981; Gommeren and Leysen, 1982) possess low Na^+ ratios, but are not opiate antagonists. These agents do, however, possess high affinities for the κ receptor (Wood et al., 1981), which again is consistent with preferential binding of [^{3}H]naloxone to κ sites in the presence of Na^+ ions.
- (c) All κ agonists possess intermediate Na^+ ratios (Wood et al., 1981) consistent with their combined high μ, δ, and κ receptor affinities.
- (d) κ Receptor binding is only slowly inactivated by NEM (Wood and Charleson, 1982). Similarly, [^{3}H]naloxone (Pasternak and Synder, 1975a), and [^{3}H]naltrexone (Simon et al., 1975) binding in the presence of Na^+ ions is much more slowly inactivated by NEM than in the absence of Na^+ ions. *Again,* this is consistent with [^{3}H]naloxone binding preferentially to κ sites in the presence of Na^+ ions.
- (e) κ Receptor sites are much more resistant to heat inactivation than are μ and δ sites (Wood and Pilapil,

1983). Similarly, [^{3}H]naltrexone binding in the presence of Na^+ ions is extremely resistant to heat (Simon et al., 1975a,b). As with the NEM studies, this is consistent with the conclusions of Loew and coworkers (1983).

In toto, these data suggest that Na^+ ions may not be distinguishing between agonist and antagonist forms of the opiate receptor, but that there is an unmasking of κ receptors. This may involve the enhanced dissociation of endogenous dynorphins from these receptors (Garzon et al., 1984).

8. Drug Classification Problems

8.1. Mu *vs* Delta Agonists

With the use of μ, δ, and κ receptor radioligands, μ and δ agonists are clearly defined by their extremely low affinity for the κ receptor. Furthermore, preference for the μ or δ site can be measured in vitro. In vivo, all CNS assay systems affected by opiates to date are influenced by both μ and δ agonists. For example, the nigrostriatal dopaminergic pathway is regulated by both μ and δ receptors that can be demonstrated to be unique receptor populations via tolerance studies (Wood et al., 1983a).

8.2. Partial Mu Agonists

Partial μ agonists (e.g., buprenorphine) differ from classical μ agonists in that they possess high affinities for μ, δ, and κ receptors (Wood et al., 1981). In vivo, however, the profile of these compounds is mainly that of a μ agonist, with most assays possessing bell-shaped dose–response curves (Table 6). This bell-shaped curve in the case of buprenorphine (Sadee et al., 1982) is clearly the result of activation of a lower affinity opiate receptor population (δ?) that opposes the action of the high affinity μ receptor population. This type of inhibition has been termed both "noncompetitive autoinhibition" (Sadee et al., 1982) and "receptor dualism" (Wood et al., 1983b).

8.3. Kappa *vs* Agonist/Antagonist Analgesics

Although κ agonists and Ag/Ant analgesics possess very similar binding profiles in vitro, these agents possess radically different in vivo pharmacological profiles (Table 6). In the case of κ

Table 6
Pharmacological Characteristics of μ, Partial μ, κ, and Agonist/Antagonist Analgesics

Parameter	Mu Ag	Partial Mu	Ag/Ant	Kappa Ag
Analgesia[a]	TF > HP	HP > TF	HP > TF	TF > HP
DA metabolites[b]	Ag curve	Bell curve	Bell curve	IA
TR_{ACh} (hippocampus)[c]	Ag curve	Bell curve	IA	IA
(cortex)	Ag curve	Bell curve	AG (ceiling)	IA
PAG analgesia[d]	Ag	—	Ag or IA	IA
Respiratory depression[e]	Ag	Ceiling	Ceiling	Ant
[^{3}H]EKC binding[f]	IA	High affinity	Medium to high affinity	High affinity

[a]Wood, 1984b.
[b]Wood, 1983b, 1984a.
[c]Wood et al., 1984a,b.
[d]Wood et al., 1983.
[e]Ling et al., 1983.
[f]Table 1.
Abbreviations: Ag, agonist; Ant, antagonist; HP, hot plate; TF, tail flick.

agonists, the agents presently available appear to be κ agonists, δ antagonists, and μ_2 antagonists (Wood et al., 1983a,b; Wood, 1984a,b), and possess the following characteristics:

(a) Potent displacers of [^{3}H]EKC.
(b) Potent analgesics.
(c) Do not alter neocortical or hippocampal TR_{ACh}, but antagonize the actions of δ agonists.
(d) Do not alter striatal DA metabolites, but antagonize the actions of δ and μ_2 agonists.

In marked contrast, the Ag/Ant agents possess extremely complex pharmacological profiles, presumably as a result of receptor dualism (Wood et al., 1983b). Clearly, the multiple receptor affinities of these agents result in dose-dependent opposing actions that are characterized by bell-shaped dose–response curves in many assays (Wood et al., 1983a). These include:

(a) Bell-shaped dose–response curve with regard to striatal DA metabolites.
(b) Agonist dose–response curve with a ceiling of activity with regard to neocortical TR_{ACh}.
(c) No effect on hippocampal TR_{ACh}.

Using these criteria, a clearcut distinction of these opioid drug classes can be obtained.

8.4. Antagonists (Selectivity?)

From Table 1 it can be clearly discerned that the opiate antagonists naloxone, MR-2266, WIN 44,441-3, MR-1464, and MR-1466 all possess a wide spectrum of antagonist actions with no major degree of receptor selectivity being evident. Therefore, the claims in the literature of selective κ antagonists (MR-2266 and WIN 44,441) are clearly unfounded (Wood, 1983a; Wood et al., 1984b).

9. Conclusions

At the present time, receptor binding studies are able to provide information about the affinities of opiates for μ_1, μ_2, δ, κ_1, and κ_2 opioid receptors. In vivo studies are further able to validate these data and provide the crucial data for the distinction between κ agonists and Ag/Ant analgesics. The in vivo validation of κ_1 vs κ_2 activities has not yet been achieved.

Appendix

Computer Programs

Analysis of radioligand-receptor binding data is usually performed with a Scatchard graph (bound/free vs bound). From linear regression analysis of the Scatchard curve, both the K_d (intercept on abscissa) and B_{max} (–1/slope) can be obtained. Program 1 will give such estimates with input of:

(a) Radioligand specific activity (Ci/mmol)
(b) Counting efficiency (%)
(c) mg wet wt/assay
(d) Total and nonspecific cpm at each radioligand concentration

The inaccuracies of the Scatchard plot that result from the clustering of data points in a small range on the abscissa, as well as from the data transformation to obtain a linear plot, can be overcome by computerized nonlinear regression analysis (Klotz, 1982; Burgisser, 1984). Of particular interest, 70–80% saturation is usually satisfactory for nonlinear regression analysis. This contrasts with Scatchard analysis in which large errors in B_{max} are evident unless saturation is achieved.

```
LIST
10 PRINT "SCATCHARD ANALYSIS"
20 PRINT
30 DIM D(60),DD(60),B(60),BF(60),Q(60)
40 DIM Z(60),ZZ(60),DI(60),A(60),W(60)
50 ZD=0
60 INPUT " SP. ACT. (CI/MMOL)=",SA
70 INPUT "COUNTING EFF. (%)  =",EF
80 INPUT " MG. WWT./ASSAY    =",MW
90 PRINT
100 EF=100/EF
110 MW=1000/MW
120 PRINT
130 INPUT "# OF CONCENTRATIONS=",ZZ
140 PRINT
150 FOR F=1 TO ZZ
160 PRINT "CONCENTRATION ";F
170 PRINT "-----------------------------------"
180 INPUT "          NM=",Q(F)
190 INPUT "TOTAL CPM =",D(F)
200 INPUT "N.S.   CPM=",DD(F)
210 B(F)=((((D(F)-DD(F))*EF)/2200)/SA)*MW
220 BF(F)=B(F)/Q(F)
230 ZZ(F)=D(F)-DD(F)
240 Z(F)=(ZZ(F)/D(F))*100
250 PRINT
260 NEXT F
270 PRINT
280 PRINT "NM"," % SP.","BOUND"," B/F"
290 PRINT
300 FOR F=1 TO ZZ
310 PRINT Q(F),INT(Z(F)),(INT(B(F)*100))/100,(INT(BF(F)*100))/100
320 NEXT F
330 PRINT
340 B=0 : C=0 : D=0 : E=0 :  I=0
350 FOR F=1 TO ZZ
360 B=B+B(F)
370 C=C+B(F)^2
380 D=D+BF(F)
390 E=E+BF(F)^2
400 I=I+(B(F)*BF(F))
410 NEXT F
```

```
420 PRINT
430 H1=C-(B^2/ZZ)
440 H2=E-(D^2/ZZ)
450 H3=I-((B*D)/ZZ)
460 J1=H3/H1
470 J2=(D-(J1*B))/ZZ
480 J3=ABS(H3^2/(H1*H2))
490 J3=SQR(J3)
500 J4=ABS((1/(ZZ-2))*(H2-J1*H3))
510 J4=SQR(J4)
520 PRINT
530 IF ZD=1 THEN 1150
540 PRINT " Y=A + BX"
550 PRINT
560 PRINT "          A=";J2
570 PRINT "          B=";J1
580 PRINT " S(Y.X)   =";J4
590 PRINT "         R =";J3
600 PRINT
610 AF=-(1/J1)
620 FA=-(J2/J1)
630 PRINT "              KD=";AF
640 PRINT " BMAX(PMOL/GM)=";FA
650 PRINT
660 PRINT "NM","OBSERVED","CALC","DIFF"
670 PRINT
680 FOR F=1 TO ZZ
690 Q1=J2+J1*B(F)
700 DI(F)=BF(F)-Q1
710 PRINT Q(F),(INT(BF(F)*100))/100,(INT(Q1*100))/100,(INT(DI(F)*100))/100
720 NEXT F
725 GOTO 1190
730 PRINT
740 INPUT " CHANGE CALCULATIONS ( Y OR N )?",V$
750 IF V$="Y" THEN 770
760 GOTO 1020
770 BP=0
780 ZF=ZZ
790 PRINT "NM","DIFF"
800 PRINT "-------------------------------------"
810 PRINT
820 FOR F=1 TO ZZ
830 BP=BP+1
840 IF BP>ZZ THEN 1000
850 PRINT Q(F),DI(F)
860 PRINT
870 INPUT "OK ( Y OR N )?",W$
880 IF W$="Y" THEN 990
890 ZF=ZF-1
900 FOR N3=F TO ZF
910 ZZ(N3)=ZZ(N3+1)
920 Q(N3)=Q(N3+1)
930 BF(N3)=BF(N3+1)
940 B(N3)=B(N3+1)
950 Z(N3)=Z(N3+1)
960 DI(N3)=DI(N3+1)
970 NEXT N3
980 F=F-1
990 NEXT F
1000 ZZ=ZF
1010 GOTO 270
1020 B=0 : C=0 : D=0 : E=0 : I=
1030 FOR F=1 TO ZZ
1040 QS=B(F)/(FA-B(F))
1050 BF(F)=(LOG(QS))/2.3026
1060 B(F)=(LOG(Q(F)))/2.3026
1070 B=B+B(F)
1080 C=C+B(F)^2
1090 D=D+BF(F)
1100 E=E+BF(F)^2
1110 I=I+(BF(F)*B(F))
1120 NEXT F
1130 ZD=1
1140 GOTO 420
1150 PRINT "HILL #=";J1
1160 PRINT "      R=";J3
1170 PRINT
1180 GOTO 1770
1190 PRINT
1200 REM X - Y PLOTS OF EXPERIMENTAL DATA
1210 INPUT " PLOT ( Y OR N )=?",L$
1220 IF L$="N" THEN 1750
1230 ESC$=CHR$(27)
1240 INPUT "X - RANGE VALUES (MIN X, MAX X)=", NX,MX
1250 INPUT "Y - RANGE VALUES (MIN Y, MAX Y)=", NY,MY
1260 XS=60/(MX-NX) : YS=20/(MY-NY)
1270 PRINT
1280 FOR I=1 TO ZZ
1290 A(I)=20-(BF(I)-NY)*YS
1300 IF (A(I)-INT(A(I)))>.5 THEN A(I)=A(I)+1
1310 W(I)=(B(I)-NX)*XS :IF (W(I)-INT(W(I)))>.5 THEN W(I)=W(I)+1
1320 NEXT I
1330 PRINT
```

```
1340 PRINT
1350 DEF FNHOME$=ESC$+"h"+ESC$+"J"
1360 PRINT FNHOME$;
1370 FOR I=1 TO 20
1380 PRINT "       I"
1390 NEXT I
1400 PRINT "       "; :FOR I=1 TO 61 :PRINT "-"; :NEXT I
1410 DEF FNHOME$=ESC$+"H"
1420 FOR I=MX/50 TO MX STEP MX/50
1430 BB=(I-NX)*XS : IF (BB-INT(BB))>.5 THEN BB=BB+1
1440 BB=BB+4 : YD=J2+J1*I
1450 CC=20-(YD-NY)*YS : IF(CC-INT(CC))>.5 THEN CC=CC+1
1460 PRINT FNHOME$;
1470 IF CC<0 OR CC>22 THEN 1500
1480 FOR J= 1 TO CC : PRINT " " : NEXT J
1490 PRINT TAB(BB);"."
1500 NEXT I
1510 FOR I=1 TO ZZ
1520 PRINT FNHOME$;
1530 IF A(I)<0 OR A(I)>22 THEN 1570
1540 FOR J=1 TO A(I) : PRINT " " : NEXT J
1550 W(I)=(B(I)-NX)*XS :IF (W(I)-INT(W(I)))>.5 THEN W(I)=W(I)+1
1560 W(I)=W(I)+4 :PRINT TAB(W(I));"*"
1570 NEXT I
1580 PRINT FNHOME$;
1590 FOR I=1 TO 20
1600 PRINT "       I"
1610 NEXT I
1620 PRINT FNHOME$;
1630 PRINT MY
1640 DEF FNHOME$=ESC$+"F"
1650 PRINT FNHOME$;
1660 PRINT "      ";NX; TAB(65);MX
1670 DEF FNHOME$=ESC$+"A"
1680 PRINT FNHOME$;
1690 PRINT FNHOME$;
1700 PRINT "   ";NY
1710 PRINT " "
1720 PRINT " "
1730 INPUT "CHANGE AXES ( Y or N )=",B$
1740 IF B$="Y" THEN 1240
1750 PRINT
1760 GOTO 730
1770 INPUT "ANOTHER CALCULATION ( Y or N )=",B$
1780 IF B$="Y" THEN 50
1790 END
Ok
```

```
RUN
SCATCHARD ANALYSIS

 SP. ACT. (CI/MMOL)=16.2
COUNTING EFF. (%)  =44
 MG. WWT./ASSAY    =20

# OF CONCENTRATIONS=7

CONCENTRATION  1
---------------------------
        NM=1.21
TOTAL CPM =1358
N.S.   CPM=321

CONCENTRATION  2
---------------------------
        NM=2.41
TOTAL CPM =2208
N.S.   CPM=541

CONCENTRATION  3
---------------------------
        NM=3.62
TOTAL CPM =3097
N.S.   CPM=719

CONCENTRATION  4
---------------------------
        NM=4.83
TOTAL CPM =3910
N.S.   CPM=1041

CONCENTRATION  5
---------------------------
        NM=7.24
TOTAL CPM =5294
N.S.   CPM=1524

CONCENTRATION  6
---------------------------
        NM=18.1
TOTAL CPM =9665
N.S.   CPM=4328
```

```
          Y=A + BX

               A= 2.964689
               B=-.1140317
     S(Y.X)    = .1097454
               R = .9908892

                  KD= 8.769488
     BMAX(PMOL/GM)= 25.9988

NM          OBSERVED     CALC       DIFF
1.21         2.73        2.58        .14
2.41         2.2         2.35       -.16
3.62         2.09        2.1        -.01
4.83         1.89        1.92       -.03
7.24         1.66        1.59        .06
18.1         .94         1.02       -9.000001E-02
36.2         .58         .53         .05
```

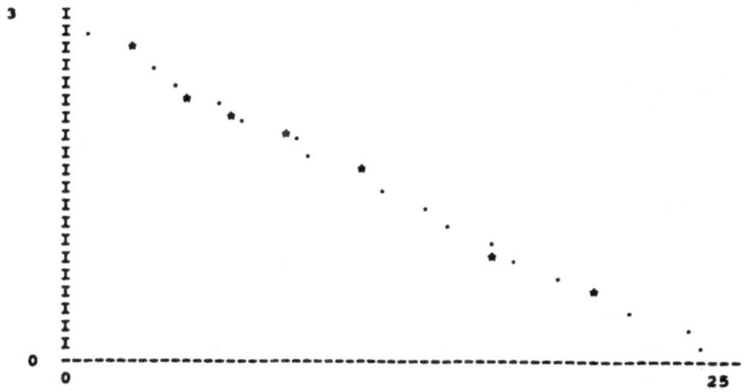


```
CHANGE AXES ( Y or N )=
CHANGE AXES ( Y or N )=N
```

```
CONCENTRATION  7                    CHANGE CALCULATIONS ( Y OR N )?N
--------------------------------
           NM=36.2
TOTAL CPM =15843                    HILL #= 1.010064
N.S.   CPM=9148                            R= .9983913

NM              % SP.               BOUND          B/F

 1.21           76                   3.3           2.73
 2.41           75                   5.31          2.2
 3.62           76                   7.58          2.09
 4.83           73                   9.140001      1.89
 7.24           71                   12.02         1.66

 18.1           55                   17.01         .94
 36.2           42                   21.34         .58
```

Program 1. This program for Scatchard analysis is written in basic for the HP-150 personal computer. The program listing, as well as a sample input/output, is presented.

Program 2 is a nonlinear analysis routine that has the same input as Program 1, but which fits the data via maximum likelihood analysis into a rectangular hyperbola according to the iterative procedures of Bliss and James (1966). A preliminary estimate of K_d is obtained from a linear least squares fit to a Woolf plot (F/B vs B). Standard deviations for both the K_d and B_{max} are also calculated.

```
LIST
10 PRINT "KD NONLINEAR ITERATIVE FIT"
20 DIM X(80), Y(80), Z(80), A(80), W(80)
30 DE=0
40 PRINT
50 PRINT "ITERATIVE FIT TO HYPERBOLA: B=(BMAX*L)/(KD+L)"
60 PRINT
70 INPUT "DATA (1=FMOL/MG PROT   2=CPM)?",DE
80 PRINT
90 PRINT
100 INPUT "      EXP. DATE=",V$
110 INPUT "   RADIOLIGAND=",M$
120 IF DE=1 THEN 180
130 INPUT "SP.ACT.(CI/MMOL)=",SA
140 INPUT "COUNTING EFF (%)=",EF
150 INPUT "MG. WWT./ASSAY=",MW
160 EF=100/EF
170 MW=1000/MW
180 PRINT
190 K1=0 : K2=0 : K3=0 : K4=0 : K5=0 : K6=0 : Y2=0
200 INPUT "            TISSUE=",Z$
210 PRINT
220 AQ=0 : N1=0 : A1=0
230 PRINT
240 AQ=AQ+1
250 INPUT "[LIGAND] (0 TO QUIT)=",X(AQ)
260 IF X(AQ)=0 THEN 430
270 IF DE=2 THEN 300
280 INPUT "SP. FMOL/MG PROT=",Y(AQ)
290 GOTO 350
300 INPUT "TOTAL CPM=",ZG
310 INPUT " N.S. CPM=",GZ
320 Y(AQ)=((((ZG-GZ)*EF)/2200)/SA)*MW
330 VV=((ZG-GZ)/ZG)*100
340 Z(AQ)=ABS(VV)
350 Y2=X(AQ)/Y(AQ)
360 K1=K1+X(AQ)
370 K2=K2+X(AQ)^2
380 K3=K3+Y2
390 K4=K4+Y2^2
400 K5=K5+X(AQ)*Y2
410 K6=K6+1
420 GOTO 230
430 L1=K2-K1^2/K6
440 L2=K4-K3^2/K6
450 L3=K5-K1*K3/K6
460 L4=L3/L1
470 L5=(K3-L4*K1)/K6
480 L6=L5/L4
490 Y9=L6
500 PRINT
510 PRINT "      EXP. DATE=";V$
520 PRINT "   RADIOLIGAND=";M$
530 PRINT "      TISSUE   =";Z$
540 IF DE=1 THEN 560
550 PRINT "N.S. DISPLACER=";T$
560 PRINT
570 PRINT "LIGAND CONC.","SP. BOUND","% SP."
580 PRINT
590 FOR JJ=1 TO K6
600 IF DE=1 THEN 630
610 PRINT X(JJ),Y(JJ),Z(JJ)
620 GOTO 640
630 PRINT X(JJ),Y(JJ)
640 NEXT JJ
650 PRINT
```

```
660 PRINT "INITIAL EST. OF KD=";(INT(L6*10000))/10000
670 PRINT
680 PRINT "B2/B1";TAB(12);"KD";TAB(24);"S.D";TAB(36);"BMAX";TAB(48);
"S.D."
690 PRINT
700 FOR N1=1 TO 10
710 A=0 : B=0 : C=0 : D=0 : E=0 : G=0
720 FOR A1=1 TO K6
730 H1=X(A1)/(X(A1)+Y9)^2
740 H2=X(A1)/(X(A1)+Y9)
750 A=A+H2^2
760 B=B+H1*H2
770 C=C+H1^2
780 D=D+H2*Y(A1)
790 E=E+H1*Y(A1)
800 G=G+Y(A1)^2
810 NEXT A1
820 W=A*C-B^2
830 J1=(D*C-E*B)/W
840 J2=(E*A-D*B)/W
850 V1=J2/J1
860 V2=Y9-V1
870 Q=(G-J1*D-J2*E)/2
880 R1=Q/J1^2
890 R2=A/W-2*(-B/W)*V1
900 R3=ABS(R2*R1)
910 R4=SQR(R3)
920 R5=C/W
930 R6=ABS(Q*R5)
940 R7=SQR(R6)
950 Y9=V2
960 PRINT (INT(V1*10000))/10000;TAB(12);(INT(V2*10000))/10000;TAB(24);(INT
(R4*10000))/10000;TAB(36);(INT(J1*10000))/10000;TAB(48);(INT(R7*10000))/
10000
970 IF N1<9 THEN 990
980 PRINT "-----------------------------------------------------------"
990 NEXT N1
1000 PRINT
1010 U1=(J1*D+J2*E)/G
1020 U2=ABS(U1)
1030 U3=SQR(U2)
1040 PRINT "R=";U3
1050 PRINT
1060 REM X - Y PLOTS OF EXPERIMENTAL DATA
1070 INPUT " PLOT ( Y OR N )=?",L$
1080 IF L$="N" THEN 1610
1090 ESC$=CHR$(27)
1100 INPUT "X - RANGE VALUES (MIN X, MAX X)=", NX,MX
1110 INPUT "Y - RANGE VALUES (MIN Y, MAX Y)=", NY,MY
1120 XS=60/(MX-NX) : YS=20/(MY-NY)
1130 PRINT
1140 FOR I=1 TO K6
1150 A(I)=20-(Y(I)-NY)*YS
1160 IF (A(I)-INT(A(I)))>.5 THEN A(I)=A(I)+1
1170 W(I)=(X(I)-NX)*XS :IF (W(I)-INT(W(I)))>.5 THEN W(I)=W(I)+1
1180 NEXT I
1190 PRINT
1200 PRINT
1210 DEF FNHOME$=ESC$+"h"+ESC$+"J"
1220 PRINT FNHOME$;
1230 FOR I=1 TO 20
1240 PRINT "       I"
1250 NEXT I
1260 PRINT "        "; :FOR I=1 TO 61 :PRINT "-"; :NEXT I
1270 DEF FNHOME$=ESC$+"H"
1280 FOR I=MX/50 TO MX STEP MX/50
1290 BB=(I-NX)*XS : IF (BB-INT(BB))>.5 THEN BB=BB+1
1300 BB=BB+4 : YD=(J1*I)/(V2+I)
1310 CC=20-(YD-NY)*YS : IF(CC-INT(CC))>.5 THEN CC=CC+1
1320 PRINT FNHOME$;
1330 IF CC<0 OR CC>22 THEN 1360
1340 FOR J= 1 TO CC : PRINT " " : NEXT J
1350 PRINT TAB(BB);"."
1360 NEXT I
1370 FOR I=1 TO K6
1380 PRINT FNHOME$;
1390 IF A(I)<0 OR A(I)>22 THEN 1430
1400 FOR J=1 TO A(I) : PRINT " " : NEXT J
1410 W(I)=(X(I)-NX)*XS :IF (W(I)-INT(W(I)))>.5 THEN W(I)=W(I)+1
1420 W(I)=W(I)+4 :PRINT TAB(W(I));"*"
1430 NEXT I
1440 PRINT FNHOME$;
1450 FOR I=1 TO 20
1460 PRINT "       I"
1470 NEXT I
1480 PRINT FNHOME$;
1490 PRINT MY
1500 DEF FNHOME$=ESC$+"F"
1510 PRINT FNHOME$;
1520 PRINT "       ";NX; TAB(65);MX
1530 DEF FNHOME$=ESC$+"A"
1540 PRINT FNHOME$;
1550 PRINT FNHOME$;
1560 PRINT "  ";NY
1570 PRINT " "
1580 PRINT " "
1590 INPUT "CHANGE AXES ( Y or N )=",B$
1600 IF B$="Y" THEN 1100
1610 PRINT
1620 INPUT "ANOTHER CALCULATION ( Y or N )=",B$
1630 IF B$="Y" THEN 30
1640 END
Ok
RUN
KD NONLINEAR ITERATIVE FIT

ITERATIVE FIT TO HYPERBOLA: B=(BMAX*L)/(KD+L)

DATA (1=FMOL/MG PROT   2=CPM)?2


     EXP. DATE=JAN 6/84
   RADIOLIGAND=[3H]5-HT
SP.ACT.(CI/MMOL)=16.2
COUNTING EFF (%)=44
MG. WWT./ASSAY=20

          TISSUE=HIPPOCAMPUS


[LIGAND] (0 TO QUIT)=1.21
TOTAL CPM=1358
```

```
 N.S. CPM=321

[LIGAND] (0 TO QUIT)=2.41
TOTAL CPM=2208
 N.S. CPM=541

[LIGAND] (0 TO QUIT)=3.62
TOTAL CPM=3097
 N.S. CPM=719

[LIGAND] (0 TO QUIT)=4.83
TOTAL CPM=3910
 N.S. CPM=1041

[LIGAND] (0 TO QUIT)=7.24
TOTAL CPM=5294
 N.S. CPM=1524

[LIGAND] (0 TO QUIT)=18.1
TOTAL CPM=9665
 N.S. CPM=4358

[LIGAND] (0 TO QUIT)=36.2
TOTAL CPM=15843
 N.S. CPM=9148

[LIGAND] (0 TO QUIT)=0

      EXP. DATE=JAN 6/84
   RADIOLIGAND=[3H]5-HT
      TISSUE   =HIPPOCAMPUS
N.S. DISPLACER=

LIGAND CONC.  SP. BOUND     % SP.

 1.21            3.306423     76.3623
 2.41            5.315147     75.49818
 3.62            7.582134     76.78398
 4.83            9.147663     73.37596
 7.24            12.02046     71.2127
 18.1            16.92111     54.90947
 36.2            21.34667     42.25841

INITIAL EST. OF KD= 9.0934

B2/B1       KD           S.D          BMAX        S.D.

 .0439      9.049401     .8456        26.2873     1.0172
 .0005      9.048801     .8382        26.2866     1.0138
 0          9.048801     .8383        26.2866     1.014
 0          9.048801     .8385        26.2866     1.0142
 0          9.048801     .8385        26.2866     1.0142
 0          9.048801     .8385        26.2866     1.0142
 0          9.048801     .8385        26.2866     1.0142
 0          9.048801     .8385        26.2866     1.0142
 0          9.048801     .8385        26.2866     1.0142
------------------------------------------------------------
 0          9.048801     .8385        26.2866     1.0142
------------------------------------------------------------

R= .9996849


 PLOT ( Y OR N )=?Y
X - RANGE VALUES (MIN X, MAX X)=0,40
Y - RANGE VALUES (MIN Y, MAX Y)=0,25
```

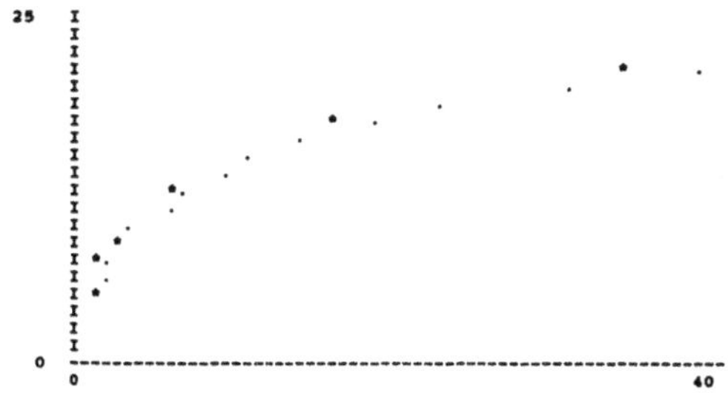


```
CHANGE AXES ( Y or N )=
```

Program 2. The nonlinear analysis of K_d/B_{max}, along with error estimates, performs 10 iterations (Line 690) to obtain the final values. The program listing, as well as an example of input/output, is presented.

For competitive displacement analysis, IC_{50} values are easily obtained from log dose–logit plots (Program 3). Pseudo Hill numbers can also be obtained with these data by regression of log (% control specific/100 – % control specific) vs log dose.

```
10 PRINT "IC50 CALCULATIONS"
20 PRINT
30 INPUT "  EXP. DATE=",A$
40 INPUT "RADIOLIGAND=",B$
50 PRINT
60 DIM LL(100),WW(100),TT(50),VV(50),UU(50),SS(50),RR(50)
70 INPUT "TOTAL CPM=",T
80 INPUT "   NS CPM=",LE
90 SB=T-LE
100 BS=(SB/T)*100
110 PRINT
120 PRINT "SP. BINDING (%)=",INT(BS)
130 PRINT
140 PRINT
150 PRINT
160 QV=1
170 N1=0
180 B=0 : C=0 : D=0 : E=0 : G=0 : I=0
190 P1=0 : P2=0 : P3=0 : DS=0
200 PRINT
210 INPUT "COLD DISPLACER=",C$
220 PRINT
230 N1=N1+1
240 INPUT "DOSE (NM  0 TO QUIT)=",WW(N1)
250 IF WW(N1)=0 THEN 510
260 INPUT "                  CPM=",AP
270 YY=((AP-LE)/SB)*100
280 IF YY>90 THEN 310
290 IF YY<10 THEN 310
300 GOTO 340
310 PRINT "DATA OUTSIDE OF LINEAR LIMITS (10 TO 90%); NEXT VALUE?"
320 PRINT
330 GOTO 240
340 VV(N1)=AP
350 Y=LOG(YY/(100-YY))
360 LL(N1)=(LOG(YY/(100-YY)))/2.3026
370 X=(LOG(WW(N1)))/2.3026
380 B=B+X
390 C=C+X^2
400 UU(N1)=X
410 D=D+Y
420 P1=P1+LL(N1)
430 P2=P2+LL(N1)^2
440 P3=P3+(LL(N1)*X)
450 E=E+Y^2
460 I=I+(X*Y)
470 PRINT
480 SS(N1)=Y
490 TT(N1)=INT(YY)
500 GOTO 230
510 PRINT
520 G=N1-1
530 H1=C-(B^2/G)
540 H2=E-(D^2/G)
550 H3=I-((B*D)/G)
560 J1=H3/H1
570 J2=(D-(J1*B))/G
580 J3=ABS(H3^2/(H1*H2))
590 J3=SQR(J3)
600 J4=ABS((1/(G-2))*(H2-J1*H3))
610 J4=SQR(J4)
620 M=-J2/J1
630 IF DS=1 THEN 1310
640 PRINT
650 PRINT "       EXP. DATE=";A$
660 PRINT "     RADIOLIGAND=";B$
670 PRINT "   % SP. BINDING=";INT(BS)
680 PRINT " COLD DISPLACER=";C$
690 PRINT
700 PRINT "DOSE";TAB(10);"CPM";TAB(21);"% CONTROL";TAB(31);"LOG DOSE";
TAB(41);"LOGIT";TAB(51);"LOG EFFECT"
710 PRINT
720 FOR NQ= 1 TO N1-1
730 PRINT WW(NQ);TAB(10);VV(NQ);TAB(21);TT(NQ);TAB(31);(INT(UU(NQ)*100))
/100;TAB(41);(INT(SS(NQ)*100))/100;TAB(51);(INT(LL(NQ)*1000))/1000
740 NEXT NQ
750 PRINT
760 PRINT "A=";J2
770 PRINT "B=";J1
780 PRINT "S(Y.X)=";J4
790 PRINT
800 PRINT "IC50=";10^M
810 PRINT "______________________________"
820 PRINT
830 PRINT "R=";J3
840 PRINT
850 INPUT "CHANGE CALCULATIONS (1=YES  0=NO)?",VD
860 IF VD>0 THEN 880
```

```
870 GOTO 1250
880 BP=0
890 ZF=N1-1
900 PRINT
910 PRINT "DISPLACER CONC.","% INHIBITION"
920 PRINT
930 FOR N2=1 TO N1-1
940 BP=BP+1
950 IF BP>G THEN 1100
960 PRINT WW(N2),INT(TT(N2))
970 PRINT
980 INPUT "OK (1=YES  0=NO)?",WD
990 IF WD>0 THEN 1090
1000 ZF=ZF-1
1010 FOR N3=N2 TO ZF
1020 WW(N3)=WW(N3+1)
1030 UU(N3)=UU(N3+1)
1040 TT(N3)=TT(N3+1)
1050 VV(N3)=VV(N3+1)
1060 SS(N3)=SS(N3+1)
1065 LL(N3)=LL(N3+1)
1070 NEXT N3
1080 N2=N2-1
1090 NEXT N2
1100 B=0 : C=0 : D=0 : E=0 : G=0 : I=0
1110 P1=0 : P2=0 : P3=0 : DS=0
1120 FOR NQ=1 TO ZF
1130 B=B+UU(NQ)
1140 C=C+(UU(NQ))^2
1150 D=D+SS(NQ)
1160 E=E+(SS(NQ))^2
1170 I=I+(UU(NQ)*SS(NQ))
1180 LL=(LOG(TT(NQ)/(100-TT(NQ))))/2.3026
1190 P1=P1+LL(NQ)
1200 P2=P2+LL(NQ)^2
1210 P3=P3+LL(NQ)*UU(NQ)
1220 NEXT NQ
1230 G=ZF
1240 GOTO 530
1250 PRINT
1260 D=P1
1270 E=P2
1280 I=P3
1290 DS=1
1300 GOTO 530
1310 PRINT "HILL #=";ABS(J1)
1320 PRINT "     R=";J3
1330 PRINT
1335 DS=0
1340 INPUT "QUIT? (1=YES   2=NEW EXP.    3=NEW DISPLACER)",JK
1350 IF JK=1 THEN 1380
1360 IF JK=2 THEN 10
1370 IF JK=3 THEN 150
1380 END

       EXP. DATE=FEB 16/84
     RADIOLIGAND=3H-DAGO
   % SP. BINDING= 89
  COLD DISPLACER=MORPHINE

 DOSE      CPM        % CONTROL LOG DOSE  LOGIT     LOG EFFECT

  31.25     863        11         1.49     -2.09     -.907
  15.625    1248       21         1.19     -1.31     -.567
  7.8125    1555       29         .89      -.87      -.378
  3.906     2080       43         .59      -.26      -.113
  1.953     2669       59         .29       .37       .164
  .9766     3236       74        -.02       1.07      .465
  .488      3661       85        -.32       1.8       .783

 A= 1.059483
 B=-2.096064
 S(Y.X)= 9.170957E-02

 IC50= 3.202307
 ______________________________

 R= .9981194

 CHANGE CALCULATIONS (1=YES  0=NO)?
 0

 HILL #= .910303
      R= .9981193
```

Program 3. This program calculates IC_{50} values for displacement curves via linear regression of log dose–logit data. An example output is presented at the end of the program listing.

References

Ahtee L. and Kaariainen I. (1973) The effect of narcotic analgesics on the homovanillic acid content of rat nucleus caudatus. *Eur. J. Pharmacol.* **22,** 206–208.

Akil H., Hewlett W. A., Barchas J. D., and Li C. H. (1980) Binding of ^{3}H-β-endorphin to rat brain membranes: Characterization of opiate receptors and interaction with ACTH. *Eur. J. Pharmacol.* **64,** 1–8.

Algeri A., Brunello N., Calderini G., and Consolazione A. (1978) Effect of enkephalins on catecholamine metabolism in rats CNS. *Adv. Biochem. Psychopharmacol.* **18,** 199–209.

Attali B., Gouardères L., Mazarguil H., Audigier Y., and Cros J. (1982) Evidence for multiple "kappa" binding sites by use of opioid peptides in the guinea pig lumbosacral spinal cord. *Neuropeptides* **3,** 53–64.

Atweh S. F. and Kuhar M. J. (1977) Autoradiographic localization of opiate receptors in rat brain. I. Spinal cord and lower medulla. *Brain Res.* **124,** 53–67.

Audigier Y., Attali B., Mazarguil H., and Cros J. (1982) Characterization of [^{3}H]-etorphine binding in guinea pig striatum after blockade of mu and delta sites. *Life Sci.* **31,** 1287–1290.

Barnard E. A. and Demoliou-Mason C. (1983) Molecular properties of opioid receptors. *Brit. Med. Bull.* **39,** 37–45.

Becket A. H. and Casey A. F. (1954) Synthetic analgesics: Stereo-chemical considerations. *J. Pharm. Pharmacol.* **6,** 986–1001.

Bliss C. I. and James A. T. (1966) Fitting the rectangular hyperbola. *Biometrics* **22,** 573–602.

Burgisser E. (1984) Radioligand-receptor binding studies: What's wrong with the Scatachard analysis? *Trends Pharmacol. Sci.* **5,** 142–144.

Chang K. J. and Cuatrecasas P. (1979) Multiple opiate receptors. Enkephalins and morphine bind to receptors of different specificity. *J. Biol. Chem.* **254,** 2610–2618.

Chang K. J., Killian A., Hazum E., Cuatrecasas P., and Chang J. K. (1981a) Morphiceptin (NH_4-Tyr-Pro-Phe-Pro-$CONH_2$): A potent and specific agonist for morphine (μ) receptors. *Science* **212,** 75–77.

Chavkin C., James I. F., and Goldstein A. (1982) Dynorphin is a specific endogenous ligand of the κ opioid receptor. *Science* **215,** 413–415.

Cheney D. L., Trabucchi M., Racagni G., Wang C., and Costa E. (1974) Effects of acute and chronic morphine on regional rat brain acetylcholine turnover rate. *Life Sci.* **15,** 1977–1990.

Childers S. R., Creese I., Snowman A. M., and Snyder S. H. (1979) Opiate receptor binding affected differentially by opiates and opioid peptides. *Eur. J. Pharmacol.* **55,** 11–18.

Costa E., Cheney D. L., Racagni G., and Zsilla G. (1975) An analysis at the synaptic level of the morphine action in striatum and N. accumbens: Dopamine and acetylcholine interactions. *Life Sci.* **17,** 1–8.

Duka T., Wuster M., Schubert P., Stoiber R., and Herz A. (1981) Selective localization of different types of opiate receptors in hippocampus as revealed by in vitro autoradiography. *Brain Res.* **205,** 181–186.

Ferrara P. and Li C. H. (1980) β-Endorphin: Characteristics of binding sites in rabbit spinal cord. *Proc. Natl. Acad. Sci. USA* **77,** 5746–5748.

Fields H. L., Emson P. C., Leigh B. K., Gibert R. F. T., and Iversen L. L. (1980) Multiple opiate receptor sites on primary afferent fibers. *Nature* (Lond.) **284,** 351–353.

Finnerty E. P. and Chan S. H. H. (1979) Morphine suppression of substantia nigra zona reticulata neurons in the rat: Implicated role for a novel striatonigral feedback mechanism. *Eur. J. Pharmacol.* **59,** 307–310.

Foote R. W. and Maurer R. (1982) Autoradiographic localization of opiate κ-receptors in the guinea pig brain. *Eur. J. Pharmacol.* **85,** 99–103.

Foote R. W. and Maurer R. (1983) Kappa opiate binding sites in the substantia nigra and bulbus olfactorius of the guinea pig as shown by in vitro autoradiography. *Life Sci.* **33,** 243–246.

Gamse R., Holzer P., and Lembeck F. (1979) Indirect evidence for presynaptic location of opiate receptors on chemosensitive primary sensory neurons. *Naunyn-Schmiedeberg's Arch. Pharmacol.* **308,** 281–285.

Gardner E. L., Zukin R. S., and Makman M. H. (1980) Modulation of opiate receptor binding in striatum and amygdala by selective mesencephalic lesions. *Brain Res.* **194,** 232–239.

Garzon J., Sanchez-Blazquez P., Gerhart J., Loh H. H., and Lee N. M. (1984) Dynorphin A: Inhibitory effect on other opiate ligand bindings in the mouse brain. *Biochem. Pharmacol.* **33,** 2609–2614.

Gauchy C., Agid Y., Glowinski J., and Cheramy A. (1973) Acute effects of morphine on dopamine synthesis and release and tyrosine metabolism in the rat striatum. *Eur. J. Pharmacol.* **22,** 311–319.

Gilbert P. E. and Martin W. R. (1976) The effects of morphine- and nalorphine-like drugs in nondependent morphine-dependent and cyclazocine-dependent chronic spinal dog. *J. Pharmacol. Exp. Ther.* **198,** 66–82.

Gillan M. G. C. and Kosterlitz H. W. (1982) Spectrum of the μ-, δ-, and κ-binding sites in homogenates of rat brain. *Brit. J. Pharmacol.* **77,** 461–469.

Gommeren W. and Leysen J. E. (1982) Binding characteristics of ^{3}H-lofentanil at the opiate receptor. *Arch. Int. Pharmacodyn.* **258,** 171–173.

Goodman R. R., Snyder S. H., Kuhar M. J., and Young W. S. (1980) Differentiation of delta and mu opiate receptor localizations by light microscopic autoradiography. *Proc. Natl. Acad. Sci. USA* **77,** 6239–6243.

Goodman R. R. and Snyder S. H. (1982) Kappa opiate receptors localized by autoradiography to deep layers of cerebral cortex: relation to sedative effects. *Proc. Natl. Acad. Sci. USA* **79,** 5703–5707.

Guidotti A., Moroni F., Gale K., and Kumakura K. (1978) Opiate Receptor Stimulation Blocks the Activation of Striatal Tyrosine Hydroxylase (TH) Induced by Haloperidol, in *Catecholamines: Basic and Clinical Frontiers* (Usdin E., Kopin I. J., and Barchas J., eds.) Pergamon Press, New York, pp. 1035–1037.

Hahn E. F., Carroll-Buatti M., and Pasternak G. W. (1982) Irreversible opiate agonists and antagonists: The 14-hydroxydihydromorphinone azines. *J. Neuroscience* **2,** 572–576.

Harris D. W. and Sethy V. H. (1980) High affinity binding of [^{3}H]-ethylketocyclazocine to rat brain homogenate. *Eur. J. Pharmacol.* **66,** 121–123.

Herling S. and Woods J. H. (1981) Discriminative stimulus effects of narcotics: Evidence for multiple receptor-mediated actions. *Life Sci.* **28,** 1571–1584.

Hiller J. M., and Simon E. J. (1980) Specific high affinity [^{3}H]-ethylketocyclazocine binding in rat central nervous system: Lack of evidence for κ receptors. *J. Pharmacol. Exp. Ther.* **214,** 516–519.

Hiller J. M. Angel L. M., and Simon E. J. (1984) Characterization of the selective inhibition of the delta subclass of opioid binding sites by alcohols. *Mol. Pharmacol.* **25,** 249–255.

Hughes J. (1975) Isolation of an endogenous compound from the brain with pharmacological properties similar to morphine. *Brain Res.* **88,** 295–308.

Itzhak Y., Hiller J. M., and Simon E. J. (1984) Solubilization and characterization of μ-, δ-, and κ-opioid binding sites from guinea pig brain: Physical separation of κ receptors. *Proc. Natl. Acad. Sci. USA* **81,** 4217–4221.

Iwamoto E. T. and Martin W. R. (1981) Multiple opioid receptors. *Med. Res. Rev.* **1,** 411–440.

Iwatsuba K. and Clouet D. H. (1977) Effects of morphine and haloperidol on the electrical activity of rat nigrostriatal neurons. *J. Pharmacol. Exp. Ther.* **202,** 429–436.

James I. F., Chavkin C., and Goldstein A. (1982) Preparation of brain membranes containing a single type of opioid receptor highly selective for dynorphin. *Proc. Natl. Acad. Sci. USA* **79,** 7570–7574.

Kelly P. D., Rance M. J., and Traynor J. R. (1982) Properties of opiate binding in the rat spinal cord. *Neuropeptides* **2,** 319–324.

Klotz I. M. (1982) Numbers of receptor sites from Scatchard graphics: Facts and fantasies. *Science* **217,** 1247–1249.

Kosterlitz H. W. and Paterson S. J. (1980) Characterization of opioid receptors in nervous tissue. *Proc. Roy. Soc. Lond.* (B.) **210,** 113–122.

Lamotte C., Pert C. B., and Snyder S. H. (1976) Opiate receptor binding in primate spinal cord: Distribution and changes after dorsal root section. *Brain Res.* **112,** 407–412.

Law P. Y., Loh H. H., and Li C. H. (1979) Properties and localization of β-endorphin receptor in rat brain. *Proc. Natl. Acad. Sci. USA* **76,** 5455–5459.

Lewis M. E., Young E. A., Houghten R. A., Akil H., and Watson S. J. (1984) Binding of [^{3}H]-dynorphin to apparent κ opioid receptors in deep layers of guinea pig cerebral cortex. *Eur. J. Pharmacol.* **98,** 149–150.

Ling G. S. F., Spiegel K., Nishimura S. L., and Pasternak G. W. (1983) Dissociation of morphine's analgesic and respiratory depressant actions. *Eur. J. Pharmacol.* **86,** 487–488.

Loew G. H., Toll L., and Keys C. (1983) Computer-assisted analysis of receptor-binding studies of [^{3}H]-naloxone and [^{3}H]-DADL: A reinterpretation of the Na^+ effect. *Life Sci.* **33,** 163–166.

Martin W. R., Eades C. G., Thompson J. M., Huppler R. E., and Gilbert P. E. (1976) The effects of morphine- and nalorphine-like drugs in the non-dependent chronic spinal dog. *J. Pharmacol. Exp. Ther.* **197,** 517–532.

Maurer R. (1982) Multiplicity of opiate receptors in different species. *Neurosc. Lett.* **30,** 303–307.

Meibach R. C. and Mayani S. (1980) Localization of naloxone-sensitive [^{3}H]-dihydromorphine binding sites within the hippocampus of the rat. *Eur. J. Pharmacol.* **68,** 175–179.

Moleman P. and Bruinvels J. (1979) Effect of morphine on dopaminergic neurons in the rat basal forebrain and striatum. *J. Neural. Transm.* **46,** 225–237.

Moroni F., Cheney D. L., and Costa E. (1977) Inhibition of acetylcholine turnover in rat hippocampus by intraseptal injections of β-endorphin and morphine. *Naunyn-Schmiedeberg's Arch. Pharmacol.* **299,** 149–153.

Moroni F., Cheney D. L., and Costa E. (1978) The turnover rate of acetylcholine in brain nuclei of rats injected intraventricularly and intraseptally with alpha and beta-endorphin. *Neuropharmacology* **17,** 191–196.

Moroni F., Peralta E., Cheney D. L., and Costa E. (1979) On the regulation of γ-aminobutyric acid neurons in caudatus, pallidus, and nigra: Effects of opioids and dopamine agonists. *J. Pharmacol. Exp. Ther.* **208,** 190–194.

Mosberg H. I., Hurst R., Hruby V. J., Gee K., Akiyama K., Yamamura M. I., Galligan J. J., and Burks T. F. (1983) Cyclic pennicillamine containing enkephalin anologs displays profound delta receptor selectivities. *Life Sci.* **33,** 447–450.

Murrin L. C., Coyle J. T., and Kuhar M. J. (1980) Striatal opiate receptors: Pre- and postsynaptic localization. *Life Sci.* **27,** 1175–1183.

Nowycky M. C., Walters J. R., and Roth R. H. (1978) Dopaminergic neurons: Effect of acute and chronic morphine administration on single cell activity and transmitter metabolism. *J. Neural. Transm.* **49,** 99–116.

Pasternak G. W. and Snyder S. H. (1975a) Opiate receptor binding: Enzymatic treatments that discriminate between agonist and antagonist interactions. *Mol. Pharmacol.* **11,** 478–484.

Pasternak G. W. and Snyder S. H. (1975b) Identification of novel high-affinity opiate receptor binding in rat brain. *Nature* (Lond.) **253,** 563–565.

Pasternak G. W. (1980) Multiple opiate receptors: [^{3}H]-Ethylketocyclazocine receptor binding and ketocyclazocine analgesia. *Proc. Natl. Acad. Sci. USA* **77,** 3691–3694.

Pasternak G. W., Carroll-Buatti M., and Spiegel K. (1981) The binding and analgesic properties of a sigma opiate, SKF 10047. *J. Pharmacol. Exp. Ther.* **219,** 192–198.

Pasternak G. W. (1982) High and low affinity opioid binding sites: Relationship to mu and delta sites. *Life Sci.* **31,** 1303–1306.

Pert C. B. and Snyder S. H. (1973) Properties of opiate-receptor binding in rat brain. *Proc. Natl. Acad. Sci. USA* **70,** 2243–2247.

Pert C. B. and Snyder S. H. (1974) Opiate receptor binding of agonists and antagonists affected differentially by sodium. *Mol. Pharmacol.* **10,** 868–879.

Pert C. B. and Taylor D. (1979) Type 1 and Type 2 Opiate Receptors; A Subclassification Scheme Based upon GTP's Differential Effects on Binding, in *Endogenous and Exogenous Opiate Agonists and Antagonists* (Way E. L., ed.) pp. 87–90, Pergamon, New York.

Pfeiffer A. and Herz A. (1981) Demonstration and distribution of an opiate binding site in rat brain with high affinity for ethylketocyclazocine and SKF 10047. *Biochem. Biophys. Res. Commun.* **101,** 38–44.

Pfeiffer A., Pasi A., Mehraein P., and Herz A. (1981) A subclassification of κ-sites in human brain by use of dynorphin 1–17. *Neuropeptides* **2,** 89–97.

Pfeiffer A. and Herz A. (1982) Discrimination of three opiate-receptor binding sites with the use of a computerized curve-fitting technique. *Mol. Pharmacol.* **21,** 266–671.

Pilapil C. and Wood P. L. (1983) [^{3}H]-SKF 10047 binding to rat brain membranes: Evidence for kappa isoreceptors. *Life Sci.* **33,** 263–265.

Pollard H., Llorens C., and Schwartz J. C. (1977) Enkephalin receptors on dopaminergic neurons in rat striatum. *Nature* (Lond.) **268,** 745–747.

Pollard H., Llorens C., Schwartz J. C., Gros C., and Dray F. (1978) Localization of opiate receptors and enkephalins in the rat striatum in relationship with the nigrostriatal dopaminergic system: Lesion studies. *Brain Res.* **151,** 392–398.

Quirion R., Hammer R. P., Herkenham M., and Pert C. B. (1981) Phencyclidine (angel dust) σ "opiate" receptor: Visualization by tritium-sensitive film. *Proc. Natl. Acad. Sci. USA* **78,** 5881–5885.

Quirion R., Bowen W. D., Herkenham M., and Pert C. B. (1982) Visualization and solubilization of rat brain opiate receptors with a "κ" ligand selectivity pattern. *Cell. Mol. Neurobiol.* **2,** 333–346.

Robson L. E. and Kosterlitz H. W. (1979) Specific protection of the binding sites of D-Ala^2-D-Leu enkephalin (δ-receptors) and dihydromorphine (μ-receptors). *Proc. Roy. Soc. Lond.* (B.) **205,** 425–432.

Robson L. E., Gillan M. G. C., McKnight A. T., and Kosterlitz H. W. (1983) [^{3}H]-Dynorphin A(1–9): Binding characteristics and degradation profile in brain homogenates. *Life Sci.* **33,** 283–286.

Roemer D., Buscher H., Hill R. C., Maurer R., Petcher T. J., Welle H. B. A., Bakel H. C. E. K., and Akkerman A. M. (1980) Bramazocine: A potent long-acting opiate kappa-agonist. *Life Sci.* **27,** 971–978.

Sadee W., Rosenbaum J. S., and Herz A. (1982) Buprenorphine: Differential interaction with opiate receptor subtypes in vivo. *J. Pharmacol. Exp. Ther.* **223,** 157–162.

Schmidt D. E. and Buxbaum D. M. (1978) Effect of acute morphine administration on regional acetylcholine turnover in the rat. *Brain Res.* **147,** 194–200.

Schubert D. P., Wuster M., Stoiber R., and Herz A. (1981) A selective distribution of different opiate receptors in certain areas of rat brain as revealed by in vitro autoradiography. *Neurosci. Lett.* **21,** 119–124.

Simantov R., Childers S. R., and Snyder S. H. (1978) The opiate receptor interaction of [^{3}H]-methionine enkephalin: An opioid peptide. *Eur. J. Pharmacol.* **47,** 319–331.

Simon E. J., Hiller J. M., and Edelman I. (1973) Stereospecific binding of the potent narcotic analgesic [^{3}H]-etorphine to rat brain homogenate. *Proc. Natl. Acad. Sci. USA* **70,** 1947–1949.

Simon E. J., Hiller J. M., Edelman I., Groth J., and Stahl K. D. (1975a) Opiate receptors and their interactions with agonists and antagonists. *Life Sci.* **16,** 1795–1800.

Simon E. J., Hiller J. M., Groth J., and Edelman I. (1975b) Further properties of stereospecific opiate binding sites in rat brain: On the nature of the sodium effect. *J. Pharmacol. Exp. Ther.* **192,** 531–537.

Slater P. and Blundell C. (1979) Effects of morphine on amphetamine-induced circling and striatal cyclic AMP in rats and mice. *Neuropharmacology* **18,** 705–708.

Smith J. R. and Simon E. J. (1980) Selective protection of sterospecific enkephalin and opiate binding against inactivation by *N*-Ethylmaleimidine: Evidence for two classes of opiate receptors. *Proc. Natl. Acad. Sci. USA* **77,** 281–284.

Snyder S. H. and Goodman R. R. (1980) Multiple neurotransmitter receptors. *J. Neurochem.* **35,** 5–15.

Spiegel K., Kourides I. A., and Pasternak G. W. (1982) Different receptors mediate morphine-induced prolactin and growth hormone release. *Life Sci.* **31,** 2177–2180.

Spiegel K. and Pasternak G. W. (1984) Meptazinol: A novel mu-1 selective opiodal analgesic. *J. Pharmacol. Exp. Ther.* **228,** 414–419.

Sugrue M. F. (1974) The effects of acutely administered analgesics on the turnover of noradrenaline and dopamine in various regions of the rat brain. *Brit. J. Pharmacol.* **52,** 159–165.

Tepper P. and Woods J. H. (1978) Changes in locomotor activity and naloxone-induced jumping in mice produced by WIN 35, 197–2 (ethylketazocine), and morphine. *Psychopharmacology* **58,** 125–129.

Trabucchi M., Poli A., Tonon G. C., and Spano P. F. (1978) Interaction Among Enkephalinergic and Dopaminergic Systems in Stiatum and Limbic Forebrain, in *Catecholamines: Basic and Clinical Frontiers* (Usdin E., Kopin I. J., and Barchas J., eds.) Pergamon Press, New York, pp. 1053–1055.

Watson S. J., Akil H., Richard C. W., and Barchas J. D. (1978) Evidence for two separate opiate peptide neuronal systems and the coexistence of β-lipotropin, β-endorphin, and ACTH immuno-reactivities in the same hypothalamus neurons. *Nature* (Lond.) **275,** 226–228.

Ward S. J., Metclaf G., and Rees J. M. H. (1978) Effects of μ-, κ-, and δ-agonists in rodents. *Develop. Neurosci.* **4,** 497–498.

Wolozin B. L., Nishimura S., and Pasternak G. W. (1982) The binding of κ-, and δ-opiates in rat brain. *J. Neuroscience* **2,** 708–713.

Wong D. T. and Horng J. S. (1977) Affinities of opiate agonists and antagonists for enkephalin receptors of rat brain. *Res. Commun. Chem. Path. Pharmacol.* **16,** 749–752.

Wood P. L., Cheney D. L., and Costa E. (1979) An investigation of whether septal γ-aminobutyrate-containing interneurons are involved in the reduction of the turnover rate of acetylcholine elicited by substance P and β-endorphin in the hippocampus. *Neuroscience* **4,** 1479–1484.

Wood P. L. and Stotland L. M. (1980) Actions of enkephalin mu and partial agonist and analgesics on acetylcholine turnover in rat brain. *Neuropharmacology* **19,** 975–982.

Wood P. L., Stotland M., Richard J. W., and Rackham A. (1980) Actions of mu, kappa, sigma, delta, and agonist/antagonist opiates on striatal dopaminergic function. *J. Pharmacol. Exp. Ther.* **215,** 697–703.

Wood P. L. and Rackham A. (1981) Actions of kappa, sigma, and partial mu narcotic receptor agonists on rat brain acetylcholine turnover. *Neurosci. Lett.* **23,** 75–80.

Wood P. L., Charleson S., Lane D., and Hudgin R. L. (1981) Multiple opiate receptors: Differential binding of mu, kappa, and delta agonists. *Neuropharmacology* **20,** 1215–1220.

Wood P. L. (1982) Multiple opiate receptors: Support of unique mu, delta, and kappa sites. *Neuropharmacology* **21,** 487–497.

Wood P. L. and Richard J. W. (1982) Morphine and nigrostriatal function in the rat and mouse: Role of nigral and striatal opiate receptors. *Neuropharmacology* **21,** 1305–1310.

Wood P. L. and Charleson S. (1982) High affinity [^{3}H]-ethylketazocine binding: Evidence for specific κ receptors. *Neuropharmacology* **21,** 215–219.

Wood P. L., Richard J. W., and Thakur M. (1982) Mu opiate isoreceptors: Differentiation with kappa agonists. *Life Sci.* **31,** 2313–2317.

Wood P. L. and Pasternak G. W. (1983) Specific μ_2 isoreceptor regulation of nigrostriatal neurons: In vivo evidence with naloxonazine. *Neurosci. Lett.* **36,** 291–293.

Wood P. L., Sanschagrin D., Richard J. W., and Thakur M. (1983a) Multiple receptor affinities of kappa and agonist/antagonist analgesics: In vivo assessment. *J. Pharmacol. Exp. Ther.* **226,** 545–550.

Wood P. L., McGuade P., Richard J. W., and Thakur M. (1983b) Agonist/antagonist analgesics and nigrostriatal dopamine metabolism in the rat: Evidence for receptor dualism. *Life Sci.* **33,** 759–762.

Wood P. L. (1983a) Opioid receptor affinities of kappa agonists, angonist/antagonists and antagonists in vitro and in vivo. *Prog. Neuro-Psychopharmacol. Biol. Psychiat.* **7,** 657–662.

Wood P. L. (1983b) Opioid regulation of CNS dopaminergic pathways: A review of methodology, receptor types, regional variations, and species differences. *Peptides* **4,** 595–601.

Wood P. L. and Pilapil C. (1983) κ-Opiate receptor sites: Unique heat stability in vitro. *Eur. J. Pharmacol.* **88,** 281–282.

Wood P. L. (1984a) Kappa agonist analgesics: Evidence for μ_2 and δ opioid receptor antagonism. *Drug Develop. Res.* **4,** 429–435.

Wood P. L. (1984b) Animal Models in Analgesic Testing, in *Analgesics: Neurochemical, Behavioral, and Clinical Perspectives.* (Kuhar M. and Pasternak G., eds.) Raven Press, New York, pp. 175–194.

Wood P. L., Stotland L. M., and Rackham A. (1984a) Opiate Receptor Regulation of Acetylcholine Metabolism: Role of Mu, Delta, Kappa, and Sigma Narcotic Receptors, in *Dynamics of Neurotransmitter Function.* (Hannin I., ed.) Raven Press, New York, pp. 99–107.

Wood P. L., Pilapil C., Thakur M., and Richard J. W. (1984b) WIN 44441: A stereospecific and long-acting narcotic antagonist. *Pharmaceut. Res.* **1,** 46–48.

Woods J. H., Fly C. L., and Swain H. H. (1978) Behavioral actions of some *N*-furyl benzomorphans and ketazocines in rhesus monkeys and mice. *Develop. Neurosci.* **4,** 403–411.

Young E., Walker J. M., Houghten R., and Akil H. (1983) [^{3}H]-Dynorphin binding to guinea pig and rat brain. *Life Sci.* **33,** 287–290.

Zsilla G., Cheney D. L., Racagni G., and Costa E. (1976) Correlation between analgesia and the decrease of acetylcholine turnover rate in cortex and hippocampus elicited by morphine, meperidine, viminol R_2, and azidomorphine. *J. Pharmacol. Exp. Ther.* **199,** 662–668.

Zsilla G., Racagni G., Cheney D. L., and Costa E. (1977) Constant rate infusion of deuterated phosphorylcholine to measure the effects of morphine on acetylcholine turnover rate in specific nuclei of rat brain. *Neuropharmacology* **16,** 25–31.

Zukin S. R. and Zukin R. S. (1979) Specific [^{3}H]-phencyclidine binding in rat central nervous system. *Proc. Natl. Acad. Sci. USA* **76,** 5372–5376.

Zukin R. S. and Zukin S. R. (1981) Multiple opiate receptors: Emerging concepts. *Life Sci.* **29,** 2681–2690.

Chapter 11
Purinergic Receptors in the CNS

Michael Williams

1. Introduction

The phenomenon of "nonadrenergic, noncholinergic" neurotransmission in the mammalian peripheral nervous system probably involves the mediation of the purines, adenosine and its nucleotide, ATP (Burnstock, 1972, 1983).

Adenosine has been known for some time to have cardiovascular (Drury and Szent-Gyorgi, 1929; Green and Stoner, 1950) and CNS (Feldberg and Sherwood, 1954) effects. However, the examination of the responses elicited by adenosine and ATP at a molecular level has only occurred in the last 15 yr.

Based on the intensive research effort directed toward elaborating the second messenger function of cyclic AMP, Kakiuchi et al. (1968) found that adenosine was potent in eliciting increases in the level of the cyclic nucleotide in brain slices. The xanthine central stimulants, caffeine and theophylline, antagonized these effects of adenosine (Sattin and Rall, 1970). Until this time, the xanthines had been classified as calcium mobilizers and cyclic nucleotide phosphodiesterase inhibitors. The use of these compounds in combination with adenosine would, therefore, have been expected to enhance, rather than inhibit, cyclic AMP formation. From these findings, Sattin and Rall made the seminal, and now widely accepted (Rall, 1980; Daly, 1982; Snyder, 1984; Williams, 1984a), observation that the xanthines are adenosine antagonists. Since these compounds were known to be central stimulants, it was suggested, and later demonstrated, that adenine nucleosides are central depressants. It was also concluded that the physiological effects of the xanthines were caused by antagonism of the effects of endogenous adenosine (Sattin and Rall, 1970; Rall, 1980).

In vivo and in vitro studies in mammalian central and peripheral tissues have shown that adenosine may play an important physiological role in CNS (Snyder, 1984; Williams, 1984a), cardiovascular (Berne et al., 1983a), renal (Osswald, 1983), pulmonary (Persson, 1984), and immune (Polmar et al., 1983) system function. Despite the skepticism related to the possibility that compounds such as adenosine and ATP might subserve a discrete physiological role in mammalian tissue because of their ubiquitous distribution and intimate involvement in all aspects of cellular energy metabolism (Arch and Newsholme, 1978), considerable evidence has been accumulated for the existence of distinct cell surface recognition sites, or receptors, for adenosine and ATP (Burnstock, 1978; Burnstock and Buckley, 1985), and for a possible localization of adenosine to neuronal elements of mammalian brain (Snyder et al., 1984).

2. Purinergic Receptors

Based on extensive pharmacological studies in a variety of peripheral tissue preparations, Burnstock (1978) proposed the existence of two types of purinergic receptor on the cell surface. The P-1 receptor is responsive to adenosine, blocked by the xanthines, and coupled to adenylate cyclase (Table 1; Fig. 1). The P-2 receptor is sensitive to ATP, had (and still has) no definitive antagonist, and may act via modulation of prostaglandin synthesis. A light-sensitive ATP antagonist arylazidoaminopropronyl ATP (AN-APP_3) has been described (Hogaboom et al., 1980), and appears to be more effective against excitatory than inhibitory actions of the nucleotide (Frew and Lundy, 1982).

Concomitantly to these studies, two groups—Londos and Wolff in Bethesda (Londos and Wolff, 1977; Londos et al., 1980) and Hamprecht et al. in Wurzburg (Van Calker et al., 1978, 1979)—studying the effects of adenosine and several of its stable analogs on cyclic AMP production in a number of cell lines and cultures, identified two subclasses of the P-1 receptor that inhibit and stimulate cyclic AMP production, respectively. In Leydig tumor cells, adipocyte, and liver cell preparations, Londos et al. (1980) noted that adenosine agonists in which the ribose moiety was unchanged could, in the nanomolar range, activate a receptor that inhibited cyclic AMP production. This receptor was designated as the Ri receptor. In the micromolar concentration range, adenosine agonists, once again with an intact ribose moiety, could stimulate cyclic AMP production via interaction with what was designated the Ra receptor. Similar results were obtained in neuroblastoma cell lines by Van Calker et al. (1978, 1979), except that the subclasses were

TABLE 1
Characteristics of Adenosine Receptor Subtypes

Receptor	Location	Agonist pharmacology[a]	Antagonist	Affinity, *M*	Comments
P-1	Cell surface	Adenosine > AMP > ADP > ATP	Alkyxanthines	10^{-6}	Activation increases cAMP formation
P-2	Cell surface	ATP > ADP > AMP = adenosine	None	—	Activation related to increased prostaglandin formation
Ri/A-1	Cell surface	CHA = R-PIA > 2-CADO > adenosine > NECA > S-PIA	Alkyxanthines	10^{-9}	Activation inhibits cAMP formation; subclass of P-1 receptor
Ra/A-2	Cell surface	NECA > 2-CADO > R-PIA > S-PIA	Alkyxanthines	10^{-6}	Activation increases cAMP formation; subclass of P-1 receptor
P	Intracellular	2′5′-Dideoxyadenosine >> adenosine	5′-Methylthioadenosine	10^{-5}	Activation inhibits cAMP formation

[a]Abbreviations: CHA = cyclohexyladenosine; PIA = phenylisopropyladenosine; 2-CADO = 2-chloroadenosine; NECA = 5′-*N*-ethylcarboxamidoadenosine.

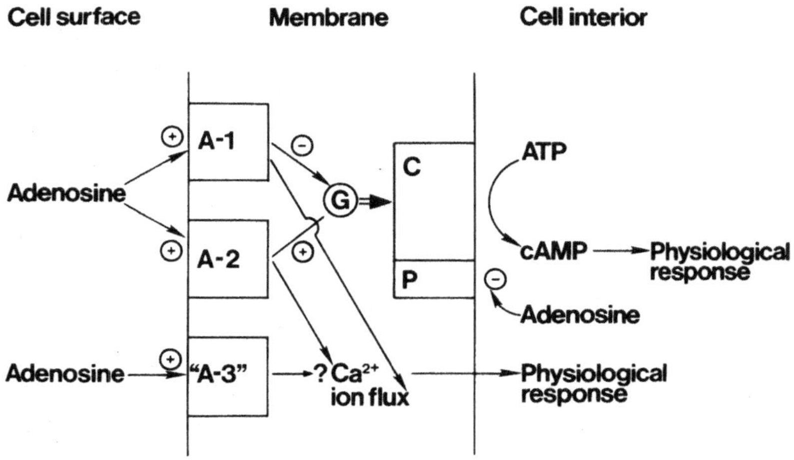

Fig. 1. Adenosine receptors in mammalian cells. Abbreviations: G = G protein of adenylate cyclase; C = Catalytic unit of adenylate cyclase; + = Activation; – = Inhibition.

designated A-1 and A-2, a nomenclature that is in more common usage than that using the R classification. The A receptor-mediated effects of adenosine are blocked by the xanthines (Table 1).

A third type of adenosine recognition site, the P-site or P-receptor, has also been described (Haslam et al., 1978). This site is not to be confused with either the P-1 or P-2 sites, and is located intracellularly on the catalytic subunit of adenylate cyclase (Londos et al., 1983). It can be activated by adenosine analogs, such as 2′,5′-dideoxyadenosine (DDA) (Fig. 2) and xylofuranosyladenosine, that are modified in the ribose ring, but not by analogs such as phenylisopropyladenosine (PIA) (Fig. 2) or 2-chloroadenosine (2-CADO) (Fig. 2), that are A-site agonists. The P site is activated by adenosine in the 10^{-5} *M* concentration range (Daly, 1982), but is not blocked by xanthines. The physiological significance of the P site has not been fully defined; however, in certain circumstances, its activation can play a role in the pharmacological actions of adenosine (Collis and Brown, 1983; Hazeki et al., 1983). There is evidence, both biochemical and behavioral (see below), to indicate the existence of subtypes of each of the two A-receptor subclasses.

The two types of A-receptor also differ markedly in their pharmacological characteristics (Daly, 1982; Fredholm, 1982a). The A-1 receptor is more sensitive to R-PIA than to NECA (5′-ethylcarboxamide adenosine) (Fig. 1), which is in turn more active than the S-diastereomer of PIA (Table 1). In contrast, the A-2 receptor shows a greater sensitivity to NECA, whereas the

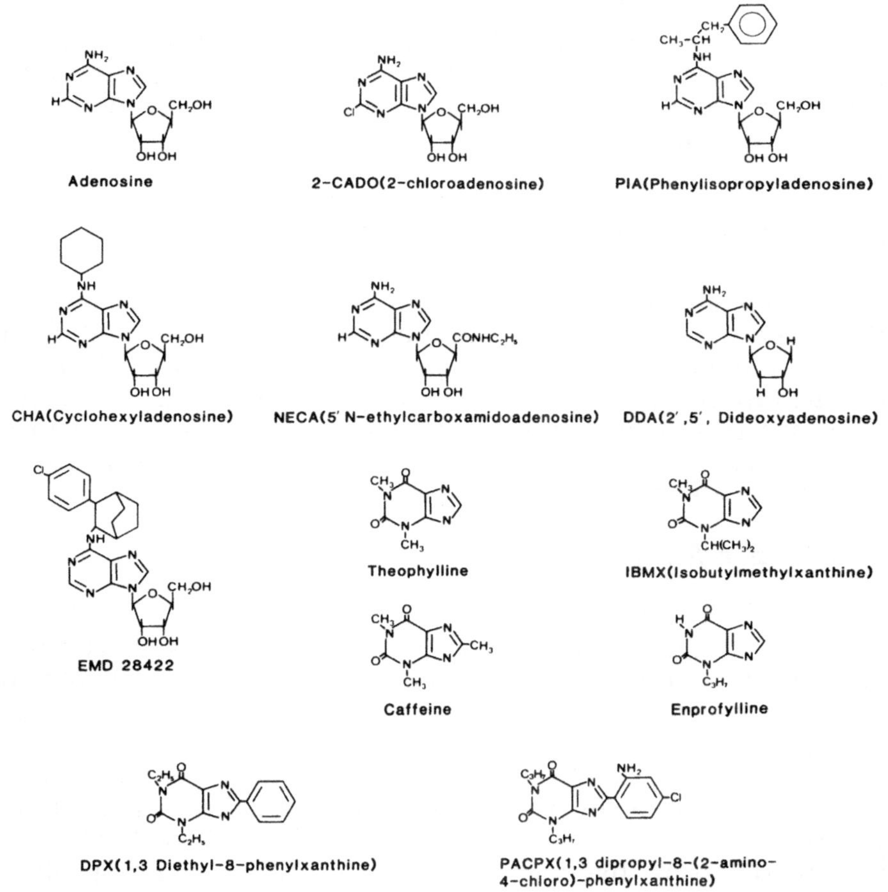

Fig. 2. Structure of adenosine agonists and antagonists.

stereospecific actions of the PIA isomers are not as marked as those seen with the A-1 site.

Other evidence has suggested the possibility that there might be a third cell surface adenosine receptor—that may tentatively be termed "A-3"—which is not linked to adenylate cyclase. The possible existence of such a site, initially suggested by Premont et al. (1979) was further developed by Phillis and Wu (1981) from electrophysiological data. Other workers (Smellie et al., 1979; Reddington and Schubert, 1979; Dolphin and Archer, 1983; Henon and McAfee, 1983) have reported that the recognition sites mediating some of the electrophysiological and biochemical effects of the purine are not identical with those mediating changes in adenylate cyclase activity, since the effects of the purine are not mimicked by cyclic nucleotides. Several analogs of cyclic AMP enhance postsynaptic potentials in olfactory slice preparations, rather than decrease them (Kuroda, 1978). Furthermore, Okada and Saito (1979) found that, although adenine nucleotides elevate cyclic AMP levels in superior collicular slices, they have no inhibitory actions on postsynaptic potential generation.

3. Adenosine Effects in Mammalian Tissues

As already mentioned, studies over the past 30–50 yr suggest that adenosine (or ATP) could affect tissue function. Although there is a high probability that many of the actions of ATP can be ascribed to the catabolic degradation of the nucleotide by ectoenzyme systems to the nucleoside, the detailed studies of Burnstock and his many colleagues (Berne et al., 1983b) have indicated that ATP itself can function as a neuromodulator.

3.1. CNS Effects

3.1.1. Electrophysiological Actions

Early studies on the electrophysiological actions of ATP and related compounds in the CNS resulted in the identification of two components—a short-acting excitatory effect and a more long-lasting depressant action (Phillis et al., 1979). This biphasic response was attributed to an initial excitatory action of the nucleotide caused by its ability to chelate calcium, or a direct effect of the polyphosphate side chain (Phillis and Wu, 1981) prior to the formation of adenosine, the nucleoside that depresses cell firing. More recently Jahr and Jessell (1983) have shown a direct effect of ATP on cell firing in dorsal root ganglia, an effect that is not blocked by theophylline. Thus it appears probable that there are P-2 receptors in nervous tissue.

The effects of adenosine in nervous tissue appear to involve both pre- and postsynaptic (Phillis and Wu, 1981; Segal, 1982; Henon and McAfee, 1983) actions and alterations in calcium currents. Adenosine reduces the spontaneous firing of central neurons (Phillis et al., 1979), as well as the amplitude of evoked excitatory postsynaptic potentials (Siggins and Schubert, 1981; Dunwiddie et al., 1981; Schubert and Mitzdorf, 1979), and inhibits interictal spiking (Dunwiddie, 1980), a phenomenon that may be related to epileptic seizure activity. These actions are blocked by theophylline and potentiated by inhibitors of adenosine uptake, such as dipyrdamole (Phillis et al., 1979), indicating the involvement of a cell surface recognition site similar to the A sites that mediate changes in cyclic AMP production. The lipid solubility of the adenosine analogs used to study electrophysiological responses is a matter of some concern (Dunwiddie et al., 1984); R-PIA is probably less effective than NECA in rat hippocampal slices because of such properties. Reddington et al. (1982) have demonstrated an excellent correlation between the effects of several adenosine analogs in depressing evoked potentials in rat hippocampal slices and their ability to displace A-1 receptor ligands. It is possible that A-1 and A-2 receptors may also be present as adenylate cyclase-independent moieties (Fig 1.) (Burnstock and Buckley, 1985). Evidence from cardiac muscle (Schrader et al., 1975), hippocampus (Schubert and Mitzdorf, 1979), and superior cervical ganglion (Henon and McAfee, 1983) indicates that adenosine can antagonize calcium conduction via a direct action on voltage-sensitive calcium channels. An indirect receptor-mediated calcium antagonist effect of adenosine has also been reported (Segal, 1982) that involves changes in K^+ conductance. Similar indirect actions of adenosine have been seen in cardiac muscle (Hartzell, 1979). Adenosine has been reported to affect Ca^{2+} uptake in K^+-depolarized synaptosomal preparations (Riberio et al., 1979), an effect that may provide a molecular basis for its mode of action. The purine has also been reported to facilitate synaptic transmission during repetitive firing (Henon and McAfee, 1983), a response dependent on extracellular calcium concentrations that, although not fully explainable, has been equated with adenosine acting as a "highpass filter." Thus, during low frequency events, adenosine reduces sensitivity and acts to enhance high frequency (repetitive) impulses. Whether these observations have physiological significance in relation to the neuromodulatory role of the purine, either in the CNS or elsewhere, or any relation to the apparent biphasic effects of the purine of transmitter release (Rabe and McGee, 1983; Spignoli et al., 1984) (*see* section 3.1.2) is a matter for further study. It is interesting, however, that although these facilitatory effects of the purine may be relevant to the enhancement of postsynaptic potentials by the dibutyryl, monobutyryl, and 8-bromo analogs of cyclic AMP in olfactory cortex and hippocampal slices (Kuroda, 1978; Dunwiddie and Hoffer, 1980), since cyclic AMP has been found to mimic the effects of adenosine (Henon and McAfee, 1983) by a mechanism that is sensitive to the catabolic enzyme adenosine deaminase (ADA), implying degradation of the cyclic nucleotide to the nucleoside, it would appear that these former effects are cyclic nucleotide- rather than adenosine-related.

The effects of adenosine on cell firing are complex, and although evidence has accumulated to suggest the likelihood that calcium rather than cyclic AMP (see above) may be involved as a second messenger for the purine, this conclusion requires further study. The physiological actions of adenosine (and no doubt ATP) are thought to reflect alterations in neurotransmitter release (*see* section 3.1.2), an explanation that could both adequately explain the depressant effects of the purine on CNS function and the potential mechanism, and involvement, of calcium. An additional complication is the finding by Katims et al. (1983) that the CNS effects of adenosine on locomotor activity (*see* section 3.1.3) involve both excitant and depressant components that may be mediated via distinct A-1 receptor subtypes.

Addition of the catabolic enzyme ADA or the xanthine adenosine antagonists to brain slice preparations can increase their spontaneous activity (Dunwiddie, 1980; Dunwiddie and Hoffer, 1980; Dunwiddie et al., 1981), providing electrophysiological evidence for the existence of "purinergic inhibitory tone" (Harms et al., 1978), the presence of endogenous adenosine proximal to cell surface recognition sites that results in modulation of cell firing. Spontaneous activity in hippocampal slice preparations has been directly correlated with endogenous adenosine levels (Fredholm et al., 1984).

3.1.2. Biochemical Effects of Adenosine

Acting via cell surface receptors, adenosine is the most potent modulator of brain slice cyclic AMP levels known (Sattin and Rall, 1970; Daly, 1977). Although the purine is a precursor of ATP, the substrate for adenylate cyclase, its effects on cyclic AMP production are not a reflection of alterations in substrate availability (Skolnick and Daly, 1974). The effects of adenosine are potentiated by the uptake inhibitor dipyridamole (Huang and Daly, 1974), indicating the involvement of cell surface recognition sites. Both A-1- and A-2-modulated adenylate cyclase activity have been demonstrated in brain membrane preparations (Premont et al., 1979; Cooper et al., 1980; Ebersolt et al., 1983). In order to demonstrate cyclase activity in broken membrane preparations, it is necessary to remove endogenous adenosine using ADA, then use a stable analog of the purine, such as 2-CADO, as a receptor agonist. The source of the endogenous adenosine in mammalian tissues has not been adequately resolved at the present time (*see* section 4).

The topographical distribution of A-1 and A-2 receptors is different (Ebersolt et al., 1983). Both subtypes are present in striatum and cerebellum, although cerebral cortex and hippocampus contain A-1, but not A-2, receptors. Studies on adenosine-modulated cyclic AMP production in brain slices and broken cell preparations has also led to the description of A-2 receptor subtypes (Daly et al., 1983). Structure–activity studies using the adenosine-stimulated cyclase of the VA 13 human fibroblast line (Bruns, 1980, 1981) have indicated that the N^6-amino group of the nucleoside is involved in the change in receptor conformation associated with enzyme activation. Also, an "anti" conformation of the purine is necessary for binding to the receptor. N^6-substituted adenosine analogs, such as CHA and R-PIA, are relatively selective for A-1 receptors, having weak agonist activity at A-2 sites (Daly, 1983). Ligands for the A-2 receptor appear to involve modifications at the 5′-position on the ribose moeity (Daly, 1983; Olssen, 1984).

Adenosine can also inhibit neurotransmitter release *in situ,* in vitro (Stone, 1981; Daly, 1982), and, by inference from its electrophysiological actions, in vivo (Phillis and Wu, 1981). The purine inhibits the release of acetylcholine (Sawynok and Jhamandas, 1976), dopamine (Harms et al., 1978; Michaelis et al., 1979), serotonin, noradrenaline (Harms et al., 1978, 1979; Fredholm and Hedqvist, 1980), and glutamate (Dolphin and Archer, 1983). Although the release of GABA has been reported to be susceptible to inhibition by the purine (Hollins and Stone, 1980), this finding has not been confirmed (Dolphin and Archer, 1983).

Transmitter release is a presynaptic event involving stimulus–secretion coupling and the mediation of calcium (Baker, 1972). In the neuromuscular junction and other ganglia, phosphodiesterase inhibitors and cyclic AMP analogs can increase transmitter release (Standaert and Dretchen, 1979; Kuba et al., 1981). Since adenosine can stimulate presynaptic cyclic AMP formation (Kobayashi et al., 1981), and the majority of the cyclic AMP-stimulated protein kinase activity in rat brain is located presynaptically (Weller, 1977), a reasonable hypothesis to explain the effects of the nucleoside on transmitter release is, as has been suggested Standaert and Dretchen (1979), that a cyclic nucleotide-dependent protein phosphorylation event regulates calcium fluxes and, consequently, stimulus–secretion coupling.

Elevated cyclic AMP levels have been reported to be associated with increases in neurotransmitter release in a number of nervous tissue preparations (Goldberg and Singer, 1969; Wooten et al., 1973; Cubeddu et al., 1975; Dunwiddie and Hoffer, 1980). In the PC-12 rat pheochromocytoma cell line, adenosine enhances stimulus-dependent, but not basal, neurotransmitter release via a mechanism that may be cyclic AMP-dependent (Rabe et al., 1980; Rabe and McGee, 1982). 2-CADO (10 n*M*) was found to stimulate [^{3}H] noradrenaline release, although the P-site agonist DDA at concentrations (100 μ*M*) that decreased cellular cyclic AMP levels by approximately 50% had no effect on K^+-evoked noradrenaline release (Rabe and McGee, 1983). The PC-12 cell line appears to have only the A-2 mediated cyclase response (Guroff et al., 1981; Rabe and McGee, 1983) and it may be argued, on the basis of the tentative involvement of cyclic AMP in the effects of adenosine on transmitter release in this cell line, that the inhibition of release normally associated with adenosine (Fredholm and Hedqvist, 1980; Phillis and Wu, 1981) is, as suggested by Daly (1982), an A-1 receptor-mediated event. In normal cells, adenosine can both inhibit and facilitate noradrenaline and dopamine release from brain vesicles (Ebstein and Daly, 1982), effects that may relate to interactions of the purine with the two receptor subclasses. The A-1 agonist CHA has been reported to inhibit acetylcholine release from rat cortical slices (Spignoli et al., 1984), whereas the A-2 agonist NECA can stimulate release. Although both effects were blocked by aminophylline, high concentrations (1 m*M*) of adenosine itself, the presumed endogenous modulator, inhibited release. Nucleoside effects on transmitter release appear to be biphasic, but require further validation before any definitive role can be ascribed to the stimulatory effect of this putative neuromodulator. Certainly, the fact that adenosine may have selective effects on both the nature (Dolphin and Archer, 1983) and direction (Ebstein and Daly, 1982; Spignoli et al., 1984) of transmitter release is promising in terms of a discrete neuromodulatory role for the purine. The effects of 2-CADO on acetylcholine turnover in rat hippocampus, parietal cortex, and striatum are regionally (Haubrich et al., 1981), as well as stereo (Murray et al., 1982), selective. Thus, the effects of adenosine on transmitter release reflect more than a "blanket" inhibition. The involvement of cyclic AMP in adenosine-related effects on cell firing and neurotransmitter release is, as discussed above, in question. Adenosine inhibition of glutamate release is not mediated via a cyclic AMP step, and in fact, 8-bromo-cyclic AMP has effects opposite to those of the purine (Dolphin and Archer, 1983).

In addition to defining the precise role (and synaptic location) of the adenosine receptors involved in transmitter release, their function within the context of discrete neurotransmitter "autoreceptors" (Carlsson, 1975) will be important to define. It is probable that the effect of adenosine on transmitter release may be more global than that of the individual transmitters themselves. Additional considerations are the potential role of the third "A-3" receptor in such events (Phillis and Wu, 1981; Rabe and McGee, 1983) and the precise nature of the calcium channels involved in CNS transmitter release. Although calcium entry blockers have been reported to interact with adenosine (Murphy and Snyder, 1982), these compounds have no apparent effect on CNS transmitter release (Shallaby et al., 1983).

3.1.3. Behavioral Actions of Adenosine

Peripheral or central administration of adenosine and its analogs, or of ATP, results in profound sedative, hypnotic, and anticonvulsant activity (Feldberg and Sherwood, 1954; Snyder et al., 1981; Coffin and Carney, 1983). Direct administration of adenosine into cat (Feldberg and Sherwood, 1954) and fowl (Marley and Nistico, 1972) brain produces a sleep-like state with an EEG profile similar to that seen in deep sleep. Parenteral administration of R-PIA can promote deep sleep and decrease wakefulness (Radulovacki et al., 1984), an effect mediated via an A-1 receptor mechanism associated with the inhibition of dopamine release (Radulovacki et al., 1984). Rats deprived of paradoxical sleep undergo an increase in brain adenosine levels, although injection of the purine into the dog lateral ventricle, or im in mice, produces hypnogenic and sedative effects (Haulica et al., 1973). Parenteral administration of the diastereomers of PIA stereospecifically depresses mouse locomotor activity (Vapaatalo et al., 1975), whereas at 0.5 mg/kg, ip, CHA and 2-CADO decrease movement, increase pause duration, and decrease mean exploratory behavior in mice (Crawley et al., 1981). Given intracerebroventricularly, NECA is more potent than R-PIA, 2-CADO, and S-PIA in depressing locomotor activity (Barraco et al., 1983), a finding that suggests the involvement of an A-2 receptor (Table 1). Muscle relaxation (Baird-Lambert et al., 1980), ataxia, and hypothermia (Dunwiddie and Worth, 1982), as well as defects in hotplate reaction time and rotorod performance (the latter an indication of ataxia), have also been noted (Crawley et al., 1981) following adenosine agonist administration. The purine can also depress breathing (Hedner et al., 1982) and supress food intake (Levin and Morley, 1983) when administered centrally. Adenosine and its analogs are antinociceptive (Yarbrough and McGuffin-Clineschmidt, 1981), whereas theophylline has been reported (Paazlow and Paazlow, 1973) to increase sensitivity to noxious stimuli. Adenosine may have diazepam-like properties and can interact with the anxiolytic both centrally (Crawley et al., 1981; Dunwiddie and Worth, 1982; Katims et al., 1983; Wagner and Katz, 1983) and peripherally (Kenakin, 1982). In keeping with the depressant actions of adenosine, theophylline has been found to increase locomotor activity (Thithapandha et al., 1972; Snyder et al., 1981). Dose-range studies on the locomotor depressant effects of R-PIA in mice have shown that the potencies of a series of alkylxanthines (administered ip) to reverse R-PIA-induced locomotor depression parallels their ability to displace [^{3}H]-CHA from A-1 receptors in vitro (Katims et al., 1983). This observation was somewhat complicated by the fact that R-PIA showed a biphasic response. At doses greater than 100 nmol/kg, ip, R-PIA depressed locomotor activity, whereas below 50 nmol/kg the adenosine analog augmented such activity, providing functional evidence for the possible existence of A-1 receptor subtypes. Similar conclusions have been reached by Coffin et al. (1984), although in this instance A-2 rather than A-1 receptors may be involved in the locomotor effects of the purines. Biphasic effects have also been observed for certain xanthine derivatives; caffeine, for instance, can reduce locomotor activity at low doses and increase it at high doses (Katims et al., 1983). The depressant actions of R-PIA can be reversed by doses of the xanthines, which are depressant, a phenomenon that may be ascribed to xanthine blockade of the lower-affinity A-1 sites through which the purine depresses locomotor activity, as a consequence "unveiling" the higher affinity sites that are responsible for the stimulatory actions of the adenosine analog (Katims et al., 1983).

Parenteral administration of adenosine can protect mice against audiogenic seizures (Maitre et al., 1974); this anticonvulsant activity has also been shown for kainic acid, strychnine, picrotoxin, and 3-mercaptopropionic acid (Dunwiddie and Worth, 1982) using the stable adenosine analogs CHA, R-PIA, and 2-CADO. Although the sedative and hypothermic effects of all three adenosine analogs could be blocked by theophylline, only the anticonvulsant effects of CHA and 2-CADO were antagonized by the xanthine. This observation, and differences in the potencies of the adenosine analogs, led Dunwiddie and Worth (1982) to suggest that the in vivo effects of R-PIA might reflect either an interaction with "nonpurinergic receptors," or that distinct adenosine receptor subtypes with differing affinities for the agonists were present in the brain.

Chronic treatment of rats or rabbits with large doses of methylxanthines results in a self-mutilation paradigm that is used as a model of Lesch-Nyhan syndrome (Nyhan, 1973). This disease is an X-linked disorder biochemically related to a deficiency in the enzyme hypoxanthine-guanine-phosphoribosyltransferase (Kelley, 1983), which may also be manifest by changes in striatal dopaminergic function (Lloyd et al., 1981). NECA has been shown to modulate striatal dopaminergic function in rats (Green et al., 1982), an effect that probably involves A-2 type receptors. In addition, NECA in combination with the dopamine agonist apomorphine can elicit a self-mutilation syndrome. From this observation, and the known biochemical lesion of purine metabolism, it has been suggested that Lesch-Nyhan syndrome may involve changes in central purinergic function that relate to a functional deficit in adenosine receptor-linked systems and a consequent alteration in associated monoamine system function, both of which may underlie the pathophysiology of this disease state (Green et al., 1982).

Central purinergic systems have also been tentatively linked to the effects of two "recreational" drugs—phencyclidine (PCP) and alcohol. Using an animal discriminative model, Browne and Welch (1982) found that R-PIA and CHA block responsiveness to PCP, a finding that may relate to an interaction between adenosine and PCP. The adenosine analog R-PIA has greater sedative and hypothermic effects in ethanol-sensitive, "long sleep" mice than in ethanol-insensitive, "short sleep" mice (Proctor and Dunwiddie, 1984). Theophylline causes an increase in the escape latency of the "long sleep" mouse group using an escape platform paradigm, without affecting significantly the performance of the second group. The possibility that ethanol might be a purinergic agonist was dismissed based on the fact that theophylline and caffeine can reduce, but not block, the effects of ethanol (Proctor and Dunwiddie, 1984). It was concluded, however, that ethanol might produce its effects via a central purinergic pathway. Theophylline can decrease the duration of ethanol-induced hypnosis, and on withdrawal of alcohol a decrease in R-PIA binding has been reported (Dar et al., 1983).

3.2. Cardiovascular Actions of Adenosine

In view of the profound effects of adenosine on cardiovascular system function (Berne, 1980), the hypotensive effects of various adenosine analogs (Vapaatalo et al., 1975), and the positive inotropic (cardiotonic) actions of the methylxanthines that have, as yet, no clearly defined molecular basis (Farah et al., 1984), it has been suggested (Phillis and Wu, 1983) that the behavioral effects of adenosine, especially those elicited by parenteral administration, might be related to a fall in blood pressure that would affect CNS function. In fact, a good correlation has been reported for the effects of a series of adenosine analogs in inhibiting neuronal cell firing and their hypotensive actions (Phillis and Wu, 1983). However, in two studies using both central and parenteral routes of administration, it was found that the hypotensive actions of adenosine agonists could be clearly distinguished from their sedative actions (Katims et al., 1983; Barraco et al., 1984), with doses approximately 10 times higher than those required to depress locomotor activity being necessary to lower blood pressure. It is possible that adenosine may produce its behavioral actions by a combination of both central and peripheral actions, or perhaps by a direct effect on cerebral blood flow (Edvinsson and Fredholm, 1983). It is equally possible that adenosine-related influences in the area of renal function (Osswald, 1983) may reflect on the central actions of compounds interacting with purinergic systems.

4. Adenosine Availability in the CNS

The evidence for the existence of a purinergic inhibitory tone (Harms et al., 1978; Dunwiddie and Hoffer, 1980; Fredholm et al., 1984) points toward a physiological role for adenosine in the CNS. Studies with exogenously applied adenosine, related analogs, adenosine uptake, and ADA inhibitors would confirm this conclusion. However, one is left with the question: What is the source of endogenous extracellular, intersynaptic adenosine? Purine release has been reported following nerve stimulation from various nerve tissue preparations (Fredholm and Hedqvist, 1980). In the CNS, the release of radiolabeled purines by electrical stimulation, elevated K^+, ouabain, and veratridine from brain slice preparations prelabeled with [^{14}C]- or [^{3}H]-adenine or adenosine has been demonstrated (Pull and McIlwain, 1974; Kuroda and McIlwain, 1973). In guinea pig cortical synaptosomes, Kuroda and McIlwain (1974) demonstrated that K^+-evoked purine release was calcium-dependent. However, in guinea pig hypothalamic synaptosomes, release was found to be calcium-independent (Fredholm and Vernet, 1978). Other workers (Stone, 1981) have reported that the release of purine(s) from cortical slices differs from that seen with other putative neurotransmitters, both temporally and quantitatively. Studies using prelabeled brain tissue are difficult to interpret (Daly, 1982), especially in terms of the specific activity of the various purine derivatives released, a problem that may be circumvented by using more specific HPLC detection procedures (Wojcik and Neff, 1982). Despite such caveats, ex vivo release of purines from hippocampal slices prepared from in vivo labeled brain tissue has been shown to be a specific event, with perforant pathway, but not fimbria stimulation, leading to labeled purine release (Lee et al., 1982). Purine release *in situ* from cat and rat cerebral cortex is primarily in the form of nucleotides (Phillis et al., 1980), and is calcium-dependent and tetrodotoxin-sensitive (Jhamandas and

Dumbrille, 1980). The calcium dependence of purine release is a matter of some concern (Stone, 1981; Snyder, 1984; Williams, 1984a), both for adenosine and ATP.

Cyclic AMP and 5′-AMP, as well as ATP, have been considered potential sources for extracellular adenosine (Newman and McIlwain, 1977; Pons et al., 1980; Potter and White, 1980; MacDonald and White, 1984), with the release of any of these adenine nucleotides, either on their own (Potter and White, 1980) or as cotransmitters (Israel et al., 1980; Burnstock, 1983), resulting in their rapid degradation by synaptic ectoenzyme systems to the nucleoside.

Adenosine can be formed in brain tissue by four enzymatic mechanisms (Nagata et al., 1984). These are the 2′- and 5′-nucleotidases and the nonspecific acid and alkaline phosphatases. Of the four enzymes, the 5′-nucleotidase—a membrane-associated entity that uses 5′-AMP as substrate (Scott, 1967)—is the most likely source of the purine under physiological conditions (Meflah et al., 1984; Nagata et al., 1984). The topographical distribution of hippocampal 5′-nucleotidase (Hunt and Kunzle, 1976) has been related (Schubert et al., 1979) to the transneuronal transfer of adenosine, a process that may be loosely related to the rheoseme concept proposed by McIlwain (1978). A correlation between the autoradiographic distribution of adenosine A-1 receptors and 5′-nucleotidase has been reported (Goodman and Snyder, 1982), suggesting a functional relationship between the enzyme and the recognition sites for its product. In this context, it is of interest that the electrophysiological effects of adenosine in hippocampal slice preparations appear to be related to A-1 receptor density rather than purine availability (Lee et al., 1983).

Although it has been suggested that 5′-nucleotidase acts intracellularly to produce adenosine (Nakamura et al., 1983; Nagata et al., 1984), the possible localization of the enzyme to external cell surfaces (Kluge et al., 1972; Newby et al., 1975) may indicate the potential for the enzyme to have relevant activity in the synaptic cleft. Using an HPLC method to measure adenosine in rat synaptosomal preparations, MacDonald and White (1984) have demonstrated the potential for the extracellular conversion of nucleotides to adenosine via an adenosine-activated ATPase (Trams and Lauter, 1974). This finding is in contrast to that reported by Pons et al. (1980). Clearly, the production of adenosine by ATPase and 5′-nucleotidase activity (Sun et al., 1976) is different from purine release mediated by stimulus–secretion coupling, and it remains to be seen what factors may influence the activity of these enzymes under appropriate physiological conditions. It should not be underemphasized that the purine may also be made available through a stimulus–secretion-dependent process that, like the 5′-nucleotidase process, may be either neuronal or glial in location. It will be of considerable interest to see if the activity of 5′-nucleotidase is subject to modulation by factors other than the cellular energy charge (Sun et al., 1976; Nagata et al., 1984) that may be susceptible to transmitter-related modulation.

The other facet of adenosine action in the CNS is the mechanism(s) available for the termination of its action at the receptor level. Adenosine uptake is one method of removing the nucleoside from its site of action. High-affinity adenosine uptake has been characterized in guinea pig (Shimizu et al., 1972) and mouse (Kuroda and McIlwain, 1974) cortical slices, and in guinea pig and rat synaptosomes (Kuroda and McIlwain, 1973; Bender et al., 1981). In the rat tissue preparation, two uptake systems have been described: one is rapid, with a K_m of 0.9 μM that is saturated within 1 min; and the other is slower, comprised of two relatively high-affinity sites (K_m = 1 and 5 μM), and requires 30 min to reach equilibrium. These latter two sites may be glial and neuronal in location, respectively. It may be noted that vasodilators such as dipyridamole produce their effects by inhibition of adenosine uptake (Moritaki, 1983), whereas a case has been made for linking the effects of some psychotropic agents to their potential, albeit weak, ability to inhibit synaptosomal adenosine uptake (Phillis and Wu, 1983; Phillis, 1984).

The major catabolic enzyme for adenosine is adenosine deaminase (ADA). This enzyme is primarily cytoplasmic (Pull and McIlwain, 1974) and would, by inference, be unlikely to be involved in the termination of the actions of adenosine unless uptake were to precede its involvement. The potent ADA inhibitors, deoxycoformycin and erythrohydroxynonyladenine (EHNA), have behavioral effects when administered parenterally (Major et al., 1981; Mendelson et al., 1983; Radulovacki et al., 1983), suggesting that ADA does play a significant role in vivo. Of interest is the immunochemical detection of a discrete plexus of ADA-containing neurons in the basal hypothalamus (Nagy et al., 1984) that may be considered a marker for those nerve cells in which adenosine is a physiological modulator. Since this group of neurons appears to innervate diverse areas of the brain, including the amygdala, striatum, and cortex, the hypothalamic ADA system may be of great significance with regard to the understanding of the neuromodulatory potential of adenosine and, together with the receptor autoradiography described below, provides evidence for a discrete role for the nucleoside in CNS function.

The use of a specific antisera to adenosine (Snyder et al., 1984) has yielded preliminary data suggesting that the distribution of immunoreactive neurons, i.e., in hippocampal pyramidal cells, parallels the distribution of A-1 receptors and 5′-nucleotidase activity. Clearly, more in-depth studies will define whether adenosine immunoreactivity is specific for unique "adenosinergic" neurons or for enzymes such as *S*-adenosyl homocysteinase that use the purine as substrate.

Although the factors responsible for the availability of extracellular adenosine are not clearly defined at the present time, it is important to consider that the majority of studies addressing adenosine availability in the CNS have focused on what may be termed "conventional" approaches to neurotransmitter function. With putative neuromodulators, such as substance P and neurotensin, it is difficult to confuse metabolic pathways for these peptides with intracellular pathways related exclusively to cellular homeostasis. If, indeed, adenosine (and ATP) function as global, as opposed to local, discrete neuromodulatory agents (Williams, 1984b), it may be necessary to consider the possibility that the general level of the extracellular purine may be a homeostatic process subject to "fine tuning" within the context of the actual availability of the receptor rather than its effector agent. Thus, adenosine may function as a neurohormone in the true sense of the definition.

5. Binding of Ligands to Purinergic Receptors

In addition to the criteria of synaptic synthesis, release, inactivation, and appropriate neuronal responses to the exogenously applied putative substance, for a compound to be considered a neurotransmitter/neuromodulator it is now necessary to also demonstrate the binding of high-specific radioactivity ligands, agonist, and antagonist, to recognition sites in appropriate subfractions of mammalian brain tissue. The delineation from the designation of a simple recognition site to that of a receptor proper requires the assignment of relatively rigorous criteria to binding data (Cuatrecasas and Hollenberg, 1976). Thus, although binding sites and subsites for any number of putative neuromodulators, their antagonists, and drugs have been described (Williams and U'Prichard, 1984), without appropriate physiological and pharmacological validation (Hrdina, 1985), such sites cannot be described as receptors. The search to identify and characterize purinergic receptors by binding techniques began in the mid 1970s and reached fruition in the 1980s with the description of the binding of stable adenosine analogs to brain tissue by several groups (Bruns et al., 1980; Williams and Risley, 1980a; Schwabe and Trost, 1980). Efforts to identify P-2 receptors by ligand binding techniques have been few to date.

5.1. ATP Binding

Although in view of the dynamic, labile nature of ATP it might appear foolhardy to use this adenine nucleotide as a radioligand, high-affinity (K_d = 20 nM) [^{3}H]-ATP binding to rabbit urinary bladder has been described (Levin et al., 1983), such binding being distinct from that involving divalent ion ATPase activity. Binding of the stable ATP analog, AppNHp, although having a dissociation constant in the nanomolar range, was found to be irreversible in rat brain membranes (Williams and Risley, 1980b). In view of the increasing evidence (discussed above) that there might be distinct P-2 receptors in the CNS (Jahr and Jessell, 1983), it seems likely that efforts to develop viable ATP radioligand binding assays will intensify.

5.2. Adenosine Agonist Binding

Adenosine and its related ADA-stable agonists have been used to label both central and peripheral adenosine receptors. In general, binding has been more successful in central than in peripheral tissues, a reflection of the relative density of adenosine receptors in the tissue preparations. To date, the majority of studies in central tissue have related to the A-1 receptor and, as might be anticipated, agonists are more potent in displacing agonist ligands than antagonist, with the converse being true for antagonist ligands.

In ascribing A-1 and A-2 characteristics to adenosine agonist recognition sites, a good deal has depended on the effects of the diastereomers of PIA. At A-1 receptors, as defined by the effects of PIA on evoked potential generation in rat hippocampus, the *R*- and *S*-isomers have a 50–100-fold difference in potency (Smellie et al., 1979; Daly, 1982). At A-2 receptors, defined by their effects on cyclic AMP formation in guinea pig hippocampal slices, only a fourfold separation in the potencies of the isomers of PIA is seen. Fredholm et al. (1982) have noted that the differences in the effects of the isomers of PIA may reflect species, as well as receptor, differences, an issue that is addressed further in section 5.6.

5.2.1. Adenosine Binding

Early studies using radiolabeled adenosine to define adenosine receptors in brain and peripheral tissues were none too successful. Binding sites with dissociation constants in the micromolar range and equivocal pharmacology (Shimizu, 1979; Schwabe et al., 1979; Newman et al., 1980; Williams, 1981) were described. In the study of Schwabe et al. (1979), the diastereomers of PIA were not stereospecific. High-affinity [^{3}H]-adenosine binding has been described in NG108CC15 cells and brain tissue by Snell and Snell (1983). In general, however, adenosine is considered to be a poor ligand for either A-1 or A-2 receptors in brain tissue (Newman, 1983).

5.2.2. Adenosine Analog Binding

The key to being able to detect high-affinity adenosine recognition sites is the use of the catabolic enzyme ADA in conjunction with adenosine analogs that are stable to degradation by this enzyme. Although this tactic had originally been used by Premont et al. (1977) to detect adenosine-stimulated adenylate cyclase activity in broken membrane preparations, the lack of availability of radioactive forms of the adenosine analogs CHA, 2-CADO, and R-PIA limited the use of this approach to the binding aspects of central adenosine receptors until 1980.

The source of endogenous adenosine in broken cell preparations has not been rigorously established. It is well known that following trauma, especially decapitation, tissue levels of adenosine are dramatically increased (Wojcik and Neff, 1982; Berne et al., 1983a). In broken cell preparations, although ADA pretreatment can dramatically increase the specific binding of radiolabeled adenosine agonists (Williams, 1981; Patel et al., 1982), cursory washing, which reduces the ADA concentration, allows significant amounts of the nucleoside to reappear. The source of this adenosine may be a sequestered form of the purine, such as *S*-adenosylhomocysteine (Daly, 1982).

5.2.2.1. CHA BINDING. Following pretreatment of tissue with ADA, CHA (Fig. 2) incubated with brain tissue for 2 h at 23°C binds with high affinity (K_d = 0.3–6.0 n*M*) to recognition sites in guinea pig and bovine brain that have characteristics of the A-1 receptor (Bruns et al., 1980). Nonspecific binding was determined using 10 μ*M*R-PIA. CHA is a weak agonist at A-2 receptors and has no affinity for the P site (Daly, 1983). Binding of CHA is: theophylline-sensitive (Table 2); enriched in synaptosomal fractions of mammalian brain; sensitive to the alkylating agent iodoacetamide (Patel et

TABLE 2
Pharmacology of Adenosine Radioligand Binding (K_i, nM)

Compound	CHA[a]	R-PIA[a]	DPX[a]	2-CADO[b]	NECA[c,d] NEM-treated striatum (A-2)	NECA[c,d] Hippocampus (A-1)
R-PIA	0.24	0.23	0.21	0.6	234	3.5
S-PIA	4.37	5.60	8.10	17.0	3045	102
2-CADO	—	—	—	1.5	—	—
DPX	1.6	3.1	1.4	—	—	—
IBMX	3837	4000	3499	2217	—	—
NECA	—	—	—	24.3	303	98
CHA	0.68	0.69	0.81	0.7	57	4.3
S/R-PIA Ratio	18	24	39	28	13	29

[a]From Murphy and Snyder, 1982; calf cortex.
[b]From Williams and Risley, 1980b; rat cortex.
[c]From Yeung and Green, 1984; rat brain.
[d]Approximate K_i values calculated.

al., 1982); and reduced by guanine nucleotides in both membrane-bound and solubilized receptor preparations (Gavish et al., 1982). CHA binding was reduced by sodium (Goodman et al., 1982; Marangos et al., 1984a) and, like other agonist ligands, labile to thermal and proteolytic degradation. CHA sites are also sensitive to irreversible denaturation by copper.

5.2.2.2. 2-CADO Binding. The binding of [^{3}H]-2-CADO to rat brain membranes has been examined both with and without ADA pretreatment. Following removal of endogenous adenosine with ADA, high-affinity (K_d = 1.3 nM) theophylline-sensitive binding has been described that has pharmacology consistent with an A-1 recognition site (Williams and Risley, 1980a). Binding was measured for 60 min at room temperature, with nonspecific binding being determined in the presence of 10 μM unlabeled 2-CADO. Specific binding was enriched in synaptosomal subfractions of rat brain and had the following regional distribution: caudate ≥ hippocampus > cerebellum > thalamus >> pons medulla ≥ cortex > hypothalamus > spinal cord, a distribution that to some extent parallels that of A-1 modulated-cyclase activity (Ebersolt et al., 1983).

Examination of 2-CADO binding in non-ADA treated tissue using standard incubation conditions of 60 min at 0°C in the presence of Mg^{2+} and Ca^{2+}, showed a high-affinity binding site (K_d = 23 n*M*) (Wu et al., 1980; Wu and Phillis, 1982). Such binding was, however, potently antagonized by adenine and inosine, a finding that was not consistent with the known structure–activity relationships demonstrated for adenosine receptor-mediated effects and with other adenosine receptor binding assays.

Further studies with 2-CADO as a radioligand were not too successful (Haubrich et al., 1981) because of the probable instability of later batches of the ligand. Recently, a 2-CADO binding assay has been reported in uterus (Fox and Kurpis, 1983). However, in general, problems with the stability of this ligand have precluded its routine use.

5.2.2.3. R-PIA BINDING. The *R*-diastereomer of the N^6-substituted analog PIA is a selective A-1 receptor ligand (Daly, 1983). In ADA-pretreated membranes, in the presence of 1 m*M* Mg^{2+}, using a 15 min incubation at 37°C and isolating receptor–ligand complexes by vacuum filtration, R-PIA binds with a K_d of 5 n*M*. Specific binding represents 60% of the total counts bound, using 100 μ*M* R-PIA to define nonspecific binding (Schwabe and Trost, 1980). R-PIA itself has a K_i of 24 n*M*, with 2-CADO having an IC_{50} of 947 n*M*. This weak effect of 2-CADO has been attributed to the assay temperature (Daly, 1973).

5.2.2.4. NECA BINDING. The adenosine 5′-carboxamide NECA and its corresponding cyclopropyl analog have the highest potencies of known adenosine analogs for the A-2 receptor (Londos et al., 1980). Binding of adenosine 5′-cyclopropylcarboxamide (Daly et al., 1979), although of high affinity (K_d = 14 n*M*), has pharmacology that is not entirely consistent with A-type receptor binding. In rat striatum, NECA binds with high affinity to both A-1 and A-2 binding sites (Yeung and Green, 1984). In hippocampal membranes, which are apparently devoid of A-2 receptors (Yeung and Green, 1983), NECA binds with a K_d of 12.3 n*M* (Yeung and Green, 1984). Using *N*-ethylmalemide (NEM) to reduce A-1 binding, NECA binding to A-2 receptors is disassociated from that occurring to A-1 receptors. A K_d of 17 n*M* for [^{3}H]-NECA has been observed in NEM-treated membranes incubated at 37°C for 15 min. Preliminary pharmacological analysis (Table 2) of NECA binding in NEM-treated striatal membranes (A-2 receptors) and hippocampus (A-1 receptors) has shown that NECA is more active at A-2 sites than are R- or S-PIA or CHA, whereas at putative A-1 sites, R-PIA and CHA are considerably more active than the carboxamide. It may be noted that there is a 13-fold separation in the activity of the diastereomers of PIA at the putative A-2 site, and a 29-fold separation at the putative A-1 site. The ratio for other A-1 ligands is: CHA, 18; 2-CADO, 28; and R-PIA, 24 (Table 2).

5.2.3. DDA Binding

2′5′-Dideoxyadenosine (DDA), an adenosine analog specific for the P site (Daly et al., 1981; Londos et al., 1983), binds to rat cortical and striatal membranes with a dissociation constant of approximately 90 n*M* (Nimit et al., 1982). Binding that was measured for 40 min at room temperature using adenosine arabinoside (100 μ*M*) to define specific binding, isolating receptor–ligand complexes by vacuum filtration, was insensitive to agonists and antagonists for A sites, including 2-CADO, R-PIA, and theophylline. DDA was, however, displaced by adenosine and 5′-methylthioadenosine. Comparison of the effects of a series of adenosine analogs on DDA binding and P site-mediated inhibition of rat brain adenylate cyclase activity has shown that the pharmacology for the two sites is not consistent; for instance, 5′-methythioadenosine, which is the most potent compound studied in terms of its inhibition of specific DDA binding, has no effect on cyclase activity (Nimit et al., 1982; Daly, 1983). It has been concluded from compound displacement curves that DDA binds to two adenosine recognition sites, one of which might be involved in adenosine uptake.

5.3. Adenosine Antagonist Binding

Attempts to use [^{3}H]-theophylline (Fig. 1) as a ligand for CNS adenosine recognition sites have not been successful (Williams, 1981). However, 1,3-diethyl-8-phenylxanthine (DPX) (Fig. 1) has been used to label adenosine receptors. Initial studies with DPX (Bruns et al., 1980) using guinea pig and bovine brain tissue have shown marked species differences. In bovine brain, a K_d of 5 n*M* was observed using a 4-h incubation period at 25°C. In guinea pig brain, binding (K_d = 70 n*M*) could only be determined at 0°C. Binding in bovine brain was consistent with A-1 receptor labeling, with the R isomer of PIA being 25 times more active than the *S* isomer. In guinea pig and avian brain, binding was initially suggested to reflect labeling of both A-1 and A-2 receptor classes (Bruns et al., 1980; Barnes and Thampy, 1982), but subsequent studies (Murphy and Snyder, 1982) suggest (Daly, 1983) that DPX binding in guinea pig brain might not be consistent with the labeling of adenosine recognition sites. In assessing species differences in DPX binding, Murphy and Snyder (1982) found that in calf, rat, and rabbit brain membranes, DPX labeled "typical"

A-1 receptors. The inability to detect significant DPX binding in guinea pig and human was attributed to the low affinity of the xanthine for the A-1 receptors present in these tissues. The fact that adenosine agonists cannot completely displace specific DPX binding (Barnes and Thampy, 1982; Marangos et al., 1984a) has been taken as evidence that the antagonist radioligand binds to additional sites that differ from those binding agonists. The number of DPX binding sites in rat brain can be increased by GppNHp or NEM pretreatment (Yeung and Green, 1983), a finding that may also point to the existence of different recognition sites.

5.4. Differences in Agonist and Antagonist Radioligand Binding

Binding of adenosine agonists and antagonists to brain tissue differs significantly in terms of thermal and proteolytic stability (Marangos et al. 1984a), thermodynamic effects, and sensitivity to guanine nucleotides and sodium ions. Agonist affinities increase at high temperatures, whereas antagonist binding is enhanced at temperatures approaching 0°C (Murphy and Snyder, 1982). Agonist binding to adenosine receptors is entropy-driven, whereas binding of antagonists appears to involve both entropy and enthalpy effects. Agonist binding is reduced by guanine nucleotides (Bruns et al., 1980; Goodman et al., 1982) and by sodium ions (Gavish et al., 1982). Agonist sites are also more sensitive than antagonist sites to thermal and proteolytic denaturation and to irreversible denaturation by copper (Marangos et al., 1984a). These differencies in agonist and antagonist binding may be used to determine such properties in novel compounds and may reflect on the different properties of adenosine receptor subclasses. It may be noted that interspecies differences in adenosine radioligand binding have been observed (*see* 5.6). These are more pronounced for antagonists than agonists. The potent xanthine, PACPX (Bruns et al., 1983a), which has picomolar affinity in bovine brain has nanomolar activity in rat and guinea pig brain (Williams and Valentine, 1985; Ferkany et al., 1986). These differences that span three orders of magnitude are not as dramatic with agonists and may be unique to adenosine recognition site(s).

5.5. Localization of Adenosine Receptors in the CNS

Direct visualization of [^{3}H]-CHA binding sites in rat brain (Goodman and Snyder, 1982) by autoradiographic techniques has shown that the highest density of sites occurs in the hippocampus, a finding consistent with the regional distribution studies mentioned above. This binding is highest in the molecular and polymorphic areas, with lower levels of binding over the pyramidal cell layer. The medial, gelatinosus, and lateral nuclei of the thalamus have high densities of binding, as do the lateral septum and medial geniculate body. Although other brain regions have intermediate densities of receptors, no binding could be observed in the spinal or pyramidal tracts, the spinal tract of the trigeminal nerve, or the superior cerebellar peduncle. It has been noted (Goodman and Snyder, 1982; Snyder, 1984) that the autoradiographic localization of CHA binding sites correlates with the histochemical localization of the enzyme 5′-nucleotidase (Hunt and Kunzle, 1976), which can generate adenosine, suggesting the possibility of a functional relationship between the enzyme and the recognition sites for its product (*see* section 4 above).

In the cerebellum of mouse neurological mutant strains, A-1 receptors are associated with granule cells (Wojcik and Neff, 1983a; Goodman et al., 1983), whereas in the rat, the removal of one eye depletes A-1 receptors in the contralateral superior colliculus, suggesting that the receptors are present on axon terminals of excitatory projections from retinal ganglion cells (Goodman et al., 1983). This localization is specific for these pathways since ablation of rat cerebral cortex or unilateral lesions of the fornix are without effect on hippocampal or thalamic receptor density. In rat striatum, A-2 receptors appear, from the effects of intrastriatal kainic acid and 6-OHDA lesions, to be partially located on cortical-striatal nerve terminals and intrinsic striatal neurons (Wojcik and Neff, 1983b,c), although A-1 receptors are primarily associated with intrinsic striatal neurons.

5.6. Adenosine Receptor Heterogeneity

Biochemical and behavioral evidence supports the existence of both A-1 and A-2 receptor subtypes. Using CHA, R-PIA, and DPX as radioligands, Murphy and Snyder (1982) studied A-1 receptor binding in rat, rabbit, human, guinea pig, and calf membranes. The affinity of CHA in calf neocortical membranes was considerably different from that seen in human (0.3 vs 5.2 n*M*), but the number of human binding sites was one-third the number of calf sites. In calf and rabbit brain, R-PIA was eight times more active in rat, guinea pig, or human brain; CHA was 2–6 times more active in calf membranes than in those from the other species; and S-PIA was 15–57 times more active in calf in displacing CHA than in the other species studies. The stereospecific effects of PIA on CHA binding in membranes from the several species studied were also different. A tenfold separation in the effects of *R*- and *S*-PIA was seen in calf, with a 34-fold separation in guinea pig and human, a 78-fold separation in rat, and a 132-fold separation in rabbit. The effects of DPX and IBMX (Fig. 2) also show species differences in their potencies in displacing CHA binding. The differences in the effects of *R*- and *S*-PIA noted by Fredholm et al. (1982) may also reflect on species differences in receptors. In calf brain, however, [^{3}H]-CHA, [^{3}H]-R-PIA, and [^{3}H]-DPX have a very similar pharmacological profile, consistent with the labeling of A-1 receptors.

The anticonvulsant profiles of *R*-PIA, CHA, and 2-CADO have also suggested the possible existence of adenosine receptor subtypes (Dunwiddie and Worth, 1982), as has the effect of *R*-PIA on mouse locomotor activity (Katims et al., 1983) (*see* 3.1.3). Guanine nucleotide and NEM treatment, reveals the existence of three affinity states of the A-1 receptor that have K_d/K_i values of 1.8, 15, and 430 n*M* (Yeung and Green, 1983). The relationship of these sites to adenylate cyclase activity remains to be determined. Other studies using adenylate cyclase activity as an endpoint have also provided evidence for adenosine receptor heterogeneity in brain slices and homogenates (Daly et al., 1983).

Recent studies with the potent xanthine antagonist PAPCX (Fig. 2) (Bruns et al., 1983a) have demonstrated distinct differences in A-1 receptors in brain and cardiac tissue (Burnstock and Hoyle, 1985), whereas the xanthine antiasthmatic enprofylline (Fig. 2) (Fredholm and Persson, 1982) also appears to provide evidence for tissue-selective receptor subtypes. In contrast, studies on CHA binding in guinea pig brain and ileal membranes have shown that the A-1 receptors present in the two tissues are very similar (Williams and Valentine, 1985).

6. In Vivo Changes in Adenosine And Related Receptors

6.1. *Animal Manipulation*

Metrazole-induced convulsions selectively decrease the number of cerebellar A-1 receptor sites by 21% (Wybenga et al., 1981), reinforcing the involvement of the purine in the seizure process (Dunwiddie and Worth, 1982). Preliminary chemical and surgical lesion studies (Williams and Wood, 1981) have shown that when the cholinergic innervation to the hippocampus is chronically removed, a 57% increase in CHA binding is observed. In contrast, Murray and Cheney (1982) found that selective neuronal lesions of serotonergic, noradrenergic, and cholinergic afferents to the hippocampus were without effect on [^{3}H]-CHA binding.

6.2. Caffeine-Elicited Changes

Administration of caffeine to animals and humans causes behavioral excitation and anxiety (Charney et al., 1984). Chronic caffeine (Fredholm, 1982b; Boulenger et al., 1983) and theophylline (Murray, 1982) administration can increase the number of adenosine receptors in rat brain. Xanthine-related changes in diazepam binding have also been reported (Boulenger et al., 1983; Wu and Coffin, 1984). Treatment of mice from birth with caffeine results in a 20–30% increase in the number of A-1 receptors (Marangos et al., 1984b). Caffeine treatment can also down-regulate β-adrenoceptors (Goldberg et al., 1982), an effect presumably caused by the xanthine antagonizing the inhibitory effects of adenosine, resulting in an increase in noradrenaline release and a consequent subsensitization of β-receptors.

To date, no literature data is available on the chronic effects of adenosine agonists.

6.3. Adenosine Effects on α_2-Adrenoceptor Binding

In early studies (Stone and Taylor, 1978) the α_2-adrenoceptor agonist clonidine was found to antagonize the electrophysiological effects of adenosine. However, it had no effect on binding to A-1 receptors (Williams and Risley, 1980a), which suggested that either the electrophysiological actions of clonidine were indirect, or that the α_2-agonist acted at an adenosine recognition site other than the A-1. Clonidine can in fact displace radiolabeled NECA binding from A-2 sites in striatum, although this effect does not appear to be concentration-dependent (E. Hicks and M. Williams, unpublished). In rat caecum, clonidine, but not the α_2-agonist guanfacine, has been shown to competitively block 2-CADO-dependent relaxation (Kulkarni and Mehta, 1984). Adenosine and 2-CADO can "unmask" α_2-adrenoceptor binding sites in rat vas deferens (Watanabe et al., 1983) without affecting α_1-binding, an effect that is blocked by theophylline (Lai et al., 1984). The physiological significance of these changes has not been elucidated. However a relationship between adenosine and α_2-receptors has been documented both in regard to cyclic AMP formation (Partington et al., 1980) and the putative involvement of the nucleoside in analgesic mechanisms. The fact that clonidine (Gold et al., 1980) is used and adenosine analogs may have potential use (Collier et al.,

1984) in the treatment of morphine withdrawal (*see* section 7.3) suggests that subtle interactions between α_2-receptors and adenosine at the molecular level may be of therapeutic significance.

6.4. Adenosine Receptors in Aged Rats

Examination of the binding of *R*-PIA in aged rats (Virus et al., 1984) has shown a trend toward a slight increase in K_d (reduction in affinity) and in the number of binding sites. Although neither of these effects is statistically significant, they might relate to changes in the amount and patterns of sleep in aged rats (Rosenberg et al., 1979), which may in turn have implications with regard to the potential use of adenosine antagonists as cognitive enhancers.

7. Therapeutic Potential of Adenosine-Related Compounds in the CNS

The advent of selective adenosine agonists, and to a lesser extent, antagonists, and of adenosine binding assays, has made the search for a role for adenosine at the molecular level in the CNS a challenging task. The ubiquitous distribution of the purine and its demonstrated effects in modulating the function of other putative neurotransmitters makes it difficult to definitively ascribe a role for the nucleoside in any given disease state. Despite this caveat, detailed studies have implicated adenosine in the mode(s) of action of several classes of psychotropic agents.

The role of ATP with regard to a therapeutic potential has not been adequately or thoroughly addressed at this point, although a role for ATP in the pathophysiology of migraine has been suggested (Burnstock, 1981).

7.1. Depression

High concentrations of classical and "atypical" antidepressants increase guinea pig brain slice cyclic AMP levels via a mechanism that can be blocked by theophylline (Sattin et al., 1978). Based on this observation and the potentiation by antidepressants of the depressant actions of adenosine (Stone and Taylor, 1979), Sattin (1981) suggested that antidepressants may cause release of the purine and thus increase cyclic AMP production. In preliminary studies, Stone (1983) demonstrated that high concentrations (500 μM) of antidepressants increased purine release.

Chronic treatment of rats with mianserin or desmethylimipramine, although decreasing β-adrenoceptor number, a well-known biochemical effect of this class of therapeutic agent (Sugrue, 1981), has no effect on the binding of CHA to A-1 receptors (Williams et al., 1983). The mood disorders associated with adrenocortical insufficiency have been tentatively linked to an adenosine-related inhibition of phosphodiesterase activity (de Mazancourt and Giudicelli, 1984); PIA at nanomolar concentrations can biphasically modulate the activity of this enzyme by a GTP-dependent mechanism. This effect of the adenosine analog may be mediated via a receptor-mediated mechanism, since theophylline (500 μM) can antagonize the effects of *R*-PIA. Although this finding remains speculative in regard to the involvement of adenosine in the etiology of depression, the potential interaction of adenosine agonists with phosphodiesterase(s) may have important implications in regard to A-1/A-2 adenylate cyclase-linked receptor definitions (Burnstock and Buckley, 1985). It will be of interest to see if the reported effects of PIA are stereospecific, and whether the effects of theophylline reflect true receptor antagonism or an alteration in the baseline activity of the cyclic nucleotide phosphodiesterase.

7.2. Anxiety

The anxiogenic actions of caffeine (Rall, 1980; Charney et al., 1984) provided impetus to examine the interactions of adenosine and related compounds with the benzodiazepine (BZ) receptor (Williams, 1983). The isolation of an endogenous inhibitor of [^{3}H]diazepam binding from brain tissue and its identification as inosine (Skolnick and Paul, 1981) was of great interest. However, the weak activity of inosine ($IC_{50} \sim 10^{-3}M$) and the potent activity of the β-carbolines ($10^{-9}M$) (Braestrup et al., 1980) at the BZ recognition site has led to general skepticism toward purines being involved in anxiety. The N^6-substituted purine, EMD 28422 (Fig. 2) (Skolnick et al., 1980) was found to have anxiolytic activity and potent interactions with the BZ receptor complex. The in vivo effects of this compound have not been confirmed (Lal et al., 1983), although R-PIA and 2-CADO have no demonstrable diazepam-like anxiolytic activity in a pentylenetetrazol-saline discrimination model of anxiety (Spencer et al., 1984). There is also no correlation between the effects of a series of BZ and non-BZ putative anxiolytics on [^{3}H]diazepam and [^{3}H]-2-CADO binding (Williams et al., 1981). The sedative and muscle relaxant, but not the anxiolytic or anticonvulsant, actions of the BZs may be adenosine-related (Bruns et al., 1983b).

BZs are also inhibitors of adenosine uptake into rat brain synaptosomes (Phillis and Wu, 1983), an effect insensitive to the BZ antagonists Ro 15-1788 and CGS 8216 (Morgan et al., 1983). The adenosine uptake inhibitor dipyridamole is a potent displacer of [^{3}H]diazepam binding (Davies et al., 1980), with an IC_{50} value of 2–3 × 10^{-7}*M*, and can also undergo an "agonist" tenfold shift in photoaffinity-labeled brain membranes (Davies et al., 1984). It has been postulated (Phillis and Wu, 1983) that inhibition of adenosine uptake may reflect the sedative and ataxic properties of the BZs. Of interest in this regard is the possibility that the acute muscle relaxant effects of the neuroleptic trifluoperazine may be mediated by inhibition of adenosine uptake (Phillis, 1984).

The role of purines in the mechanisms of action of anxiolytics is therefore somewhat confused, despite the demonstrated anxiogenic actions of the methylxanthines. There is, however, still considerable interest in an adenosine–anxiolytic interaction at the molecular level (Wagner and Katz, 1983), either directly at the receptor level or via the intermediacy of other putative neurotransmitters, the function of which is amenable to modulation by the purine. The BZ partial antagonist Ro 15-1788 can antagonize caffeine-induced seizures in mice (Velluci and Webster, 1984), whereas aminophylline can antagonize diazepam sedation (Arvidsson et al., 1982).

7.3. Analgesia

Alkylxanthines antagonize the antinociceptive action of morphine and can also elicit a "quasimorphine withdrawal syndrome" (QMWS) (Collier et al., 1974, 1981), a response which, like that occurring with the opiate antagonist naloxone, is associated with an increase in central noradrenaline turnover via α_2-receptor activation (Reinhard et al., 1983; Galloway and Roth, 1983a).

The QMWS effects of the xanthine IBMX (Fig. 2) can be antagonized by the α_2-agonist, clonidine (Grant and Redmond, 1982), which can also antagonize xanthine-related increases in noradrenaline metabolism (Galloway and Roth, 1983b), an effect blocked by morphine. Naloxone-precipitated withdrawal increases locus ceruleus cell firing, an effect also blocked by clonidine. Adenosine can suppress withdrawal phenomenon in guinea pig ileum and opiate-dependent rats (Collier and Tucker, 1983; Collier et al., 1984). It has been suggested (Collier et al., 1984) that an adenosine agonist such as 2-CADO may have potential in the management of opiate withdrawal in humans. Clonidine has been used to treat withdrawal (Gold et al., 1980), and it seems likely that withdrawal involves hyperactive noradrenergic pathways. Xanthines can also antagonize the effects of endogenous opiates (Elliott et al., 1983) and can reverse their respiratory depressant actions (Arnanada and Turmen, 1979).

The antinociceptive actions of the $GABA_B$ agonist baclofen have been reported to be potentiated by theophylline (Sawynok, 1983). This effect contrasts to the reported nociceptive actions of the xanthine (Paazlow and Paazlow, 1973) and may reflect a dual mechanism of action for the compound. Caffeine has well-documented actions as an analgesic adjuvant (Laska et al., 1984) and, in contrast to its anxiogenic actions (Charney et al., 1984), has anxiolytic effects (White et al., 1980). In view of the biphasic actions of xanthines on mouse locomotor activity (Katims et al., 1983), it is quite possible that the apparent conflicting roles of caffeine and other xanthines in regard to analgesia may reflect interactions with adenosine receptor subtypes.

7.4. Miscellaneous

Mention has already been made (section 3.1.3) of the potential involvement of adenosine in Lesch-Nyhan syndrome (Green et al., 1982), as a possible factor in the effects of PCP (Browne and Welch, 1982) and ethanol (Proctor and Dunwiddie, 1984), and as a potential somnificant (Radulovacki et al., 1984). The purine may also be an aggrevating factor in the pathophysiology of myasthenia gravis, a disease associated with impairment of neuromuscular transmission (Riberio, 1982). The effects of adenosine (and ATP) on cerebral blood flow may also be factors that contribute to alterations in CNS function that can be termed "purine-dependent."

8. Adenosine, Purinergic Tone, and CNS Function

Receptor density (Lee et al., 1983) and regional formation (Goodman and Snyder, 1982; Snyder et al., 1984) and degradation (Nagy et al., 1984), rather than stimulus–secretion-dependent availability, may possibly determine the function of adenosine, and to an unknown extent, ATP, in the CNS. This possibility raises questions, however, as to the extent to which purine-related neuromodulation may reflect a global, as opposed to local, phenomenon. The sedative and hypnotic effects of adenosine may reflect a generalized modulation of neuronal circuits and, indeed, a decreased "awareness" of the environment may explain the putative anxiolytic-like effects seen with the nucleoside. Conversely, the CNS stimulant, anxiogenic, and convulsant actions of the xanthines methyl, alkyl, and, presumably, aryl probably results from a nonspecific, but progressive antagonism of CNS purinergic tone. It is possible that adenosine antagonists may have potential use as analeptics or in the treatment of coma.

In peripheral tissues, the adenosine produced as the result of hypoxia (Berne et al., 1983a) is generally thought to subserve a homeostatic role, antagonizing the β-receptor mediated stimulatory actions of noradrenaline to prevent damage to the heart (Belardinelli et al., 1983). In congestive heart failure, the negative inotropic actions of adenosine may be counteracted by the use of the xanthine, aminophylline, as a cardiotonic agent (Farah et al., 1984).

Decapitation dramatically increases brain adenosine levels (Wojcik and Neff, 1982; Fredholm et al., 1984). This effect may be compared to the changes in the nucleoside level in heart tissue during hypoxia. It has been suggested (Williams, 1984b) that adenosine may act as a central homeostatic modulator, preventing tissue "damage" during insults to the CNS. In this context, decapitation is the ultimate insult. However, pathophysiological states associated with chemical imbalances that reflect minor deviations from a "norm," i.e., mental illness that is not currently associated with gross organic changes, may be amenable to rectification by alterations in the purinergic tone by endogenous mechanisms as yet undefined. When such states require medication, it may be plausible, as has been suggested by Phillis (1984), that many psychotropic agents produce certain of their effects as weak inhibitors of adenosine uptake. As an extension of these considerations, it will be of interest to see if the reported differences in human brain A-1 receptors, as compared to those of other species (Murphy and Snyder, 1982), are of evolutionary and behavioral significance.

These speculations and the issues surrounding the role(s) of adenosine (and ATP) as a local neuromodulatory agent require a better understanding of the mechanism(s) controlling the presence of the purines in the extracellular milieu, including glial adenosine stores (Phillis and Wu, 1981).

The availability of novel, nonxanthine adenosine antagonists (Davies et al., 1983), better ligands for the A-2 receptor, better understanding of the importance of the putative "A-3" receptor, and more physiologically related studies on the roles of both adenosine and ATP in the CNS should allow a more focused understanding of the role of purines in the pathophysiology of the various CNS states discussed above and will provide a more realistic framework for the development of novel therapeutic agents in the area.

Note added in proof: Discussion of the receptor classification for adenosine receptor at a recent symposium (Stone, 1985) has led to the recommended adoption of the A-1/A-2 terminology (Hamprecht and Van Calker, 1985) independent of the effects of the purine on adenylate cyclase formation. Thus the nomenclature is pharmacological rather than biochemical.

Acknowledgments

The author would like to thank Charlene Kirby for her excellent help in preparing the manuscript and Holly Roth Williams for the preparation of the figures.

References

Arch J. R. S. and Newsholme E. A. (1978) The control of metabolism and the hormonal role of adenosine. *Essays Biochem.* **14,** 82–123.

Arnanada J. V. and Turmen T. (1979) Methylxanthines in apnea of prematurity. *Clin. Perinatol.* **6,** 87–108.

Arvidsson S. B., Ekstrom-Jodal B., Martinell S. A. G., and Niemand D. (1982) Aminophylline antagonizes diazepam sedation. *Lancet* **ii,** 1467.

Baird-Lambert J., Marwood J. F., Davies L. P., and Taylor K. M. (1980) l-methylisoguanosine: an orally active marine natural product with skeletal muscle and cardiovascular effects. *Life. Sci.* **26,** 1069–1077.

Baker P. F. (1972) Transport and metabolism of calcium ions in nerve. *Prog. Biophys. Mol. Biol.* **24,** 177–224.

Barnes E. M. and Thampy K. G. (1982) Subclasses of adenosine receptors in brain membranes from adult tissue and from primary cultures of chick embryo. *J. Neurochem.* **39,** 647–652.

Barraco R. A., Coffin V. L., Altman H. J., and Phillis J. W. (1983) Central effects of adenosine analogs and locomotor activity in mice and antagonism by caffeine. *Brain Res.* **272,** 392–395.

Barraco R. A., Aggarwal A. K., Phillis J. W., Moore M. A., and Wu P. H. (1984) Dissociation of the locomotor and hypotensive effects of adenosine analogs in the rat. *Neurosci. Lett.* **48,** 139–144.

Belardinelli L., West A., Crampton R., and Berne R. M. (1983) Chronotropic and Dromotropic Effects of Adenosine, in *Regulatory Function of Adenosine.* (Berne R. M., Rall T. W., and Rubio R., eds.) pp. 377–398, Martinus Nijhoff, Boston.

Bender A. S., Wu P. H., and Phillis J. W. (1981) The rapid uptake and release of [^{3}H]-adenosine by rat cerebral cortical synaptosomes. *J. Neurochem.* **36,** 651–660.

Berne R. M. (1980) The role of adenosine in the regulation of coronary blood flow. *Circ. Res.* **46,** 807–813.

Berne R. M., Knabb R. M., Ely S. W., and Rubio R. (1983a) Adenosine in the local regulation of blood flow: a brief overview. *Fed. Proc.* **42,** 3136–3142.

Berne R. M., Rall T. W., and Rubio R. (1983b) *Regulatory Function of Adenosine.* Martinus Nijhoff, Boston.

Boulenger J-P., Patel J., Post R. M., Parma A., and Marangos P. J. (1983) Chronic caffeine consumption increases the number of brain adenosine receptors. *Life Sci.* **32,** 1135–1142.

Braestrup C., Nielsen M., and Olsen C. E. (1980) Urinary and brain β-carboline-3-carboxylates as potent inhibitors of brain benzodiazepine receptors. *Proc. Natl. Acad. Sci. USA* **77,** 2288–2292.

Browne R. G. and Welch W. M. (1982) Steroselective antagonism of phencyclidine's discriminative properties by adenosine receptor agonists. *Science* **217,** 1157–1158.

Bruns R. F. (1980) Adenosine receptor activation in human fibroblasts: Nucleoside agonists and antagonists. *Can. J. Physiol. Pharmacol.* **58,** 673–691.

Bruns R. F. (1981) Adenosine antagonism by purines, pteridines, and benzopteridines in human fibrobasts. *Biochem. Pharmacol.* **30,** 325–333.

Bruns R. F., Daly J. W., and Snyder S. H. (1980) Adenosine receptors in brain membranes: binding of N^6-cyclohexyl-[^{3}H]-adenosine and 1,3-diethyl-8-[^{3}H]-phenylxanthine. *Proc. Natl. Acad. Sci. USA* **77,** 5547–5551.

Bruns R. F., Daly J. W., and Snyder S. H. (1983a) Adenosine receptor binding: structure activity analysis generates extremely potent xanthine antagonists. *Proc. Natl. Acad. Sci. USA* **80,** 2077–2080.

Bruns R. F., Katims J. J., Annau Z., Snyder S. H., and Daly J. W. (1983b) Adenosine receptor interactions and anxiolytics. *Neuropharmacol.* **22,** 1523–1529.

Burnstock G. (1972) Purinergic nerves. *Pharmacol Rev.* **24,** 509–581.

Burnstock G. (1978) A Basis for Distinguishing Two Types of Purinergic Receptors, in *Cell Membrane Receptors for Drugs and Hormones.* (Bolis L, and Straub R. N., eds.) pp. 107–118, Raven Press, New York.

Burnstock G. (1981) Pathophysiology of migraine: a new hypothesis. *Lancet* **ii,** 1397–1399.

Burnstock G. (1983) A Comparison of Receptors for Adenosine and Adenine Nucleotides, in *Regulatory Function of Adenosine.* (Berne R. M., Rall T. W., and Rubio R., eds.) pp. 49–62, Martinus Nijhoff, Boston.

Burnstock G. and Buckley N. J. (1985) The Classification of Receptors for Adenosine and Adenine Nucleotides, in *Methods in Pharmacology 6, Methods Used In Adenosine Research.* (Paton D. M., ed.) Plenum, New York.

Burnstock G. and Hoyle C. H. V. (1985) PACPX—a substituted xanthine—antagonizes both the A_1 and A_2 subclasses of the P1 purinoceptor: antagonism of the A_2 subclass is competitive but antagonism of the A_1 subclass is not. *Brit. J. Pharmacol.*, **85,** 291–296.

Carlsson A. (1975) Receptor-Mediated Control of Dopamine Metabolism, in *Pre-and Postsynaptic Receptors.* (Usdin E. and Bunney W. E., Jr., eds.) pp. 49–67, Marcel Dekker, New York.

Charney P. S., Galloway M. P., and Heninger G. R. (1984) The effects of caffeine on plasma MHPG, subjective anxiety, autonomic symptoms, and blood pressure in healthy humans. *Life Sci.* **35,** 135–144.

Coffin V. C. and Carney J. M. (1983) Behavioral Pharmacology of Adenosine Analogs, in *Physiology and Pharmacology of Adenosine Derivatives.* (Daly J. W., Kuroda Y., Phillis J. W., Shimizu H., and Ui M., eds.) pp. 267–274, Raven, New York.

Coffin V. C., Taylor J. A., Phillis J. W., Altman H. J., and Barraco R. A. (1984) Behavioral interaction of adenosine and methylxanthines on central purinergic systems. *Neurosci. Lett.*, **47,** 91–98.

Collier H. O. J. and Tucker J. F. (1983) Novel form of drug-dependence—on adenosine in guinea pig ileum. *Nature* (Lond.) **302,** 618–621.

Collier H. O. J., Francis D. L., Henderson G., and Schnieder C. (1974) Quasi-morphine abstinence syndrome. *Nature* (Lond.) **249,** 471.

Collier H. O. J., Cuthbert N. J., and Francis D. L. (1981) Character and meaning of quasi-morphine withdrawal phenomena elicited by methylxanthines. *Fed. Proc.* **40,** 1513–1518.

Collier H. O. J., Plant N. T., Tucker J. F., and Von Uexull A. (1984) Inhibition with adenosine derivatives of opiate withdrawal effects. *Brit. J. Pharmacol.* **81,** 131P.

Collis M. G., and Brown C. M. (1983) Adenosine relaxes the aorta by interacting with an A_2-receptor and an intracellular site. *Eur. J. Pharmacol.* **96,** 61–69.

Cooper D. M. F., Londos C., and Rodbell R. (1980) Adenosine receptor-mediated inhibition of rat cerebral cortical adenylate cyclase by a GTP-dependent process. *Mol. Pharmacol.* **18,** 598–601.

Crawley J. N., Patel J., and Marangos P. J. (1981) Behavioral characterization of two long lasting adenosine analogs: sedative properties and interaction with diazepam. *Life Sci.* **29,** 2623–2630.

Cuatrecasas P. and Hollenberg M. (1976) Membrane receptor and hormone action. *Adv. Protein Chem.* **30,** 251–451.

Cubeddu L., Barnes E., and Winer N. (1975) Release of norepinephrine and dopamine hydroxylase by nerve stimulation. IV. An evaluation of a role for cyclic AMP. *J. Pharmacol. Exp. Ther.* **193,** 105–127.

Daly J. W. (1973) Role of ATP and Adenosine Receptors in Physiologic Processes: Summary and Prospectus, in *Physiology and Pharmacology of Adenosine Derivatives* (Daly J. W., Kuroda Y., Phillis J. W., Shimizu H., and Ui M., eds). pp. 275–290, Raven, New York.

Daly J. W. (1977) *Cyclic Nucleotides in the Nervous System.* Plenum, New York.

Daly J. W. (1982) Adenosine receptors: target sites for drugs. *J. Med. Chem.* **25,** 197–207.

Daly J. W. (1983) Binding of Radioactive Ligands to Adenosine Receptors in the Central Nervous System, in *Regulatory Function of Adenosine.* (Berne R. M., Rall T. W., and Rubio R., eds.) pp. 97–113, Martinus Nijhoff, Boston.

Daly J. W., Bruns R. F., and Snyder S. H. (1981) Adenosine receptors in the central nervous system: Relationship to the central actions of methylxanthines. *Life Sci.* **28,** 2083–2097.

Daly J. W., Butts-Lamb P., and Padgett W. (1983) Subclasses of adenosine receptors in the central nervous system: interaction with caffeine and related methylxanthines. *Cell. Mol. Neurobiol.* **3,** 69–80.

Daly J. W., Mimitkitpaisan Y., Pons F., Bruns R. F., Smellie F., and Skolnick P. (1979) Binding sites for adenosine analogs: Possible relationship to cyclic-AMP generating systems in brain tissue. *Pharmacologist* **21,** 253.

Dar M. S., Mustafa S. J., and Wooles W. R. (1983) Possible role of adenosine in the CNS. Effects of ethanol. *Life Sci.* **33,** 1363–1374.

Davies L. P., Cook A. F., Poonian M., and Taylor K. M. (1980) Displacement of [^{3}H]-diazepam binding in rat brain by dipyridamole and by 1-methylisoguanosine, a marine natural product with muscle-relaxant activity. *Life Sci.* **26,** 1089–1094.

Davies L. P., Brown D. J., Chen Chow S., and Johnston G. A. R. (1983) Pyrazolo [3,4-D]-pyrimidines, a new class of adenosine antagonists. *Neurosci. Lett.* **41,** 189–194.

Davies L. P., Chow S. C., and Johnston G. A. R. (1984) Interaction of purines and related compounds with photoaffinity-labeled benzodiazepine receptors in rat brain membranes. *Eur. J. Pharmacol.* **97,** 325–329.

de Mazancourt P. and Giudicelli Y. (1984) N^6-Phenylisopropyladenosine stimulates in normal and inhibits in adrenalectomized rats the low Km cyclic AMP phosphodiesterase in brain. *Brain Res.* **300,** 211–217.

Dolphin A. C. and Archer E. R. (1983) An adenosine agonist inhibits and a cyclic AMP analogue enhances the release of glutamate but not GABA from slices of rat dentate gyrus. *Neurosci. Lett.* **43,** 49–54.

Drury A. N. and Szent-Gyorgi A. (1929) The physiological activity of adenine compounds with especial reference to their actions upon the mammalian heart. *J. Physiol.* (Lond.) **68,** 213–237.

Dunwiddie T. V. (1980) Endogenously released adenosine regulates excitability in the in vitro hippocampus. *Epilepsia* **21,** 541–548.

Dunwiddie T. V. and Hoffer B. J. (1980) Adenine nucleotides and synaptic transmission in the in vitro rat hippocampus. *Brit. J. Pharmacol.* **69,** 59–68.

Dunwiddie T. V., Hoffer B. J. and Fredholm B. B. (1981) Alkylxanthines elevate hippocampal excitability: evidence for a role for endogenous adenosine. *Naunyn-Schmeideberg's Arch. Pharmacol.* **316,** 326–330.

Dunwiddie T. V. and Worth T. (1982) Sedative and anticonvulsant effects of adenosine analogs in mouse and rat. *J. Pharmacol. Exp. Ther.* **220,** 70–76.

Dunwiddie T. V., Basile A. S., and Palmer M. R. (1984) Electrophysiological responses to adenosine analogs in rat hippocampus and cerebellum. Evidence for mediation by adenosine receptors of the A-1 subtype. *Life Sci.* **34,** 34–47.

Ebersolt C., Premont J., Prochiantz A., Perez M., and Bockaert J. (1983) Inhibition of brain adenylate cyclase by A_1-adenosine receptors: pharmacological characteristics and location. *Brain Res.* **267,** 123–124.

Ebstein R. P. and Daly J. W. (1982) Release of norepinephrine and dopamine from brain vesicular preparations: effects of adenosine analogs. *Cell Mol. Neurobiol.* **2,** 193–204.

Edvinsson L. and Fredholm B. B. (1983) Characterization of adenosine receptor in isolated cerebral arteries of cat. *Brit. J. Pharmacol.* **80,** 631–637.

Elliott J., Jhamandas K., Notman H., and Sutak M. (1983) Antagonism of enkephalin action on acetylcholine release by methylxanthine: lack of a purine link. *Brit. J. Pharmacol.* **80,** 727–734.

Farah A. E., Alousi A. A., and Schwarz, Jr. R. P. (1984) Positive inotropic agents. *Ann. Rev. Pharmacol. Toxicol.* **24,** 275–328.

Feldberg W. and Sherwood S. L. (1954) Injections of drugs into lateral ventricle of the cat. *J. Physiol.* (Lond.) **123,** 148–167.

Ferkany J. W., Valentine H., Stone G., and Williams M. (1986) Adenosine A-1 receptors in mammalian brain: Species differences in their interactions with agonists and antagonists. *J. Neurochem.*, submitted.

Fox I. H. and Kurpis L. (1983) Binding characteristics of an adenosine receptor in human placenta. *J. Biol. Chem.* **258,** 6952–6955.

Fredholm B. B. (1982a) Adenosine receptors. *Med. Biol.* **60,** 289–293.

Fredholm B. B. (1982b) Characterization of adenosine receptors in the brain. *Acta Physiol. Scand.* Suppl. **508,** 31.

Fredholm B. B. and Vernet L. (1978) Morphine increases depolarization-induced purine release from rat cortical slices. *Acta Physiol. Scand.* **104,** 502–504.

Fredholm B. B. and Hedqvist P. (1980) Modulation of neurotransmission by purine nucleotides and nucleosides. *Biochem. Pharmacol.* **29,** 1635–1643.

Fredholm B. B. and Persson C. G. A. (1982) Xanthine derivatives as adenosine receptor antagonists. *Eur. J. Pharmacol.* **81,** 673–676.

Fredholm B. B., Jonzon B., Lindgren E., and Lindstrom K. (1982) Adenosine receptors mediating cyclic AMP production in the rat hippocampus. *J. Neurochem.* **39,** 165–175.

Fredholm B. B., Dunwiddie T. V., Bergman B., and Lindsham, K. (1984) Levels of adenosine and adenine nucleotides in slices of rat hippocampus. *Brain Res.* **295,** 127–136.

Frew R. and Lundy P. M. (1982) Effect of arylazido aminopropionyl ATP ($ANAPP_3$), a putative ATP antagonist, on ATP responses of isolated guinea pig smooth muscle. *Life Sci.* **30,** 259–267.

Galloway M. P. and Roth R. H. (1983a) Clonidine prevents methylxanthine stimulation of norepinephrine metabolism in rat brain. *J. Neurochem.* **40,** 246–251.

Galloway M. P. and Roth R. H. (1983a) Clonidine prevents methylxanthine stimulation of norepinephrine metabolism in rat brain. *J. Neurochem.* **40,** 246–251.

Galloway M. P. and Roth R. H. (1983b) Neuropharmacology of 3-isobutylmethylxanthine: effect on central noradrenergic systems in vivo. *J. Pharmacol. Exp. Ther.* **227,** 1–8.

Gavish M., Goodman R. R., and Snyder S. H. (1982) Solubilized adenosine receptors in the brain: regulation by guanine nucleotides. *Science* **215,** 1633–1635.

Gold M. S., Pottash A. C., Sweeney D. R., and Kleber H. D. (1980) Opiate withdrawal using clonidine. A safe, effective, and rapid non-opiate treatment. *J. Am. Med. Assoc.* **243,** 343–346.

Goldberg A. L. and Singer J. J. (1969) Evidence for a role of cyclic AMP in neuromuscular transmission. *Proc. Natl. Acad. Sci. USA* **64,** 134–141.

Goldberg M. R., Curatolo P. W., Tung C. S., and Robertson D. (1982) Caffeine down-regulates beta adrenoceptors in rat forebrain. *Neurosci. Lett.* **31,** 47–52.

Goodman R. R. and Snyder S. H. (1982) Autoradiographic localization of adenosine receptors in rat brain using [^{3}H]cyclohexyladenosine. *J. Neurosci.* **2,** 1230–1241.

Goodman R. R., Cooper M. J., Gavish M., and Snyder S. H. (1982) Guanine nucleotide and cation regulation of the binding of [^{3}H]-cyclohexyladenosine and [^{3}H]diethylphenylxanthine to adenosine receptors in brain membranes. *Mol. Pharmacol.* **211,** 329–335.

Goodman R. R., Kuhar M. J., Hester L., and Snyder S. H. (1983) Adenosine receptors: autoradiographic evidence for their location on axon terminals of excitatory neurons. *Science* **220,** 967–969.

Grant S. J. and Redmond D. E. (1982) Methylxanthine activation of noradrenergic unit activity and reversal by clondine. *Eur. J. Pharmacol.* **85,** 1054–109.

Green H. N. and Stoner H. B. (1950) *Biological Actions of the Adenosine Nucleotides.* Lewis, London.

Green R. D., Proudfit H. K., and Yeung S-M. H.(1982) Modulation of striatal dopaminergic function by injection of 5′-N-ethylcarboxamide adenosine. *Science* **218,** 58–61.

Guroff G., Dickens G., End D., and Londos C. (1981) The action of adenosine analogs on PC 12 cells. *J. Neurochem.* **37,** 1431–1439.

Hamprecht B. and Van Calker D. (1985) Nomenclature of adenosine receptors. *Trends Pharmacol. Sci.* **6,** 153–154.

Harms H. H., Wardeh G., and Mulder A. H. (1978) Adenosine modulates depolarization-induced release of [^{3}H]noradrenaline from slices of rat brain neocortex. *Eur. J. Pharmacol.* **49,** 305–308.

Harms H. H., Wardeh G., and Mulder A. H. (1979) Effects of adenosine depolarization-induced release of various radiolabeled neurotransmitters from slices of rat corpus striatum. *Neuropharmacol.* **18,** 577–880.

Hartzell C. H. (1979) Adenosine receptor in frog sinus venosus: slow inhibitory potentials produced by adenine compounds and acetylcholine. *J. Physiol.* (Lond.) **293,** 23–49.

Haslam R. J., Davidson M. M. L., and Desjardins J. W. (1978) Inhibition of adenylate cyclase by adenosine analogs in preparations of broken and intact human platelets. *Biochem. J.* **176,** 83–95.

Haubrich D. R., Williams M., Yarbrough G. G., and Wood P. L. (1981) 2-chloroadenosine inhibits brain acetylcholine turnover. *Can. J. Physiol. Pharmacol.* **59,** 897–1020.

Haulica I., Aradei L., Bransteau D., and Topoliceanu F. (1973) Preliminary data on the possible hypnogenic role of adenosine. *J. Neurochem.* **21,** 1019–1020.

Hazeki O., Katada T., Kurose H., and Ui M. (1983) Effect of Adenosine on Cyclic AMP Accumulation in Cardiac and Other Cells and Its Modification by Islet-Activating Protein, in *Physiology and Pharmacology Of Adenosine Derivatives.* (Daly J. W., Phillis J. W., Kuroda Y., Shimizu H., and Ui M., eds.) pp. 41–49, Raven, New York.

Hedner T., Hedner J., Wessberg P., and Jonason J. (1982) Regulation of breathing in the rat: indications for a role of central adenosine mechanisms. *Neurosci. Lett.* **33,** 147–151.

Henon B. A. and McAfee D. A. (1983) The ionic basis of adenosine receptor actions of postsynaptic neurones in the rat. *J. Physiol.* (Lond.) **336,** 607–620.

Hogaboom G. K., O'Donnell J. P., and Fedan J. S. (1980) Purinergic receptors: photoaffinity analog of adenosine triphosphate is a specific adenosine triphosphate antagonist. *Science* **208,** 1273–1276.

Hollins C. and Stone T. W. (1980) Adenosine inhibition of γ-aminobutyric acid release from slices of rat cerebral cortex. *Brit. J. Pharmacol.* **69,** 107–112.

Hrdina P. D. (1985) Introduction in *Neuromethods IV: Receptor Binding.* (Boulton A. A., Baker G. B., and Hrdina, P. D., eds.) Humana, Clifton, New Jersey.

Huang, M. and Daly J. W. (1974) Adenosine-elicited accumulation of cyclic AMP in brain slices: potentiation by agents which inhibit uptake of adenosine. *Life Sci.* **14,** 489–503.

Hunt S. P. and Kunzle H. (1976) Bidirectional movement of label and transneuronal transport phenomena after injection of [^{3}H]adenosine in the central nervous system. *Brain Res.* **112,** 127–132.

Israel M., Lesbats B., Manaranche R., Muenier F. M., and Franchon P. (1980) Retrograde inhibition of transmitter release by ATP. *J. Neurochem.* **34,** 923–932.

Jahr C. E. and Jessel T. M. (1983) ATP excites a subpopulation of rat dorsal horn neurons. *Nature* (Lond.) **304,** 730–732.

Jhmandas K. and Dumbrille A. (1980) Regional release of [^{3}H]adenosine derivatives from rat brain in vivo: effect of excitatory amino acids, opiate agonists, and benzodiazepines. *Can. J. Physiol. Pharmacol.* **58,** 1262–1278.

Kakiucki S., Rall T. W., and McIlwain H. (1968) The effect of electrical stimulation upon the accumulation of adenosine 3′,5′-phosphate in isolated cerebral tissue. *J. Neurochem.* **16,** 485–491.

Katims J. J., Annau Z., and Snyder S. H. (1983) Interactions in the behavioral effects of methylxanthines and adenosine derivatives. *J. Pharmacol. Exp. Ther.* **227,** 167–173.

Kelley W. N. (1983) Disorders of Nucleic Acid Metabolism, in *Principles of Internal Medicine.* (Petersdorf R. G., Adams R. A., Braunwald E., Isselbacher K. J., Martin J. B., and Wilson J. D., eds.) pp. 517–524, McGraw-Hill, New York.

Kenakin T. P. (1982) The potentiation of cardiac responses to adenosine by benzodiazepines. *J. Pharmacol. Exp. Ther.* **222,** 752–758.

Kluge H., Hartmann W., Wieczorek V., and Zahlten W. (1972) Kinetic properties of cerebral 5′-nucleotidase. *J. Neurochem.* **19,** 1409–1411.

Kobayashi K., Kuroda Y., and Yoshioka M. (1981) Change of cyclic AMP level in synaptosomes from cerebral cortex: increase by adenosine derivatives. *J. Neurochem.* **36,** 86–91.

Kuba M. L., Kato E., Kumamoto E., Koketsu K., and Hirai K. (1981) Sustained potentiation of transmitter release by adrenaline and dibutyryl cyclic AMP in sympathetic ganglia. *Nature* (Lond.) **291,** 654–656.

Kulkarni S. K. and Mehta A. K. (1984) P_1-purinoceptor antagonism by clonidine in the rat caecum. *Life Sci.* **34,** 2273–2277.

Kuroda Y. (1978) Physiological roles of adenosine derivatives which are released during neurotransmission in mammalian brain. *J. Physiol.* (Paris) **74,** 463–470.

Kuroda Y. and McIlwain H. (1973) Subcellular localization of [^{14}C]adenine derivative newly formed in cerebral tissues and the effects of electrical excitation. *J. Neurochem.* **21,** 889–900.

Kuroda Y. and McIlwain H. (1974) Uptake and release of [^{14}C]adenine derivatives at beds of mammalian cortical synaptosomes in a superfusion system. *J. Neurochem.* **22,** 691–699.

Lai R. T., Watanabe Y., Kamino Y., and Yoshida H. (1984) Interaction between 2-chloroadenosine and α-adrenoceptors in rat vas deferens. *Life Sci.* **34,** 409–418.

Lal H., Gherezghiher T., and Carney J. (1983) Ineffectiveness of a purine analogue, EMD 28422, in two animal tests of anxiolytic action. *Drug Dev. Res.* **3,** 75–79.

Laska E. M., Sunshine A., Mueller F., Elvers W. B., Siegel C., and Rubin A. (1984) Caffeine as an analgesic adjuncant. *J. Am. Med. Assoc.* **251,** 1711–1718.

Lee K., Schubert P., Gribkoff V., Sherman B., and Lynch G. (1982) A combined in vivo/in vitro study of the presynaptic release of adenosine derivatives in the hippocampus. *J. Neurochem.* **28,** 80–83.

Lee K. S., Reddington M., Schubert P., and Kreutzberg G. (1983) Regulation of the strength of adenosine modulation in the hippocampus by a differential distribution of the density of A_1 receptors. *Brain Res.* **260,** 156–159.

Levin A. S. and Morley J. E. (1983) Effect of intraventricular adenosine on food intake in rats. *Pharmacol. Biochem. Behav.* **19,** 23–26.

Levin R. M., Jacoby R., and Wein A. J. (1983) High-affinity, divalent ion-specific binding of [^{3}H]-ATP to homogenate derived from rabbit urinary bladder. *Mol. Pharmacol.* **23,** 1–7.

Lloyd K. G., Hornykiewicz O., Davidson L., Shannak K., Farley I., Goldstein M., Shibuya M., Kelley W. N., and Fox I. (1981) Biochemical evidence of dysfunction of brain neurotransmitters in the Lesch-Nyhan syndrome. *N. Eng. J. Med.* **305,** 1106–1111.

Londos C. and Wolff J. (1977) Two distinct adenosine-sensitive sites on adenylate cyclase. *Proc. Natl. Acad. Sci. USA* **74,** 5482–5486.

Londos C., Cooper D. M. F., and Wolff J. (1980) Subclasses of adenosine receptors. *Proc. Natl. Acad. Sci. USA* **77,** 2551–2554.

Londos C., Wolff J., and Cooper D. M. F. (1983) Adenosine Receptors and Adenylate Cyclase Interactions, in *Regulatory Functions of Adenosine.* (Berne R. M., Rall T. W., and Rubio R., eds.) pp. 17–32, Martinus Nijhoff, Boston.

MacDonald W. F. and White T. D. (1984) Adenosine released from synaptosomes is derived from the extracellular dephosphorylation of released ATP. *Prog. Neuropsychopharmacol. & Biol. Psychiat.* **8,** 487–494.

Maitre M., Ciesielski L., Lehmann A., Kempt E., and Mandell P. (1974) Protective effect of adenosine and nicotinamide against audogenic seizures. *Biochem. Pharmacol.* **23,** 2807–2816.

Major P. P., Agarawal R. P., and Kufe D. W. (1981) Deoxycoformycin: neurological toxicity. *Cancer Chemother. Pharmacol.* **5,** 193–196.

Marangos P. J., Patel P., Martino A. M., Dilli M., and Boulenger J-P

(1984a) Differential binding properties of adenosine receptor agonists and antagonists in brain. *J. Neurochem.* **41,** 365–374.

Marangos P. J., Boulenger J-P, and Patel J. (1984b) Effects of chronic caffeine on brain adenosine receptors: regional and ontogenetic studies. *Life Sci.* **34,** 899–907.

Marley E. and Nistico G. (1972) Effects of catecholamines and adenosine derivatives given into the brain of fowls. *Brit. J. Pharmacol.* **46,** 619–636.

McIlwain H. (1978) Synaptic mediation and the structuring of cerebral activity. *Prog. Neurobiol.* **11,** 189–203.

Meflah K., Harb J., Duflos Y., and Bernard, S. (1984) 5′-Nucleotidase from bovine caudate nucleus synaptic plasma membranes: specificity for substrates and cations; study of the carbohydrate moiety by glycosidases. *J. Neurochem.* **42,** 1107–1115.

Mendelson W. B., Kuruvilla A., Watlington T., Goehl K., Paul S. M., and Skolnick P. (1983) Sedative and electroencephalographic actions of erythro-9-(2-hydroxy-3-nonyl) adenine (EHNA): Relationship to inhibition of brain adenosine deaminase. *Psychopharmacol.* **79,** 126–129.

Michaelis M. L., Michaelis E. K., and Myers S. L. (1979) Adenosine modulation of synaptosomal dopamine release. *Life Sci.* **24,** 2083–2092.

Morgan P. F., Lloyd H. G. E., and Stone T. W. (1983) Benzodiazepine inhibition of adenosine uptake is not prevented by benzodiazepine antagonists. *Eur. J. Pharmacol.* **87,** 121–126.

Moritaki H. (1983) Possible Mechanism of Potentiation of the Action of Adenosine by Some Vasodilators, in *Physiology and Pharmacology of Adenosine Derivatives* (Daly J. W., Kuroda Y., Phillis J. W., Shimizu H., and Ui M., eds.) pp. 197–207, Raven, New York.

Murphy K. M. M. and Snyder S. H. (1982) Heterogeneity of adenosine A_1 receptor binding in brain tissue. *Mol. Pharmacol.* **21,** 250–257.

Murray T. F. (1982) Upregulation of rat cortical adenosine receptors following chronic administration of theophylline. *Eur. J. Pharmacol.* **82,** 113–114.

Murray T. F. and Cheney D. C. (1982) Neuronal location of N^6-cyclohexyl-[^{3}H]-adenosine binding sites in rat and guinea pig brain. *Neuropharmacol.* **21,** 575–580.

Murray T. F., Blaker W. D., Cheney D. L., and Costa E. (1982) Inhibition of acetylcholine turnover rate in rat hippocampus and cortex by intraventricular injection of adenosine analogs. *J. Pharmacol. Exp. Ther.* **222,** 550–554.

Nagata H., Mimori Y., Nakamura S., and Kameyama M. (1984) Regional and subcellular distribution in mammalian brain of the enzymes producing adenosine. *J. Neurochem.* **42,** 1001–1007.

Nagy J. I., Labella F. S., Buss M., and Daddona P. E. (1984) Immunohistochemistry of adenosine deaminase: implications for adenosine neurotransmission. *Science* **224,** 166–168.

Nakumura S., Mimori Y., Iijima S., Nagata H., Yamao S., and Kameyama M. (1983) Distribution of Adenosine-Producing Enzymes in the Brain, in *Physiology and Pharmacology of Adenosine Derivatives.* (Daly J. W., Kuroda Y., Phillis J. W., Shimiza H., and Ui M., eds.) pp. 21–29, Raven, New York.

Newby A. C., Luzio J. P., and Hales C. N. (1975) The properties and extracellular location of 5′-nucleotidase of the rat fat cell plasma membrane. *Biochem. J.* **146,** 625–633.

Newman M. E. (1983) Adenosine binding sites in brain: relationship to endogenous levels of adenosine and its physiological and regulatory roles. *Neurochem. Inter.* **5,** 21–35.

Newman M. E. and McIlwain H. (1977) Adenosine as a constituent of the brain and of isolated cerebral tissues, and its relationship to the generation of adenosine 2′,5′-cyclic monophosphate. *Biochem. J.* **164,** 131–137.

Newman M. E., Patel J., and McIlwain H. (1980) The binding of adenosine to synaptosomal and other preparations from mammalian brain. *Biochem. J.* **194,** 611–620.

Nimit Y., Law J., and Daly J. W. (1982) Binding of 2′,5′-dideoxyadenosine to brain membranes. Comparison to P-site inhibition of adenylate cyclase. *Biochem. Pharmacol.* **31,** 3279–3287.

Nyhan W. C. (1973) The Lesch-Nyhan Syndrome. *Ann. Rev. Med.* **24,** 41–60.

Okada Y. and Saito M. (1979) Inhibitory actions of adenosine, 5HT (serotonin), and GABA (γ-aminobutyric acid) on the postsynaptic potential (PSP) of slices from olfactory cortex and superior colliculus in correlation to the level of cyclic AMP. *Brain Res.* **160,** 368–371.

Olssen R. A. (1984) Structure of the coronary artery adenosine receptor. *Trends Pharmacol. Sci.* **5,** 113–116.

Osswald H. (1983) Adenosine and Renal Function, in *Regulatory Function of Adenosine* (Berne R. M., Rall T. W., and Rubio R., eds.) pp. 399–415, Martinus Nijhoff, Boston.

Paazlow G. and Paazlow C. (1973) The effects of caffeine and theophylline on nociceptive stimulation in the rat. *Acta. Pharmacol. Toxicol.* **32,** 22–32.

Partington C. R., Edwards M. N., and Daly J. W. (1980) Calcium-dependent desensitization of adenylate cyclase in rat cerebral cortical slices. *J. Neurochem.* **34,** 76–82.

Patel J., Marangos P. J., Stivers J., and Goodwin F. K. (1982) Characterization of adenosine receptors in brain using N^6-cyclohexyl-[^{3}H]adenosine. *Brain Res.* **237,** 203–214.

Persson C. G. A. (1984) The Pharmacology of Antiasthmatic Xanthines and the Role of Adenosine, in *Asthma Reviews* (Moley J., ed.) Academic Press, in press.

Phillis J. W. (1984) Antagonism of the actions of psychoactive agents by caffeine. *Prog. Neurobiol. and Biol. Psychiatry.* **8,** 494–502.

Phillis J. W. and Wu P. H. (1981) The role of adenosine and its nucleotides in central synaptic transmission. *Progr. Neurobiol.* **16,** 187–193.

Phillis J. W. and Wu P. H. (1983) The Role of Adenosine in Central Neuromodulation, in *Regulatory Function of Adenosine.* (Berne R. M., Rall T. W., and Rubio R., eds.) pp. 419–437, Martinus Nijhoff, Boston.

Phillis J. W., Edstrom J. P., Kostopoulos G. K., and Kirkpatrick J. R. (1979) Effects of adenosine and adenine nucleotides on synaptic transmission in the cerebral cortex. *Can. J. Physiol. Pharmacol.* **57,** 1289–1312.

Phillis J. W., Jiang Z. G., Chelack B. J., and Wu P. H. (1980) The effect of morphine on purine and acetylcholine release from rat cerebral cortex: Evidence for a purinergic component in morphine's action. *Pharmacol. Biochem. Behav.* **13,** 421–427.

Polmar S. H., Birch R. E., Mandle R., and Rudolph S. A. (1983) Adenosine and Cyclic Nucleotides in Modulation of Immune Responses, in *Regulatory Function of Adenosine.* (Berne R. M., Rall T. W., and Rubio R., eds.) pp. 249–259, Martinus Nijhoff, Boston.

Pons F., Bruns R. F., and Daly J. W. (1980) Depolarization-evoked accumulation of cyclic AMP in brain slices: the requisite intermediate adenosine is not derived from hydrolysis of released ATP. *J. Neurochem.* **34,** 1319–1323

Potter P. and White T. D. (1980) Release of adenosine 5′-triphosphate from synaptosomes from different regions of rat brain. *Neuroscience* **5,** 1351–1356.

Premont J., Perez M., and Bockaert J. (1977) Adenosine-sensitive adenylate cyclase in rat striatal homogenates and its relationship to dopamine and Ca^{2+}-sensitive adenylate cyclase. *Mol. Pharmacol.* **13,** 662–670.

Premont J., Perez M., Blanc G., Tarsin J. P., Thierry A. M., Herve D., and Bockaert J. (1979) Adenosine-sensitive adenylate cyclase in rat brain homogenates: kinetic characteristics, specificity, topographical, subcellular, and cellular distribution. *Mol. Pharmacol.* **16,** 790–804.

Proctor W. R. and Dunwiddie T. V. (1984) Behavioral sensitivity to purinergic drugs parallels ethanol sensitivity in selectivity bred mice. *Science* **224,** 519–521.

Pull I. and McIlwain H. (1974) Rat cerebral cortex adenosine deaminase activity and its subcellular distribution. *Biochem. J.* **144,** 37–41.

Rabe C. S., and McGee, Jr. R. (1982) The Effects of Methylation Inhibitors on Depolarization-Dependent Exocytosis, Phospholipid Methylation, and Protein Carboxymethylation in Clonal Pheochromocytoma Cells, in *The Biochemistry of S-Adenosyl-Methionine and Related Compounds.* (Usdin E., Borchardt R. T., and Creveling C. R., eds.) MacMillan, London.

Rabe C. S. and McGee, Jr. R. (1983) Regulation of depolarization-dependent release of neurotransmitter by adenosine: cyclic AMP-dependent enhancement of release from PC 12 cells. *J. Neurochem.* **41,** 1623–1634.

Rabe C. S., Williams T. P., and McGee Jr. R. (1980) Enhancement of depolarization-dependent release of norepinephrine by inhibition of S-adenosylmethionine-dependent transmethylation. *Life Sci.* **27,** 1573–1579.

Radulovacki M., Virus R. M., Djuricic-Nedelson M., and Green R. D. (1983) Hypnotic effects of deoxycoformycin in rats. *Brain Res.* **271,** 392–395.

Radulovacki M., Virus R. M., Djuricic-Nedelson M., and Green R. D. (1984) Adenosine analogs and sleep in rats. *J. Pharmacol. Exp. Ther.* **228,** 268–274.

Rall T. W. (1980) Central Nervous Stimulants: The Xanthines, in *The Pharmacological Basis of Therapeutics* (6th Ed.). (Gilman A. G., Goodman L. S., and Gilman A., eds.) pp. 592–607, MacMillan, New York.

Reddington M. and Schubert P. (1979) Parallel investigations of the effects of adenosine on evoked potentials and cyclic AMP accumulation in hippocampus slices of rat. *Neurosci Lett.* **14,** 37–42.

Reddington M., Lee K., and Schubert P. (1982) An A-1 adenosine receptor, characterized by [^{3}H]cyclohexyladenosine binding, mediates the depression of evoked potentials in a rat hippocampal slice preparation. *Neurosci. Lett.* **28,** 275–279.

Reinhard, Jr. J. F., Galloway M. P., and Roth R. H. (1983) Noradrenergic stimulation of serotonin synthesis and metabolism. II. Stimulation by 3-isobutyl-1-methylxanthine. *J. Pharmacol. Exp. Ther.* **226,** 764–769.

Riberio J. A. (1982) The decrease of neuromuscular transmission by adenosine depends on previous neuromuscular depression. *Arch. Int. Pharmacodyn.* **255,** 59–67.

Riberio J. A., Sa-Almedia A. M., and Namordo J. M. (1979) Adenosine and adenosine triphosphates decrease $^{45}Ca^{2+}$ uptake by synaptosomes stimulated by potassium. *Biochem. Pharmacol.* **28,** 1297–1300.

Rosenberg R. S., Zepelin H., and Rechtschaffen A. (1979) Sleep in young and old rats. *J. Gerontol.* **34,** 525–532.

Sattin A. (1981) Adenosine as a Mediator of Antidepressant Treatment, in *Chemisms in the Brain.* (Rodnight R., Bachelard H. S., and Stahl W. S., eds.) pp. 265–275, Churchill-Livingston, Edinburgh.

Sattin A. and Rall T. W. (1970) The effect of adenosine and adenine nucleotides on the cyclic adenosine 3',5'-phosphate content of guinea pig cerebral cortex slices. *Mol. Pharmacol.* **6,** 12–23.

Sattin A., Stone T. W., and Taylor D. A. (1978) Biochemical and electrophysiological studies with tricyclic antidepressants in rat and guinea pig cerebral cortex. *Life Sci.* **23,** 2321–2626.

Sawynok J. (1983) Theophylline-induced potentiation of the antinociceptive action of baclofen. *Brit. J. Pharmacol.* **78,** 353–357.

Sawynok J. and Jhamandas K. H. (1976) Inhibition of acetylcholine release from cholinergic nerves by adenosine, adenine nucleotides, and morphine: antagonism by theophylline. *J. Pharmacol. Exp. Ther.* **197,** 379–390.

Schrader J., Rubio R., and Berne R. M. (1975) Inhibition of slow action potentials of guinea pig atrial muscle by adenosine. A possible effect on Ca^{2+} influx. *J. Mol. Cell. Cardiol.* **7,** 427–433.

Schubert P. and Mitzdorf V. (1979) Analysis and quantitative evaluation of the depressive effect of adenosine on evoked potentials in hippocampal slices. *Brain Res.* **172,** 186–190.

Schubert P., Komp W., and Kreutzberg G. W. (1979) Correlation of 5′-nucleotidase activity and selective transneuronal transfer of adenosine in the hippocampus. *Brain Res.* **168,** 419–424.

Schwabe U., Kiffe H., Puchstein C., and Trost T. (1979) Specific binding of [^{3}H]adenosine to rat brain membranes. *Naunyn-Schmiedeberg's Arch. Pharmacol.* **313,** 59–67.

Schwabe U. and Trost T. (1980) Characterization of adenosine receptors in rat brain by (–)-[^{3}H]-N^6-phenylisopropyladenosine. *Naunyn-Schmeideberg's Arch. Pharmacol.* **313,** 179–188.

Scott T. G. (1967) The distribution of 5′-nucleotidases in the brain of the mouse. *J. Comp. Neurol.* **129,** 97–113.

Segal M. (1982) Intracellular analysis of a postsynaptic action of adenosine in the rat hippocampus. *Eur. J. Pharmacol.* **79,** 193–199.

Shallaby I. A., Freedman S. B., and Miller R. J. (1983) The effect of organic calcium antagonists on release of [^{3}H]dopamine from aggregate cultures of midbrain neurons. *Soc. Neurosci. Abstr.* **9,** 565.

Shimizu H., Tanaka S., and Kodama T. (1972) Adenosine kinase of mammalian brain: partial purification and its role for the uptake of adenosine. *J. Neurochem.* **19,** 687–698.

Shimizu H. (1979) Biochemical Characterization of Adenosine Receptors in the Brain, in *Physiological and Regulatory Function of Adenosine and Adenine Nucleotides* (Baer H. P. and Drummond G. I., eds.) pp. 243–248. Raven, New York.

Siggins G. R. and Schubert P. (1981) Adenosine depression of hippocampal neurons in vitro: an intracellular study of dose-dependent actions on synaptic and membrane potentials. *Neurosci. Lett.* **23,** 55–60.

Skolnick P. and Daly J. W. (1974) The accumulation of adenosine 3′,5′-monophosphate in cerebral cortical slices of the quaking mouse, a neurological mutant. *Brain Res.* **73,** 513–525.

Skolnick P. and Paul S. M. (1981) The mechanism(s) of action of the benzodiazepines. *Med. Res. Rev.* **1,** 3–22.

Skolnick P., Lock K. C., Paul S. M., Marangos P. J., Jones R., and Irmscher K. (1980) Increased benzodiazepine receptor number elicited in vitro by a novel purine EMD 28422. *Eur. J. Pharmacol.* **67,** 179–186.

Smellie F. W., Davis C. W., Daly J. W., and Wells J. N. (1979) Alkylxanthines—inhibition of adenosine-elicited accumulation of cyclic-AMP in brain slices and of brain phosphodiesterase activity. *Life Sci.* **24,** 2475–2481.

Snell P. H. and Snell C. R. (1983) [^{3}H]Adenosine binding sites on 108CC15 neuroblastoma × glioma hybrid cell line and rat brain membranes. *Neurochem. Int.* **5,** 245–249.

Snyder S. H. (1985) Adenosine as a neuromodulator. *Ann Rev. Neurosci.,* **8,** 103–124.

Snyder S. H., Katims J. J., Annau Z., Bruns R. F., and Daly J. W. (1981) Adenosine receptors and behavioral actions of methylxanthines. *Proc. Natl. Acad. Sci. USA* **78,** 3260–3764.

Snyder S. H., Braas K. M., and Newby A. C. (1984) Immunocytochemical localization of adenosine-containing neurons in rat brain. *Proc. Intl. Life Sci. Inst. Meeting,* Mexico, June, 1984, in press.

Spencer D. G., Gherezghiher T., and Lal H. (1984) Inosine and N^6-substituted adenosine analogs lack anxiolytic activity in the pentylenetetrazol discrimination model of anxiety. *Drug Develop. Res.* **4,** 201–206.

Spignoli G., Pedata F., and Pepeu G. (1984) A_1 and A_2 adenosine receptors modulate acetylcholine relase from brain slices. *Eur. J. Pharmacol.* **97,** 341–342.

Standaert F. G. and Dretchen K. L. (1979) Cyclic nucleotides and neuromuscular transmission. *Fed. Proc.* **38,** 2183–2192.

Stone T. W. (1981) Physiological roles for adenosine and adenosine 5′-triphosphate in the nervous system. *Neurosci.* **6,** 523–525.

Stone T. W. (1983) Interaction of Adenosine with Other Agents, in *Regulatory Function of Adenosine* (Berne R. M., Rall T. W., and Rubio R., eds.) pp. 467–477, Martinus Nijhoff, Boston.

Stone T. W. (1985) *Purines: Pharmacology and Physiological Roles.* MacMillan, London.

Stone T. W. and Taylor D. A. (1978) Clonidine as an adenosine agonist. *J. Pharmacol.* **30,** 792–793.

Stone T. W. and Taylor D. A. (1979) Antidepressant drugs potentiate supression by adenosine of neuronal firing in rat cerebral cortex. *Neurosci. Lett.* **11,** 93–97.

Sugrue M. F. (1981) Chronic antidepressant administration and adaptive changes in central monoaminergic systems, in *Antidepressants: Neurochemical, Behavioral, and Clinical Perspectives* (Enna S. J., Malick J. B., and Richelson E., eds.) pp. 13–30, Raven, New York.

Sun M-C, McIlwain H., and Pull I. (1976) The metabolism of adenine derivatives in different parts of the brain of the rat and their release from hypothalamic preparations in excitation. *J. Neurobiol.* **7,** 109–122.

Thithapandha A., Maling H. M., and Gillette G. R. (1972) Effects of caffeine and theophylline on activity of rats in relation to brain xanthine concentration. *Proc. Soc. Exp. Biol. Med.* **139,** 582–586.

Trams E. G. and Lauter C. J. (1974) On the sidedness of plasma membrane enzymes. *Biochem. Biophys. Acta.* **345,** 180–197.

Van Calker D., Muller M., and Hamprecht B. (1978) Adenosine inhibits the accumulation of cyclic AMP in cultured brain cells. *Nature* (Lond.) **276,** 839–841.

Van Calcker D., Muller M., and Hamprecht B. (1979) Adenosine regulates via two different types of receptors, the accumulation of cyclic AMP in cultured brain cells. *J. Neurochem.* **33,** 999–1005.

Vapaatalo H., Onken D., Neuvonen P., and Westerman E. (1975) Stereospecificity in some central and circulatory effects of phenylisopropyladenosine (PIA). *Arnzeim Forsch.* **25,** 407–410.

Vellucci S. V. and Webster R. A. (1984) Antagonism of caffeine-induced seizures in mice by Ro 15-1788. *Eur. J. Pharmacol.* **97,** 289–293.

Virus R. M., Baglajewski T., and Radulovacki M. (1984) [^{3}H]-N^6-(L-Phenylisopropyl) adenosine binding in brain from young and old rats. *Neurobiol. Aging* **5,** 61–62.

Wagner J. A. and Katz R. J. (1983) Purinergic control of anxiety: direct behavioral evidence in rat. *Neurosci. Lett.* **43,** 333–337.

Watanabe Y., Lai R. T., and Yoshida H. (1983) Increase of [^{3}H]clonidine binding sites induced by adenosine receptor agonists in rat vas deferens in vitro. *Eur. J. Pharmacol.* **86,** 265–269.

Weller M. (1977) Evidence for the presynaptic location of adenylate cyclase and the cyclic-AMP stimulated protein kinase which is bound to synaptic membranes. *Biochem. Biophys. Acta.* **409,** 350–345.

White C., Lincoln C. A., and Pearce N. W. (1980) Anxiety and muscle tension as consequences of caffeine withdrawal. *Science* **209,** 1547–1548.

Williams M. (1981) Purinergic Receptors in Mammalian Brain: Characterization by Ligand Binding Techniques, in *Chemisms in the Brain.* (Rodnight R., Bachelard H. S., and Stahl W. S., eds.) pp. 88–99, Churchill-Livingstone, Edinburgh.

Williams M. (1983) Anxioselective anxiolytics. *J. Med. Chem.* **26,** 619–628.

Williams M. (1984a) Mammalian Central Adenosine Receptors, in *Handbook of Neurochemistry.* vol. 6 (Lajtha A., ed.) pp. 1–26, Plenum, New York.

Williams M. (1984b) Adenosine—a selective neuromodulator in the CNS? *Trends Neurosci.* **7,** 164–168.

Williams M. and Risley E. A. (1980a) Biochemical characterization of putative purinergic receptors by using 2-chloro-[^{3}H]adenosine, a stable analog of adenosine. *Proc. Natl. Acad. Sci. USA* **77,** 6892–6896.

Williams M., and Risley E. A. (1980b) Binding of ^{3}H-adenyl-5′-imidodiphosphate (AppNHp) to rat brain synaptic membranes. *Fed. Proc.* **39,** 1009.

Williams M. and U'Prichard D. C. (1984) Drug discovery at the molecular level: a decade of radioligand binding in retrospect. *Ann. Rep. Med. Chem.* **19,** 283–292.

Williams M. and Valentine H. (1985) Binding of [^{3}H]cyclohexyladenosine (CHA) to adenosine recognition sites in guinea pig ileal membranes: comparison with binding in brain membranes. *Neurosci. Lett.* **57,** 79–83.

Williams M. and Wood P. L. (1981) Effect of chronic chemical and surgical denervation on A-1 adenosine receptor density in rat brain. *Pharmacologist* **23,** 184.

Williams M., Risley E. A., and Huff J. R. (1981) Interaction of putative anxiolytic agents with central adenosine receptors. *Can. J. Physiol. Pharmacol.* **59,** 897–900.

Williams M., Risley E. A., and Robinson J. L. (1983) Chronic in vivo treatment with desmethylimipramine and mianserin does not alter adenosine A-1 radioligand binding in rat cortex. *Neurosci. Lett.* **35,** 47–51.

Wojcik W. J. and Neff N. H. (1982) Adenosine measurement by a rapid HPLC-fluorometric method: Induced change of adenosine content in regions of rat brain. *J. Neurochem.* **39,** 280–282.

Wojcik W. J. and Neff N. H. (1983a) Adenosine A_1 receptors are associated with cerebellar granule cells. *J. Neurochem.* **41,** 759–763.

Wojcik W. J. and Neff N. H. (1983b) Differential location of adenosine A_1 and A_2 receptors in striatum. *Neurosci. Lett.* **41,** 55–60.

Wojcik W. J. and Neff N. H. (1983c) Location of adenosine release and adenosine A_2 receptors to rat striatal neurons. *Life Sci.* **33,** 755–763.

Wooten F. G., Thoa N. B., Kopin I. J., and Axelrod J. (1973) Enhanced release of dopamine-β-hydroxylase and norepinephrine from sympathetic nerves by dibutyryl cyclic adenosine monophosphate and theophylline. *Mol. Pharmacol.* **9,** 178–183.

Wu P. H. and Coffin V. L. (1984) Up-regulation of brain [^{3}H]diazepam binding sites in chronic caffeine-treated rats. *Brain Res.* **294,** 186–189.

Wu P. H. and Phillis J. W. (1982) Adenosine receptors in rat brain membranes: characterization of high affinity binding of [^{3}H]-2-chloroadenosine. *Int. J. Biochem.* **14,** 399–404.

Wu P. H., Phillis J. W., Balls K., and Rinaldi B. (1980) Specific binding of 2-[^{3}H]-chloroadenosine to rat brain cortical membranes. *Can. J. Physiol. Pharmacol.* **58,** 576–579.

Wybenga M. P., Murphy M. G., and Robertson H. A. (1981) Rapid changes in cerebellar adenosine receptors following experimental seizures. *Eur. J. Pharmacol.* **75,** 79–80.

Yarbrough G. G. and McGuffin-Clineschmidt J. (1981) In vivo behavioral assessment of central nervous system purinergic receptors. *Eur. J. Pharmacol.* **76,** 137–144.

Yeung S. M. H. and Green R. D. (1983) Agonist and antagonist affinities for inhibitory adenosine receptors are reciprocally affected by 5′-guanylylimidodiphosphate or *N*-ethylmaleimide. *J. Biol. Chem.* **258,** 2334–2339.

Yeung S. M. H. and Green R. D. (1984) [^{3}H]-5′-*N*-ethylcarboxamide adenosine binds to both Ra and Ri adenosine receptors in rat striatum. *Naunyn-Schmiedeberg's Arch. Pharmacol.* **325,** 218–225.

Chapter 12
The Benzodiazepine Receptor

Ian L. Martin

1. Introduction

The 1,4-benzodiazepines have become the most frequently prescribed of all psychotropic drugs since the introduction of the first member of this series, chlordiazepoxide, into clinical practice in 1960. The compounds are used as anxiolytics, sedative/hypnotics, anticonvulsants, and muscle relaxants. This unique spectrum of activity, together with their large therapeutic index and essential lack of disturbing peripheral side effects, has assured their popularity and today over 25 analogs are available for prescription worldwide.

Like many drug discoveries, that of the benzodiazepines owes much to serendipity (Sternbach, 1978). Chlordiazepoxide was originally thought to be a quinazoline-*N*-oxide (Fig. 1.1) and its subsequent pharmacological testing showed a novel activity profile, now uniquely associated with drugs of this chemical class. This in turn led to the synthesis of further analogs, and correction of the original structural assignment (Fig. 1.2). Structure–activity studies resulted in the synthesis of more potent compounds possessing the same activity profile and the second compound in the series, diazepam (Fig. 1.3), was introduced to the market in 1963.

The first clues to the mechanism of action of the compounds came from electrophysiological studies with the observation by Schmidt et al. (1967) that diazepam increased primary afferent depolarization in the spinal cord. It is now clear that this was because of presynaptic inhibition, an event mediated by the inhibitory neurotransmitter γ-aminobutyric acid (GABA) (*see* Levy, 1977). Since that time many other electrophysiological studies have shown that the benzodiazepines appear to facilitate GABA-mediated transmission in many areas of the mammalian CNS; the phenomenon appears to be quite general for those events mediated through $GABA_A$ receptors (bicuculline-sensitive, baclofen-

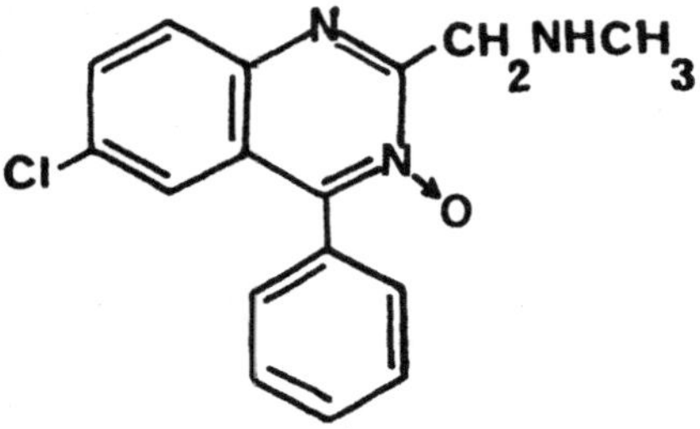

1.1

quinazoline-N-oxide

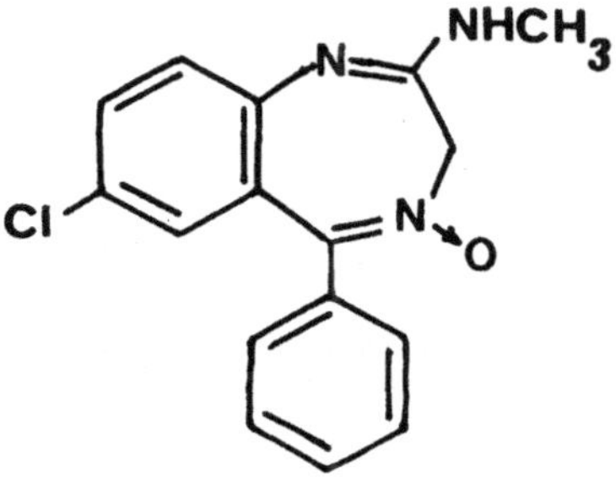

1.2

chlordiazepoxide

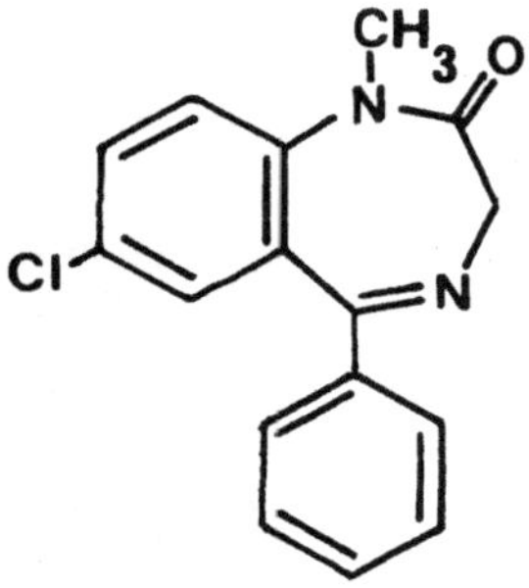

1.3

diazepam

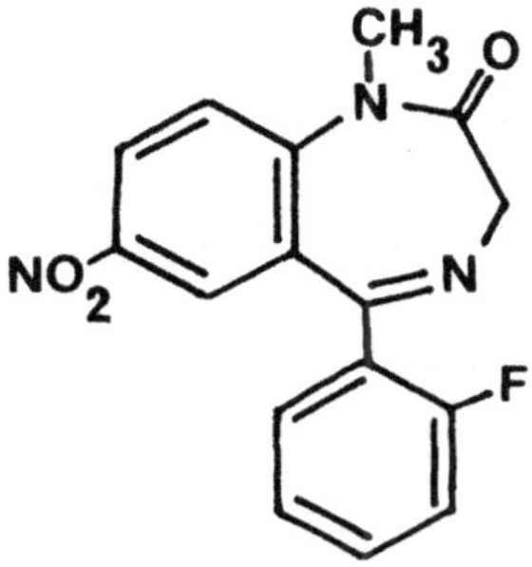

1.4

flunitrazepam

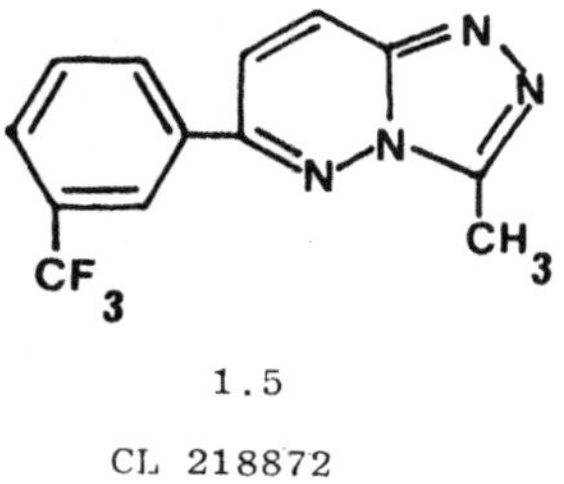

1.5

CL 218872

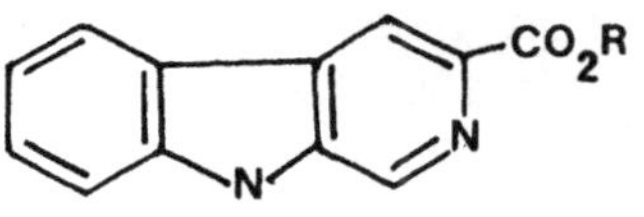

1.6

$R = CH_3$

$R = CH_2CH_3$

$R = CH_2CH_2CH_3$

β-carboline-3-carboxylic acid esters

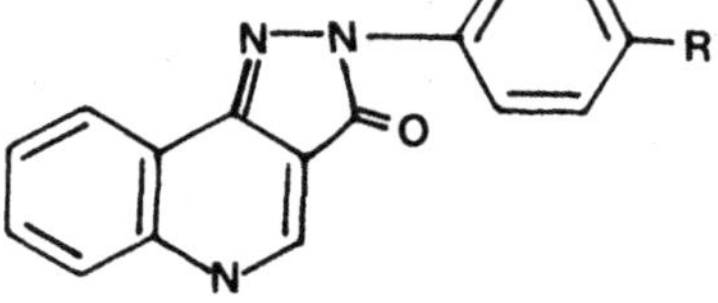

1.7

R = H : CGS 8216

$R = OCH_3$: CGS 9895

R = Cl : CGS 9896

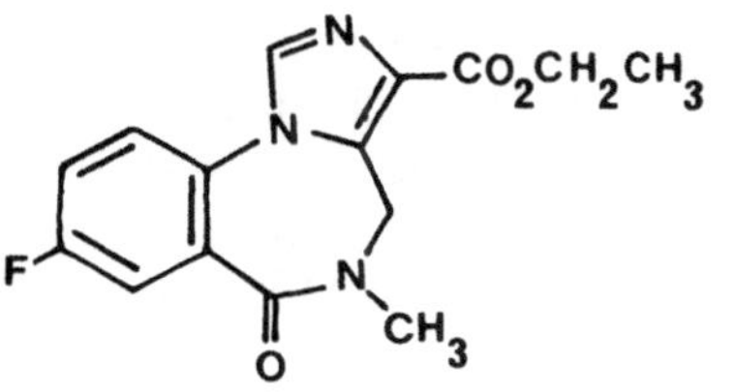

1.8
Ro 15-1788

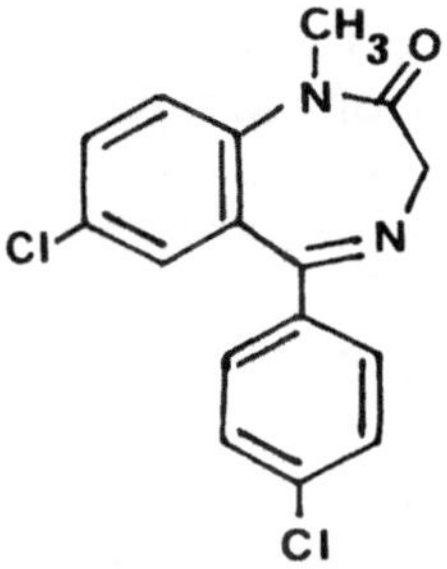

1.9
Ro 5-4864

Fig. 1. Chemical structures of drugs discussed in this chapter.

insensitive) (*see*, for example, Haefely, 1978, 1984). The action of GABA at this receptor subtype causes an increase in the probability of chloride channel opening in the cell membrane; the benzodiazepines appear to increase the frequency of this channel opening in response to a given GABA stimulus, but do not affect the channel lifetime (Study and Barker, 1981).

The early electrophysiological studies produced invaluable information concerning benzodiazepine actions at the neuronal level, but the precise mechanism by which this occurred remained unknown. It was with the discovery of specific high-affinity binding sites for the benzodiazepines in the mammalian CNS that the first link in this chain of events became apparent. These binding sites proved to have characteristics that suggested that they were, in fact, part of the pharmacological receptor through which these compounds exerted their effects.

2. Characterization

2.1. Biochemical Characterization

2.1.1. Benzodiazepine Ligands

Initial equilibrium studies with [^{3}H]-diazepam revealed that membranes prepared from the rat CNS were able to bind this ligand in a saturable manner with high affinity. Attempts to displace the bound ligand with a range of other compounds suggested that the interaction was quite specific for benzodiazepine analogs (Squires and Braestrup, 1977; Möhler and Okada, 1977). The binding characteristics of [^{3}H]-diazepam indicated that the interaction took place with a single homogeneous population of sites, and further that it obeyed simple mass action kinetics. Although the equilibrium dissociation constant remained essentially constant in different regions of the rat CNS, it was clear that the binding capacity of diazepam was unequally distributed regionally, being highest in the cortex and hippocampus and lowest in the spinal cord (Squires and Braestrup, 1977; Möhler and Okada, 1977; Müller et al., 1978).

The initial studies with [^{3}H]-diazepam were carried out at 0°C and little specific binding was found with this ligand if the temperature was raised to 37°C. The reason for such an observation is that the dissociation rate constant for diazepam is much higher at 37°C than 0°C with the result that during the separation of the bound ligand from free ligand and subsequent washing, in standard filtration assay procedures, much of the bound ligand was lost. The use of [^{3}H]-flunitrazepam (Fig. 1.4) appears largely to overcome such technical limitations, because of its slower dissociation constant (at both 0 and 37°C). Studies carried out to compare the two ligands suggest that they bind to the same population of sites (Braestrup and Squires, 1978).

That these binding sites were in fact part of the pharmacological receptor through which these drugs exert their effects was suggested by the marked correlation observed between the K_i value for a number of benzodiazepines in displacing [^{3}H]-diazepam from its binding sites in CNS membranes with a series of pharmacological tests thought to be predictive of the activity profile of these compounds.

In general terms the initial conclusions retain their credence, but the detailed investigations of this receptor system over the past 7 yr have shown that the situation is a good deal more complex. Although in equilibrium binding experiments the benzodiazepines appear to interact with their binding sites according to simple mass action kinetics, association and dissociation kinetics suggest that the mechanism of this interaction is more complex (Quast and Mahlman, 1982; Chiu et al., 1982). It is also clear from careful studies in different brain regions that the K_d values of [^{3}H]-flunitrazepam do differ significantly in, for example, the hippocampus and cerebellum (Supavilai and Karobath, 1980). This and additional data have been used to support arguments for a heterogeneity in the benzodiazepine receptor population in the mammalian CNS, although the precise nature of such heterogeneity is still the subject of some argument. (This will be discussed in more detail in later sections.)

2.1.2. Nonbenzodiazepine Ligands

Initially it was thought that the benzodiazepine binding site was specific only for benzodiazepines. However, it is now known that there are a number of classes of nonbenzodiazepinoid compounds that are able to potently and completely displace the benzodiazepines from their binding sites on mammalian CNS membranes.

The first such compounds to be reported were the triazolopyridazines, of which CL 218872 (Fig. 1.5) has received most attention. It was found that this compound was able to completely displace either [^{3}H]-diazepam or [^{3}H]-flunitrazepam from rat brain membranes with an apparent K_i value of about 70 nM; however, the Hill coefficient obtained from such a displacement analysis was significantly less than unity (~0.7) (Squires et al., 1979). It was pointed out by the authors that such an observation was consistent with the hypothesis of two populations of benzodiazepine receptors, each displaying the same affinity for the classical 1,4-benzodiazepines, but exhibiting different affinities for CL 218872. Such an observation gained further interest from the report that this compound exhibited anxiolytic and anticonvulsant activity with little or no sedative effects (Lippa et al., 1979), in contrast to the benzodiazepines. Subsequent studies, however, suggest that such a differentiation is not marked (Oakley et al., 1984).

The first member of the second class of compounds to be studied was ethyl-β-carboline-3-carboxylate (β-CCE) (Fig. 1.6). This material was found to displace [^{3}H]-flunitrazepam completely from rat brain membranes although the potency of the compound in the cerebellum (K_i ~0.9 nM) was about 4-fold greater than that found in the hippocampus. It was also shown that the Hill coefficient in the hippocampus was about 0.75, whereas that in the cerebellum was close to unity (Nielsen and Braestrup, 1980). Again the data were rationalized in terms of the multiple-receptor concept with the high-affinity β-CCE site comprising essentially all sites in the cerebellum and about 50% of the sites in the hippocampus. Subsequently, β-CCE, together with its methyl and *n*-propyl ester analogs, have become available labeled with tritium, and it has been confirmed that indeed these ligands do appear to preferentially label a subpopulation of benzodiazepine receptors with high affinity (Braestrup and Nielsen, 1980; 1981a; Nielsen et al., 1981). The detailed characterization of their lower-affinity interaction has proved difficult, using saturation analysis, because of the rapid increase of nonspecific binding at higher-ligand concentrations (Mitchell and Wilson, 1984), though displacement analyses are consistent with the hypothesis (Toll et al., 1984). However, such data do not show conclusively the existence of two independent subpopulations of benzodiazepine receptors, and alternative explanations of the data are equally viable (*see* Martin et al., 1983; Chiu and Rosenberg, 1983).

The third group of compounds that displays very high potency in displacing the benzodiazepines from their binding sites are the pyrazoloquinolinones (Czernik et al., 1982; Yokoyama et al., 1982). The first of these compounds upon which binding data were reported was CGS 8216 (Fig. 1.7). A series of benzodiazepines displaced [^{3}H]-CGS 8216 binding in the same rank order as that found for displacement of [^{3}H]-flunitrazepam binding, suggesting that the two ligands bound to the same population of sites; however, in a series of saturation experiments at various temperatures the B_{max} values obtained for CGS 8216 were consistently below those obtained for [^{3}H]-flunitrazepam. It is also clear that the inhibition of [^{3}H]-flunitrazepam binding by CGS 8216 is of the mixed type and not strictly competitive. Hill coefficients obtained in both hippocampal and cerebellar membranes in competition experiments with [^{3}H]-flunitrazepam were significantly greater than unity, indicating a complexity in the interaction (Martin and Baker, unpublished).

The imidazodiazepine, Ro 15-1788, (Fig. 1.8) also displays a high potency in displacement of [^{3}H]-flunitrazepam binding. The interaction appears to be strictly competitive in nature and displays an almost complete congruence with the distribution of [^{3}H]-benzodiazepines (Hunkeler et al., 1981).

2.1.3. Irreversible Labeling

It has been found that the benzodiazepine receptor can be irreversibly labeled using flunitrazepam (Battersby et al., 1979; Möhler et al., 1980). If flunitrazepam is allowed to come into equilibrium with its binding site at 0°C and then subsequently the preparation is irradiated with UV light, a fragment of the ligand becomes irreversibly attached to the benzodiazepine binding protein in the vicinity of the ligand recognition site. This observation has proved extremely useful, not only for the localization of flunitrazepam binding sites in tissue sections, but also to probe further the mechanisms by which the different ligand classes interact with this receptor complex.

The precise nature of the irreversible interaction of flunitrazepam with the benzodiazepine binding site is presently unclear. The situation has been complicated by the fact that individual groups have used widely differing procedures to accomplish the irreversible labeling, and this makes generalization difficult. A number of conclusions, however, can be drawn from the data available. Only those benzodiazepines with a nitro- function at position 7 are able to act as photoactivated irreversible agents (Johnson and Yamamura, 1979; Sieghart and Möhler, 1982). The irreversible attachment can be prevented by the inclusion of compounds that will compete with the photosensitive benzodiazepine for its specific binding site. Subsequent to photoaffinity labeling and removal of the excess–unused ligand, saturation analysis with either [^{3}H]-flunitrazepam or [^{3}H]-diazepam reveals that up to 90% of the high-affinity sites are lost; it has not proved possible to reduce this capacity factor further. It is also clear that the binding of some ligands such as the β-carboline-3-carboxylic acid esters and Ro 15–1788 are not so compromised after photoaffinity modification of the benzodiazepine receptor population (Brown and Martin, 1982; Gee and Yamamura, 1982; Hirsch, 1982; Karobath and Supavilai, 1982; Möhler, 1982).

It has been possible to use this observation to show that subsequent to photoaffinity modification of the receptor the [^{3}H]-β-CCE or [^{3}H]-Ro 15–1788 bound may be completely displaced by the benzodiazepines, although with a much reduced affinity. Such an observation indicates that the photoaffinity attachment of a fragment of flunitrazepam to the site does not leave the site in an "on" conformation, since it does not preclude binding of other ligands; under control conditions the different types of ligand bind in a mutually exclusive manner. This has now received support from the electrophysiological studies of Chan et al. (1983). The binding of other ligand classes is compromised to a lesser extent; CL 218872 shows a 2-fold reduction in affinity (Möhler, 1982; Brown and Martin, unpublished), whereas that for methyl 6,7-dimethoxy-4-ethyl-β-carboline-3-carboxylate (DMCM) is doubled (Braestrup et al., 1983).

The stoichiometry of the interaction is difficult to ascertain. It does appear that a greater number of sites are affected than would be suggested from the amount of flunitrazepam irreversibly attached to the membrane fragments (Thomas and Tallman, 1981). Problems of nonspecific labeling, which occur under certain conditions, and the effect of UV irradiation alone on the receptor population (Martin and Baker, unpublished observations) make meaningful calculations difficult. The fact that this stoichiometry is clearly not 1:1, however, is consistent with cooperative interactions within the binding site population.

The benzodiazepines clearly function as agonists at the benzodiazepine receptor; they produce a clear functional response. The interaction of an agonist with its receptor comprises an initial recognition process, then subsequently, or in some concerted fashion, a conformational change occurs in the protein as the first stage in the initiation of the effector sequence of events that leads to an overt response (Franklin, 1980). Marked loss of affinity of the benzodiazepines subsequent to photoaffinity modification of the receptor population could clearly represent a change in either the recognition site process or the effector sequence. It has been proposed that the increase in affinity of compounds for the benzodiazepine receptor in the presence of GABA is predictive of an agonist profile (*see* section 4.1). It has now been shown that the GABA facilitation of benzodiazepine binding remains unimpaired after photoaffinity labeling, suggesting that the modification that does occur involves the recognition site properties of the receptor and not its effector sequence (Brown and Martin, 1984b).

Attempts have also been made to produce alkylating ligands for irreversible attachment to the benzodiazepine receptor and these appear to have met with some success under certain conditions (Rice et al., 1979; Williams et al., 1980).

2.1.4. Thermodynamics of the Binding Interaction

A study of the thermodynamics of the interaction between a ligand and its receptor can give further insights into the mechanism by which such a process occurs. Indeed, such studies with the β-receptor have led Weiland et al. (1979) to suggest that a certain generality may exist in thermodynamic parameters exhibited by agonist–receptor interactions that can be differentiated from those of the antagonists.

The equilibrium dissociation constant K_d is related to the free energy change (ΔG) on ligand receptor interaction by the equation:

$$\Delta G = RT \ln K_d$$

where R is the gas constant and T absolute temperature. This free energy change can be resolved into an enthalpic (ΔH) and entropic (ΔS) component with the Gibbs equation. Thus:

$$\Delta G = \Delta H - T\Delta S$$

hence,

$$\ln K_d = \frac{\Delta H}{RT} - \frac{\Delta S}{R}$$

By plotting ln K_d agonist $1/T$ (van't Hoff plot), estimates of both enthalpic and entropic components of the interaction can be estimated. Hence studies of the temperature dependence of the K_d values for various ligands will suggest the type of forces involved in their binding-site interaction.

The majority of benzodiazepine receptor ligands show a marked decrease in affinity with increase in temperature (Speth et al., 1979; Quast et al., 1982; Doble, 1983), although the reverse was true for triazolam (Kochman and Hirsch, 1982). The van't Hoff plots for flunitrazepam (Speth et al., 1979; Quast et al., 1982; Doble, 1983) and clonazepam (Möhler and Richards, 1981) show discontinuities. This does not appear to be a general observation with the agonist benzodiazepines, since it is not shown by diazepam or *N*-desmethyldiazepam (Quast et al., 1982). β-CCE, with a clearly distinct pharmacological profile, again shows such a discontinuity (Doble, 1983), although the imidazobenzodiazepine Ro 15–1788 does not (Möhler and Richards, 1981). Clearly no generalization can be made about the type of interaction occurring with agonists as opposed to antagonists in this receptor system, and the reason for the occurrence of this discontinuity remains unexplained. At temperatures above the discontinuities in each case (and throughout for those ligands displaying linear van't Hoff plots) the binding interaction is largely enthalpically driven, although at lower temperatures a significant entropic component appears.

Estimates of the temperature dependence of the rate of reaction have been made in only one study (Quast et al., 1982), and assuming a simple bimolecular interaction it is suggested that activation energy for diazepam with the receptor complex is approximately 80 kJ/mol, whereas for flunitrazepam the figure is about 40 kJ/mol.

2.1.5. Multiple Receptors

A unique receptor must be defined, using the strictest criteria, in terms of its primary chemical structure. However, in the absence of such detailed information currently available about the benzodiazepine receptor, a number of second-order criteria have been used to argue for a multiplicity of this receptor system. These criteria derived from behavioral, electrophysiological, or receptor binding studies have produced a considerable amount of information, but unfortunately much confusion in the literature.

Evidence obtained solely from work carried out with the 1,4-benzodiazepines provides very little evidence to suggest that this receptor system is heterogeneous. There are many drugs of this class and the behavioral profile for each appears to be almost invariant. Although there are differences between members of the class in terms of their pharmacokinetics, pharmacodynamically their properties appear uniform throughout the series. Each is anxiolytic, sedative/hypnotic, anticonvulsant, and muscle relaxant; furthermore each displays these properties in approximately the same ratio. Electrophysiologically the compounds seem to be uniform in their properties; they facilitate GABA-mediated transmission in the mammalian CNS, and mechanistically their actions in this respect can be differentiated from those of the barbiturates (Study and Barker, 1981). Initial receptor binding studies with these compounds, under equilibrium conditions, were consistent with the conclusion that the benzodiazepines react with a single homogeneous population of binding sites according to simple mass-action kinetics. There was, therefore, no evidence to suggest any type of heterogeneity in this receptor population.

The discovery of compounds with distinct chemical structures that were able to displace the benzodiazepines from their binding sites did cast doubts on this simple picture.

The binding data of CL 218872 and β-CCE, representatives of the first two such chemical structures discovered, have already been discussed. Their binding characteristics were very similar. However, behaviorally the compounds were quite distinct. CL 218872 was an anxiolytic with anticonvulsant properties, and β-CCE has now been shown to be proconvulsant and anxiogenic (Tenen and Hirsch, 1980; Cowan et al., 1981; Dorow et al., 1983). The lack of coherence between the binding and behavioral data has since been extended. Within the β-carboline-3-carboxylates, the methyl ester is an overt convulsant, the ethyl ester a proconvulsant, as already discussed, and the *n*-propyl ester is a weak anticonvulsant (Oakley and Jones, 1982; Jensen et al., 1983). Each, however, has very similar characteristics in its ability to displace the classical benzodiazepines from their binding sites. It seems, therefore, that the differences observed in the behavioral characterization of compounds that appear to interact at the benzodiazepine receptor are unrelated to the complexities that occur in the binding phenomenology.

As pointed out earlier, the displacement characteristics of the benzodiazepines by the β-carboline-3-carboxylate analogs were not uniform throughout the mammalian brain. Competition curves were shallow in the hippocampus, but not in the cerebellum (Nielsen and Braestrup, 1980). It was suggested that two subpopulations of binding site existed, so called Bz_1 and Bz_2. The benzodiazepines were not able to distinguish between the two, whereas β-CCE exhibited an approximately 10-fold higher affinity for Bz_1 than Bz_2 sites. The cerebellum consisted largely of Bz_1

subtype, whereas in the hippocampus the population was equally divided between Bz_1 and Bz_2. Such a hypothesis is quite valid if it can be shown that each ligand obeys simple mass-action kinetics at each subtype, but this appears not to be the case (Martin et al., 1983). More detailed studies indicated that the situation was a good deal more complex than first thought. Both the association and dissociation kinetics of the benzodiazepines are biphasic, and it has been suggested that, subsequent to the formation of the initial ligand–receptor complex, a further step occurs to form some activated complex (Quast and Mahlmann, 1982). The kinetics of dissociation in the cerebellum and hippocampus are different for flunitrazepam, although the differences are small (Brown and Martin, unpublished). Nevertheless it suggests that the precise mechanism by which these drugs interact with their receptor in these two regions is distinct. However, similar studies with the imidazodiazepine Ro 15–1788 show very pronounced differences in these two brain regions; in the hippocampus the dissociation kinetics are clearly biphasic, whereas in the cerebellum they are monophasic. Differences in the interaction at the mechanistic level are therefore apparent (Brown and Martin, 1984c). In a series of very detailed studies over several years it has also been shown that the polypeptide chains that constitute the benzodiazepine receptor are apparently different in these two, and other, brain regions (Sieghart and Karobath, 1980; Sieghart and Drexler, 1983). These techniques have relied on the photoaffinity labeling of the receptor population with [^{3}H]-flunitrazepam, the subsequent solubilization of the affinity labeled material under denaturing conditions, and the separation of these labeled fragments. The cerebellum was found to contain largely material of molecular weight 51,000 daltons, whereas in the hippocampus another major additional band was observed at 55,000 daltons.

It is clear from a number of distinct types of observation, therefore, that a heterogeneity exists in the benzodiazepine receptor population in different regions of the brain. The details of the complexity underlying these observations have yet to be elucidated.

2.1.6. Solubilization

In the final analysis, the concept of receptor multiplicity will be resolved only when these receptors have been isolated and their primary structure ascertained, for it is this parameter and this only that can be used as the criterion for true identity. Attempts are already being made to answer this particular question and the solubilization of the receptor from the CNS membranes is the first stage in such studies.

Several different procedures have been adopted to solubilize the benzodiazepine receptor from cell membranes. Most studies have included attempts to ascertain whether or not the neurochemical characteristics of the receptor have been retained in the solubilized material. The loss of any particular attribute can be ascribed either to the "freezing" of the receptor protein into a particular conformation or its degradation during the solubilization, with the irretrievable loss of these characteristics.

The receptor has been solubilized using 0.2% sodium deoxycholate containing 1*M* potassium chloride in approximately 30% yield (Asano and Ogasawara, 1980); this material showed the characteristic GABA facilitation of benzodiazepine binding and later studies by the same authors pointed out the fact that high-affinity GABA binding also appeared to be copurified with the benzodiazepine binding sites (Asano and Ogasawara, 1981). The GABA facilitation found by these authors is by no means universal using sodium deoxycholate, since in a number of similar solubilization procedures it has not been found (Sigel et al., 1982, 1983), whereas in others it has (Sherman-Gold and Dudai, 1980; Schoch and Möhler, 1983). Undoubtedly the precise conditions, particularly relating to the concentration of the agent, are important.

Triton X-100 (Braestrup et al., 1981), digitonin (Gavish et al., 1979; Gavish and Snyder, 1980), and Lubrol (Yousoufi et al., 1979) have also been used with varying degrees of success. However, undoubtedly one of the most suitable agents appears to be the zwitterionic detergent 3-[(cholamidopropyl)-dimethylammonio]propanesulfonate (CHAPS) (Stephenson and Olsen, 1982). A maximum yield was obtained with 20 m*M* CHAPS, with which 67% of the total membrane bound [^{3}H]-flunitrazepam binding activity was obtained. Under the conditions used, both GABA and barbiturate facilitation of [^{3}H]-flunitrazepam binding was found.

This method has been extended to the purification of this receptor material using both affinity and ion-exchange chromatography (Sigel et al., 1982, 1983; Sigel and Barnard, 1984) and the essentially purified receptor partially characterized. The material, which exhibits high-affinity binding sites for both GABA and the benzodiazepines on the same physical structure, has a molecular weight of about 230,000 daltons. This is in close agreement with the figure obtained by the same group using radiation-inactivation analysis either in membrane preparations or in solution (Chang and Barnard, 1982). Similar purification studies have also been carried out by Martini et al. (1982).

2.2. Localization

2.2.1. Peripheral vs Central Type Sites

High-affinity binding sites for both [^{3}H]-diazepam and [^{3}H]-flunitrazepam are found not only in the CNS, but also in peripheral tissues such as liver, kidney, and lung (Braestrup and Squires, 1977; Möhler and Okada, 1977; Gallager et al., 1981). However, it is apparent that the binding sites in peripheral tissues exhibit quite distinct recognition characteristics from those in the CNS; clonazepam, with a very high affinity for the central type site, exhibits an extremely low affinity at peripheral sites, whereas the reverse is true for Ro 5–4864 (Fig. 1.9). No function has yet been proposed for these peripheral-type sites. The benzodiazepines also display a high-capacity, low-affinity binding to human serum albumin, but again the recognition properties here are quite distinct from those found in the CNS (Sjodin et al., 1976; Lucek and Coutinho, 1976; Albec-Kolbah et al., 1979; Möhler and Okada, 1977). The characteristics of these sites can be differentiated with relative ease, since defining nonspecific [^{3}H]-flunitrazepam or [^{3}H]-diazepam binding with clonazepam will yield specific binding only to central-type sites. [^{3}H]-Ro 5–4864 has been used directly to investigate peripheral-type sites (Marangos et al., 1982; Schoemaker et al., 1983).

2.2.2. Regional Distribution

Using standard binding assay procedures on dissected brain tissue it has been shown that benzodiazepine receptors are unevenly distributed in the brain of both rat (Squires and Braestrup, 1977; Möhler and Okada, 1977; Müller et al., 1978) and man (Braestrup et al., 1977; Möhler and Okada, 1978; Speth et al., 1978). The receptor capacity is highest in the cortical regions and the cerebellum (approximately 1 pmol/mg protein), intermediate in such regions as the hypothalamus, thalamus, and caudate nucleus, whereas in the medulla oblongata and spinal cord the levels are low (approximately 20% of the levels found in the cortex). The receptor is almost absent from white matter. The affinity of the receptor is essentially uniform throughout the CNS, although minor differences do occur (Supavilai and Karobath, 1980). Within these gross regions there are variations, e.g., the molecular layer of the cerebellum contains more binding sites than the granule cell layer in bovine brain (Candy and Martin, unpublished). As opposed to these in vitro neurochemical procedures, the distribution of this receptor has been revealed in much greater detail by light microscopic autoradiographic techniques (Kuhar, 1978). It is clear from such studies that not only are there marked subregional variations of this receptor, but subtle interspecies differences also become apparent (Young and Kuhar, 1979, 1980; Biscoe et al., 1984). An extension of this technique by the use of tritium-sensitive film and its quantitation using computer-assisted densitometry has also allowed quantitative receptor pharmacology to be carried out in areas too small to study easily using conventional techniques (Palacios et al., 1981; Penney et al., 1981; Unnerstall et al., 1982).

Central type benzodiazepine receptors have also been identified on bovine adrenal chromaffin cells (Kataoka et al., 1984) and in the rat adenohypophysis, which is of ectodermal origin (Brown and Martin, 1984a). This latter study also shows, paradoxically, that peripheral type binding sites are most dense in the neurally derived posterior pituitary.

2.2.3. Cellular Localization

Lesioning experiments, both physically and with toxic agents, have proved of enormous value in the localization of neuronal pathways within the CNS. Such studies have therefore been used to provide evidence for a neuronal localization of these receptors, although the results have frequently proved difficult to interpret. Kainic acid injected into the corpus striatum, substantia nigra, or cerebellum resulted in an approximately 50% decrease in the benzodiazepine binding sites (Braestrup and Squires, 1978; Braestrup et al., 1979; Chang et al., 1980; Biggio et al., 1980), whereas ibotenic acid injected directly into the hippocampus produced only a 30% decrease in this receptor population (Fuxe et al., 1981). Such studies provide evidence of a degree of neuronal localization, although the ubiquitous occurrence of this receptor on many different cell types within a given region is undoubtedly responsible for the low depletions seen.

An alternative approach has been to study the benzodiazepine receptor in a number of mutant strains of mice in which there is a distinct loss of a particular cell type. Such studies indicate that whereas the cerebellar Purkinje cells do possess benzodiazepine receptors, cerebellar granule cells do not, although incomplete loss of receptors in the presence of complete Purkinje cell loss means that other cell types in the cerebellum express the benzodiazepine receptor (Braestrup et al., 1979; Chang et al., 1980; Skolnick et al., 1979; Speth and Yamamura, 1979; Lippa et al., 1978). There is, of course some additional difficulty associated with such investigations on mutant strains, since the mutation may also be responsible for changes other than cell loss that would clearly confuse the interpretation of such data. This is apparent in the mutant strain "spastic," in which distinct changes in the benzodiazepine-GABA receptor population are apparent with no known histological abnormalities (Biscoe et al., 1981).

Although none of this work suggests that the binding sites for the benzodiazepine receptors are strictly located on neurons, because of the incomplete losses observed, subcellular distribution studies are consistent with the neuronal localization of the receptor population. The crude synaptosomal fraction contained larger amounts of specific [^{3}H]-diazepam binding sites than those of the crude nuclear or microsomal fractions, whereas the osmotically shocked crude synaptic membrane fraction also contained large concentrations of binding activity (Möhler and Okada, 1978).

However, despite this plethora of inconclusive data, it is generally believed that the central-type benzodiazepine receptor obtains a restricted neuronal localization, whereas the peripheral-type sites are present only on glial cells in the CNS. The dogma, however, is supported by tissue culture studies in which primary cultures of both neurons and glia display the appropriate type of binding sites (Braestrup et al., 1978; Syapin and Skolnick, 1979; Baraldi et al., 1979; Huang et al., 1980; McCarthy and Harden, 1981; White et al., 1981).

2.2.4. *Phylogeny and Ontogeny*

Phylogenetically the benzodiazepine receptor appears to have developed rather late. It cannot be identified in invertebrate species and its binding characteristics in those species prior to the bony fishes is atypical (Nielsen et al., 1978; Fernholm et al., 1979). This is distinct from the high-affinity GABA binding sites that are present in cyclostomes and chondrichthyes (Mann and Enna, 1980) in which benzodiazepine binding sites are absent. This is generally in agreement with electrophysiological studies in which responses to GABA can be found in invertebrates (e.g., Takeuchi and Takeuchi, 1975), although these responses are not potentiated by chlordiazepoxide (Mathews and Wickelgren, 1979).

Ontogenetically the benzodiazepine receptor appears to develop early; specific [^{3}H]-diazepam binding could be detected 8 d before birth in P_2 fraction from the rat whole brain. At birth the levels were about 35% of the adult values, which were reached at about 4 wk postnatally (Braestrup and Nielsen, 1978). These results were largely confirmed in a postnatal study, although a difference in the rates of cortical and cerebellar development was noted (Candy and Martin, 1979b). Autoradiographic studies have recently provided much more detailed information (Richards et al., 1983; Schlumpf et al., 1983). The results are broadly in agreement with the previous neurochemical studies, although it is clear that the development of benzodiazepine binding sites in different brain regions parallels axonal growth and synaptogenesis, occurring much earlier in the spinal cord than in the cerebral cortex or cerebellum. It is interesting, however, that an autoradiographic study of the ontogenic development of the high-affinity GABA receptor was quite distinct (Palacios and Kuhar, 1982). Here, for example, the GABA binding sites in the cerebellum followed largely the granule cell development, whereas those of the benzodiazepine binding sites were closely associated with the development of the Purkinje cell dendritic tree in the molecular layer (Richards et al., 1983; Schlumpf et al., 1983). Despite this, it has been shown that modulation of [^{3}H]-flunitrazepam binding by GABA appears to be relatively constant throughout development (Palacios et al., 1979; Mallorga et al., 1980; Regan et al., 1980).

There have been several studies concerning the possible differential development of the putative multiple benzodiazepine sites, and it appears that the 55,000-dalton fragment labeled in [^{3}H]-flunitrazepam photoaffinity experiments develops more rapidly than the 51,000-dalton material (Sieghart and Mayer, 1982).

3. Benzodiazepine Receptor Function

3.1. *In Vivo Receptor Binding*

Although the in vitro binding studies with benzodiazepine ligands yield a considerable amount of information, it is clear that the situation is completely artificial. However, it has proved possible to carry out similar studies in vivo using both [^{3}H]-diazepam (Williamson et al., 1978; Tallman et al., 1979) or [^{3}H]-flunitrazepam (Chang and Snyder, 1978; Duka et al., 1979). The latter ligand has proved more suitable because the dissociation rate for this ligand is much slower than that of diazepam, with the result that equilibria obtained in vivo change less rapidly postmortem. The technique involves the iv injection of [^{3}H]-flunitrazepam. At some suitable timepoint the animals are killed, and the brains removed, dissected, homogenized, and filtered to yield a measure of radioligand bound to the membrane-associated receptor sites. The definition of nonspecific binding cannot be carried out precisely, although a high dose of an unlabeled benzodiazepine prior to the injection of the radiolabel provides a measure of displaceable binding.

The characteristics of this in vivo binding compare favorably with those expected from the in vitro studies in terms of both distribution and ligand specificity (Blanchard et al., 1979; Chang and Snyder, 1978). No estimates have been made of either equilibrium dissociation constants or capacity factors in vivo, and thus all results assume that such tracer doses of [^{3}H]-flunitrazepam bind to a representative sample of the receptor population. Comparisons are made, therefore, in qualitative rather than quantitative terms. Despite these restrictions, a great deal of useful information can be obtained by using such an approach. One of these that has proved of considerable value concerns the determination of efficacy at the benzodiazepine receptor.

3.2. Efficacy at the Benzodiazepine Receptor

The classical 1,4-benzodiazepines all appear to be full agonists at the benzodiazepine receptor. Small doses of a given benzodiazepine produce marked anticonvulsant effects, whereas higher doses produce anticonflict effects and much higher doses still are required to produce muscle relaxation. This information, in conjunction with the in vivo binding assay studies mentioned in the previous section, has shown that the ED_{50} dose for antagonism of pentylenetetrazole-induced seizures occupies 20–25% of the benzodiazepine receptor population (Duka et al., 1979), whereas anticonflict effects are elicited at about 60% occupancy (Petersen and Buus-Lassen, 1981); muscle relaxant effects require still greater occupancy (~80%).

This gradation of receptor occupancy can be turned to advantage. Partial agonists would be expected to require greater receptor occupancy to produce a given effect, thus even at near maximal occupancy such compounds would be unable to produce those effects associated with muscle relaxation. The development of such compounds has taken place both in the benzodiazepine and β-carboline series, i.e., the application of these principles has allowed the spectrum of activity normally associated with benzodiazepine receptor occupation to be limited to those effects that occur at relatively low receptor occupancy by full agonists (Braestrup et al., 1984).

No suitable irreversible ligands have yet been obtained for in vivo use and thus the estimation of the receptor reserve has not been possible.

3.3. The Three-State Model

There is, however, a further complexity associated with this receptor system. The 1,4-benzodiazepines possess a clearly defined profile of action as anticonvulsants, sedative/hypnotics, anxiolytics, and muscle relaxants. It is now apparent that certain compounds that appear to interact specifically with the benzodiazepine receptor can produce diametrically opposed effects, and yet others can antagonize both types of action (Tenen and Hirsch, 1980; Mitchell and Martin, 1980; Cowan et al., 1981; Dorow et al., 1983).

It is possible to rationalize the data by assuming that an endogenous ligand is present for the benzodiazepine receptor that imposes some tone upon the system. The benzodiazepines are able to displace such an endogenous material and act as agonists on the system, increasing the tone. Those compounds with opposite effects to the benzodiazepines would displace the endogenous ligand and release that tone that it imposed upon the system; we could classify these as antagonists that effectively decrease the endogenous tone. Such an explanation, with the endogenous ligand as a partial agonist, is perfectly logical, but has not received widespread acceptance, probably because no endogenous ligand has yet been identified.

Instead, the three-state model has been adopted. This model has been put forward in essentially the same form by three groups (Polc et al., 1982; Nutt et al., 1982; Jensen et al., 1983; see also Prado de Carvalho et al., 1983). The benzodiazepine receptor is envisaged to exist in three energy states. The first, when occupied, results in no effect being observed on the GABA activation of Cl^- channel opening; the antagonist state. The second is that in which occupation by the benzodiazepines allows the receptor–ligand complex to relax into such a conformation that results in an increased coupling between the GABA stimulus and its effector mechanism. This so called "agonist" state can also be induced by certain nonbenzodiazepine ligands. The third state is obtained when the receptor is occupied by a compound such as methyl β-carboline-3-carboxylate with behavioral actions opposite to those of the benzodiazepines; this ligand produces overt convulsions. Hence the receptor–ligand complex relaxes into another conformation that decreases the coupling between the GABA stimulus and its effector sequence; the so-called inverse or contra agonist state. Such a model is thermodynamically quite sound, but its essential novelty has caused some disquiet among a number of pharmacologists.

It is clear that this receptor system is atypical in the sense that effects are simply modulatory via the GABA receptor. As such, the three-state model provides a putative mechanism for "fine-tuning" the GABA stimulus–effector coupling.

4. Interactions With the Benzodiazepine Receptor

4.1. γ-Aminobutyric Acid

There is compeling evidence, mainly from electrophysiological studies, that the benzodiazepines produce their effects by modulation of GABA-mediated transmission (*see*, e.g., Haefely, 1978, 1984). However, neurochemically it was not until 1982 that a direct interaction was shown. It is now clear that the benzodiazepines are able to facilitate the binding of GABA to a low-affinity receptor in mammalian CNS membranes (Skerritt et al., 1982). The reverse interaction was demonstrated much earlier (Tallman et al., 1978; Martin and Candy, 1978). Again this interaction appears to take place via a low affinity GABA receptor as high concentrations of this neurotransmitter must be included in in vitro binding assay systems to demonstrate the effect. The presence of GABA affects both the on and off rates of the benzodiazepines at their binding sites and the result is an increased affinity of the receptor for its agonist ligands. Although the effect has not been adequately rationalized, apart from its classification as an allosteric modulation, it has proved of considerable practical value. It appears (with only one exception published so far) that the affinity of all agonist ligands is increased in the presence of GABA, whereas the affinity of inverse agonists is decreased (Braestrup and Nielsen, 1981b; Möhler and Richards, 1981; Doble et al., 1982; Fujimoto et al., 1982; Skolnick et al., 1982; Braestrup et al., 1982). Those compounds that display antagonist profiles of action suffer no affinity modulation in the presence of GABA.

It has also been possible to demonstrate the phenomenon in vivo, since the use of pharmacological agents that increase the GABA content in the CNS produce significant increases in binding of the appropriate ligands to the benzodiazepine receptor (Tallman et al., 1979).

4.2. Effect of Various Ions

In direct extension of the modulatory effects of GABA on benzodiazepine binding, and because of the chloride channel gating involved in GABA-mediated transmission, attempts were made to investigate the effects of various ions on the binding characteristics of the benzodiazepine receptor. It was initially shown that chloride ions produced a synergistic effect with GABA, increasing the affinity of this binding site for [^{3}H]-diazepam (Martin and Candy, 1978). The effect was later shown to generalise to other anions, which were able to reverse GABA inhibitory postsynaptic potentials electrophysiologically (Costa et al., 1979; Martin and Candy, 1980). It was argued that such an observation was caused by a close connection between the benzodiazepine receptor and chloride ion channels (Costa et al., 1979), although the limitations of such an argument were pointed out (Candy and Martin, 1979a). It has since been shown that certain distilbene derivatives can at least partially block the facilitation of agonist binding caused by chloride ions, although the effects are complex (Costa et al., 1981).

Both Ni^{2+} and Cd^{2+} ions have also been shown to increase the affinity of the benzodiazepine receptor for [^{3}H]-diazepam (Mackerer et al., 1978).

4.3. Barbiturates

Although the overt pharmacological actions of the barbiturates have some similarities to those of the benzodiazepines, at the molecular level their mechanism of interaction is different. Although the benzodiazepines produce an increase in the frequency of chloride channel opening in response to a given GABA stimulus, the major effect of the barbiturates is on channel lifetime (Study and Barker, 1981). The barbiturates display no direct interaction with the benzodiazepine receptor in vitro; their locus of action appears to be at a recognition site that can be labeled with dihydropicrotoxinin (Ticku and Olsen, 1978). This site does interact, however, with the benzodiazepine binding site as the anesthetic barbiturates have been shown to produce a chloride-dependent enhancement of benzodiazepine binding (Leeb-Lundberg et al., 1980). This is because of an affinity change at the receptor with no alteration in capacity. The interactions also appear to be somewhat complex, with, for example, the anticonvulsant phenobarbital producing no direct effect on benzodiazepine receptors alone, but able to block the facilitatory effects of pentobarbital (Leeb-Lundberg et al., 1981; Olsen, 1981).

The benzodiazepine receptor, therefore, appears to suffer modulation by a number of allosteric sites, such as those for GABA and the barbiturates. Agents that interact with those sites can produce changes in the affinity of the benzodiazepine receptor for its ligands. These ligands, of course, have been classed as exhibiting positive and negative efficacy, and as a general rule the modulatory effects produced by these sites are opposite on these two classes of benzodiazepine ligand (Braestrup et al., 1983; Honore et al., 1983). The fact that there are a number of exceptions to this general rule must indicate that our understanding of these actions remains far from complete (Honore et al., 1984).

A new ligand, *t*-butylbicyclophosphorothionate, has recently become available (Squires et al., 1983). It appears to bind to a site very similar to that for dihydropicrotoxinin. This ligand, unlike dihydropicrotoxinin, has excellent binding characteristics and is responsive to occupation of many of the recognition sites presently thought to be involved in this complex (Supavilai and Karobath, 1984).

4.4. Others

In addition to those agents mentioned above, a number of other compounds have been reported to modulate binding at the benzodiazepine receptor, though their mechanisms of interaction are even less well characterized.

The pyrazolopyridines tracazolate, etazolate, and cartazolate produce a chloride-dependent affinity change at the benzodiazepine receptor, with characteristics similar to that produced by the hypnotic barbiturates (Beer et al., 1978; Williams and Risley, 1979; Supavilai and Karobath, 1979, 1981; Meiners and Salama, 1982).

The avermectin B_{1a} (Williams and Yarbrough, 1979) and ivermectin (Williams and Risley, 1982, 1984) appear to affect both affinity and capacity factors at the benzodiazepine receptor. Both compounds are macrolide anthelmintics that appear to have effects on chloride channels at the lobster neuromuscular junction (Fritz et al., 1979).

A number of adenosine analogs (EMD 28422, EMD 35993, and EMD 41717), as well as the diaryltriazine L 781067 (Wong et al., 1983) and the H_2-receptor antagonist cimetidine (Lakoski et al., 1983) also modify benzodiazepine binding (Skolnick et al., 1980, 1983).

5. Endogenous Ligands for the Benzodiazepine Receptor

The identification of the opioid peptides as natural ligands for opiate receptors inevitably set in motion attempts to identify compounds that would serve the same function at the benzodiazepine receptor. Although a good deal of effort has been expended in this direction, arguments have been made concerning the necessity to envisage the existence of such a ligand (Möhler, 1981).

Although many endogenous compounds have been found that show some ability to displace the benzodiazepines from their binding sites on CNS membranes, few have any reasonable affinity for the site (for review, *see* Martin, 1980). Davis and Cohen (1980) and Davis et al. (1981) reported isolation of a relatively small molecular weight material (original estimate ~3000 daltons) from rat or bovine brain; this material competitively displaced the benzodiazepines from their binding sites in vitro and also produced EEG changes similar to those caused by diazepam in animals. No further information has been published on this material since that time.

A second material, named nepenthin by Wolff and Nixon (1981), has been isolated from the bile duct of rats and purified to apparent homogeneity. This material, with a molecular mass of about 16,000 daltons, has been successfully used to raise antibodies that in turn have been used to identify cells containing immunologically similar material in deep cortical regions of the rat forebrain.

The most promising candidate to date, however, would appear to be that originally isolated in 1978 by Guidotti et al. This material, with a molecular weight of approximately 11,000 daltons, displaces the benzodiazepines from their binding sites in a competitive manner (Massotti and Guidotti, 1980; Guidotti et al., 1981). When given icv to rats, it produces effects characteristic of a partial inverse agonist (Guidotti et al., 1983). The material so far has not been completely characterized, although cyanogen bromide treatment yields three fragments, two of which, with a total of 45 amino acids, have been completely sequenced. The structure of the third polypeptide, which retains the biological activity of the parent material, has not so far been characterized (Guidotti et al., 1983).

6. Structure–Activity Relationships at the Benzodiazepine Receptor

The problem of defining the structural parameters important in the interaction between the benzodiazepine receptor and its ligands is an extremely intriguing one.

The initial studies of Sternbach et al. (1968), in which correlations were sought between overt behavioral activity and individual compounds and their various structural modifications, led to a number of empirical rules. Activity was increased by the inclusion of electron-withdrawing substituents at position 7 and 2′, whereas groups larger than methyl at position 1 decreased activity. Further it was shown that the replacement of the carbonyl function at position 2 by hydrogen atoms, reduction of the *N*4–C5 double bond or the introduction of halo-substituents at 3′ or 4′ positions drastically reduced activity.

Later studies making use of theoretical calculations and using anticonvulsant potency as the measure of biological activity, led to suggestions that significant correlations could be found with the charge on *N*4 (Sarrazin et al., 1976), the charge on C2 carbonyl oxygen and the total molecular dipole moment (Blair and Webb, 1977), the difference between energies of the lowest empty and highest occupied molecular orbitals (Gilli et al., 1977), or the electron density on the p_y orbital at C10 (Lucek et al., 1979). All these studies, however, suffered from the fact that the measure of biological activity used was dependent not only in the recognition properties of the pharmacological receptor, responsible for the actions of the benzodiazepines, but also on the pharmacokinetic properties of the individual compounds.

Attempts have since been made to circumvent these difficulties by using as the measure of biological activity the affinity of the individual compounds for their CNS specific benzodiazepine receptor.

In a series of 28 benzodiazepine molecules it has been found that the appearance of lowest empty molecular orbital (a measure of overall electron affinity of the molecule) in correlations suggests that an important part of the interaction between the benzodiazepine receptor and its ligands may be charge–transfer complex formation. The dipole moment (μ_x; parallel to C10–C11 bond) also appears to play a significant role in such correlations; this may be important in the alignment of the molecule prior to the initial interaction with its binding site. The third most important parameter appears to relate to the geometry around the *N*4–C5 region, though at this stage it is not clear whether this is an electronic or geometric factor (Butcher et al., 1984).

It has also become clear that substituents at position 3 of the classical 1,4-benzodiazepines exhibit pronounced effects on the affinity of these compounds for their receptor. It has now been shown that it is the R-conformation of the benzodiazepines that exhibits the highest affinity for the receptor (Sunjic et al., 1979; Blount et al., 1983).

Clearly the topography of the recognition site for the benzodiazepines must also be able to accommodate the other ligands that appear to produce their effects through the pharmacological receptor that includes this site. Attempts have been made to identify this by a number of approaches. The first, the distance–geometry method of Crippen (1982) is purely theoretical, although the more classical methods using as their basis X-ray crystallographic and molecular orbital calculations have been more popular.

The structures of zopiclone, CL 218872, the β-carboline-3-carboxylic acid analogs and the pyrazoloquinolinone CGS 8216 have all been determined and attempts have been made to obtain structural commonalities by the overlay of these structures (Gilli, 1984).

In addition it has also been possible to show a marked correlation, in a series of imidazobenzodiazepines, between the agonist and inverse agonist profiles of these compounds (Fryer et al., 1984). This latter study suggests a relationship between the separation of the benzene ring A and an associated charge is predictive of such activity. This model is also able to accommodate the activity profiles of the nonbenzodiazepine ligands for this receptor (Fryer, 1983).

Considerable progress has been made in explaining the detail of the interaction of this receptor with its ligands, but only time will permit an adequate refinement of these important hypotheses.

7. Binding Assay Technology

There are many ways of preparing tissue for membrane binding assays and it is important to recognize these differences in the interpretation of results obtained by different groups.

The most common is the homogenization of tissue in 20 vol of hypotonic buffer (pH 7–7.5; Tris-HCl, Tris-citrate, or phosphate based buffers). We have also found it important in certain cases to fragment membranes using the Polytron homogenizer to allow reproducible data to be obtained; e.g., cerebellar tissue tends to "clump" if care is not taken. The homogenized membranes are then centrifuged (normally ~30,000*g* for 20 min), the supernatant removed, and the pellet resuspended in fresh buffer. This is again followed by centrifugation and the procedure is repeated three times. The final pellet is then suspended in 20 vol of buffer and appropriate aliquots are taken, and frozen at –20°C until required. For most mammalian CNS tissue this is then diluted to give a final 1:80 homogenate, 100 μL of which is used in the assay. To this is added the appropriate radiolabeled ligand, which is then allowed to come to equilibrium. The bound radioactivity is then separated from free radioligand by filtration through previously wetted glass fiber filters (usually GF/B) under mild vacuum. The filters are then rapidly washed with 15 mL of the same buffer. This procedure, using well washed membranes essentially free from endogenous GABA, is very reproducible. The lower the incubation temperature (in general), the longer it will take for equilibrium to be attained; this is normally 20–60 min, depending on the ligand.

One alternative procedure, that of using unwashed membranes in which the tissue is merely homogenized, centrifuged, and resuspended for use, contains considerable amounts of GABA and thus, depending on the ligand used, will be subject to the appropriate GABA modulation. Similarly, the ionic content of the buffer may induce a number of effects.

The third approach, that in fact used initially, was to prepare P_2 fraction of CNS material using density gradient centrifugation. However, the results from such studies, although likely to be more specific for the appropriate synaptic membrane fractions, suffer from a number of disadvantages. The fractions are clearly much more time-consuming to prepare and it has been found that the recovery of binding activity is dependent, at least partially, on the brain region used (Nielsen et al., 1981). The specificity and the low level of nonspecific binding, if defined with the appropriate ligand and concentration, generally obviate the use of such membrane fractions, unless of course that particular type of study is intended.

8. The Future

Our understanding of the mechanism by which the benzodiazepines produce their array of pharmacological effects has progressed considerably over the years since their introduction into clinical practice. Many questions, however, remain to be answered. The radioligand binding assays have given a vast array of information. However, to interpret this data adequately it is imperative that methods be developed for the determination of efficacy at this receptor complex. Sophisticated electrophysiological techniques have been used in this regard, but rapid neurochemical techniques to study chloride flux still elude us. Modern techniques of molecular biology may soon provide answers to the structural intricacies of the receptor protein itself and perhaps give new insights to the underlying mechanisms of the disease states that these drugs are used to treat. Whether such understanding will allow us to design drugs with a more restricted, and therefore more useful, spectrum of activity remains for the future.

NOTE ADDED IN PROOF: Since the original manuscript for this article was submitted, two papers of considerable interest have appeared.

The first concerns the development of 22 monoclonal antibodies for the $GABA_A$–benzodiazepine receptor complex. The authors state that these allow four different epitopes of the $GABA_A$–benzodiazepine receptor complex to be distinguished. Two of the monoclonal antibodies have been described in some detail and their use in the immunological delineation of this receptor protein in rat brain is described and compared with the autoradiographic distribution of [^{3}H]-Ro 15-1788 and [^{3}H]-muscimol (Schoch et al., 1985).

The second concerns further work on the putative endogenous ligand for the benzodiazepine receptor (Guidotti et al., 1978). Polyclonal antibodies have now been raised to this purified peptide and the distribution of the immunoreactive material carried out using standard radioimmunoassay techniques on microdissected tissue, together with immunocytochemical localization in thin sections of glutaraldehyde fixed tissue (Alho et al., 1985).

References

Albic-Kolbah T., Kajfez F., Rendic S., Sunjic V., Konowal A., and Snatzke G. (1979) Circular dichroism and gel filtration study of binding of prochiral and chiral 1,4-benzodiazepin-2-ones to human serum albumin. *Biochem. Pharmacol.* **28,** 2457–2464.

Alho H., Costa E., Ferrero P., Fujimoto M., Cosenza-Murphy D., and Guidotti A. (1985) Diazepam binding inhibitor: A neuropeptide located in selected neuronal populations of rat brain. *Science,* **229** 179–182.

Asano T. and Ogasawara N. (1980) Solubilisation of γ-aminobutyric acid receptor from rat brain. *Life Sci.* **26,** 1131–1137.

Asano T. and Ogasawara N. (1981) Soluble gamma-aminobutyric acid and benzodiazepine receptors from rat cerebral cortex. *Life Sci.* **29,** 193–200.

Baraldi M., Guidotti A., Schwartz J. P., and Costa E. (1979) GABA receptors in clonal cell lines: A model for study of benzodiazepine action at molecular level. *Science* **205,** 821–823.

Battersby M. K., Richards J. G., and Möhler H. (1979) Benzodiazepine receptor: photoaffinity labeling and localization. *Eur. J. Pharmacol.* **57,** 277–278.

Beer B., Klepner C. A., Lippa A. S., and Squires R. F. (1978) Enhancement of ^{3}H-diazepam binding by SQ 65396: A novel antianxiety agent. *Pharmacol. Biochem. Behav.* **9,** 849–851.

Biggio G., Corda M. G., de Montis G., and Gessa G. L. (1980) Differential Effects of Kainic Acid on Benzodiazepine Receptors, GABA Receptors, and GABA-Modulin in the Cerebellar Cortex, in *Receptors for Neurotransmitters and Peptide Hormones.* (Pepeu G., Kuhar M. J., and Enna S. J., eds.), pp. 265–270, Raven, New York.

Biscoe T. J., Fry J. P., Martin I. L., and Rickets C. (1981) Binding of GABA and benzodiazepine receptor ligands in the spinal cord of the spastic mouse. *J. Physiol.* (Lond.) **317,** 32–33P.

Biscoe T. J., Fry J. P., and Rickets C. (1984) Autoradiography of benzodiazepine receptor binding in the central nervous system of the normal C57BL6J mouse. *J. Physiol.* (Lond.) **352,** 495–508.

Blair T. and Webb G. A. (1977) Electronic factors in the structure–activity relationship of some 1,4-benzodiazepine-2-ones. *J. Med. Chem.* **20,** 1206–1210.

Blanchard J. C., Boireau A., Garret C., and Joulou L. (1979) In vitro and in vivo inhibition by zopiclone of benzodiazepine binding to rodent brain receptors. *Life Sci.* **24,** 2417–2420.

Blount J. F., Fryer R. I., Gilman N. W., and Todaro L. J. (1983) Quinazolines and 1,4-benzodiazepines: 92. Conformational recognition of the receptor by 1,4-benzodiazepines. *Mol. Pharmacol.* **24,** 425–428.

Braestrup C., Albrechtsen R., and Squires R. F. (1977) High densities of benzodiazepine receptors in human cortical areas. *Nature* (Lond.) **269,** 702–704.

Braestrup C., Honore T., Nielsen M., Petersen E. N., and Jensen L. J. (1984) Ligands for benzodiazepine receptors with positive and negative efficacy. *Biochem. Pharmacol.* **33,** 859–862.

Braestrup C. and Nielsen M. (1978) Ontogenetic development of benzodiazepine receptors in the rat brain. *Brain Res.* **147,** 170–173.

Braestrup C. and Nielsen M. (1980) Multiple benzodiazepine receptors. *Trends Neurosci.* **3,** 301–303.

Braestrup C. and Nielsen M. (1981a) ^{3}H-propyl β-carboline-3-carboxylate as a selective radioligand for the Bz_1 benzodiazepine receptor subclass. *J. Neurochem.* **37,** 333–341.

Braestrup C. and Nielsen M. (1981b) GABA reduces binding of ^{3}H-methyl β-carboline-3-carboxylate to brain benzodiazepine receptors. *Nature* (Lond.) **294,** 472–474.

Braestrup C., Nielsen M., Biggio G., and Squires R. F. (1979) Neuronal localisation of benzodiazepine receptors in cerebellum. *Neurosci. Lett.* **13,** 219–224.

Braestrup C., Nielsen M., and Honore T. (1983) Binding of [^{3}H]-DMCM, a convulsive benzodiazepine ligand, to rat brain membranes: Preliminary studies. *J. Neurochem.* **41,** 454–465.

Braestrup C., Nielsen M., Skovbjerg H., and Gredal O. (1981) β-Carboline-3-Carboxylates and Benzodiazepine Receptors, in *GABA and Benzodiazepine Receptors*. (Costa E., Di Chiara G., and Gessa G. L., eds.), pp. 147–155, Raven, New York.

Braestrup C., Nissen C., Squires R. F., and Schousboe A. (1978) Lack of brain specific benzodiazepine receptors on mouse primary astroglial cultures. *Neurosci. Lett.* **9,** 45–49.

Braestrup C., Schmiechen R., Neff G., Nielsen M., and Pettersen E. N. (1982) Interaction of convulsive ligands with benzodiazepine receptors. *Science* **216,** 1241–1243.

Braestrup C., and Squires R. F. (1977) Specific benzodiazepine receptors in rat brain characterised by high affinity [^{3}H]-diazepam binding. *Proc. Natl. Acad. Sci. USA* **74,** 3805–3809.

Braestrup C. and Squires R. F. (1978) Brain specific benzodiazepine receptor. *Brit. J. Psychiatry.* **133,** 249–260.

Brown C. L. and Martin I. L. (1982) Photoaffinity labeling of the benzodiazepine receptor does not occlude the βCCE binding site. *Brit. J. Pharmacol.* **77,** 312P.

Brown C. L. and Martin I. L. (1984a) Autoradiographic localization of benzodiazepine receptors in the rat pituitary gland. *Eur. J. Pharmacol.* **102,** 563–564.

Brown C. L. and Martin I. L. (1984b) Photoaffinity labeling of the benzodiazepine receptor compromises the recognition site but not its effector mechanism. *J. Neurochem.* **43,** 272–273.

Brown C. L. and Martin I. L. (1984c) Kinetics of [^{3}H]-Ro 15-1788 binding to membrane-bound rat brain benzodiazepine receptors. *J. Neurochem.* **42,** 918–923.

Butcher H. J., Chananont P., Hamor T. A., and Martin I. L. (1984) Structure activity relationships in a series of 5-phenyl-1,4-benzodiazepines. *Proc. 8th Int. Symp. Med. Chem.*, Uppsala, Sweden.

Candy J. M. and Martin I. L. (1979a) Is the benzodiazepine receptor coupled to a chloride anion channel? *Nature* (Lond.) **280,** 172–173.

Candy J. M. and Martin I. L. (1979b) The postnatal development of the benzodiazepine receptor in the cerebral cortex and cerebellum of the rat. *J. Neurochem.* **32,** 655–658.

Chan C. Y., Gibbs T. T., Borden L. A., and Farb D. H. (1983) Multiple embryonic benzodiazepine binding sites: evidence for functionality. *Life Sci.* **33,** 2061–2069.

Chang L-R. and Barnard E. A. (1982) The benzodiazepine/GABA receptor complex: molecular size in brain synaptic membranes and in solution. *J. Neurochem.* **39,** 1507–1518.

Chang R. S. L. and Snyder S. H. (1978) Benzodiazepine receptors: labeling in intact animals with [^{3}H]-flunitrazepam. *Eur. J. Pharmacol.* **48,** 213–218.

Chang R. S. L., Tran V. T., and Snyder S. H. (1980) Neurotransmitter receptor localisations: brain lesion-induced alterations in benzodiazepine, GABA, β-adrenergic and histamine H1–receptor binding. *Brain Res.* **190,** 95–110.

Chiu T. H., Dryden D. M., and Rosenberg H. C., (1982) Kinetics of ^{3}H-flunitrazepam binding to membrane-bound benzodiazepine receptors. *Mol. Pharmacol.* **21,** 57–65.

Chiu T. H. and Rosenberg H. C. (1983) Multiple conformational states of benzodiazepine receptors. *Trends Pharmacol. Sci.* **4,** 348–350.

Costa T., Rodbard D., and Pert C. (1979) Is the benzodiazepine receptor coupled to a chloride anion channel? *Nature* (Lond.) **277,** 315–317.

Costa T., Russell L., Pert C. B., and Rodbard D. (1981) Halide and γ-aminobutyric acid-induced enhancement of diazepam receptors in rat brain. Reversal by olisulfonic acid stilbene blockers of anion channels. *Mol. Pharmacol.* **20,** 470–476.

Cowan P. J., Green A. R., Nutt D. J., and Martin I. L. (1981) Ethyl β-carboline carboxylate lowers seizure threshold and antagonises flurazepam induced sedation in rats. *Nature* (Lond.) **290,** 54–55.

Crippen G. M. (1982) Distance geometry analysis of the benzodiazepine binding site. *Mol. Pharmacol.* **22,** 11–19.

Czernik A. J., Petrack B., Kalinsky H. J., Psychoyos S., Cash W. D., Tsai C., Rinehart R. K., Granat F. R., Lovell R. A., Brundish D. E., and Wade R. (1982) CGS 8216; receptor binding characteristics of a potent benzodiazepine antagonist. *Life Sci.* **30,** 363–372.

Davis L. G. and Cohen R. K. (1980) Identification of an endogenous peptide ligand for the benzodiazepine receptor. *Biochem. Biophys. Res. Commun.* **92,** 141–148.

Davis L. G., McIntosh H., and Reker D. (1981) An endogenous ligand to the benzodiazepine receptor: preliminary evaluation of its bioactivity. *Pharmacol. Biochem. Behav.* **14,** 839–844.

Doble A. (1983) Comparative thermodynamics of benzodiazepine receptor ligand interactions in rat neuronal membranes. *J. Neurochem.* **40,** 1605–1612.

Doble A., Martin I. L., and Richards D. A. (1982) GABA modulation predicts biological activity of ligands for the benzodiazepine receptor. *Brit. J. Pharmacol.* **76,** 238P.

Dorow R., Horowski R., Paschelke, G. Amin, M., and Braestrup C. (1983) Severe anxiety induced by FG7142, a β-carboline ligand for benzodiazepine receptors. *Lancet* **ii,** 98–99.

Duka T., Höllt V., and Herz A. (1979) In vivo receptor occupation by benzodiazepines and correlation with pharmacological effect. *Brain Res.* **179,** 147–156.

Fernholm B., Nielsen M., and Braestrup C. (1979) Absence of brain specific benzodiazepine receptors in cyclostomes and elasmobranchs. *Comp. Biochem. Physiol.* **C62,** 209–211.

Franklin T. J. (1980) Binding energy and the activation of hormone receptors. *Biochem. Pharmacol.* **29,** 853–856.

Fritz L. C., Wang C. C., and Gorio A. (1979) Avermectin Bla irreversibly blocks postsynaptic potential at the lobster neuromuscular junction by reducing muscle membrane resistance. *Proc. Natl. Acad. Sci. USA* **76,** 2062–2066.

Fryer R. I. (1983) Benzodiazepine ligand–receptor interactions, in The Benzodiazepines: From Molecular Biology to Clinical Practice, (Costa E., ed), pp. 7–20, Raven, New York.

Fryer R. I., Gilman N. W., Madison V., and Walser A. (1984) Conformational differences for agonists and antagonists at the benzodiazepine receptor. *Proc. 8th Int. Symp. Med. Chem.* Uppsala, Sweden.

Fujimoto M., Hirai K., and Okabayashi T. (1982) Comparison of the effects of GABA and chloride ion on the affinities of ligands for the benzodiazepine receptor. *Life Sci.* **30,** 51–57.

Fuxe K., Koehler C., Agnati L. F., Andersson K., Oegren S. O., Eneroth P., Perez de la Mora M., Karobath M., and Krogsgaard-Larsen P. (1981) GABA and benzodiazepine receptors. Studies on their Localisation in the Hippocampus and Their Interaction With Central Dopamine Neurons in the Rat Brain, in: *GABA and Benzodiazepine Receptors* (Costa E., Di Chiara G., and Gessa G. L., eds.), pp. 61–76, Raven, New York.

Gallager D. W., Mallorga P., Oertel W., Henneberry R., and Tallman J. (1981) ^{3}H-diazepam binding in mammalian central nervous system: a pharmacological characterization. *J. Neurosci.* **1,** 218–225.

Gavish M., Chang R. S. L., and Snyder S. H. (1979) Solubilisation of histamine H-1, GABA and benzodiazepine receptors. *Life Sci.* **25,** 783–790.

Gavish M. and Snyder S. H. (1980) Soluble benzodiazepine receptors: GABAergic regulation. *Life Sci.* **26,** 579–582.

Gee K. W. and Yamamura H. I. (1982) Differentiation of benzodiazepine receptor agonist and antagonist: sparing of ^{3}H-benzodiazepine antagonist binding following the photolabeling of benzodiazepine receptors. *Eur. J. Pharmacol.* **82,** 239–241.

Gilli G., Borea P. A., Bertolasi V., and Sacerdoti M. (1977) Stereochemical and electronic properties of benzodiazepines. Correlation with their anti-convulsant activity. *Proc. 4th Eur. Crystallography Meet.* Oxford, p. 38.

Gilli G. (1984) Crystal structure and properties of molecules. *Acta Cryst. A40 suppl.* C–6.

Guidotti A., Ebstein B., and Costa E. (1981) Purification and characterisation of an endogenous brain peptide that competes with ^{3}H-diazepam binding. *Soc. Neurosci. Abs.* **7,** 634.

Guidotti A., Forchetti C. M., Corda M. G., Konkel D., Bennett C. D., and Costa E. (1983) Isolation, characterisation, and purification to homogeneity of an endogenous polypeptide with agonistic action on benzodiazepine receptors. *Proc. Natl. Acad. Sci. USA* **80,** 3531–3535.

Guidotti A., Toffano G., and Costa E. (1978) An endogenous protein modulates the affinity of GABA and benzodiazepine receptors in rat brain. *Nature* (Lond.) **275,** 553–555.

Haefely W. (1978) Central actions of benzodiazepines: general introduction. *Brit. J. Psychiatry* **133,** 231–238.

Haefely W. (1984) Actions and Interactions of Benzodiazepine Agonists and Antagonists at GABAergic Synapses, in *Actions and Interactions of GABA and Benzodiazepines,* (Bowery N. G., ed.), pp. 263–285, Raven, New York.

Hirsch J. D. (1982) Photolabeling of benzodiazepine receptors spares [^{3}H]propyl β-carboline binding. *Pharmacol. Biochem. Behav.* **16,** 245–248.

Honore T., Nielsen M., and Braestrup C. (1983) Binding of ^{3}H-DMCM to benzodiazepine receptors; chloride–dependent allosteric mechanisms. *J. Neural. Transm.* **58,** 83–98.

Honore T., Nielsen M., and Braestrup C. (1984) Barbiturate shift as a tool for determination of efficacy of benzodiazepine receptor ligands. *Eur. J. Pharmacol.* **16,** 245–248.

Huang A., Barker J. L., Paul S. M., Moncada V., and Skolnick P. (1980) Characterisation of benzodiazepine receptors in primary cultures of fetal mouse brain and spinal cord neurons. *Brain Res.* **190,** 485–491.

Hunkeler W., Möhler H., Pieri L., Polc P., Bonetti E. P., Cumin R., Schaffner R., and Haefely W. (1981) Selective antagonists of benzodiazepines. *Nature* (Lond.) **290,** 514–516.

Jensen L. H., Peteresen E. N., and Braestrup C. (1983) Audiogenic seizures in DBA/2 mice discriminate sensitively between low efficacy benzodiazepine receptor agonists and inverse agonists. *Life Sci.* **33,** 393–399.

Johnson R. W. and Yamamura H. I. (1979) Photoaffinity labeling of the benzodiazepine receptor in bovine cerebral cortex. *Life Sci.* **25,** 1613–1620.

Karobath M. and Supavilai P. (1982) Distinction of benzodiazepine agonists from antagonists by photoaffinity labeling of benzodiazepine receptors in vitro. *Neurosci. Lett.* **31,** 65–69.

Kataoka Y., Gutman Y., Guidotti A., Panula P. Wroblewski J., Cosenza–Murphy D., Wu J. Y., and Costa E. (1984) Intrinsic GABAergic system of adrenal chromaffin cells. *Proc. Natl. Acad. Sci. USA* **81,** 3218–3222.

Kochman R. L. and Hirsch J. D. (1982) Thermodynamic changes associated with benzodiazepine and alkyl β-carboline-3-carboxylate binding to rat brain homogenates. *Mol. Pharmacol.* **22,** 335–341.

Kuhar M. J. (1978) Histochemical Localisation of Neurotransmitter Receptors, in: *Neurotransmitter Receptor Binding* (Yamamura H. I., Enna S. J., and Kuhar M. J., eds.), pp. 113–126, Raven, New York.

Lakoski J. M., Aghajanian G. K., and Gallager D. W. (1983) Interaction of histamine H_2-receptor antagonists with GABA and benzodiazepine binding sites in the CNS. *Eur. J. Pharmacol.* **88,** 241–245.

Leeb-Lundberg F., Snowman A., and Olsen R. W. (1980) Barbiturate receptors are coupled to benzodiazepine receptors. *Proc. Natl. Acad. Sci. USA* **77,** 7468–7472.

Leeb-Lundberg F., Snowman A., and Olsen R. W. (1981) Perturbation of benzodiazepine receptor binding by pyrazolopyridines involves picrotoxinin/barbiturate receptor sites. *J. Neurosci.* **1,** 471–477.

Levy R. A. (1977) The role of GABA in primary afferent depolarization. *Prog. Neurobiol.* **9,** 211–267.

Lippa A. S., Critchett D. J., Sano M. C., Klepner C. A., Greenblat F. N., Coupet J., and Beer B. (1979) Benzodiazepine receptors: cellular and behavioural characteristics. *Pharmacol. Biochem. Behav.* **10,** 831–843.

Lippa A. S., Sano M. C., Coupet J., Klepner C. A. and Beer B. (1978) Evidence that benzodiazepine receptors reside on cerebellar Purkinje cells: Studies with "nervous" mutant mice. *Life Sci.* **23,** 2213–2218.

Lucek R. W. and Coutinho C. B. (1976) The role of substituents in the hydrophobic binding of the 1,4-benzodiazepines by human plasma proteins. *Mol. Pharmacol.* **12,** 612–619.

Lucek R. W., Garland W. A., and Dairman W. (1979) CNDO/2 molecular orbital study of selected 1,3-dihydro-5-phenyl-1,4-benzodiazepin-2-ones. *Fed. Proc.* **38,** 541.

Mackerer C. R. and Kochman R. L. (1978) Effects of cations and anions on the binding of [^{3}H]-diazepam to rat brain. *Proc. Soc. Exp. Biol. Med.* **158,** 393–397.

Mallorga P., Hamburg M., Tallman J. F., and Gallager D. W. (1980) Ontogenetic changes in GABA modulation of brain benzodiazepine binding. *Neuropharmacology* **19,** 405–408.

Mann E. and Enna S. J. (1980) Phylogenetic distribution of bicuculline sensitive γ-aminobutyric acid (GABA) receptor binding. *Brain Res.* **184,** 367–373.

Marangos P. J., Patel J., Boulenger J. P., and Clark-Rosenberg R. (1982) Characterisation of peripheral-type benzodiazepine binding sites in brain using [^{3}H]Ro 5–4864. *Mol. Pharmacol.* **22,** 26–32.

Martin I. L. (1980) Endogenous ligands for benzodiazepine receptors. *Trends Neurosci.* **3,** 299–301.

Martin I. L., Brown C. L., and Doble A. (1983) Multiple benzodiazepine receptors: structures in the brain or structures in the mind? A critical review. *Life Sci.* **32,** 1925–1933.

Martin I. L. and Candy J. M. (1978) Facilitation of benzodiazepine binding by sodium chloride and GABA. *Neuropharmacology* **17,** 993–998.

Martin I. L. and Candy J. M. (1980) Facilitation of specific benzodiazepine binding in rat brain membrane fragments by a number of anions. *Neuropharmacology* **19,** 175–179.

Martini C., Lucacchini A., Ronca G., Hrelia S., and Rossi C. A. (1982) Isolation of putative benzodiazepine receptors from rat brain membranes by affinity chromatography. *J. Neurochem.* **38,** 15–19.

Massotti M. and Guidotti A. (1980) Endogenous regulators of benzodiazepine recognition sites. *Life Sci.* **27,** 847–854.

Mathews G. and Wickelgren W. O. (1979) Glycine, GABA and synaptic inhibition of reticulospinal neurones of lamprey. *J. Physiol.* (Lond.) **293,** 393–415.

McCarthy G. D. and Harden T. K. (1981) Identification of two benzodiazepine binding sites on cells cultured from rat cerebral cortex. *J. Pharmacol. Exp. Ther.* **216,** 183–191.

Meiners B. A. and Salama A. J. (1982) Enhancement of benzodiazepine and GABA binding by the novel anxiolytic tracazolate. *Eur. J. Pharmacol.* **78,** 315–322.

Mitchell P. R. and Martin I. L. (1980) Ethyl β-carboline-3-carboxylate antagonises the effect of diazepam on a functional GABA receptor. *Eur. J. Pharmacol.* **68,** 513–514.

Mitchell P. R. and Wilson L. (1984) Investigation of the effects of muscimol on different components of ^{3}H-propyl β-carboline-3-carboxylate binding to rat hippocampal and cerebellar membranes. *Eur. J. Pharmacol.* **97,** 315–319.

Möhler H. (1981) Benzodiazepine receptors: are there endogenous ligands in brain? *Trends Pharmacol. Sci.* **1,** 116–119.

Möhler H. (1982) Benzodiazepine receptors: differential interaction of benzodiazepine agonists and antagonists after photoaffinity labeling with flunitrazepam. *Eur. J. Pharmacol.* **80,** 435–436.

Möhler H., Battersby M. K., and Richards J. G. (1980) Benzodiazepine receptor protein identified and visualized in brain tissue by a photoaffinity label. *Proc. Natl. Acad. Sci. USA* **77,** 1666–1670.

Möhler H. and Okada T. (1977) Benzodiazepine receptor: demonstration in the central nervous system. *Science* **198,** 849–851.

Möhler H. and Okada T. (1978) The benzodiazepine receptor in normal and pathological human brain. *Brit. J. Psychiatry* **133,** 261–268.

Möhler H. and Richards J. G. (1981) Agonist and antagonist benzodiazepine receptor interaction in vitro. *Nature* (Lond.) **294,** 763–765.

Müller W. E., Schläfer U., and Wollert U. (1978) Benzodiazepine receptor binding in rat spinal cord membrane. *Neurosci. Lett.* **9,** 239–243.

Nielsen M. and Braestrup C. (1980) Ethyl β-carboline-3-carboxylate shows differential benzodiazepine receptor interaction. *Nature* (Lond.) **286,** 606–607.

Nielsen M., Braestrup C., and Squires R. F. (1978) Evidence for a late evolutionary appearance of brain-specific benzodiazepine receptors: an investigation of 18 vertebrate and 5 invertebrate species. *Brain Res.* **141,** 342–346.

Nielsen M., Schou H., and Braestrup C. (1981) [^{3}H]-Propyl β-carboline-3-carboxylate binds specifically to brain benzodiazepine receptors. *J. Neurochem.* **36,** 276–285.

Nutt D. J., Cowen P. J., and Little H. J. (1982) Unusual interactions of benzodiazepine receptor antagonists. *Nature* (Lond.) **295,** 436–438.

Oakley N. R. and Jones B. J. (1982) Differential pharmacological effects of β-carboline-3-carboxylic acid esters. *Neuropharmacology* **21,** 587–589.

Oakley N. R., Jones B. J., and Straughan D. W. (1984) The benzodiazepine receptor ligand CL 218872 has both anxiolytic and sedative properties in rodents. *Neuropharmacology* **23,** 797–802.

Olsen R. W. (1981) GABA-benzodiazepine-barbiturate receptor interactions. *J. Neurochem.* **37,** 1–13.

Palacios J. M. and Kuhar M. J. (1982) Ontogeny of high affinity GABA and benzodiazepine receptors in rat cerebellum: an autoradiographic study. *Devel. Brain Res.* **2,** 531–539.

Palacios J. M., Niehoff D. L., and Kuhar M. J. (1979) Ontogeny of GABA and benzodiazepine receptors: effects of Triton X-100, bromide and muscimol. *Brain Res.* **179,** 390–395.

Palacios J. M., Niehoff D. L., and Kuhar M. J. (1981) Receptor autoradiography with tritium-sensitive film: potential for computerised densitometry. *Neurosci. Lett.* **25,** 101–105.

Penney J. B., Frey K., and Young A. B. (1981) Quantitative autoradiography of neurotransmitter receptors using tritium-sensitive film. *Eur. J. Pharmacol.* **72,** 421–422.

Petersen E. N. and Buus-Lassen J. (1981) A water-lick conflict paradigm using drug-experienced rats. *Psychopharmacology* **75,** 236–239.

Polc P., Bonetti E. P., Schaffner R., and Haefely W. (1982) A three-state model of the benzodiazepine receptor explains the interactions between the benzodiazepine antagonist Ro 15–1788, benzodiazepine tranquilizers, β-carbolines and phenobarbitone. *Naunyn Schmiedeberg's Arch. Pharmacol.* **321,** 260–264.

Prado de Carvalho L., Grecksch G., Chapouthier G., and Rossier J. (1983) Anxiogenic and non-anxiogenic benzodiazepine antagonists. *Nature* (Lond.) **301,** 64–66.

Quast U. and Mahlmann H. (1982) Interaction of ^{3}H-flunitrazepam with the benzodiazepine receptor: evidence for a ligand-induced conformation change. *Biochem. Pharmacol.* **31,** 2761–2768.

Quast U., Mahlmann H., and Vollmer K.-O. (1982) Temperature dependence of the benzodiazepine-receptor interaction. *Mol. Pharmacol.* **22,** 20–25.

Regan J. W., Roeske W. R., and Yamamura H. I. (1980) The benzodiazepine receptor: Its development and its modulation by γ-aminobutyric acid. *J. Pharmacol. Exp. Ther.* **212,** 137–143.

Rice K. C., Brossi A., Tallman J., Paul S. M., and Skolnick P. (1979) Irazepine, a non-competitive, irreversible inhibitor of ^{3}H-diazepam binding to benzodiazepine receptors. *Nature* (Lond.) **278,** 854–855.

Richards J. G., Schlumpf M., Lichtensteiger W., and Möhler H. (1983) Ontogeny of benzodiazepine binding sites in fetal rat brain: An in vitro autoradiographic study. *Monogr. Neural. Sci.* **9,** 111–118.

Sarrazin M., Bourdeaux-Pontier M., and Briand C. (1976) Etude de quelques relations structure-activite de benzodiazepines. *Ann. Phys. Biol. Med.* **9,** 211–220.

Schlumpf M., Richards J. G., Lichtensteiger W., and Möhler H. (1983) An autoradiographic study of the pre-natal development of benzodiazepine-binding sites in rat brain. *J. Neurosci.* **3,** 1478–1487.

Schmidt R. F., Vogel E., and Zimmerman M. (1967) Die Wirkung von Diazepam auf die präsynaptische Hemmung und andere Rückenmarks reflexe. *Naunyn Schmiedeberg's Arch. Pharmacol.* **258,** 69–82.

Schoch P., and Möhler H. (1983) Purified benzodiazepine receptor retains modulation by GABA. *Eur. J. Pharmacol.* **95,** 323–324.

Schoch P., Richards J. G., Häring P., Takacs B., Stähli C., Staehelin T., Haefeley W. and Möhler H. (1985) Co-localisation of $GABA_A$ receptors and benzodiazepine receptors in the brain by monoclonal antibodies. *Nature* (Lond.) **314,** 168–171.

Schoemaker H., Boles R. G., Horst W. D., and Yamamura H. I. (1983) Specific high affinity binding sites for [^{3}H]-Ro 5-4864 in rat brain and kidney. *J. Pharmacol. Exp. Ther.* **225,** 61–69.

Sherman-Gold R. and Dudai Y. (1980) Solubilisation and properties of a benzodiazepine receptor from calf cortex. *Brain Res.* **198,** 485–490.

Sieghart W. and Drexler G. (1983) Irreversible binding of [^{3}H]flunitrazepam to different proteins in various brain regions. *J. Neurochem.* **41,** 47–55.

Sieghart W. and Karobath M. (1980) Molecular heterogeneity of benzodiazepine receptors. *Nature* (Lond.) **286,** 285–287.

Sieghart W. and Mayer A. (1982) Postnatal development of proteins irreversibly labelled with [^{3}H]-flunitrazepam. *Neurosci. Lett.* **31,** 71–74.

Sieghart W. and Möhler H. (1982) ^{3}H-Clonazepam, like ^{3}H-flunitrazepam, is a photoaffinity label for the central type of benzodiazepine receptors. *Eur. J. Pharmacol.* **81,** 171–173.

Sigel E. and Barnard E. A. (1984) A γ-aminobutyric acid/benzodiazepine receptor complex from bovine cerebral cortex. Improved purification with preservation of regulatory sites and their interactions. *J. Biol. Chem.* **259,** 7219–7223.

Sigel E., Mamalaki C., and Barnard E. A. (1982) Isolation of a GABA receptor from bovine brain using a benzodiazepine affinity column. *FEBS Lett.* **147,** 45–48.

Sigel E., Stephenson F. A., Mamalaki C., and Barnard E. A. (1983) A γ-aminobutyric acid/benzodiazepine receptor complex in bovine cerebral cortex. *J. Biol. Chem.* **258,** 6965–6971.

Sjödin T., Roosdorp N., and Sjöholm I. (1976) Studies on the binding of benzodiazepines to human serum albumin by circular dichroism measurements. *Biochem. Pharmacol.* **25,** 2131–2140.

Skerritt J. H., Willow M., and Johnston G. A. R. (1982) Diazepam enhancement of low affinity GABA binding to rat brain membranes. *Neurosci. Lett.* **29,** 63–66.

Skolnick P., Lock K-L., Paugh B., Marango F., Windsor R., and Paul S. (1980) Pharmacologic and behavioural effects of EMD 28422; a novel purine which enhances [^{3}H]-diazepam binding to brain benzodiazepine receptors. *Pharmacol. Biochem. Behav.* **12,** 685–689.

Skolnick P., Paul S., Crawley J., Lewin E., Lippa A., Clody D., Irmscher K., Saiko O., and Minck K. (1983) Antagonism of the anxiolytic action of diazepam and chlordiazepoxide by the novel imidazopyridines EMD 39593 and EMD 41717. *Eur. J. Pharmacol.* **88,** 319–327.

Skolnick P., Schweri M. M., Williams E. F., Moncada V. Y., and Paul S. M. (1982) An in vitro binding assay which differentiates benzodiazepine agonists and antagonists. *Eur. J. Pharmacol.* **78,** 133–136.

Skolnick P., Syapin P. J., Paugh B. A., and Paul S. M. (1979) Reduction in benzodiazepine receptors associated with Purkinje cell degeneration in nervous mutant mice. *Nature* (Lond.) **277,** 397–399.

Speth R. C., Wastek C. J., Johnson P. C., and Yamamura H. I. (1978) Benzodiazepine binding in human brain: demonstration using ^{3}H-flunitrazepam. *Life Sci.* **22,** 859–866.

Speth R. C., Wastek G. W., and Yamamura H. I. (1979) Benzodiazepine receptors: temperature dependence of ^{3}H-flunitrazepam binding. *Life Sci.* **24,** 351–358.

Speth R. C. and Yamamura H. I. (1979) Benzodiazepine receptors: alterations in mutant mouse cerebellum. *Eur. J. Pharmacol.* **54,** 397–399.

Squires R. F., Benson D. I., Braestrup C., Coupet J., Klepner C. A., Myers V., and Beer B. (1979) Some properties of brain specific benzodiazepine receptors: New evidence for multiple receptors. *Pharmacol. Biochem. Behav.* **10,** 825–830.

Squires R. F. and Braestrup C. (1977) Benzodiazepine receptors in rat brain. *Nature* (Lond.) **266,** 732–734.

Squires R. F., Casida J. E., Richardson N. M., and Saederup E. (1983) [^{35}S]-*t*-Butylbicyclophosphorothionate binds with high affinity to brain specific sites coupled to γ-aminobutyric acid-A and ion recognition sites. *Mol. Pharmacol.* **23,** 326–336.

Stephenson F. A. and Olsen R. W. (1982) Solubilisation by CHAPS detergent of barbiturate-enhanced benzodiazepine-GABA receptor complex. *J. Neurochem.* **39,** 1579–1586.

Sternbach L. H. (1978) The benzodiazepine story. *Prog. Drug Res.* **22,** 229–266.

Sternbach L. H., Randall L. O., Banziger R., and Lehr H. (1968) in Structure Activity Relationships in the 1,4-Benzodiazepine Series, *Drugs Affecting the Central Nervous System.* vol. 2, (Burger A., ed.), pp. 237–264. Marcel Dekker, New York.

Study R. E. and Barker J. L. (1981) Diazepam and (–)-pentobarbital: fluctuation analysis reveals different mechanisms for potentiation of gamma-aminobutyric acid responses in cultured central neurons. *Proc. Natl. Acad. Sci. USA* **78,** 7180–7184.

Sunjic V., Lisini A., Sega A., Kovac T., Kajfez F., and Ruscic B. (1979) Conformation of 7-chloro-5-phenyl-d_5-3(S)-methyl-dihydro-1,-4-benzodiazepin-2-one in solution. *J. Heterocyclic Chem.* **16,** 757–761.

Supavilai P. and Karobath M. (1979) Stimulation of benzodiazepine receptor binding by SQ 20009 is chloride-dependent and picrotoxin-sensitive. *Eur. J. Pharmacol.* **60,** 111–113.

Supavilai P. and Karobath M. (1980) Heterogeneity of benzodiazepine receptors in rat cerebellum and hippocampus. *Eur. J. Pharmacol.* **64,** 91–93.

Supavilai P. and Karobath M. (1981) Action of pyrazolpyridines as modulators of ^{3}H-flunitrazepam binding to the GABA/benzodiazepine receptor complex in the cerebellum. *Eur. J. Pharmacol.* **70,** 183–193.

Supavilai P. and Karobath M. (1984) [^{35}S]-t-butylbicyclophosphorothionate binding sites are constituents of the γ-aminobutyric acid benzodiazepine receptor complex. *J. Neurosci.* **4,** 1193–1200.

Syapin P. J. and Skolnick P. (1979) Characterisation of benzodiazepine binding sites in cultured cells of neural origin. *J. Neurochem.* **32,** 1047–1051.

Takeuchi A. and Takeuchi N. (1975) The structure-activity relationship for GABA and related compounds in the crayfish muscle. *Neuropharmacology* **14,** 627–634.

Tallman J. F., Thomas J. W., and Gallager D. W. (1978) GABAergic modulation of benzodiazepine binding site sensitivity. *Nature* (Lond.) **274,** 383–385.

Tallman J. F., Thomas J. W., and Gallager D. W. (1979) Identification of diazepam binding in intact animals. *Life Sci.* **24,** 873–880.

Tenen S. S., and Hirsch J. D. (1980) Antagonism of diazepam activity by β-carboline-3-carboxylic acid ethyl ester. *Nature* (Lond.) **288,** 609–610.

Thomas J. W. and Tallman J. F. (1981) Characterisation of photoaffinity labelling of benzodiazepine binding sites. *J. Biol. Chem.* **256,** 9838–9842.

Ticku N. K. and Olsen R. W. (1978) Interaction of barbiturates with dihydropicrotoxinin binding sites related to the GABA receptor-ionophore system. *Life Sci.* **22,** 1643–1652.

Toll L., Keys C., Spangler D., and Loew G. (1984) Computer-assisted determination of benzodiazepine receptor heterogeneity. *Eur. J. Pharmacol.* **99,** 203–209.

Unnerstall J. R., Niehoff D. L., Kuhar M. J., and Palacios J. M. (1982) Quantitative receptor autoradiography using ^{3}H-Ultrofilm: application to multiple benzodiazepine receptors. *J. Neurosci. Methods* **6,** 59–73.

Weiland G. A., Minneman D. P., and Molinoff P. B. (1979) Fundamental differences between the molecular interaction of agonists and antagonists with the β-adrenergic receptor. *Nature* (Lond.) **281,** 114–117.

White W. F., Dichter M. A., and Snodgrass S. R. (1981) Benzodiazepine binding and interactions with the GABA receptor complex in living cultures of rat cerebral cortex. *Brain Res.* **215,** 162–176.

Williams E. F., Rice K. C., Paul S. M., and Skolnick P. (1980) Heterogeneity of benzodiazepine receptors in the central nervous system demonstrated with kenazepine, an alkylating benzodiazepine. *J. Neurochem.* **35,** 591–597.

Williams M. and Risley E. A. (1979) Enhancement of binding to rat brain membranes in vitro by SQ 20009, a novel anxiolytic agent, gamma-aminobutyric acid and muscimol. *Life Sci.* **24,** 833–841.

Williams M. and Risley E. A. (1982) Interactions of avermectins with [^{3}H]β-carboline-3-carboxylate ethyl ester and ^{3}H-diazepam binding sites in rat brain cortical membranes. *Eur. J. Pharmacol.* **77,** 307–312.

Williams M., and Risley E. A. (1984) Ivermectin interactions with benzodiazepine receptors in rat cortex and cerebellum in vitro. *J. Neurochem.* **42,** 745–753.

Williams M. and Yarbrough G. G. (1979) Enhancement in in vitro binding and some of the pharmacological properties of diazepam by a novel anthelmintic agent avermectin B_{1a}. *Eur. J. Pharmacol.* **56,** 273–276.

Williamson M. J., Paul S. M., and Skolnick P. (1978) Labelling of benzodiazepine receptors in vivo. *Nature* (Lond.) **275,** 551–553.

Wong D. T., Rathbun R. C., Bymaster F. P., and Lacefield W. B. (1983) Enhanced binding of radioligands to receptors of γ-aminobutyric acid and benzodiazepines by a new anti-convulsant agent, LY81067. *Life Sci.* **33,** 917–923.

Woolf J. H. and Nixon J. C. (1981) Endogenous effector of the benzoidazepine binding site: Purification and characterisation. *Biochemistry* **20,** 4263–4269.

Yokoyama N., Ritter B., and Neubert A. D. (1982) α-Arylpyrazolo-[4,3-c]quinolin-3-ones: novel agonist, partial agonist and antagonist of benzodiazepines. *J. Med. Chem.* **25,** 337–339.

Young W. S. III and Kuhar M. J. (1979) Autoradiographic localisation of benzodiazepine receptors in the brains of humans and animals. *Nature* (Lond.) **280,** 393–395.

Young W. S. III and Kuhar M. J. (1980) Radiohistochemical localisation of benzodiazepine receptors in rat brain. *J. Pharmacol. Exp. Ther.* **212,** 337–346.

Yousufi M. A. K., Thomas J. W., and Tallman J. F. (1979) Solubilisation of benzodiazepine binding site from rat cortex. *Life Sci.* **25,** 463–470.

Chapter 13
Binding Sites for Antidepressants

Pavel D. Hrdina

1. Introduction

Although antidepressant drugs have been successfully used in the treatment of depression for more than two decades, the mechanism(s) by which these drugs produce their therapeutic effect is not understood. Tricyclic antidepressants have been shown to inhibit the neuronal uptake of norepinephrine (noradrenaline; NA) and serotonin, (5-hydroxytryptamine; 5-HT) (Glowinski and Axelrod, 1964) and their therapeutic effect has been related to this activity (Schildkraut, 1965). However, the effect of these drugs on the uptake of biogenic amines occurs immediately and does not correlate with the time course of the therapeutic response that usually sets in after only 1–3 wk of treatment. Moreover, there are other antidepressants, such as iprindole and mianserin, that do not block 5-HT and/or NA reuptake, but have been shown to lessen the symptoms of depression. Other drugs, such as cocaine, block the uptake of biogenic amines, but are not clinically effective antidepressants (Post et al., 1974). Some tricyclic antidepressants have also been shown to interact with the in vitro binding of specific radioactive ligands to several classes of neurotransmitter receptors in brain membrane preparations: α- and β-adrenergic (U'Prichard et al., 1978), serotonergic (Peroutka and Snyder, 1980), H_1-histaminergic (Tran et al., 1981), and muscarinic cholinergic (Snyder and Yamamura, 1977) receptors. It is, however, unclear whether the interaction with any of these receptors is directly related to the primary antidepressant effect or only to some of the side effects of these drugs. Recently, long-term, but not short-term, treatment with antidepressants was shown to reduce the number of β-adrenergic receptor sites (Banerjee et al., 1977) and the activity of an NA-stimulated adenylate cyclase (Vetulani et al., 1976) in brain tissue. It has been suggested that the down-regulation of β-adrenergic receptors is caused by increased concentration of NA in the synaptic gap following inhibition of NA uptake by tricyclic antidepressant drugs. However, antidepressants (e.g., iprindole and trazodone) that do not appreciably block NA uptake have also been reported to reduce β-adrenoceptor density after repeated administration (Clements-Jewery, 1978).

The recent discovery of specific high-affinity binding sites for ^{3}H-imipramine (IMI) in rat brain (Raisman et al., 1979a,b), as well as in human brain and platelets (Briley et al., 1979; Rehavi et al., 1980), and for ^{3}H-desipramine (desmethylimipramine; DMI) in brain (Rehavi et al., 1981; Hrdina, 1981) has offered a new approach to investigation of site and mode of action of antidepressants, and possibly also a new biological marker of depression. The ^{3}H-IMI sites in brain are located on serotonergic nerve terminals (Sette et al., 1981; Gross et al., 1981; Hrdina et al., 1982) and are associated with the neuronal reuptake of serotonin (Langer et al., 1980a; Hrdina, 1982), but not identical to the substrate recognition site for serotonin uptake (Sette et al., 1983a). In contrast, binding sites for ^{3}H-DMI appear to be associated with the neuronal uptake of NA (Lee and Snyder, 1981; Rehavi et al., 1982), are localized, at least in part, on noradrenergic neurons (Hrdina et al., 1981), and can be regulated by the synaptic concentration of the transmitter NA (Lee et al., 1983). The high-affinity ^{3}H-IMI binding in platelets, which has similar characteristics to that in the brain tissue, has recently been found to be altered in depressed patients (Briley et al., 1980; Suranyi-Cadotte et al., 1982). Both ^{3}H-IMI and ^{3}H-DMI binding sites fulfill most criteria for a recognition site of a pharmacological receptor.

The aim of this chapter is to examine the experimental findings on specific binding of ^{3}H-IMI and ^{3}H-DMI, particularly from the point of view of their possible physiological significance in regulation of biogenic amine uptake, or as potential biological markers of depression.

2. Characterization of ^{3}H-Imipramine Binding in Brain

The specific high-affinity binding sites for ^{3}H-IMI were first described and characterized by Raisman et al. (1979a,b, 1980) in the rat brain, and then reported by several groups of investigators in the brain of other species, including humans (Rehavi et al., 1981; Gross et al., 1981; Langer et al., 1981a; Briley et al., 1982; Hrdina, 1982; Plenge and Mellerup, 1982; Reith et al., 1983).

2.1. Tissue Preparation and the Binding Assay

The binding assay used in most reports is based on the original procedure employed by Raisman et al. (1979a) for cerebral cortex. The brain region to be studied is homogenized in 25 or 50 vol of ice cold buffer (50 mM Tris/HCl, 100 m*M* NaCl, 5 m*M* KCl, pH 7.4) using a Brinkman Polytron (setting 6, 15 s) or other type of homogenizer. The homogenate is then centrifuged at 30,000*g* for 10 min and the pellet washed twice. Finally, the pellet is resuspended in the original volume or in 2/3 of original volume. At this point, an aliquot is taken for protein determination by the method of Lowry et al. (1951). It appears that the initial volume or the method of homogenization have little, if any, effect on the degree of specific binding of ^{3}H-IMI as long as the final protein concentration of the membrane suspension is kept within a range of 1.4–3.4 mg/mL.

For the determination of specific ^{3}H-IMI binding, aliquots (180 μL) of membrane suspension are incubated in a total volume of 250 μL, with varying concentrations of ^{3}H-IMI with or without an excess of unlabeled displacing drug. Desipramine (10 or 100 μ*M*) or IMI (0.4 μ*M*) have been used in most studies as displacers to define the nonspecific binding. After incubation for 60 min at 0°C, a 100-μL aliquot of suspension from each tube is diluted into 5 mL of ice-cold buffer and rapidly filtered under reduced pressure through Whatman GF/F glass fiber filters that are then washed three times with 5 mL of the same ice-cold buffer. Filters are placed into empty scintillation vials and dried (e.g., in an oven at 110°C). A scintillation mixture is then added to the vials and the radioactivity is measured in a liquid scintillation counter. Although in most reports on ^{3}H-IMI binding GF/F type filters have been used to separate the bound and free ligand, some investigators have also employed GF/B and GF/C types of glass fiber filters (Kinnier et al., 1981; Gross et al., 1981; Plenge and Mellerup, 1982). It is of interest that in these experiments the estimates of B_{max} for ^{3}H-IMI binding was usually substantially higher and the ratio of specific/total binding lower than in those employing GF/F filters.

The specific binding of ^{3}H-IMI is calculated as the difference between the total (without displacer) and nonspecific (with displacer) binding, and is expressed in fmol/mg protein. Examples of binding curves for total, nonspecific, and specific ^{3}H-IMI binding in rat cerebral cortex are given in Fig. 1. The specific binding of ^{3}H-IMI at nanomolar concentrations of the ligand usually represents 60–80% of the total binding.

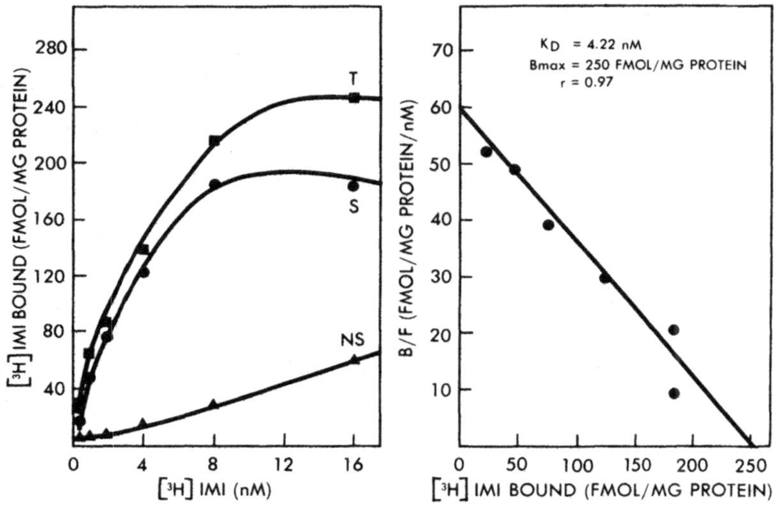

Fig. 1. Saturation curve (left) and Scatchard plot (right) of specific ^{3}H-imipramine binding to rat cerebrocortical membranes. Incubations were carried out at 0°C for 60 min at ^{3}H-imipramine concentrations of 0.5–16.0 n*M* (with or without 100 μ*M* desipramine), as described in section 2.1. Total binding (T), specific binding (S), and nonspecific binding (NS) are plotted as a function of ^{3}H-imipramine concentration. Points are means of triplicate determination of both total and nonspecific binding from a single typical experiment. The line for Scatchard analysis was fitted by linear regression using the method of least squares. Reproduced from Hrdina (1984b) by permission of Elsevier Biomedical Press, Amsterdam.

2.2. Saturation and Kinetics of ^{3}H-Imipramine Binding

When binding of ^{3}H-IMI to brain membrane preparations is measured in a low-concentration range (0.5–16 n*M*), the specific binding appears to be saturable, whereas the nonspecific binding increases linearly (Raisman et al., 1980; Rehavi et al., 1980; Hrdina, 1984b) (Fig. 1). Scatchard analysis of the saturation isotherm illustrated in Fig. 1 resulted in a linear plot (r = 0.97) that yielded an equilibrium dissociation constant (K_d) of 4.2 n*M* and a density of binding sites (B_{max}) of 250 fmol/mg protein. Under these experimental conditions, the Scatchard analysis of data suggests a single population of non-interacting binding sites. The Hill coefficient high-affinity ^{3}H-IMI binding in rat cerebral cortex or hippocampus is close or equal to unity (Raisman et al., 1980; Kinnier et al., 1981), suggesting lack of cooperation. The parameters of the high-affinity ^{3}H-IMI binding in various brain regions of different

TABLE 1

High-Affinity ^{3}H-Imipramine Binding in Brain

Region	Species	^{3}H-Imipramine K_d, nM	^{3}H-Imipramine B_{max}, fmol/mg protein	References
Cortex	Human	7.6 ± 1.9[a]	301 ± 52	Langer et al. (1981a)
	Rat	4.0 ± 0.5	249 ± 23	Raisman et al. (1980)
		7.5 ± 0.5	328 ± 14	Dumbrille-Ross et al. (1981)
		6.6 ± 1.7	490 ± 60	Kinnier et al. (1980)
		4.1 ± 0.2	374 ± 10	Hrdina (1982)
		5.5 ± 0.6	810 ± 34	Plenge and Melleup (1982)
	Mouse	14	830	Reith et al. (1983)
Hypothalamus	Human	6.7 ± 1.0	500 ± 57	Reith et al. (1983)
		1.7 ± 0.1	16.2 ± 2.9[b]	Rehavi et al. (1980)
	Cat	4.0 ± 0.9	400 ± 30	Briley et al. (1982)
	Rat	5.2 ± 3.4	279 ± 47	Raisman et al. (1980)
		4.1 ± 0.8	317 ± 50	Palkovits et al. (1981)
Hippocampus	Human	4.2 ± 1.2	263 ± 50	Langer et al. (1981a)
	Rat	3.3 ± 0.2	143 ± 26	Palkovits et al. (1981)
		8.3 ± 1.5	800 ± 60	Kinnier et al. (1980)
		4.1 ± 0.7	360 ± 43	Brunello et al. (1982b)
		4.5 ± 0.6	353 ± 43	Hrdina et al. (1982)
Caudate n.	Human	4.6 ± 1.5	219 ± 13	Langer et al. (1981a)
Striatum	Rat	2.8 ± 0.5	164 ± 11	Raisman et al. (1980)
Midbrain	Rat	2.4 ± 0.7	156 ± 35	Palkovits et al. (1981)
Pons medulla	Rat	7.8 ± 0.6	600 ± 34	Hrdina et al. (1982)
Spinal cord	Rat	4.0 ± 1.0	199 ± 63	Palkovits et al. (1980)
Cerebellum	Rat	8.0 ± 3.1	60 ± 11	Raisman et al. (1980)

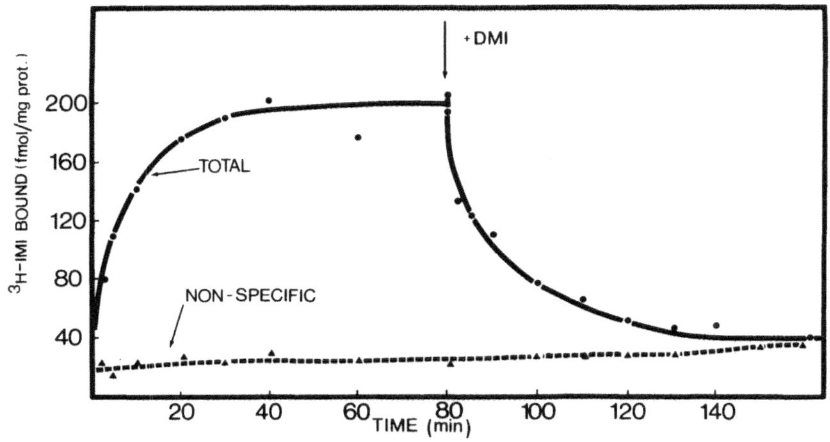

Fig. 2. Time course of total (●) and nonspecific (▲) ^{3}H-imipramine binding. Membrane suspension from rat cerebral cortex was prepared and incubated with 4.0 nM ^{3}H-imipramine (as described in section 2.1) and binding was measured at various time intervals. At equilibrium (80 min), unlabeled desipramine (DMI) was added in excess (100 μM) to displace the specific ^{3}H-imipramine binding. The values are the means of triplicate determination. ^{3}H-imipramine binding reaches equilibrium rapidly with an association $t_{1/2}$ of about 5 min. The rate constant for association (K_1) was calculated to be $1.27 \times 10^7 M^{-1}$/min. Dissociation of ^{3}H-imipramine from its binding sites after addition of DMI appeared to be a first-order process with a dissociation rate constant (K_2) of 0.0404/min and $t_{1/2}$ of 22 min. The equilibrium dissociation constant (K_d) calculated from this experiment (K_2/K_1 ratio) was 3.23 nM, which is close to the value (4.2 nM) obtained by Scatchard analysis of saturation isotherm illustrated in Fig. 1.

species as reported by various investigators are summarized in Table 1.

The binding of ^{3}H-IMI to brain membrane preparation is rapid even at low temperatures (0–4°C). Equilibrium is reached at 30–60 min of incubation with a $t_{½}$ of about 5 min, and addition of an excess of displacer (100 μM desipramine) causes a rapid decrease in binding (Fig. 2). The analysis of the time course of binding shown in Fig. 2 yielded an association rate constant (K_1) of 1.27×10^7 M^{-1}/min. The dissociation rate constant (K_2) calculated from the plot of log specific binding vs time was 0.0404/min. The equilibrium dissociation constant (K_d) derived from this kinetic analysis is 3.19 nM, a value close to that (4.2 nM) obtained from Scatchard analysis of the binding isotherm.

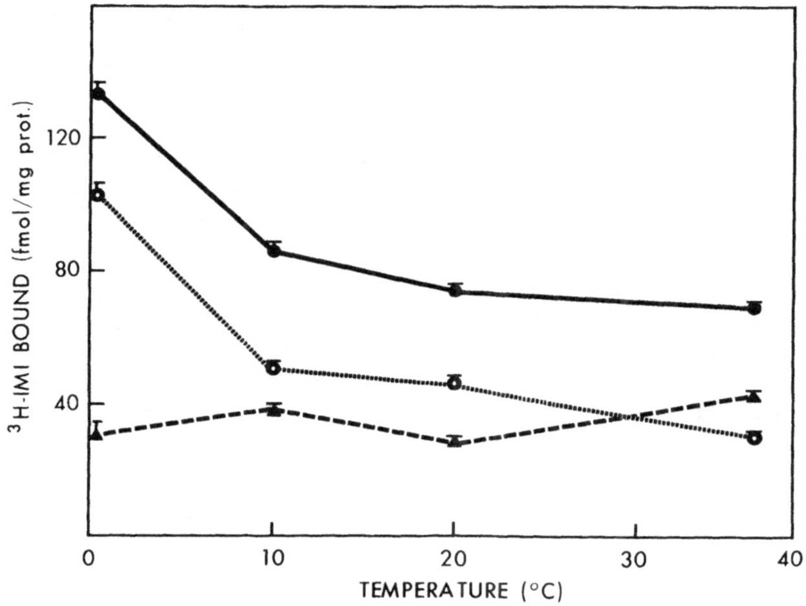

Fig. 3. Temperature dependence of ^{3}H-imipramine binding (4.0 n*M*) to membrane suspension from rat cerebral cortex. Total (●) and nonspecific (▲) binding were measured, and specific binding (○) calculated as described in section 2.1. Values are means ± SEM from four separate experiments carried out in triplicate. Note that the specific (and total) binding of ^{3}H-imipramine is highest at 0°C. The nonspecific binding did not change remarkably except at 37°C when it became higher than the specific binding.

2.3. Temperature and Ion Dependence

Binding of ^{3}H-IMI to rat cerebrocortical membranes is temperature-dependent and highest at 0°C (Fig. 3). Kinnier et al. (1981) found that the number of specific binding sites for ^{3}H-imipramine in membrane preparation from rat hippocampus is rapidly and irreversibly reduced by increasing the incubation temperature from 0 to 37°C. On the other hand, Rehavi et al. (1980) observed no significant changes in maximum number of binding sites (B_{max}) for ^{3}H-IMI in membranes from human hypothalamus incubated at either 0–4°C or at physiological (37°C) temperature. They found, however, a small but significant increase in the apparent K_d (from 1.7 to 4.0 n*M*) when incubation was carried out at 37°C. It appears that an incubation temperature of 0–4°C is optimal for measurement of ^{3}H-IMI binding, since the dissociation of ^{3}H-antidepressants from their binding sites is much faster at higher temperatures (Lee et al., 1982), and the time limitation in separating the free from bound ligand may not allow an appropriate estimate of binding at an equilibrium state.

^{3}H-IMI binding to membranes from rat brain cortex and hippocampus is sensitive to sodium ions. In the absence of sodium, the dissociation constant (K_d) of binding to cortical membranes was found to be four times greater (Briley and Langer, 1981) and the specific binding in hippocampus about 50% lower (Kinnier et al., 1981) than in the presence of 120 m*M* NaCl. The binding sites for ^{3}H-IMI in brain membranes are also sensitive to treatment with hydrolytic enzymes, such as phospholipase A_2 and pronase (Kinnier et al., 1981), which considerably inhibit the specific binding of the ligand.

2.4. Subcellular Distribution of Binding

Initial work on subcellular localization of ^{3}H-IMI binding sites suggested that these sites are located in euronal terminals since synaptosomal membranes were found to have the greatest amount (267% enrichment) of specific binding (Kinnier et al., 1981). However, Laduron et al. (1982) reported that the subcellular distribution profile of ^{3}H-IMI binding differs markedly from that of 5-HT uptake since it shows highest enrichment in nuclear fraction. They concluded that IMI binding sites are not likely to be localized on nerve terminals. More recently, Sette et al. (1983b) investigated the high affinity ^{3}H-IMI binding in subcellular fractions of the rat cerebral cortex, before and after chemical denervation of serotonergic neurons. In contrast to Laduron's report, they found the highest proportion of ^{3}H-IMI binding sites (45%) in the heavy mitochondrial fraction, but also found surprisingly high proportion (36%) in the microsomal fraction. Only 3% of binding sites were found in the nuclear fraction. After chemical denervation with 5,7-dihydroxytryptamine (5,7-DHT), the density of ^{3}H-IMI binding sites in the microsomal fraction was decreased by 80%, but in the mitochondrial fraction by only about 50%. The presence of residual ^{3}H-IMI binding after chemical denervation would indicate that a certain proportion of ^{3}H-IMI binding sites may be located either on nonserotonergic neurons or postsynaptic membranes; neither of these would be destroyed by chemical damage to presynaptic serotonergic terminals. The significance of ^{3}H-IMI sites in the microsomal fraction remains to be clarified.

Rehavi et al. (1983) pointed out that the subcellular distribution profiles of ^{3}H-IMI binding and ^{3}H-serotonin uptake in rat forebrain are very similar when the specific binding is defined by a low concentration (500 n*M*) of displacer, which blocks the high-affinity, but not the low-affinity, binding of ^{3}H-IMI. They found saturable, high-affinity binding sites for ^{3}H-IMI only in the synaptosomal fraction; in contrast, the specific binding of ^{3}H-IMI in the nuclear and the myelin fraction of the homogenate were nonsaturable, and Scatchard analysis of the binding failed to yield a determinable slope. Results of this study emphasize the importance of distinguishing the high-affinity ^{3}H-IMI binding from the low-affinity sites that might be located on structures other than synaptosomes and may have a different function (Hrdina, 1984a; Reith et al., 1983).

2.5. Regional Distribution and Ontogenesis of ^{3}H-Imipramine Binding in Brain

Rehavi et al. (1980) were first to demonstrate the presence of specific high-affinity binding sites for ^{3}H-IMI in human brain samples obtained at autopsy. Using a binding assay similar to that for rat brain tissue, they obtained an apparent equilibrium dissociation constant (K_d) of 1.7 n*M* that was the same in the hypothalamus and the cerebral and cerebellar cortex. In contrast, marked regional differences were found in the maximum number of binding sites (B_{max}), the hypothalamus being the richest region (Langer et al., 1981a) (Table 1). Kinetic rate constants of the ^{3}H-IMI binding in human hypothalamus were similar to those for cerebral cortex of rat except that the dissociation half-life (28 min) was much longer (Rehavi et al., 1980).

In the rat brain, the highest density of ^{3}H-IMI binding sites is also found in the hypothalamus, and the lowest in the cerebellum. The density of imipramine sites appears to be highly correlated with the regional distribution of endogenous serotonin (Palkovits et al., 1981). A study of ontogenic development of ^{3}H-IMI binding sites in the rat brain has revealed the presence of high-affinity sites at the 5th day after birth, whereas the high-affinity component of serotonin uptake was only detected from the 9th day on (Mocchetti et al., 1982). This different time course in development of ^{3}H-IMI binding sites and serotonin uptake sites would indicate that the two sites are probably not identical, but can be functionally related.

3. Pharmacological Profile and Physiological Significance of ^{3}H-Imipramine Binding Sites

3.1. Pharmacological Characterization of ^{3}H-Imipramine Binding

The binding of ^{3}H-IMI to membranes from human and rat brain shows many characteristics of a specific drug-recognition site: high-affinity, saturability, reversibility, and uneven distribution. The question of whether the high-affinity IMI binding site has any physiological or pharmacological significance was first investigated by determining the displacement profile of the binding using a wide variety of pharmacological agents and biogenic compounds. The initial studies (Raisman et al., 1979a,b) have revealed that the ^{3}H-IMI binding sites are distinct from the recognition sites of adrenergic, serotonergic, dopaminergic, cholinergic, or peptidergic receptors and unrelated to any known neurotransmitter receptors. As expected, the binding was selectively displaced by low concentrations of chemically related tricyclic antidepressants, but not of other classes of psychotropic drugs. An interesting possibility arose that the high-affinity ^{3}H-IMI binding site may represent a common antidepressant recognition site, activation of which would trigger a chain of events resulting in the therapeutic effect of antidepressants. However, the low potency of non-tricyclic, but clinically effective, antidepressants (e.g., iprindole, mianserin, maprotiline) in displacing the ^{3}H-IMI binding (Raisman et al., 1979b; Langer et al., 1980a; Hrdina, 1982, 1984b) has made this possibility unlikely. Displacement curves of various tri- and tetracyclic antidepressants and other related compounds, and their IC_{50} values for ^{3}H-IMI binding in rat cerebral cortex, are shown in Figs. 4A, B and Table 2. Among tricyclics, the most effective was clomipramine, followed by IMI itself and nortriptyline. A complete displacement of binding was virtually achieved over two orders of magnitude and the Hill coefficients were close to 1.0 (Hrdina, 1984b). In contrast, tetracyclic antidepressants maprotiline and mianserin had a very weak effect and a low Hill coefficient (0.61). Fluoxetine, a nontricyclic serotonin uptake blocker displaced ^{3}H-IMI in a complex manner (Sette et al., 1983a; Hrdina, 1984b). Among other drugs tested, the antihistamine pyribenzamine, which is a potent serotonin-uptake blocker, also effectively displaced ^{3}H-IMI binding. The displacement studies have revealed a highly significant correlation between the potency of various compounds tested to displace the high-affinity ^{3}H-IMI binding and

TABLE 2
Inhibition of High-Affinity ^{3}H-Imipramine and ^{3}H-Desipramine Binding in Membranes from Rat Cerebral Cortex

Drug	IC$_{50}$, n*M*[a]	
	^{3}H-Imipramine	^{3}H-Desipramine
Antidepressants		
Clomipramine	6.8	1,000
Imipramine	18.0	590
Amitriptyline	26.0	800
Nortriptyline	145	10
Desipramine	250	10
Doxepin	280	50
Maprotiline	32,000	800
Mianserin	>40,000	2,200
Neurotransmitters		
Serotonin	1,700	>10,000
Noradrenaline	>10,000	>10,000
Dopamine	>10,000	>10,000
Other drugs		
Fluoxetine	50.0	3,000
Pyribenzamine	52.0	700
Chlorpromazine	500	420
Nisoxetine	500	20
Diphenhydramine	7,000	—
Pargyline	>10,000	>10,000
Phentolamine	>10,000	>10,000
Atropine	>10,000	>10,000

[a]The IC$_{50}$ values are concentrations of drugs required to inhibit the specific ^{3}H-imipramine or ^{3}H-desipramine binding (at 4.0 n*M* concentration, with 100 μ*M* desipramine as displacer) by 50%.

to inhibit the neuronal uptake of serotonin (Fig. 5), but not of NA (Langer et al., 1980a; Rehavi et al., 1980; Hrdina, 1982; Reith et al., 1983). Langer et al. (1980b), using pairs of stereoisomers of 10-hydroxyamitriptyline, zimelidine, and of their demethylated derivatives, were able to demonstrate that the binding of ^{3}H-IMI in rat cerebral cortex is also stereospecific. The Z and E isomers of 10-hydroxyamitriptyline and 10-hydroxynortriptyline showed 7–8-fold differences, and those of zimelidine and norzimelidine ap

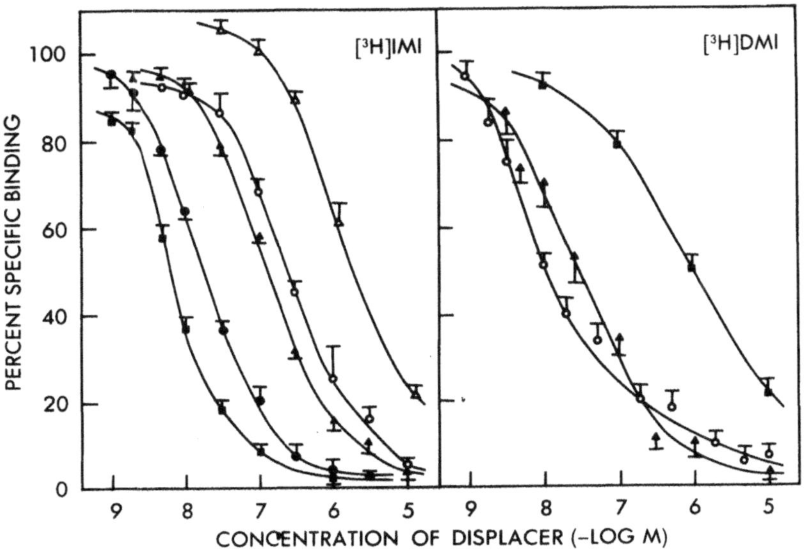

Fig. 4. Displacement of specific ^{3}H-imipramine and ^{3}H-desipramine binding in rat cerebral cortex by various drugs. Specific binding was determined as the difference between the binding at 4.0 n*M* concentration of the labeled ligand in the absence and presence of 100 μ*M* desipramine. Results are expressed as percent of specific binding remaining after adding different concentrations of displacing drugs. Values are ±SEM of three separate experiments done in triplicate for each concentration. The binding assays were performed as described in sections 2.1 and 5.1. (A) Displacement of specific ^{3}H-imipramine (left) and ^{3}H-

(*continued on next page*)

proximately 50–70-fold differences in their affinity for the high-affinity ^{3}H-IMI binding site, the Z isomer being the more active in both instances.

3.2. Localization of ^{3}H-Imipramine Binding Sites

The results of displacement experiments have suggested that the high affinity ^{3}H-IMI binding site may be closely associated with the serotonin-uptake mechanism at presynaptic serotonergic nerve terminals. This assumption is further supported by the fact that the density of ^{3}H-IMI binding sites in human (Langer et al., 1981a) and rat (Palkovits et al., 1981) brain regions tends to parallel the endogenous levels of serotonin. More direct evidence for localization of IMI binding sites on serotonergic nerve terminals has been provided by lesion studies. Sette et al. (1981) and Dumbrille-Ross et al. (1981) demonstrated that electrolytic lesions of dorsal raphe, which resulted in a significant decrease in the

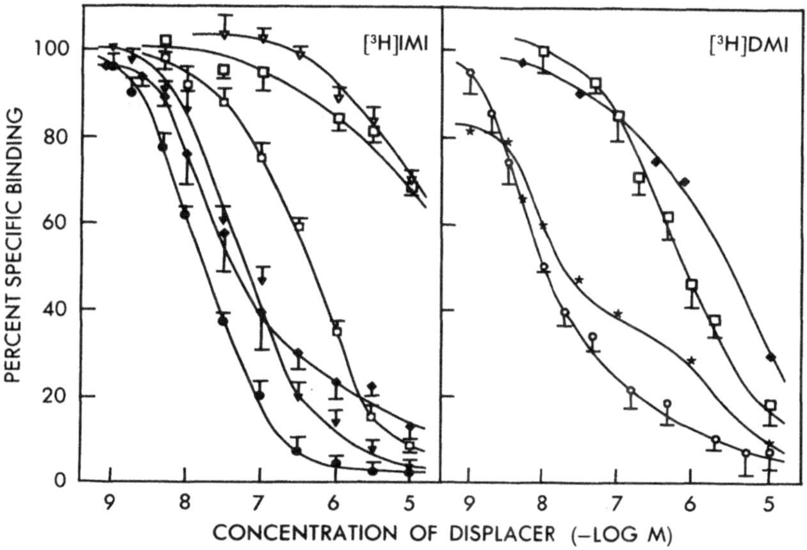

(*continued*)

desipramine (right) binding by tricyclic antidepressants: clomipramine (■), imipramine (●), nortriptyline (▲), desipramine (○), and serotonin (△). (B) Displacement of specific ^{3}H-imipramine (left) and ^{3}H-desipramine (right) binding by a serotonin-uptake inhibitor, fluoxetine (◆), a noradrenaline-uptake inhibitor, nisoxetine (*), an antihistamine, pyribenzamine (▼), a tricyclic analog, chlorpromazine (■), and tetracyclic antidepressants, maprotiline (□) and mianserin (▽), compared to displacement curves with imipramine (●) and desipramine (○), respectively (from Hrdina, 1982; Hrdina, 1984b).

endogenous levels of serotonin, produced a reduction in the density (B_{max}) of ^{3}H-IMI binding sites in rat hypothalamus and cortex without a change in the affinity constant (K_d). High-affinity ^{3}H-IMI binding sites were also shown to be reduced in cortex and hypothalamus of rats after extensive destruction of serotonergic neurons by intracerebral injection of 5,7-DHT (Gross et al., 1981; Brunello et al., 1982a) and in rat hippocampus after surgical lesioning of specific afferent pathways including serotonergic inputs (Brunello et al., 1982b). In contrast, the specific high-affinity binding of ^{3}H-IMI to cortical membranes from rats with selective destruction of noradrenergic neurons by intracerebral injection of 6-hydroxydopamine remained virtually unaffected (Hrdina et al., 1981; Lee and Snyder, 1981), as illustrated in Fig. 6. Results from our laboratory (Hrdina et al., 1982) support the notion that ^{3}H-IMI recognition sites in the rat brain are indeed related to the functioning of serotonergic neurons. It is known that neonatal 5,7-DHT

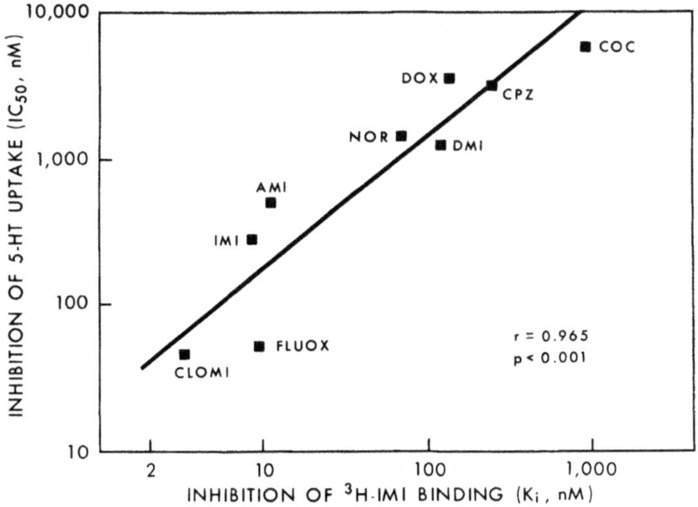

Fig. 5. Correlation between the inhibition of specific ^{3}H-imipramine binding and neuronal uptake of serotonin by various drugs. CLOMI; clomipramine: FLUOX; fluoxetine: IMI; imipramine: AMI; amitriptyline: DMI; desipramine: DOX; doxepin: CPZ; chlorpromazine: COC; cocaine. ^{3}H-Imipramine binding was determined at 4.0 n*M* concentration using 100 μ*M* desipramine to define the nonspecific binding, as described in section 2.1. Data for serotonin-uptake inhibition are from Horn (1976), Koe (1975), and Wong et al. (1975) (from Hrdina, 1982, with permission of Pergamon Press, Oxford).

treatment produces marked suppression of serotonin uptake and content in some brain regions, but a significant increase (after an initial decrease) in others (Sachs and Jonsson, 1975). We found that in the hippocampus of rats treated neonatally with 5,7-DHT, the depletion of serotonin was paralleled by a significant decrease in ^{3}H-IMI binding sites (B_{max}) (Fig. 7). In contrast, both serotonin content and ^{3}H-IMI binding were significantly enhanced by about the same degree in the pons-medulla region. These results suggest that an up-regulation of ^{3}H-IMI binding sites could be associated with increased serotonergic function.

3.3. High- and Low-Affinity ^{3}H-Imipramine Binding

Determination of specific ^{3}H-IMI binding to brain membranes over a wide concentration range (up to 900 n*M*) has revealed, in addition to a well-documented high-affinity binding, the presence of a low-affinity component in ^{3}H-IMI binding (Hrdina, 1983, 1984b; Reith et al., 1983). The Scatchard plot of binding data shows

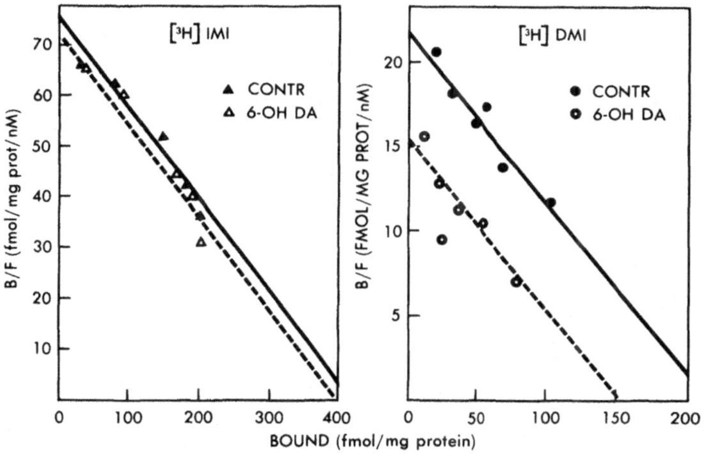

Fig. 6. High-affinity binding of ^{3}H-imipramine (^{3}H-IMI) and ^{3}H-desipramine (^{3}H-DMI) in rat cerebral cortex after selective lesion of noradrenergic neurons by intracerebral injection of 6-hydroxydopamine (6-OHDA). Saturation isotherms of ^{3}H-imipramine and ^{3}H-desipramine binding were carried out as described in sections 2.1 and 5.1 using 100 μM desipramine as displacer. Representative Scatchard plots indicate that 6-OHDA lesion decreased the B_{max} of ^{3}H-desipramine, but not of ^{3}H-imipramine binding (adapted from Hrdina et al., 1981).

an upward convex curve, indicating that under these experimental conditions ^{3}H-IMI binds to more than one population of binding sites. The specific binding could best be resolved into two distinct components: high-affinity binding with a K_d of 6.9 nM and a B_{max} of 292 fmol/mg protein, and low-affinity binding with a mean K_d value (292 ± 46 nM) about 50 times greater than for the high-affinity sites and a density of binding sites approximately ten times higher (B_{max} = 2459 ± 428 fmol/mg protein) (Fig. 8). These results and similar findings of Reith et al. (1983) in mouse cerebral cortex indicate that the parameters of high-affinity ^{3}H-IMI binding determined at free ^{3}H-IMI concentrations of up to approximately 20 nM (as in most studies) are likely to be influenced by the proportion of the low-affinity sites present. From the binding parameters of high- and low-affinity binding, one can calculate that at 4 nM concentration of free ^{3}H-IMI, about 25% of specific binding is to low-affinity sites when 100 μM DMI is used as a displacer (Hrdina, 1984b).

The low-affinity ^{3}H-IMI sites are pharmacologically distinct from the high-affinity sites. Whereas the order of potencies of various drugs tested in displacing high-affinity ^{3}H-IMI binding is correlated with the effect of these drugs on serotonin uptake, all

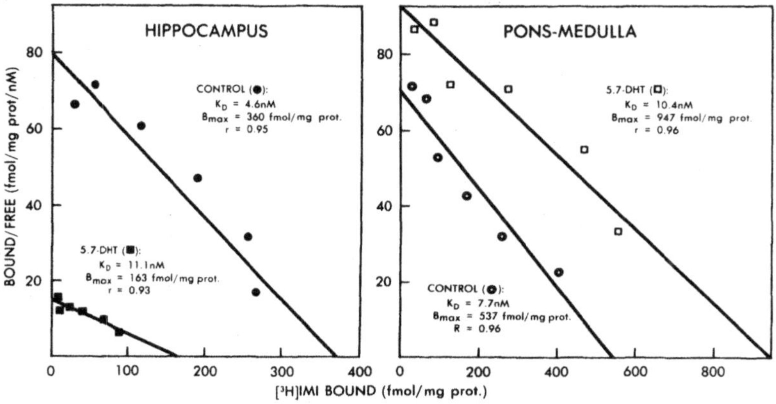

Fig. 7. High-affinity ^{3}H-imipramine binding in rat hippocampus and pons-medulla region after selective lesions of serotonergic neurons by neonatal treatment with 5,7-dihydroxytryptamine (5,7-DHT). Saturation isotherm of ^{3}H-imipramine binding was carried out as described in section 5.1. Representative Scatchard plots illustrate that neonatal treatment with 5,7-DHT markedly suppressed ^{3}H-imipramine binding in the hippocampus, but increased the density of binding sites in the pons–medulla region (adapted from Hrdina et al. 1982).

antidepressants tested were virtually equally effective in inhibiting the low-affinity ^{3}H-IMI binding (Hrdina, 1984b). Moreover, there seems to be no relationship between the low-affinity ^{3}H-IMI sites and the serotonin-uptake mechanism (Reith et al., 1983). The nature and localization of low-affinity binding for imipramine remain to be clarified. It could be related to some postsynaptic receptors, extraneuronal sites such as the glial cells, or unidentified sites in the brain that could play a role in the therapeutic effect of antidepressants. Two classes of ^{3}H-IMI binding sites have also been found in human brain (frontal cortex), with parameters of low-affinity binding (K_d = 300 nM; B_{max} = 3000 fmol/mg protein) very similar to those observed in rat cortex (Cash et al., 1984). One should be reminded that the clinically effective therapeutic range of antidepressants in plasma is far in excess of concentrations needed to saturate the high-affinity ^{3}H-IMI sites, but close to the concentrations of these drugs shown to effectively displace the low-affinity ^{3}H-IMI binding. Specific binding of ^{3}H-DMI (Hrdina, 1981) and ^{3}H-doxepin (Tran et al., 1981) in brain have also been shown to have a low-affinity component with an affinity (K_d) in low micromolar range. In view of a recent report on possibly clinically relevant benzodiazepine receptors with micromolar affinity in

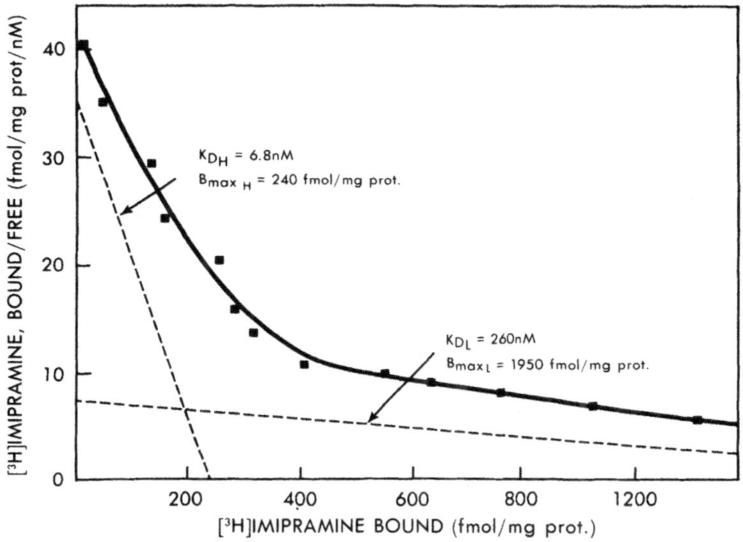

Fig. 8. Scatchard plot of specific ^{3}H-imipramine binding to rat cerebrocortical membranes. Specific ^{3}H-imipramine binding was determined at concentrations of 0.25–200 n*M* as described in section 2.1. Each point represents a mean of triplicate determination. The data were resolved into two components of binding (dotted lines) using graphical analysis procedure according to Rosenthal (1967). The experiment was replicated four times (from Hrdina, 1984b, by permission of Elsevier Biomedical Press, Amsterdam).

brain (Bowling and DeLorenzo, 1982), further study of low-affinity ^{3}H-antidepressant binding sites seems to be warranted.

3.4. High-Affinity ^{3}H-Imipramine Sites and Serotonin Uptake Regulation

Experimental evidence reviewed in section 3.2 indicates that a major proportion of the high-affinity binding sites for ^{3}H-IMI is localized at presynaptic serotonergic terminals. Their possible physiological significance, in particular their relationship to serotonin uptake sites at the same terminals and to the level of serotonergic transmission, has been a matter of intensive investigation. Table 3 summarizes the effect of various experimental treatments and conditions on the simultaneously measured ^{3}H-IMI binding (only B_{max} values are given since K_d in most instances does not change significantly) and serotonin uptake. On reviewing these data, it becomes obvious that the only condition when paral-

TABLE 3
Effect of Various Treatments on ^{3}H-Imipramine Binding and Serotonin Uptake in Rat Brain

Treatment, condition	^{3}H-Imipramine binding, B_{max}	Serotonin uptake	Reference
Drug effect "in vitro"	Similar rank order of potency		Brunello et al. (1982a); Gross et al. (1981); Hrdina (1982); Hrdina et al. (1981), (1982); Langer et al. (1980a); Reith et al. (1983); Rehavi et al. (1980); Sette et al. (1981).
5,7-DHT lesion	↓	↓	Brunello et al. (1982a); Gross et al. (1981); Hrdina et al. (1982); Sette et al. (1981).
Neonatal 5,7-DHT (hippocampus)	↓	↓	Hrdina et al. (1982).
Neonatal 5,7-DHT (pons-medulla)	↑	↑	Hrdina et al. (1982).
6-OH DA lesion	0	0	Hrdina et al. (1981); Gross et al. (1981).
Raphe lesion	↓	↓	Dumbrille-Ross et al. (1981).
Reserpine (acute)	0	↓	Hrdina et al. (1985b)
(chronic)	0	↓	Lee et al. (1983).
Imipramine, Desipramine (chronic)	↓	↑	Barbaccia et al. (1983a).
Mianserin (chronic)	0	0	Barbaccia et al. (1983a); Hrdina (1984a).
Deprenyl (chronic)	↑		Zsilla et al. (1983).
Haloperidol (chronic)	↓	0	Hrdina (1984a).
Lithium (chronic)	↓	0	Plenge and Mellerup (1982).
Fawn-hooded rats	↓	↑	Arora et al. (1983).

lel changes occur in both functions is a physical alteration of presynaptic terminals—destruction by chemical denervation, mechanical lesion to cell bodies, or a compensatory sprouting of terminals after initial chemical lesion (Sachs and Jonsson, 1975). These results, along with the pharmacological evidence (similar order in potency of various drugs to influence the two functions), have led to the suggestion (Langer et al., 1980a) that ^{3}H-IMI binding sites in brain are closely associated with serotonin-uptake sites at serotonergic presynaptic nerve terminals.

Later experiments have, however, shown that in some conditions changes in ^{3}H-IMI binding do not parallel alterations in serotonin uptake and vice versa. For instance, after treatment with reserpine, there is a profound suppression of serotonin uptake, but no changes are found in ^{3}H-IMI binding (Lee et al., 1983; Hrdina et al., 1985b). On the other hand, chronic administration of lithium (Plenge and Mellerup, 1982) or haloperidol (Hrdina, 1984a) down-regulates ^{3}H-IMI binding sites, but neither of these drugs is known to significantly inhibit serotonin uptake.

Important evidence in support of a functional association between ^{3}H-IMI binding sites and serotonin uptake in terms of modulatory role of ^{3}H-IMI sites on uptake of serotonin was recently provided by the work of Costa's group (Barbaccia et al., 1983a,b). They have succeeded in isolating from brain a nonpeptide constituent that is capable of displacing ^{3}H-IMI from its high-affinity sites and of inhibiting serotonin uptake in a dose-dependent manner (Barbaccia et al., 1983b). Presence of endogenous substances in partially purified extracts from rat brain and blood that inhibit both ^{3}H-IMI binding and serotonin uptake was subsequently reported by other groups (Rehavi et al., 1984; Angel et al., 1984). Once the chemical nature of this endogenous ligand for ^{3}H-IMI recognition sites, which may act as a modulator of serotonin uptake, is determined, the substance will become an important tool in studies of mode of action of antidepressants. Furthermore, Barbaccia et al. (1983a) have shown that down-regulation of ^{3}H-IMI binding after chronic administration of IMI and DMI was accompanied by an increase in serotonin uptake in rat hippocampal slices. This finding further indicates that the ^{3}H-IMI recognition site is distinct from the serotonin-uptake site. However, as Barbaccia et al. (1983a) hypothesize, it could be a regulatory site that is a part of a supramolecular presynaptic receptor complex regulating serotonin uptake and, consequently, the level of serotonergic transmission. According to this model, the receptor complex would function as a gating mechanism for serotonin uptake by presynaptic serotonergic terminals. To accommodate the experimental findings, the time course of action of tricyclic antidepressants is assumed to be biphasic. Initially, IMI or DMI, mimicking the action of an endogenous ligand at the ^{3}H-IMI regulatory site would suppress serotonin uptake and increase the amount of transmitter available for synaptic action. With time, repeated administration of these drugs would result, by mechanisms at present unknown, in a down-regulation of ^{3}H-IMI sites with subsequent facilitation of serotonin uptake and a decrease in its synaptic function. The concept of a reciprocal functional relationship between ^{3}H-IMI binding sites and serotonin uptake is indirectly supported by a recent finding (Arora et al., 1983) that in the brain of Fawn-hooded rats, a strain with a hereditary platelet serotonin deficiency, density of ^{3}H-IMI binding sites is significantly decreased, whereas the V_{max} of serotonin uptake is higher when compared to the Sprague-Dawley strain of rats.

Although the above hypothetical model helps to explain the possible role of ^{3}H-IMI recognition sites in the mode of action of antidepressants, the functional relationship of these sites to serotonin uptake under various other conditions (e.g., in reserpinized animals) when both functions seem to change independently, will have to be elucidated by further experiments.

3.5. Regulation of ^{3}H-Imipramine Binding by Drugs

The evidence reviewed in previous sections indicates that the high-affinity binding site for ^{3}H-IMI may represent a specific drug recognition site located at serotonergic neurons and related, in a manner yet to be clarified, to the regulation of their function. Some indication that the ^{3}H-IMI binding site may have a physiological and pharmacological significance came from the studies on the effect of various in vivo treatments on the high-affinity specific binding. The assumption here was that treatments that lead to clinical improvement of depression (e.g., administration of various antidepressants, electroconvulsive therapy, REM sleep deprivation) would significantly influence the high-affinity ^{3}H-IMI binding. It has been shown that partial or total REM sleep deprivation or chronic electroshock treatment in rats indeed produces a significant decrease in the number (B_{max}) of ^{3}H-IMI binding sites (Mogilnicka et al., 1980). Furthermore, chronic treatment (for 22 d) of rats with DMI also resulted in a decrease in the number of specific binding sites for ^{3}H-IMI in rat cerebral cortex (Raisman et al., 1980). In these experiments, the animals were killed without any period of withdrawal from the drug. However, acute treatment with DMI 2 hr before the sacrifice caused no changes in B_{max} of K_d of ^{3}H-IMI binding, indicating that the alterations in binding seen after chronic administration were not caused by the presence of residual drugs in the membrane preparation. Kinnier et al. (1980) found a down-regulation of ^{3}H-IMI binding in the hippocampus, but not in the cerebral cortex, of rats treated with IMI for 10 d. In addition, they reported that treatment with a nontricyclic antidepressant, iprindole, was ineffective in altering ^{3}H-IMI binding in either cortex or hippocampus. Results from this laboratory (Hrdina, 1984a) indicate that although chronic treatment with the tricyclic antidepressant nortriptyline results in down-regulation of high-affinity ^{3}H-IMI binding sites in rat hippocampus, repeated injections (for 21 d) of the "atypical" tetracyclic antidepressant mianserin fail to alter either the K_d or B_{max} of specific ^{3}H-IMI binding in brain (Table 3).

It thus appears that among antidepressant drugs tested so far, only the tricyclics are capable of down-regulating ^{3}H-IMI binding sites after chronic administration. The mechanism of this effect is not clear. Tricyclics have been reported to down-regulate β-adrenergic (Banerjee et al., 1977), α_2-adrenergic (Smith et al., 1981), and S_2-serotonergic (Peroutka and Snyder, 1980) receptors in the brain. This is thought to result from accumulation of serotonin and NA in synapses subsequent to the block of their reuptake by tricyclics. However, the same mechanism may not account for down-regulation of ^{3}H-IMI sites, since chronic treatment with IMI was shown, under the same experimental conditions, to decrease the density of β-adrenergic receptors, but not of ^{3}H-IMI recognition sites (Kinnier et al., 1980; Briley et al., 1982). It should be noted that there is a parallel between the ability of various compounds to displace high-affinity ^{3}H-IMI "in vitro" and to down-regulate the density of ^{3}H-IMI binding sites after chronic "in vivo" treatment. For example, tricyclic compounds such as IMI, DMI, and nortriptyline, which down-regulate ^{3}H-IMI sites, are also very effective (IC_{50}, of 20–300 n*M*) in displacing ^{3}H-IMI binding (Raisman et al., 1980; Langer et al., 1980a; Hrdina, 1982, 1984b). In contrast, mianserin and iprindole, which are very weak displacers of ^{3}H-IMI binding "in vitro" (Raisman et al., 1980; Hrdina, 1982) also fail to down-regulate ^{3}H-IMI binding sites "in vivo" (Kinnier et al., 1980; Hrdina, 1984a).

When evaluating the changes in specific ^{3}H-IMI binding after chronic in vivo treatment with drugs known to act on this binding in vitro, one has to take into account the possibility that a portion of the drug that remains after chronic treatment in the synaptosomal membranes could influence the result of the binding assay. The failure to recognize this may lead to errors in the interpretation of results. The time interval after the cessation of chronic treatment may be critical for the outcome of the binding assay. It is known that after repeated administration of DMI to rats, the concentration of the drug in various brain regions is 3–3.5 times higher than that observed at the same time interval after a single dose, and that the drug preferentially accumulates in the synaptosome-containing (P_2) fraction of the brain homogenate (Hrdina and Dubas, 1981). For instance, even at 48 h after the last of repeated doses of DMI (10 mg/kg/d), the P_2 fraction contains about 2 ng/mg protein of the drug. This is about 15–25 times more than the maximum number of ^{3}H-IMI binding sites derived from a standard binding assay, or the actual total binding at 8–10 n*M* concentration of ^{3}H-IMI. In a recent study, Briley et al. (1982) investigated the effect of chronic IMI treatment (7.5 mg/kg, 20 d) on ^{3}H-IMI binding in the hypothalamus of cats sacrificed at various time intervals (16, 40, and 88 h) after the last dose of the drug. They found a decrease in B_{max} of ^{3}H-IMI binding at 16- and 40-hr, but not at 88-hr time interval, whereas β-adrenergic receptor binding (^{3}H-DHA) was decreased at all time intervals. Moreover, chronic administration of a lower dose of IMI (2.5 mg/kg) failed to modify ^{3}H-IMI binding, but down-regulated ^{3}H-DHA binding in the cat brain. The authors have acknowledged that the decrease in the B_{max} of ^{3}H-IMI binding observed after the higher dose could be caused by the high concentration of the drug at the membrane level, which could modify in an irreversible manner the physicochemical properties of the binding site. Plenge and Mellerup (1982) also found that the effect of chronic IMI treatment on high-affinity ^{3}H-IMI binding was dependent on the length of withdrawal period, and concluded that the observed changes in ^{3}H-IMI binding parameters were most likely caused by competition in the assay procedure of IMI remaining in the membrane preparation. Clearly, the question of whether chronic treatment with tricyclic antidepressants truly down-regulates the binding sites for ^{3}H-IMI is not without controversy and will have to be answered by future experiments that take into account the possible interference of the residual drug with the binding assay itself.

Unexpectedly, chronic administration of haloperidol (1 mg/kg for 21 d) to rats also produced a lowering in the number of high-affinity ^{3}H-IMI binding sites in the hippocampus (Hrdina, 1984a). Furthermore, 5 wk of lithium treatment produced the same effect in rat cerebral cortex (Plenge and Mellerup, 1982). These observations indicate that down-regulation of ^{3}H-IMI recognition sites is not limited to tricyclic antidepressants and suggest that the influence of the residual drug may not be the only or determinant factor involved in the finding of decreased ^{3}H-IMI sites after chronic treatment with tricyclics. In the case of haloperidol, the drug's ability to down-regulate ^{3}H-IMI binding after "in vivo" administration apparently does not parallel a similar effect "in vitro" since its congener, spiroperidol, was found to be ineffective in displacing specific ^{3}H-IMI binding "in vitro" (Raisman et al., 1980).

4. High-Affinity ^{3}H-Imipramine Binding in Platelets

4.1. Characterization of Binding

Specific ^{3}H-IMI binding sites have also been found in platelets from different species, including man (Briley et al., 1979; Langer et al., 1980c; Paul et al., 1980; Dumbrille-Ross and Tang, 1981; Briley et al., 1982). This was not unexpected since platelets have been known to possess a serotonin uptake mechanism.

The characteristics of ^{3}H-IMI binding to human platelet membranes (high affinity, saturability, reversibility) are similar to those found in human or rat brain tissue. The binding assay for determination of ^{3}H-IMI sites in platelet membrane preparation is virtually the same as for membranes prepared from brain tissue. In most studies measuring ^{3}H-IMI binding, the platelet membranes are prepared according to the procedure described by Briley et al. (1979). Blood samples (20–30 mL) are collected into plastic tubes with sodium citrate (0.38%, final concentration) as anticoagulant. The blood is centrifuged at 280*g* for 10 min using a swinging bucket rotor. The platelet-rich plasma is carefully separated, measured, and then centrifuged at 16,000*g* for 10 min using a fixed-angle rotor. The plasma is decanted and the resulting platelet pellet is washed twice with 50 m*M* Tris/HCl buffer (pH 7.5) containing 150 m*M* NaCl and 20 m*M* EDTA. To prepare membranes, the platelets are then lysed and homogenized in a low ionic strength buffer (5 m*M* Tris/HCl, 5 m*M* EDTA, pH 7.5). Lysed membranes are separated by centrifugation at 39,000 g for 10 min, washed twice with 70 m*M* Tris/HCl buffer (pH 7.5), and finally resuspended in a buffer solution containing 50 m*M* Tris/HCl, 5 m*M* KCl, and 120 m*M* NaCl (pH 7.5). Aliquots of this platelet membrane suspension are then used to determine the specific ^{3}H-IMI binding.

The platelets have a slightly higher affinity ($K_D \sim 1.4$ n*M*) for ^{3}H-IMI binding than do rat brain membranes, but about the same affinity as the membrane preparation from human brain. A typical saturation isotherm and Scatchard analysis of high-affinity binding to human platelet membrane preparation is illustrated in Fig. 9. It appears that, in contrast to brain membranes, platelets only possess one set of specific binding sites for ^{3}H-IMI. The reported maximum number of binding sites in human platelets ranges from 450 to about 1010 fmol/mg protein (Table 4). This large variation may be caused in part by interindividual variations, but also by methodological differences in preparing the platelet membranes. Since B_{max} is usually expressed in terms of fmol/mg of protein, higher purity of membrane preparation will yield higher B_{max} values. Indeed, membranes prepared by sonication give B_{max} values for ^{3}H-IMI binding as high as 1500 fmol/mg membrane protein (Anderson et al., 1984). Protein content of platelet membrane suspensions used by various investigators to determine ^{3}H-IMI binding varies from as low as 0.17 mg/mL (Mellerup et al., 1982) to as high as 3.5 mg/mL (Paul et al., 1981). Friedl et al. (1983) found that mechanical disruption of platelets in hypotonic buffer can yield, because of incomplete lysis, preparations with varying proportions of intracellular and membrane protein. Therefore, the B_{max} values expressed on the basis of membrane protein may show considerable variation. They have shown that sonication yields pure membrane preparations that, because of a low protein content, yield the highest B_{max} values when ^{3}H-IMI binding is expressed on a per mg protein basis. To facilitate comparison between population studies of ^{3}H-IMI binding, Friedl et al. (1983) suggested the use of intact platelets that could provide a more reliable protein reference. There are, however, some disadvantages to this suggestion, such as possible disruption and changes in shape, aggregation of platelets, effect of released endogenous serotonin, and so on. An alternate suggestion is to use membranes prepared by sonication, which yields a pure membrane preparation.

A study of ^{3}H-IMI binding in the platelets of healthy volunteers (Langer et al., 1980c) has revealed no sex differences in either the B_{max} or K_d of binding. Moreover, the binding parameters did not change at subsequent determinations up to a period of 4 wk. Whereas the study by Langer et al. (1980c) reported a decrease in the number of ^{3}H-IMI binding sites with age, Asarch et al. (1980) were unable to find a significant correlation between age and either B_{max} or K_d of IMI binding in platelets of normal subjects.

A comparison of the IC_{50} values of various drugs for inhibition of ^{3}H-IMI binding in platelets and brain has revealed a highly significant correlation (Briley et al., 1979; Paul et al., 1980). In addition, a good correlation was found between the IC_{50} values of a series of drugs for inhibition of ^{3}H-IMI binding and of ^{3}H-serotonin uptake in human platelets (Paul et al., 1981). The similarity of binding characteristics and inhibitory potencies of a displacing

TABLE 4

Platelet ^{3}H-Imipramine Binding in Untreated Depressives

Reference	Normal controls K_d, nM	Normal controls B_{max}, fmol/mg protein	n	Depressed patients K_d, nM	Depressed patients B_{max}, fmol/mg protein	n
Briley et al. (1980)	2.6 ± 0.4^{a}	604 ± 42	21^{c}	2.5 ± 0.3	275 ± 24^{e}	16^{c}
Asarch et al. (1980)	2.0 ± 0.7^{b}	764 ± 243	7(M)	2.2 ± 0.8	608 ± 184^{e}	11(M)
	1.8 ± 0.8^{b}	650 ± 159	9(F)	1.8 ± 0.9	519 ± 154^{e}	12(F)
Paul et al. (1981)	1.5 ± 0.2^{a}	450 ± 23	28	0.9 ± 0.1	318 ± 20^{e}	14
Mellerup et al. (1982)	0.6 ± 0.2^{b}	1010 ± 280	33	0.6 ± 0.1	1190 ± 230^{e}	19
Berrettini et al. (1982)	3.1 ± 1.6^{b}	440 ± 168	12	3.0 ± 1.2	450 ± 151	12^{d}
Langer and Raisman (1983)	1.6 ± 1.6^{b}	562 ± 250	70	2.4 ± 2.0	318 ± 124^{e}	48
Suranyi-Cadotte et al. (1983)	3.0 ± 0.1^{a}	658 ± 33	17	2.4 ± 0.3	372 ± 34^{e}	6
Whitaker et al. (1984)	1.6 ± 0.3^{b}	764 ± 468	16	1.9 ± 0.4	798 ± 484	16
Baron et al. (1983)	3.4 ± 1.2^{b}	991 ± 347	15	2.5 ± 0.7^{e}	776 ± 287	15
Hrdina et al. (1985a)	1.3 ± 0.3^{b}	1000 ± 201	10	2.6 ± 1.1^{e}	894 ± 332	20

[a] ±SEM.
[b] ±SD.
[c] Females only.
[d] Euthymic bipolar patients.
[e] Significantly different from controls.

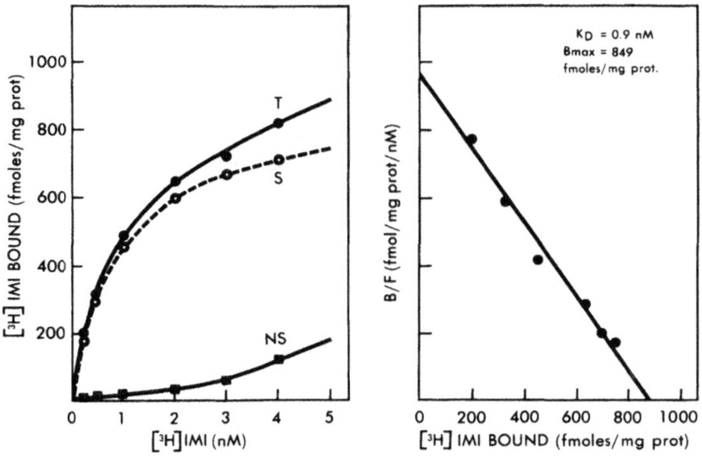

Fig. 9. Saturation curve (left) and Scatchard plot (right) of specific ^{3}H-imipramine binding to human platelet membranes prepared as described in section 4.1. High-affinity ^{3}H-imipramine binding was measured at concentrations of 0.25–4.0 n*M*, as described in section 2.1. Total binding (T), specific binding (S), and nonspecific binding (NS) are plotted as a function of ^{3}H-imipramine concentration. Points are means of triplicate determination of both total and nonspecific binding from a single typical experiment. Mean values of K_d and B_{max} (±SEM) for ^{3}H-imipramine binding to platelets from 10 healthy volunteers were 1.34 ± 0.09 n*M* and 1000 ± 64 fmol/mg protein, respectively (adapted from Hrdina et al., 1985a).

drug suggests that the high-affinity binding sites of ^{3}H-IMI in platelets are identical with those present in the brain, and that the platelets can serve as a useful model for elucidating the pharmacological and physiological significance of these binding sites.

4.2. ^{3}H-Imipramine Binding as a Potential Biological Marker of Depression

The demonstration of high-affinity binding of ^{3}H-IMI in human platelets has offered a possibility of studying the significance of this binding site for various antidepressant treatments and the etiology of the depressive illness. During the past few years, several groups of investigators reported the results of studies comparing the binding of ^{3}H-IMI in platelets of depressed patients in various stages of illness with that found in age-matched healthy subjects. Briley et al. (1980) were the first to report that the maximal number of ^{3}H-IMI binding sites (B_{max}) was significantly lower in untreated depressed subjects in comparison to age-matched controls. They found, however, no relationship between the severity of depression and the parameters of ^{3}H-IMI binding. They suggested that binding of ^{3}H-IMI in human platelets may represent a new biological marker of depression, possibly reflecting similar changes in brain. Subsequently, the same group of investigators (Raisman et al., 1981) has shown that reduction of ^{3}H-IMI binding sites occurs in both endogenous and reactive types of depression. However, they found virtually no change in B_{max} values of ^{3}H-IMI binding in platelets of depressed patients during a course of treatment with antidepressants, despite the clinical improvement of symptoms. These results were independently confirmed by Asarch et al. (1980) and Paul et al. (1981) who also found a significant decrease in the number of ^{3}H-IMI binding sites in a sample of 23 mixed and 14 endogenous depressives, respectively, in comparison with age-matched controls. In addition, the latter authors reported an important preliminary observation that ^{3}H-IMI binding remained low in four of their patients even following recovery. This, taken together with previous data, suggested that the decreased binding of ^{3}H-IMI to platelets might not be a reflection of the depressed state but rather a trait phenomenon that may be genetically regulated.

A subsequent study by Suranyi-Cadotte et al. (1982) has shown that the decrease in platelet ^{3}H-IMI binding noted in depressed patients on admission returns to normal levels after clinical recovery because of treatment with tricyclic antidepressants. This observation is at variance with the suggestion that decreased ^{3}H-IMI binding in depression is a trait marker, since it was state-dependent. In support of the latter is the finding of no difference in ^{3}H-IMI binding between medication-free euthymic bipolar patients and normal volunteers (Berrenttini et al., 1982). To further complicate the issue, Mellerup et al. (1982), in a large study including 47 bipolar depressed patients without medication and 39 control subjects, found that depressed bipolar patients had a higher B_{max} value for platelet ^{3}H-IMI binding than normal control subjects or euthymic bipolar patients. Finally, Whitaker et al. (1984) reported no difference in ^{3}H-IMI binding parameters between a population of medication-free depressed patients and healthy volunteers. They found, however, significant seasonal variation in platelet ^{3}H-IMI binding in both depressed and normal subjects and suggested that failure to control the effect of this annual rhythm could have accounted for apparent changes in ^{3}H-IMI binding sites reported by some group of investigators. Published findings on ^{3}H-IMI binding in platelets of depressed patients in comparison with control subjects are summarized in Table 4. More recent findings (Meltzer et al., 1984; Hrdina et al. 1985a; Lewis and McChesney, 1985) suggest that decreased platelet ^{3}H-IMI binding could be a characteristic of certain subgroups of depressives (e.g., psychotic, bipolar) or a predictor of poor response to antidepressant treatment.

The exact relationship between the ^{3}H-IMI binding sites and serotonin-uptake mechanism in platelets also remains to be clarified. Although both the density of imipramine binding sites and serotonin uptake (Tuomisto and Tukiainen, 1976) have been shown to be lower in depressed patients, it is not clear whether these two changes are coupled and occur in parallel. For example, in nondepressed patients with alcoholic cirrhosis or in untreated hypertensive subjects, the uptake of ^{3}H-serotonin into platelets was found to be significantly reduced, whereas the ^{3}H-IMI binding was unchanged in comparison with controls (Ahtee et al., 1981; Kamal et al., 1984). Lack of association between ^{3}H-IMI binding sites and uptake of serotonin in control and depressed subjects was also reported by Wood et al. (1983).

A number of questions will have to be answered before it is established that binding of ^{3}H-IMI in platelets reflects the binding in the brain and that the recognition site for ^{3}H-IMI in the brain is indeed a physiologically relevant receptor site involved in the pharmacological effect of antidepressants or in modulation of mood. It should be noted, for instance, that chronic treatment with desipramine or imipramine was found to down-regulate ^{3}H-IMI binding in rat brain membranes, but not in platelets of depressed subjects (Raisman et al., 1980; Briley et al., 1980). This apparent discrepancy may be caused by the differences in the properties of binding sites in the normal vs the diseased state. The situation is further complicated by the fact that in experimental animals, chronic treatment with some antidepressants (e.g., iprindole, mianserin) did not result in decreased ^{3}H-IMI binding. It is also possible that two concurrent changes occur in platelet ^{3}H-IMI binding of depressed subjects during antidepressant treatment: down-regulation of the binding caused by antidepressant drug administration and an increase in the number of binding sites caused by clinical improvement. The observed lack of apparent alterations in platelet ^{3}H-IMI binding might stem from one of these two concurrent changes masking the other. Longitudinal studies in depressed patients successfully treated by nondrug intervention (e.g., electroconvulsive therapy) and determination of platelet ^{3}H-IMI binding in families of depressed patients should help to clarify the question of whether the alteration in high-affinity binding of ^{3}H-IMI could be a biological or genetic marker of depression. At the time this manuscript was submitted, a World Health Organization (WHO) multicenter collaborative study had been initiated with the aim of answering the above question by using a standardized protocol of assessment methods and techniques. In order to determine whether the decrease in ^{3}H-IMI binding observed in platelets of patients with depressive illness is an indication of the same changes in the brain, one would have to examine ^{3}H-IMI binding in brain samples from deceased patients with a diagnosis of depression. Recent reports of lower binding of ^{3}H-IMI in brains of suicides (Stanley et al., 1982), patients with depressive illness (Perry et al., 1983), and patients with Parkinson's disease (Cash et al., 1984) suggest that such studies may be feasible.

5. Noradrenaline Uptake Sites and ^{3}H-Desipramine Binding

The tertiary amine tricyclic antidepressants (e.g., IMI and amitriptyline) are metabolized in vivo to their respective demethylated derivatives (e.g., desipramine and nortriptyline). These metabolites contribute significantly to the antidepressant effect of their parent compounds and are also used as antidepressant drugs in their own right. Although both tertiary and secondary amine tricyclic antidepressants block the neuronal uptake of biogenic amines, the secondary amines, such as DMI, are relatively more potent inhibitors of NA than of serotonin uptake, whereas the reverse is true for tertiary amines, such as IMI (Ross and Renyi, 1975). High-affinity ^{3}H-IMI binding in brain has been shown to be associated with serotonin uptake and to be inhibited more by tertiary than secondary amine tricyclic antidepressant compounds (Langer et al., 1980a; Rehavi et al., 1980; Hrdina, 1982). It was therefore of interest to investigate whether specific binding sites for ^{3}H-DMI also exist in the brain and are associated with amine uptake mechanisms.

5.1. Kinetics and Distribution of ^{3}H-Desipramine Binding

The demonstration of high-affinity binding of ^{3}H-DMI to rat brain membranes has been reported independently from several laboratories (Rohavi et al., 1981; Hrdina, 1981; Langer et al., 1981b; Lee and Snyder, 1981). Prior to these reports, Biegon and Samuel (1979) had demonstrated a low-affinity binding of ^{3}H-DMI in rat brain.

The binding assay used for ^{3}H-DMI is essentially the same as

TABLE 5
High-Affinity ^{3}H-Desipramine Binding in Rat Brain

Region	^{3}H-Desipramine K_d, nM	B_{max}, fmol/mg protein	Reference
Cerebral cortex	2.3 ± 0.6[a]	115 ± 8	Langer et al. (1981b)
	4.5	70	Hrdina (1981)
	2.4 ± 0.5	101 ± 12	Raisman et al. (1982)
	1.5 ± 0.2	240 ± 30	Rehavi et al. (1982)
Hypothalamus	2.6 ± 1.2	156 ± 22	Raisman et al. (1982)
	0.9 ± 0.2	220 ± 25	Rehavi et al. (1982)
Striatum	3.4 ± 0.8	29 ± 4	Raisman et al. (1982)
Cerebellum	1.5 ± 0.3	80 ± 8	Raisman et al. (1982)
	1.1 ± 0.3	100 ± 20	Rehavi et al. (1982)
Septum	1.7 ± 0.3	320 ± 40	Rehavi et al. (1982)
Medulla	1.0 ± 0.3	100 ± 20	Rehavi et al. (1982)

[a]Mean value ± SEM.

for ^{3}H-IMI, except that either the drug itself (DMI), nortriptyline, or nisoxetine are used as displacers. The specific binding of ^{3}H-DMI to membranes prepared from cerebral cortex of rats was found to have at least two components (high and low affinity) (Rehavi et al., 1981; Hrdina, 1981; Lee et al., 1982). In contrast to ^{3}H-IMI, the amount of ^{3}H-DMI specific binding to brain membranes is only about 50% of total binding. The equilibrium dissociation constant (K_d) of the high-affinity binding in brain was found to be in low nanomolar range, and the maximum number of binding sites (B_{max}) was seen to vary from 70 to 320 fmol/mg protein (Table 5).

The saturability of specific high-affinity ^{3}H-DMI binding has been difficult to demonstrate when relatively high concentrations

(10 or 100 μ*M*) of displacers were used. This is most likely because specific ^{3}H-DMI binding involves at least two components. Recently, Lee et al. (1982) reported that when only 0.1 μ*M* DMI is used as a displacing cold drug to define the nonspecific binding, the high-affinity ^{3}H-DMI binding reflects a single, almost saturable component with an equilibrium dissociation constant (K_d) in a low nanomolar range (2–4 n*M*). Pretreatment of brain membranes with 0.3*M* KCl increases the ratio of high-affinity to low-affinity specific ^{3}H-DMI binding. Why KCl treatment results in this effect is not clear. However, even with this experimental manipulation, the amount of displaceable ^{3}H-DMI binding is still only about 50%.

The binding of ^{3}H-DMI is rapidly displaceable by either desipramine, nortriptyline, or nisoxetine. Kinetic analysis of high-affinity ^{3}H-DMI binding has revealed an association rate constant (K_i) in a range of 0.57–6.4 × 10^7 *M*/min (Rehavi et al., 1982; Lee et al., 1982; Hrdina, unpublished observation). The dissociation rate constant (K_2) is temperature-dependent and was calculated to be 0.021 and 0.078/min at 0 and 20°C, respectively (Lee et al., 1982). ^{3}H-DMI binding to membranes from rat cerebral cortex was also found to be sodium-dependent (Rehavi et al., 1982; Lee and Snyder, 1981; Lee et al., 1982). Temperature dependence of binding was reported by some (Rohavi et al., 1982; Lee and Snyder, 1981), but not other (Raisman et al., 1982) workers. The regional distribution of high-affinity ^{3}H-DMI binding was found to be uneven, with the highest density of binding sites in the septum and hypothalamus, and the lowest in the cerebellum and striatum (Table 5). This tends to parallel that of NA uptake (Lee et al., 1982).

5.2. Pharmacological Characterization of ^{3}H-Desipramine Sites

Studies examining the effect of various tricyclic and tetracyclic antidepressants and relatively specific blockers of neuronal uptake of serotonin and NA on ^{3}H-DMI binding revealed displacement profiles that were markedly different from those seen with ^{3}H-IMI binding (Rehavi et al., 1981; Hrdina, 1981, 1982; Langer et al., 1981b). The secondary amine tricyclic antidepressants DMI and nortriptyline and the specific NA-uptake inhibitor nisoxetine inhibited ^{3}H-DMI binding most effectively (in nanomolar concentration range), whereas chlorimipramine and fluoxetine, potent inhibitors of serotonin uptake and ^{3}H-IMI binding, were least effective (Hrdina, 1982) (Fig. 4, Table 2). A highly significant correlation was reported by several authors (Rehavi et al., 1981; Hrdina, 1982; Langer et al., 1981b; Raisman et al., 1982) between the potency of

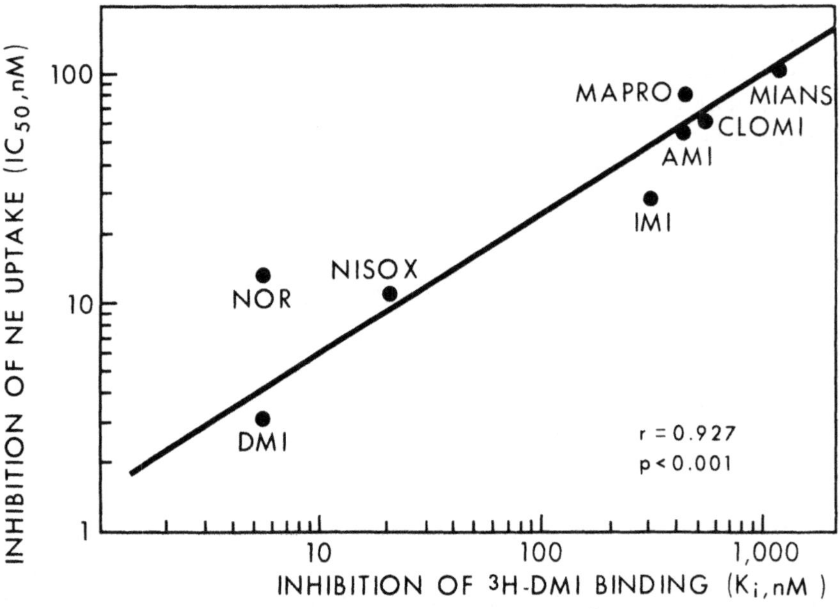

Fig. 10. Correlation between the inhibition of ^{3}H-desipramine binding and neuronal uptake of noradrenaline (A) and serotonin (B) by various drugs. DMI; desipramine: NOR; nortriptyline: NISOX; nisoxetine: IMI; imipramine: AMI; amitriptyline: CLOMI; clomipramine: MAPRO; maprotiline: MIANS; mianserin: DOX; doxepin: CPZ; chlorpromazine: FLUOX; fluoxetine. Specific ^{3}H-desipramine binding was determined at

(continued on next page)

various compounds to inhibit NA uptake and high-affinity ^{3}H-DMI binding in rat cerebral cortex (Fig. 10A) or hypothalamus. In contrast, no correlation was found between the displacement of ^{3}H-DMI binding and of serotonin uptake produced by tested drugs (Fig. 10B). Also, no ^{3}H-DMI binding sites were detected in platelets, which posses an active transport system for serotonin but not for NA. These results suggested that ^{3}-DMI binding is (a) distinct from that of ^{3}H-IMI and (b) associated with neuronal uptake of NA. This notion was further supported by the demonstration of high-affinity ^{3}H-DMI binding in peripheral tissues that contain rich adrenergic innervation, such as the heart and vas deferens (Langer et al., 1981b; Raisman et al., 1982). The displacement by drugs of ^{3}H-DMI binding in the heart and cerebral cortex has been shown to be virtually the same (Langer et al., 1981b).

5.3. Localization of ^{3}H-Desipramine Binding Sites and Noradrenaline Uptake

Direct evidence for a presynaptic localization of ^{3}H-DMI binding sites on noradrenergic nerve terminals was initially provided

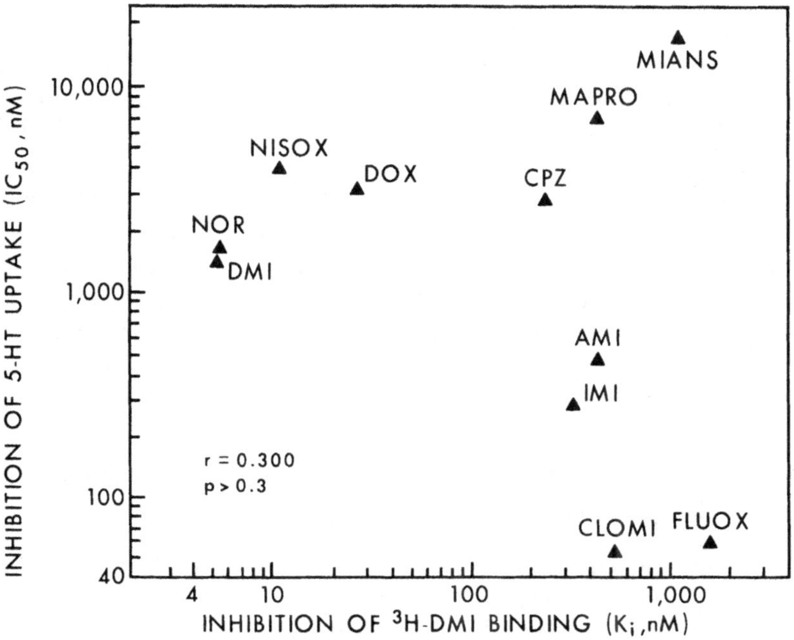

(*continued*)

4.0 n*M* concentration, as described in section 5.1. Data for noradrenaline uptake inhibition from Hall and Ögren (1981) and Langer et al. (1980a); for serotonin uptake inhibition from Horn (1976), Koe (1975), and Wang et al. (1975). Fig. 10A reproduced from Hrdina (1982) by permission of Pergamon Press, Oxford; Fig. 10B adapted from Hrdina (1982).

by studies (Lee and Snyder, 1981; Hrdina et al., 1981) that examined the high-affinity binding of ^{3}H-DMI after selective destruction of NA-containing neurons by 6-hydroxydopamine (6-OHDA). Selective lesioning of central noradrenergic neurons by intracerebral (Hrdina et al., 1981) or intraventricular (Lee and Snyder, 1981) injection of 6-OHDA in rats resulted in marked parallel decreases of both NA content or uptake and high-affinity ^{3}H-DMI binding. In contrast, ^{3}H-IMI binding was not affected by the same treatments (Fig. 6). Similar findings were subsequently reported by Rehavi et al. (1982). However, the magnitude of decrease in NA uptake in 6-OHDA-treated rats was much greater (87%) than that of ^{3}H-DMI binding (36%) (Rehavi et al., 1982). DSP-4 [*N*-(2-chloroethyl)-*N*-ethyl-2-bromobenzylamine] is a neurotoxin that selectively destroys NA neurons and alkylates NA uptake sites (Ross, 1976). When high-affinity ^{3}H-DMI binding was measured in KCl-pretreated membranes with 0.1 μ*M* DMI as displacer (Lee et al., 1982), both the NA uptake and density of ^{3}H-DMI binding sites after treatment with DSP-4 were reduced to a similar extent. Low-affinity ^{3}H-DMI binding remained uneffected by DSP-4 treatment.

These experiments provided strong evidence for association of high-affinity ^{3}H-DMI binding sites with the NA-uptake mechanism. Effect of various treatments on simultaneously determined ^{3}H-DMI binding and NA uptake in rat brain is summarized in Table 6.

The distribution of high-affinity binding sites for ^{3}H-DMI, as determined by quantitative autoradiography, was found to be remarkably similar to the distribution of noradrenergic terminals in the rat brain (Biegon and Rainbow, 1982). The highest density was seen in the locus coeruleus and in nuclei of the hypothalamus and limbic systems, whereas the lowest was in the caudate-putamen region.

5.4. Regulation and Possible Physiological Significance of ^{3}H-Desipramine Sites

In contrast to ^{3}H-IMI sites, high-affinity ^{3}H-DMI binding in rat cerebral cortex was found to be unaffected by chronic treatment with IMI, DMI, or iprindole, but appears to be regulated by the actual concentration of the neurotransmitter NA (Lee et al., 1983). Depletion of NA (and serotonin) by reserpine resulted in a significant reduction of ^{3}H-DMI binding; on the other hand, accumulation of amines (NA and serotonin) after treatment with an MAO inhibitor (iproniazid) was accompanied by a modest up-regulation of ^{3}H-DMI recognition sites. In both instances, however, the changes in ^{3}H-DMI binding were much less pronounced than alterations in NA levels. Although reserpine depletes brain of not only NA, but also of serotonin and dopamine, association of decreases in ^{3}H-DMI binding with NA depletion is strengthened by findings that sympathetic denervation of rat heart and submaxillary gland or reserpine treatment also resulted in marked concomitant decreases in both endogenous NA levels and ^{3}H-DMI binding sites in these peripheral tissues (Langer et al., 1981b; Raisman et al., 1982; Lee et al., 1983).

Specific high-affinity binding sites for ^{3}H-DMI have thus been demonstrated in several brain regions and also in adrenergically innervated peripheral tissues of the rat. These binding sites are distinct from those for ^{3}H-IMI, have a characteristic pharmacological profile, appear to be located, at least in part, on noradrenergic nerve terminals, and are related to neuronal uptake of NA. However, several problems remain to be clarified concerning the nature and localization of ^{3}H-DMI binding sites and their physiological significance. ^{3}H-DMI binding appears to be multiphasic. The significance of low-affinity binding sites and their relationship to the

TABLE 6

Effect of Various Treatments on ^{3}H-Desipramine Binding and Noradrenaline Uptake in Rat Brain

Treatment	^{3}H-Desipramine binding, density	Noradrenaline uptake	Reference
Drugs added "in vitro"	Similar rank order of potency		Hrdina (1981, 1982); Langer et al. (1981b); Lee et al. (1981, 1982); Raisman et al. (1982); Rehavi et al. (1981, 1982)
6-OH DA lesion	↓	↓ ↓	Hrdina et al. (1981); Hrdina (1982); Lee et al. (1981), Rehavi et al. (1982)
DSP-4 lesion	↓	↓	Lee et al. (1983)
Reserpine	↓	↓ ↓	Lee et al. (1983)
Antidepressants (imipramine, desipramine, iprindole-chronic)	0	0	Lee et al. (1983)
Iproniazid	↑	↑	Lee et al. (1983)

high-affinity sites is at present unclear. The exact relationship of high-affinity ^{3}H-DMI binding sites to the NA-uptake mechanism is not without controversy. According to Lee et al. (1983), ^{3}H-DMI labels NA uptake sites that are regulated by transmitter concentration at the synapse. This is certainly different from the more complex relationship between serotonin uptake and the ^{3}H-IMI recognition site. The reason for this difference is not clear. Recently, Laduron (1984) pointed out that some very potent NA-uptake inhibitors, such as D- and L-imafen, do not display the same high potency in displacing ^{3}H-DMI binding. In addition, Laduron et al. (1982) also found that the subcellular distribution profile of ^{3}H-DMI binding differed markedly from that of NA uptake. On the basis of these experiments, they concluded that ^{3}H-DMI does not bind to the sites associated with the uptake of NA. It is possible that ^{3}H-DMI does not label the substrate recognition site for the NA uptake but rather an unidentified membrane subunit that would modulate the neuronal uptake of the amine. Noradrenergic nerve terminals were shown to possess receptors modulating the release of the neurotransmitter (Langer, 1980). It is thus possible that analogous regulatory sites exist at presynaptic nerve terminals for NA uptake and are labeled by ^{3}H-DMI. The endogenous ligand for such a hypothetical regulating unit remains to be identified. It also remains to be determined whether chronic treatment with NA-uptake inhibitors will be able to affect the high-affinity ^{3}H-DMI binding. Finally, it will be important to examine whether high-affinity ^{3}H-DMI binding sites are present in human brain and whether they are altered by some diseased states. ^{3}H-DMI binding may prove to be an important tool in studies of the regulation of NA neuronal uptake and of the biochemical bases of antidepressant drug action.

6. Conclusions

High-affinity recognition sites for ^{3}H-IMI have been described and characterized in brain and platelets of several species, including man. Experimental evidence indicates that these sites are located at the serotonergic nerve terminals in brain and are, in a manner not yet completely understood, associated with serotonin uptake process in both brain and platelets. ^{3}H-IMI sites in brain may be of physiological and pharmacological importance in regulation of serotonin uptake and consequently of the level of serotonergic transmission in the CNS. A putative endogenous ligand from brain has recently been isolated that inhibits both

^{3}H-IMI binding to its recognition sites and serotonin uptake. Future studies aimed at characterization of endogenous ligand(s) should help to elucidate the role of this site. Recently, it has been shown that ^{3}H-IMI binding in brain also has a low-affinity component that is distinct from high-affinity sites, not related to serotonin uptake, and affected in a similar fashion by different types of antidepressants. It remains to be seen whether these low-affinity ^{3}H-IMI sites are related to the clinical antidepressant effect. ^{3}H-IMI binding in platelets may be a useful biological marker of altered serotonergic function since it has been found to be reduced in some groups of depressed patients.

On the other hand, high-affinity ^{3}H-DMI binding sites appear to subserve a different function since they are located, at least in part, at noradrenergic nerve terminals and regulated by the synaptic level of the neurotransmitter. No endogenous ligand for the ^{3}H-DMI site has yet been identified.

Binding of ^{3}H-antidepressants has served as a useful tool in studies of physiological and biochemical aspects of serotonin and noradrenaline function in the brain. Their use as possible biological markers of disturbances in functioning of central biogenic amines underlying mood disorders is of potential clinical importance in psychiatry.

Acknowledgments

This work was supported by the Ontario Mental Health Foundation grant #871 and Medical Research Council of Canada grant #MA 6775. I am grateful to Dr. David Peters for reading the manuscript and for helpful comments. I thank Diane McNeil for typing the manuscript.

References

Ahtee L., Briley M., Raisman R., Lebrec D., and Langer S. Z. (1981) Reduced uptake of serotonin but unchanged ^{3}H-Imipramine binding in the platelets from cirrhotic patients. *Life Sci.* **29,** 2323–2329.

Anderson G. M., Minderaa R. B., van Bentheem P.-P.G., Volkmar F. R., and Cohen D. J. (1984) Platelet imipramine binding in autistic subjects. *Psychiatry Res.* **11,** 133–141.

Angel I., Goldman M. E., and Paul S. M. (1984) Inhibition of synaptosomal (^{3}H)-serotonin uptake and (^{3}H)-imipramine binding by an endogenous substance(s) isolated from rat and human blood. *14th CINP Congress* (Abstracts), June 19–23, 1984, Florence, Italy, p. 283.

Arora R. C., Tong C., Jackman H. L., Stoff D., and Meltzer H. Y. (1983) Serotonin uptake and imipramine binding in blood platelets and brain of Fawn-hooded and Sprague-Dawley rats. *Life Sci.* **33,** 437–442.

Asarch K. B., Shih J. C., and Kulcsar A. (1980) Decreased ^{3}H-imipramine binding in depressed males and females. *Commun. Psychopharmacol.* **4,** 425–432.

Benerjee S. P., Kung L. S., Riggi S. J., and Chanda S. K. (1977) Development of β-adrenergic receptor subsensitivity by antidepressants. *Nature* (Lond.) **268,** 455–456.

Barbaccia M. L., Brunello N., Chuang D. M., and Costa E. (1983a) On the mode of action of imipramine: relationship between serotonergic axon terminal function and down-regulation of β-adrenergic receptors. *Neuropharmacol.* **22,** 373–383.

Barbaccia M. L., Gandolfi O., Chuang D. M., and Costa E. (1983b) Modulation of neuronal serotonin uptake by a putative endogenous ligand of imipramine recognition sites. *Proc. Natl. Acad. Sci. USA* **80,** 5134–5138.

Baron M., Barkai A., Gruen R., Kowalik S., and Quitkin F. (1983) ^{3}H-Imipramine platelet binding sites in unipolar depression. *Biol. Psychiat.* **18,** 1403–1409.

Berrettini W. H., Nurnberger Jr., J. I., Post R. M., and Gershon E. S. (1982) Platelet ^{3}H-imipramine binding in euthymic bipolar patients. *Psychiatry Res.* **7,** 215–219.

Biegon A. and Samuel D. (1979) Binding of a labeled antidepressant to rat brain tissue. *Biochem. Pharmacol.* **28,** 3361–3366.

Biegon A. and Rainbow T. C. (1982) Quantitative autoradiography of (^{3}H)-desipramine binding sites in rat brain. *Eur. J. Pharmacol.* **82,** 245–246.

Bowling A. C. and DeLorenzo R. J. (1982) Micromolar affinity benzodiazepine receptors: identification and characterization in central nervous system. *Science* **216,** 1247–1250.

Briley M. and Langer S. Z. (1981) Sodium dependency of (^{3}H)-imipramine binding in rat cerebral cortex. *Eur. J. Pharmacol.* **72,** 377–380.

Briley M. S., Raisman R., and Langer S. Z. (1979) Human platelets possess a high-affinity binding site for (^{3}H)-imipramine. *Eur. J. Pharmacol.* **58.** 347–348.

Briley M. S., Langer S. Z., Raisman R., Sechter D., and Zarifian E. (1980) Tritiated imipramine binding sites are decreased in platelets of untreated depressed patients. *Science* **209,** 303–305.

Briley M. S., Raisman R., Arbilla S., Casadamont M., and Langer S. S. (1982) Concomitant decrease in (^{3}H)-imipramine binding in cat brain and platelets after chronic treatment with imipramine. *Eur. J. Pharmacol.* **81,** 309–314.

Brunello N., Chuang D. M., and Costa E. (1982a) Different synaptic location of mianserin and imipramine binding sites. *Science* **215,** 1112–1115.

Brunello N., Chuang D. M., and Costa E. (1982b) Specific binding of (^{3}H)-mianserin and (^{3}H)-imipramine to structures of rat hippocampus. *Eur. J. Pharmacol.* **78,** 383–384.

Cash R., Raisman R., Ploska A., and Agid Y. (1984) ^{3}H-Imipramine binding in brain of Parkinsonian patients. *14th CINP Congress,* June 19–23, 1984, Florence, Italy, p. 17.

Clements-Jewery S. (1978) The development of cortical β-adrenoceptor subsensitivity in the rat by chronic treatment with trazodone, doxepin, and mianserin. *Neuropharmacology* **17,** 779–781.

Dumbrille-Ross A. and Tang S. W. (1981) Absence of high-affinity (^{3}H)-imipramine binding in platelets and cerebral cortex of Fawn-hooded rats. *Eur. J. Pharmacol.* **72,** 137–138.

Dumbrille-Ross A., Tang S. W., and Coscina D. W. (1981) Differential binding of ^{3}H-imipramine and ^{3}H-mianserin in rat cerebral cortex. *Life Sci.* **29,** 2049–2058.

Friedl W., Propping P., and Weck B. (1983) ^{3}H-Imipramine binding in platelets: influence of varying proportions of intact platelets in membrane preparations on binding. *Psychopharmacol.* **80,** 96–99.

Glowinski J. and Axelrod J. (1964) Inhibition of uptake of tritiated noradrenaline in the intact rat brain by imipramine and structurally related compounds. *Nature* (Lond.) **204,** 1318–1319.

Gross G., Gothert M., Ender H. P., and Schumann H.-J. (1981) ^{3}H-Imipramine binding sites in the rat brain. Selective localization on serotonergic neurons. *Naunyn-Schmiedeberg's Arch. Pharmacol.* **317,** 310–314.

Hall H. and Ögren S.-O. (1981) Effects of antidepressant drugs on different receptors in the brain. *Eur. J. Pharmacol.* **70,** 393–407.

Horn A. S. (1976) The interaction of tricyclic antidepressants with the biogenic amine uptake systems in the central nervous system. *Postgrad. Med. J.* **52,** (Suppl) 25–30.

Hrdina P. D. (1981) Pharmacological characterization of (^{3}H)-desipramine binding in rat. cerebral cortex. *Prog. Neuro-Psychopharmacol.* **5,** 553–557.

Hrdina P. D. and Dubas T. C. (1981) Brain distribution and kinetics of desipramine in the rat. *Can. J. Physiol. Pharmacol.* **59,** 163–167.

Hrdina P. D., Elson-Hartman K., Roberts D. C. S., and Pappas B. A. (1981) High-affinity (^{3}H)-desipramine binding in rat cerebral cortex decreases after selective lesion of noradrenergic neurons with 6-hydroxydopamine. *Eur. J. Pharmacol.* **73,** 375–376.

Hrdina P. D. (1982) Tricyclic Antidepressants: Differences in High-Affinity Binding of (^{3}H)-Desipramine and (^{3}H)-Imipramine in the Brain, in *New Vistas in Depression* (Langer S. Z., Takahashi R., and Briley M., eds.) pp. 125–132, Pergamon, London.

Hrdina P. D., Pappas B. A., Bialik R. J., and Ryan C. L. (1982) Regulation of (^{3}H)-imipramine binding sites in rat brain regions: effect of neonatal 5,7-dihydroxytryptamine treatment. *Eur. J. Pharmacol.* **82,** 343–344.

Hrdina P. D. (1983) Antidepressant binding: implications for the mode of action of antidepressants and the biology of depression. *Progr. Neuropsychopharmacol. Biol. Psychiat.* **7,** 457–462.

Hrdina P. D. (1984a) ^{3}H-Imipramine binding sites in brain down-regulated by chronic nortriptyline and haloperidol but not mianserin treatment. *Psychiatry Res.* **11,** 271–278.

Hrdina P. D. (1984b) Differentiation of two components of specific (^{3}H)-imipramine binding in rat brain. *Eur. J. Pharmacol.* **102,** 481–488.

Hrdina P. D., Lapierre Y. D., Horn E., et al. (1985a) Platelet ^{3}H-imipramine binding: A possible predictor of response to antidepressant treatment. *Prog. Neuropsycho-pharmacol. Biol. Psychiat.* **9,** in press.

Hrdina P. D., Pappas B. A., Roberts D. C. S., Bialik R. J., and Ryan C. L. (1985b) Relationship between levels and uptake of serotonin and high affinity [^{3}H]imipramine recognition sites in the rat brain. *Can. J. Physiol. Pharmacol.* **63,** 1239–1244.

Kamal L. A., Raisman R., Meyer P., and Langer S. Z. (1984) Reduced V_{max} of ^{3}H-serotonin uptake but unchanged ^{3}H-imipramine binding in the platelets of untreated hypertensive subjects. *Life Sci.* **34,** 2083–2088.

Kinnier W. J., Chuang D.-M., Gwynn G., and Costa E. (1980) Down-regulation of dihydroalprenolol and impramine binding sites in brain of rats repeatedly treated with imipramine. *Eur. J. Pharmacol.* **67,** 289–291.

Kinnier W. J., Chuang D.-M., Gwynn G., and Costa E. (1981) Characteristics and regulation of high-affinity (^{3}H)-imipramine binding to rat hippocampal membranes. *Neuropharmacol.* **20,** 411–419.

Koe B. K. (1975) Effects of Antidepressant Drugs on Brain Catecholamines and Serotonin, in *Antidepressants* (Fielding S. and Lal H., eds.) pp. 143–180, Futura Publishing, Mount Kisco, NY.

Laduron P. M. (1984) Criteria for receptor sites in binding studies. *Biochem. Pharmacol.* **33,** 833–839.

Laduron P. M., Robbyns M., and Schotte A. (1982) (^{3}H)-Desipramine and (^{3}H)-imipramine binding are not associated with noradrenaline and serotonin uptake in the brain. *Eur. J. Pharmacol.* **78,** 491–493.

Langer S. Z. (1980) Presynaptic regulation of the release of catecholamines. *Pharmacol. Rev.* **32,** 337–362.

Langer S. Z. and Raisman R. (1983) Binding of (^{3}H)-imipramine and (^{3}H)-desipramine as biochemical tools for studies in depression. *Neuropharmacol.* **22,** 407–413.

Langer S. Z., Moret C., Raisman R., Dubocovich M. L., and Briley M. (1980a). High-affinity (^{3}H)-imipramine binding in rat hypothalamus:

association with uptake of serotonin but not of norepinephrine. *Science* **210,** 1133–1135.

Langer S. Z., Raisman R., and Briley M. S. (1980b) Stereoselective inhibition of ^{3}H-imipramine binding by antidepressant drugs and their derivatives. *Eur. J. Pharmacol.* **64,** 89–90.

Langer S. Z., Briley M. S., Raisman R., Henry J.-F., and Morselli P. L. (1980c) Specific ^{3}H-imipramine binding in human platelets. Influence of age and sex. *Naunyn-Schmiedeberg's Arch. Pharmacol.* **313,** 189–194.

Langer S. Z., Javoy-Agid F., Raisman R., Briley M., and Agid Y. (1981a) Distribution of specific high-affinity binding sites for (^{3}H)-imipramie in human brain. J. *Neurochem.* **37,** 267–271.

Langer S. Z., Raisman R., and Briley M. (1981b) High-affinity (^{3}H)-DMI binding is associated with neuronal noradrenaline uptake in the periphery and the central nervous system. *Eur. J. Pharmacol.* **72,** 423–424.

Lee C.-M. and Snyder S. H. (1981) Norepephrine neuronal uptake binding sites in rat brain membranes labeled with (^{3}H)-desipramine. *Proc. Natl. Acad. Sci. USA* **78,** 5250–5254.

Lee C. M., Javitch J. A., and Snyder S. H. (1982) Characterization of (^{3}H)-desipramine binding associated with neuronal norepinephrine uptake sites in rat brain membrenes. *J. Neurosci.* **2,** 1515–1525.

Lee C. M., Javitch J. A., and Snyder S. H. (1983) Recognition sites for norepinephrine uptake: regulation by neurotransmitter. *Science* **220,** 626–629.

Lewis D. A. and McChesney C. (1985) Tritiated imipramine binding distinguishes among subtypes of depression. *Arch. Gen. Psychiat.* **42,** 485–488.

Lowry D. H., Rosebrough N. J., Farr A. L., and Randall R. J. (1951) Protein measurement with the Folin phenol reagent. *J. Biol. Chem.* **192,** 265–275.

Mellerup E. T., Plenge P., and Rosenberg R. (1982) ^{3}H-Imipramine binding sites in platelets from psychiatric patients. *Psychiatry Res.* **7,** 221–227.

Meltzer H. Y., Arora R. C., Robertson A., and Lowy M. (1984) Platelet ^{3}H-imipramine binding and platelet 5-HT uptake in affective disorders and schizophrenia. *Clin. Neuropharmacol.* **7** (suppl. 1) S169.

Mocchetti I., Brunello N., and Racagni G. (1982) Ontogenic study of (^{3}H)-imipramine binding sites and serotonin uptake system: indication of possible interdependence. *Eur. J. Pharmacol.* **83,** 151–152.

Mogilnicka E., Arbilla S., Depoortere H., and Langer S. Z. (1980) Rapid eye movement sleep deprivation decreases the density of ^{3}H-dihydroalprenolol and ^{3}H-imipramine binding sites in the rat cerebral cortex. *Eur. J. Pharmacol.* **65,** 289–292.

Palkovits M., Raisman R., Briley M., and Langer S. Z. (1981) Regional distribution of (^{3}H)-imipramine in rat brain. *Brain Res.* **210,** 493–498.

Paul S. M., Rehavi M., Skolnick P., and Goodwin F. K. (1980) Demonstration of specific high-affinity binding sites for (^{3}H)-imipramine in human platelets. *Life Sci.* **26,** 953–959.

Paul S. M., Rehavi M., Skolnick P., Ballenger J., and Goodwin F. K. (1981) Depressed patients have decreased binding of tritiated imipramine to platelet serotonin "transporter." *Arch. Gen. Psychiat.* **38,** 1315–1317.

Peroutka S. J. and Snyder S. H. (1980) Long-term antidepressant treatment decreases spiroperidol-labeled serotonin receptor binding. *Science* **210,** 88–90.

Perry E. K., Marshall E. F., Blessed G., Tomlinson B. E., and Perry R. H. (1983) Decreased imipramine binding in the brains of patients with depressive illness. *Brit. J. Psychiat.* **142,** 188–192.

Plenge P. and Mellerup E. T. (1982) ^{3}H-Imipramine high-affinity binding sites in rat brain. Effects of imipramine and lithium. *Psychopharmacol.* **77,** 94–97.

Post R. M., Kotin J., and Goodwin F. K. (1974) The effect of cocaine on depressed patients. *Am. J. Psychiat.* **131,** 511–517.

Raisman R., Briley M., and Langer S. Z. (1979a) High-affinity ^{3}H-imipramine binding in rat cerebral cortex. *Eur. J. Pharmacol.* **54,** 307–308.

Raisman R., Briley M., and Langer S. Z. (1979b) Specific tricyclic antidepressant binding sites in rat brain. *Nature* (Lond.) **281,** 148–150.

Raisman R., Briley M. S., and Langer S. Z. (1980) Specific tricyclic antidepressant binding sites in rat brain characterized by high-affinity ^{3}H-imipramine binding. *Eur. J. Pharmacol.* **61,** 373–380.

Raisman R., Sechter D., Briley M. S., Zarifian E., and Langer S. Z. (1981) High-affinity ^{3}H-imipramine binding in platelets from untreated and treated depressed patients compared to healthy volunteers. *Psychopharmacol.* **75,** 368–371.

Raisman R., Sette M., Pimoule C., Briley M., and Langer S. Z. (1982) High-affinity (^{3}H)-desipramine binding in the peripheral and central nervous system: a specific site associated with the neuronal uptake of noradrenaline. *Eur. J. Pharmacol.* **78,** 345–351.

Rehavi M., Paul S. M., Skolnick P., and Goodwin F. K. (1980) Demonstration of specific high-affinity binding sites for (^{3}H)-imipramine in human brain. *Life Sci.* **26,** 2273–2279.

Rehavi M., Skolnick P., Hulihan B., and Paul S. M. (1981) "High affinity" binding of (^{3}H)-desipramine to rat cerebral cortex: relationship to tricyclic antidepresssant-induced inhibition of norepinephrine uptake. *Eur. J. Pharmacol.* **70,** 597–599.

Rehavi M., Skolnick P., Brownstein M. J., and Paul S. M. (1982) High-affinity binding of (^{3}H)-desipramine to rat brain: a presynaptic marker for noradrenergic uptake sites. *J. Neurochem.* **38,** 889–895.

Rehavi M., Skolnick P., and Paul S. M. (1983) Subcellular distribution of high-affinity (^{3}H)-imipramine binding and (^{3}H)-serotonin uptake in rat brain. *Eur. J. Pharmacol.* **87,** 335–339.

Rehavi M., Ventura I., and Sarne Y. (1984) Endogenous regulator of imipramine recognition site. *14th CINP Congress* (Abstracts) June 19–23, 1984, Florence, Italy, p. 282.

Reith M. E. A., Sershen H., Allen D., and Lajtha A. (1983) High- and low-affinity binding of (^{3}H)-imipramine in mouse cerebral cortex. *J. Neurochem.* **40,** 389–395.

Rosenthal E. H. (1967) A graphic method for the determination and presentation of binding parameters in a complex system. *Anal. Biochem.* **20,** 525–532.

Ross S. B. (1976) Long-term effects of *N*-2-chloroethyl-*N*-ethyl-2-bromobenzylamine hydrochloride on noradrenergic neurons in the rat brain and heart. *Brit. J. Pharmacol.* **58,** 521–527.

Ross S. B. and Renyi A. (1975) Tricyclic antidepressant agents. I. Comparison of the inhibition of the uptake of ^{3}H-noradrenaline and ^{14}C-5-hydroxytryptamine in slices and crude synaptosome preparation of the midbrain–hypothalamus region of the rat brain. *Acta Pharmacol. Toxicol.* **36,** 382–394.

Sachs C. and Jonsson G. (1975) 5,7-Dihydroxytryptamine induced changes in the postnatal development of central 5-hydroxytryptamine neurons. *Med. Biol.* **53,** 156–164.

Schildkraut J. J. (1965) The catecholamine hypothesis of affective disorders: a review of the supporting evidence. *Am. J. Psychiat.* **122,** 509–522.

Sette M., Raisman R., Briley M., and Langer S. Z. (1981) Localization of tricyclic antidepressant binding sites on serotonin nerve terminals. *J. Neurochem.* **37,** 40–42.

Sette M., Briley M. S., and Langer S. Z. (1983a) Complex inhibition of (^{3}H)-imipramine binding by serotonin and nontricyclic serotonin uptake blockers. *J. Neurochem.* **40,** 622–628.

Sette M., Ruberg M., Raisman R., Scatton B., Zivkovic B., Agid Y., and Langer S. Z. (1983b) (^{3}H)-Imipramine binding in subcellular fractions of the rat cerebral cortex after chemical lesion of serotonergic neurons. *Eur. J. Pharmacol.* **95,** 41–51.

Smith C. B., Garcia-Sevilla J. A., and Hollingsworth P. J. (1981) α_2-Adreno-receptors in rat brain are decreased after long-term tricyclic antidepressant treatment. *Brain Res.* **210,** 413–418.

Snyder S. H. and Yamamura H. I. (1977) Antidepressants and the muscarinic acetylcholine receptors. *Arch. Gen. Psychiat.* **34,** 236–239.

Stanley M., Virgilio J., and Gershon S. (1982) Tritiated imipramine binding sites are decreased in the frontal cortex of suicides. *Science* **216,** 1337–1339.

Suranyi-Cadotte B. E., Wood P. L., Nair N. P. V., and Schwartz G. (1982) Normalization of platelet ^{3}H-imipramine binding in depressed patients during remission. *Eur. J. Pharmacol.* **85,** 357–358.

Suranyi-Cadotte B. E., Wood P. L., Schwartz G., and Nair N. P. V. (1983) Altered platelet (^{3}H)-imipramine binding in schizo-affective and depressive disorders. *Biol. Psychiatry* **18,** 923–927.

Tran V. T., Lebovitz R., Toll L., and Snyder S. H. (1981) (^{3}H)-Doxepin interactions with histamine H_1-receptors and other sites in guinea pig and rat brain homogenates. *Eur. J. Pharmacol.* **70,** 501–509.

Tuomisto J. and Tukiainen E. (1976) Decreased uptake of 5-hydroxytryptamine in blood platelets from depressed patients. *Nature* (Lond.) **262,** 596–598.

U'Prichard D. C., Greenberg D. A., Sheehan P. R., and Snyder S. H. (1978) Tricyclic antidepressants: therapeutic properties and affinity for α-noradrenergic receptor binding sites in the brain. *Science* **199,** 197–199.

Vetulani J., Stawarz R. J., Dingell J. V., and Sulser F. (1976) A possible common mechanism of action of antidepressant treatments—reduction in the sensitivity of noradrenergic cyclic AMP generating system in the rat limbic forebrain. *Naunyn-Schmiedeberg's Arch. Pharmacol.* **293,** 109–114.

Whitaker P. M., Warsh J. J., Stancer H. C., Persad E., and Vint C. K. (1984) Seasonal variation in platelet ^{3}H-imipramine binding: comparable values in control and depressed populations. *Psychiatry Res.* **11,** 127–131.

Wong D. T., Bymaster F. P., Horng J. S., and Molloy B. B. (1975) A new selective inhibitor for uptake of serotonin into synaptosomes of rat brain: 3-(*p*-trifluoromethylphenoxy)-*n*-methyl-3-phenylpropylamine. *J. Pharmacol. Exp. Therap.* **193,** 804–811.

Wood P. L., Suranyi-Cadotte B. E., Nair N. P. V., LaFaille F., and Schwartz G. (1983) Lack of association between (^{3}H)-imipramine binding sites and uptake of serotonin in control, depressed, and schizophrenic patients. *Neuropharmacol.* **22,** 1211–1214.

Zsilla G., Barbaccia M. L., Gandolfi O., Knoll J., and Costa E. (1983) (–)-Deprenyl a selective MAO "B" inhibitor increases (^{3}H)-imipramine binding and decreases β-adrenergic receptor function. *Eur. J. Pharmacol.* **89,** 111–117.

Chapter 14
Phencyclidine (PCP) and Amphetamine Receptor Binding Sites

Remi Quirion

1. Introduction

Both phencyclidine [1-(1-phenylcyclohexyl)piperidine, PCP, angel dust] and amphetamine are widely abused in North America (Biel and Bopp, 1978; Petersen and Stillman, 1978). Among their various effects, it is well known that PCP induces sedation, sensory isolation, and hallucinations. Morever, drug abusers have also described strong feelings of depersonalization that make PCP an impressive drug model of schizophrenia (Domino, 1981; Snyder, 1980). These disturbances in thinking have been described as "loosening of association, overinclusive thoughts, delusional thinking, and general disorganization, all strikingly reminiscent of thought disorders of schizophrenics" (Snyder, 1980). Thus, PCP's pharmacological profile is unlike that of any other psychoactive drug and may provide a good animal model of schizophrenia (Domino, 1981; Quirion and Pert, 1983a; Pert and Quirion, 1983).

Amphetamine and related phenylethylamine derivatives possess pronounced behavioral actions, including central nervous system stimulation, anorexia, vasoconstriction, and hyperthermia (Biel and Bopp, 1978). Possible use of amphetamines as anorectic and stimulant drugs has generated great scientific interest and is likely to be related to its abuse. The *precise* mechanism of the action of amphetamine is still unknown, but is probably related to biogenic amines (Biel and Bopp, 1978).

Thus, PCP and amphetamine are unique CNS-acting drugs that possess very peculiar spectra of actions. However, much remains to be known about their respective molecular mechanisms of action. In this review, I will describe recent evidence suggesting the existence of specific brain receptor binding sites for both PCP and amphetamine. The possible physiological significance of the presence of specific receptors for these two drugs of abuse is also discussed.

2. Neurochemical Pharmacology of PCP

Over the past 10–15 yr, various groups have reported on the interaction of PCP with various neurotransmitter systems. It seems appropriate to briefly review here some of the evidence that should place in proper context the subsequent search for PCP receptor binding sites.

2.1. Effects on Cholinergic Systems

It is well known that PCP and related analogs interact in various ways with cholinergic systems (Table 1). PCP inhibits (at micromolar to millimolar doses) acetylcholine release, choline uptake, acetylcholinesterase activity, muscarinic and nicotinic (peripheral and invertebrates) receptors, and cholinergic-induced contractions. Acetylcholine turnover is also stimulated by high doses of PCP in rat brain (Domino and Wilson, 1972; Murray, 1983). Thus, PCP appears to be a relatively weak antagonist of peripheral and central cholinergic systems. Moreover, the discordant results of behavioral studies (Adams, 1980; Chait and Balster, 1978; Pinchasi et al., 1978) do not allow any definitive conclusions concerning the physiological significance of the action of PCP on cholinergic neurons in the brain.

Finally, recent reports have demonstrated that the ligand selectivity pattern of [^{3}H]-PCP binding to nicotinic receptors present in invertebrate electric organs is not correlated to its behavioral activities (Eldefrawi et al., 1982a). The same group has also reported that [^{3}H]-PCP binds with high affinity to its own sites in the guinea pig ileum and only with a much lower potency to muscarinic receptors (El-Fakahany et al., 1984a). All these data indicate that PCP interactions with cholinergic systems are unlikely to be associated with the relevant CNS effects of PCP.

2.2. Effects on Dopaminergic Systems

Both PCP (Table 2) and amphetamine act on dopaminergic systems, often in a similar manner. For example, PCP and amphetamine decrease tyrosine hydroxylase activity in vivo (Smith et al.,

TABLE 1
Effects of PCP and Derivatives on Various Cholinergic Parameters

Parameter	PCP action, μ*M*–m*M*	References
Acetylcholine turnover	Stimulation	Domino and Wilson (1972) Murray (1983) Murray and Cheney (1981)
Acetylcholine release	Inhibition	Leventer and Johnson (1983) Wolverton et al. (1980)
Choline uptake	Inhibition	Haggerty et al. (1983)
Acetylcholinesterase activity	Inhibition	Maayani et al. (1974) Pinchasi et al. (1977)
Cholinergic-induced contraction	Inhibition	Aronstam et al. (1980) Gabrielovitz et al. (1980) Maayani et al. (1974)
Muscarinic receptor	Inhibition	Aronstam et al. (1980) Boggan et al. (1982) Fosset et al. (1979) Vincent et al. (1978)
Nicotinic receptor	Inhibition	Albuquerque et al. (1980)

1977; Walters and Roth, 1976) and dopamine-stimulated adenylate cyclase activity (Smith et al., 1981). However, PCP decreases striatal DOPAC and homovanillic acid (HVA) levels. Thus, these data would not suggest an increased release of striatal dopamine (Doherty et al., 1980; Meltzer et al., 1981).

Behaviorally, PCP induces locomotor hyperactivity, stereotypy, and ataxia in the rat (Castellani and Adams 1981a,b; Meltzer et al., 1981). PCP-induced hyperactivity and stereotypy, like that of amphetamine, can be inhibited by pretreatment with neuroleptics (Castellani and Adams, 1981b; Freed et al., 1984; Murray and Horita, 1979; Sturgeon et al., 1981) and by selective lesions of dopaminergic pathways (French and Vantini, 1984; Nabeshima et al., 1983a,b). However, recent biochemical data indicate that PCP does not directly alter brain levels of tyrosine, dopamine, noradrenaline, or HVA. PCP is only able to potentiate haloperidol-induced decreases in dopamine levels following amphetamine treatment, as well as haloperidol-induced elevation of HVA (Johnson and Oeffinger, 1981). These effects are similar to those reported for the "nonamphetamine-type" stimulants, such as cocaine, methyl-

TABLE 2
Effects of PCP and Derivatives on Various Dopaminergic Parameters

Parameter	PCP action, μ*M*–m*M*	References
Tyrosine hydroxylase activity	Inhibition (in vivo)	Smith et al. (1977)
	Stimulation (in vitro)	Vickroy and Johnson (1981)
Dopamine-stimulated adenylate cyclase activity	Inhibition	Smith et al. (1981)
Levels of dopamine metabolites	Decrease	Doherty et al. (1980) Meltzer et al. (1981)
Dopamine release	Stimulation	Ary and Komisky (1980) Bagchi (1981) Doherty et al. (1980) Johnson and Oeffinger (1981)
Dopamine uptake	Inhibition	Doherty et al. (1980) Johnson and Oeffinger (1981) Vignon and Lazdunski (1984)
Caudate neurons	Inhibition	Johnson et al. (1984)
Activity		Quirion et al. (1981a)
dopamine receptors	No effect	Vincent et al. (1978)
Amphetamine stereotypy	Potentiation	Balster and Chait (1978)
PCP-induced behaviors: locomotor activity stereotypy	Antagonism by neuroleptics; reduction following lesions	Castellani and Adams (1981b) Freed et al. (1984) Murray and Horita (1979) Sturgeon et al. (1981) French and Vantini (1984) Nabeshima et al. (1983a,b)

phenidate, and amfonelic acid (Shore, 1976). Thus, it appears that the effects of PCP on dopaminergic systems are complex and do not precisely fit into any class of drugs interacting with this important neurotransmitter.

2.3. Effects on Serotonergic Systems

Much less is known about interactions between PCP and serotonergic systems. It has been shown that PCP blocks serotonin (5-hydroxytryptamine, 5-HT) synthesis and metabolism, probably by inhibiting serotonin uptake (Table 3). Behaviorally, PCP induces in rats a biphasic response consisting of the serotonergic syndrome, *followed* by stereotyped behavior and ataxia (Martin et al., 1979). However, it is clear that the PCP-induced stereotyped behavior and ataxia only partly involve serotonergic systems (Martin et al., 1979; Freed et al., 1984). Thus, it appears unlikely that PCP induces most of its behavioral activities by acting through serotonergic mechanisms.

2.4. Effects on Noradrenergic and Adrenergic Systems

Many of the effects of PCP on noradrenergic and adrenergic systems are similar to those reported for dopamine (Tables 2 and 4). PCP is able to block catecholamine reuptake, decrease total brain catecholamine content, and stimulate catecholamine release (Table 4). Moreover, PCP-induced electrophysiological depressions of hippocampal and cerebellar neurons appear to involve actions on presynaptic noradrenergic terminals (Bickford et al., 1981; Marwaha et al., 1980; Palmer et al., 1983; Raja and Guyenet, 1980). Moreover, the α_1 antagonist, prazosin, and the α_2 agonist, clonidine, block various behavioral and biochemical effects of PCP (Vonvoigtlander et al., 1983). For example, the ability of PCP to disrupt Y-maze performance involving light–dark discrimination is blocked by prazosin and clonidine. Biochemically, the marked enhancement of glucose metabolism induced by PCP in corticolimbic regions (Meibach et al., 1979; Hammer et al., 1982) is completely blocked by prazosin and clonidine (Vonvoigtlander et al., 1983). However, it is known that PCP does not directly act on catecholaminergic receptors (Table 4), and the peculiar pharmacological profile of PCP is certainly not easily explained by exclusive actions on catecholaminergic pathways.

TABLE 3
Effects of PCP and Derivatives on Various Serotonergic Parameters

Parameter	PCP action, μM–mM	References
Serotonin uptake	Inhibition	Arora and Meltzer (1980) Martin et al. (1982) Smith et al. (1977)
Serotonin synthesis and metabolism	Inhibition	Martin and Smith (1982)
Serotonin receptors	No effect	Quirion et al. (1981a) Vincent et al. (1978)
Serotonergic syndrome	Induced	Martin et al. (1979)
PCP-induced behavior	Antagonism by neuroleptics-	Freed et al. (1984)
	reduction following lesions.	Nabeshima et al. (1983b)

TABLE 4
Effects of PCP and Derivatives on Various Catecholaminergic Parameters

Parameter	PCP action, μM–mM	References
Catecholamine uptake	Inhibition	Ary and Kowinskey (1980) Smith et al. (1977) Taube et al. (1975)
Catecholamine release	Stimulation	Bayorh and Kopin (1982) Marwaha (1982)
Catecholamine content	Decrease	Leonard and Tonge (1969)
Catecholamine receptors	No effects	Quirion et al. (1981a) Vincent et al. (1978)
Action on catechol-aminergic neurons	Inhibition	Bickford et al. (1981) Marwaha (1982) Marwaha et al. (1980) Palmer et al. (1983) Raja and Guyenet (1980)
PCP-induced behaviors	Blockade by α-adrenergic antagonists	vonVoigtlander et al. (1983)

2.5. Effects on Amino Acids, Peptides, and Ion Channels

In the past few years, the interaction of PCP with amino acid neurotransmitters has generated much interest. Even if PCP/GABA interactions appear to be fairly limited (Table 5), recent observations have demonstrated that PCP selectively reduces the excitatory effects of another amino acid transmitter, *N*-methyl-D,L-aspartate (NMA) on spinal neurons (Berry et al., 1983; Lodge and Anis, 1982). NMA is a selective agonist for one of the three classes of excitatory amino acid receptors (L-glutamate and L-aspartate) in the brain. Moreover, the relative potency of various PCP derivatives in reducing the excitatory action of NMA correlates well with behavioral and binding data (Berry et al., 1983). However, NMA and PCP do not appear to bind to the same substrate to trigger their primary events (Table 5).

Recent data also suggests that certain adenosine receptor agonists such as N^6-cyclohexyladenosine and L-phenylisopropyladenosine block the discriminative properties of PCP (Browne and Welch, 1982). However, it is unlikely that this effect is through receptor blockade since PCP is a very weak inhibitor of adenosine receptors (Table 5). Further studies are needed to precisely determine the mechanism involved in PCP–adenosine agonists interaction.

Very few studies related to possible actions of PCP on neuropeptides have been done. Only opioid peptides and their receptors have been investigated thus far (Table 5). Vincent et al. (1978) have reported that PCP is a weak inhibitor (K_i = 26 μM) of [^{3}H]-dihydromorphine binding. Two other groups have shown that PCP alters enkephalin contents in the brain (Fratta et al., 1983; Nabeshima et al., 1983c,d). However, opioid peptide levels were differentially modified following acute vs chronic PCP treatment (Table 5).

Most PCP behaviors are not blocked by various opiate antagonists such as naloxone and naltrexone (Table 1). Only a few reports have shown that PCP-related behavioral effects could be either potentiated (Contreras et al., 1981; Wagner et al., 1984) or blocked (Glick and Guido, 1982) by opioid receptor antagonists. Thus, it appears that PCP interactions with opioid systems are limited, although further investigations are needed.

PCP acts on various ion channels (Table 6). For example, certain calcium channel antagonists, such as verapamil and D-600, are able to displace [^{3}H]-PCP binding in the low micromolar range (El-Fakahany et al., 1984b; Eldefrawi et al., 1982b; Quirion and Pert, 1982a). PCP-induced cerebrovasospasms and inotropic action are also blocked by certain calcium channel antagonists (Table 6). However, PCP does not seem to significantly block Ca^{2+} channels in various electrophysiological assays and it is unlikely that behavioral actions of PCP are related to Ca^{2+} channel blockade.

PCP also interacts with sodium channels (Table 6), but its most potent action is certainly on potassium channels. It has been reported that PCP, in the nanomolar range, blocks depolarization-activated K^+ channels in rat brain synaptosomes (Albuquerque et al., 1981; Blaustein and Ickowicz, 1983) and neuroblastoma cell lines (Tourneur et al., 1982). Moreover, the potencies of various PCP derivatives in blocking the K^+ channel correlate relatively well with their respective activities in PCP-behavioral assays (Aguayo et al., 1984; Tourneur et al., 1982). Thus, it appears that PCP blockade of K^+ channels could be related to some CNS effects of this drug.

3. PCP Receptor Binding Sites

3.1. Early Evidence

As described above, various pieces of evidence suggest that PCP interacts with many different neurotransmitters (acetylcholine, dopamine, serotonin, noradrenaline, aspartate, GABA, opiate) and ion channels (K^+, Na^+, Ca^{2+}). However, none of these interactions fully explain the unique pharmacological profile of PCP. Similarly, multiple mechanisms of action were also suggested for morphine and congeners before the discovery of highly selective opiate receptors in brain (Pert and Snyder, 1973).

In 1979, two groups independently reported on the existence of specific [^{3}H]-PCP receptors in rat brain and suggested that these sites were the major target sites that mediated PCP actions (Vincent et al., 1979; Zukin and Zukin, 1979).

Vincent et al. (1979) demonstrated that [^{3}H]-PCP binds to rat brain synaptic membranes in a saturable, reversible, and selective manner with high affinity (250 n*M*) to a maximal number of sites (2.4 pmol/mg protein). [^{3}H]-PCP binding was concentrated in cerebral cortex and striatum. Only PCP-related analogs were able to bind to the PCP receptor. Other neurotransmitters and/or neuromodulators such as acetylcholine, glycine, glutamate, epinephrine, serotonin, dopamine, substance P, and enkephalins were unable to compete for [^{3}H]-PCP binding sites. Moreover, there was a good correlation between the apparent affinities of PCP analogs in the binding assay and their activities in the rotarod behavioral assay, suggesting that [^{3}H]-PCP binding sites were relevant to the pharmacological effects of PCP.

TABLE 5
Interactions of PCP and Derivatives With Various Amino Acids and Peptides in the Brain

Compound	PCP action, μ*M*–m*M*	References
Amino Acids		
Glutamic acid decarboxylase	Stimulation Inhibition	Hsu et al. (1980) Peat and Gibb (1983)
GABA content[a]	Decrease	Tonge and Leonard (1971)
GABA receptors	No effect	Quirion et al. (1981a)
Glutamate and aspartate electro-physiological effects (*N*-methyl-D,L-aspartate)	Inhibition	Berry et al. (1983) Lodge and Anis (1982) Raja and Guyenet (1982)
N-methyl-D,L-aspartate receptors	No effect	Quirion (unpublished results)
PCP-discriminative stimulus	Blockade by adenosine agonists	Browne and Welch (1982) Browne et al. (1983)

Adenosine receptor	Inhibition at high doses	Browne et al. (1983)
Peptides		
Enkephalin contents	Decrease (acute)	Fratta et al. (1983)
		Nabeshima et al. (1983c)
	Increase (chronic)	Nabeshima et al. (1983d)
Opioid receptors	Inhibition at high doses	Vincent et al. (1978)
PCP-induced behaviors	No effect of opioid antagonists	Browne (1982)
		Brady and Balster (1982)
		Holtzman (1980)
		Shannon (1981)
		Vaupel (1983)
		Verebey et al. (1981)
	Potentiation by opioid antagonists	Contreras et al. (1981)
		Wagner et al. (1984)
	Blockade by opioid antagonists	Glick and Guido (1982)

[a]GABA = gamma-aminobutyric acid.

TABLE 6
Interactions of PCP and Derivatives With Various Ion Channels

Channel	PCP action, n*M*–μ*M*	References
Potassium (K^+)	Blockade	Aguayo et al. (1984) Albuquerque et al. (1981) Blaustein and Ickowicz (1983) Tourneur et al. (1982)
Sodium (Na^+)	Blockade	Tourneur et al. (1982)
Calcium (Ca^{2+}) antagonists	Inhibition of [^{3}H]-PCP binding	El-Fakahany et al. (1984b) Eldefrawi et al. (1982b) Quirion and Pert (1982a)
	Inhibition of PCP-induced cerebro-vasospasms iontropic action	Altura and Altura (1983) Temma et al. (1983)
Ca^{2+} Translocation processes	Blockade	Calixto et al. (1983)

Similar results were obtained by Zukin and Zukin (1979). They reported that [^{3}H]-PCP binds specifically and with high-affinity (150 n*M*) to a single class of saturable binding sites in rat brain membranes. Bound [^{3}H]-PCP was displaced by PCP and by a series of derivatives with relative potencies that paralleled those determined in the rotarod and the discriminative stimulus behavioral assays. [^{3}H]-PCP binding was concentrated in the hippocampus and cortical areas. Trypsin and pronase reduced specific [^{3}H]-PCP binding, demonstrating the proteinous nature of the binding site. Thus, they suggested that PCP exerts its effects on the central nervous system via binding to its own brain receptor sites.

3.2. Controversy

Just a few months after the publication of the reports by Vincent et al. (1979) and Zukin and Zukin (1979), Maayani and Weinstein (1980) reported that the specific binding of [^{3}H]-PCP was only an artifact of the rapid filtration method, the binding technique used in the first two studies. They reported that [^{3}H]-PCP binds to GF/B glass fiber filters in a manner undistinguishable from binding to membrane preparations. Moreover, binding of [^{3}H]-PCP to the P_2 fraction of rat brain membranes was not affected by denaturing tissues in boiling water for 5 min. They also suggested that the very large degree of filter binding of [^{3}H]-PCP was the main reason for differences in the density of PCP binding found in various brain regions. They finally concluded that "the artifacts created in this case by the rapid filtration method render it unsuitable for labeling the pharmacologically relevant binding sites of [^{3}H]-PCP" (Maayani and Weinstein, 1980), and suggested that if PCP receptors exist, methods other than rapid filtration should be used to reveal them.

Naturally, Maayani and Weinstein's report raises important doubts on the existence of PCP receptors in the brain and was widely publicized in the scientific literature. In retrospect, it seems that this report had at least the merit of stimulating further research in PCP binding sites, as well as showing that great care must be taken in the interpretation of receptor binding data.

Following that report (Maayani and Weinstein, 1980), the first two groups involved in the characterization of [^{3}H]-PCP binding sites performed an extensive series of experiments that clearly demonstrated the existence of specific [^{3}H]-PCP binding sites in brain, using the rapid filtration method (Kamenka et al., 1982; Lazdunski et al., 1983; Vignon et al., 1982; Vincent et al., 1980a,b, 1981, 1983; Zukin and Zukin, 1981; Zukin et al., 1983, 1984).

Lazdunski and coworkers performed multiple controls to ensure that [^{3}H]-PCP binding sites represent a unique biochemical and pharmacological entity. First, they demonstrated that it was possible to determine [^{3}H]-PCP binding by rapid filtration (Vincent et al., 1980a) when substracting specific [^{3}H]-PCP binding to filters alone from binding in presence of membranes. They clearly showed that binding to filters displaceable by unlabeled PCP is always much lower than specific binding to rat brain homogenate. [^{3}H]-PCP binding was also dependant on the membrane concentration present in the assay. Moreover, it was shown that [^{3}H]-PCP binding should be done in medium of very low ionic strength. Under these conditions, the affinity of the [^{3}H]-PCP binding sites was markedly increased (from 250 to 63 n*M* at low ionic strength). Also, [^{3}H]-PCP binding sites present in the brain are different from those found in peripheral tissues such as the liver and the kidney; with only the ligand selectivity pattern of [^{3}H]-PCP brain binding correlated with data obtained in the rotarod behavioral assay (Vincent et al., 1980b).

In a subsequent study (Vignon et al., 1982), [^{3}H]-PCP binding sites were characterized in greater detail. It was shown that [^{3}H]-PCP binding sites were rapidly heat-inactivated at temperatures over 50°C (5–10 min) and destroyed by proteases like trypsin, pronase, and papain, suggesting a proteinous nature. A good correlation was found between the pharmacological activity (rotarod assay) of at least 30 PCP analogs and their respective affinity in the binding assay. Finally, various ions markedly decreased [^{3}H]-PCP binding, revealing the high sensitivity of PCP binding sites to ionic strength. Thus, these data demonstrate that under proper conditions it is possible to characterize [^{3}H]-PCP brain receptors using the rapid membrane filtration technique.

Similar studies were also conducted by Zukin and coworkers (Zukin et al., 1983, 1984). They demonstrated that the presoaking of filters in 0.1% poly-L-lysine eliminated displaceable [^{3}H]-PCP binding to filters. Under these conditions, K_d values obtained at equilibrium were in agreement with kinetic data derived from association and dissociation rate constants. The stereospecificity of the PCP receptor was demonstrated by showing that (+)-ketamine was four times more potent than (–)-ketamine in displacing specifically bound [^{3}H]-PCP. Similar data were also obtained using centrifugation assays. Again, these data demonstrate that the rapid filtration method is adequate to characterize [^{3}H]-PCP binding sites when performed under appropriate conditions.

Various other groups have now reported on the characterization of [^{3}H]-PCP binding sites using either the rapid filtration method (Geary et al., 1983; Hampton et al., 1982; Jasinski et al., 1979; McQuinn et al., 1981; Mendelsohn et al., 1984; Murray and Leid, 1984), or an in vitro autoradiographic technique (Kozlowski and Lebel, 1983; Quirion et al., 1981a,b, 1982a,b, 1983; Zukin and Sircar, 1984). As shown in Table 7, it is generally agreed that [^{3}H]-PCP binds to an apparent saturable class of sites with an affinity between 25 and 250 n*M* and a number of sites B_{max} between 0.87 and 3.0 pmol/mg protein. Various groups have also reported that enzymes and heat denaturation treatments markedly decrease [^{3}H]-PCP binding to brain membranes (Table 7). Moreover, presoaking glass fiber filters in either polylysine, *tert*-amyl alcohol or polyethyleneimine drastically reduces displaceable [^{3}H]-PCP binding to filters (Hampton et al., 1982; McQuinn et al., 1981; Murray and Leid, 1984; Quirion et al., 1984; Zukin et al., 1983). Under these conditions, it is possible to demonstrate that the relative potencies of PCP and its derivatives in binding assays correlate very well with their respective activities in various behavioral assays (Table 7). Thus, an important body of literature reported from many independent laboratories has clearly demonstrated the existence of [^{3}H]-PCP binding sites in mammalian brain (Table 7).

3.3. Autoradiographic Visualization and Distribution of PCP Binding Sites

Another approach used to study receptor binding sites is the in vitro receptor autoradiography technique. With this method, it is possible to pharmacologically (kinetics, ligand selectivity) and anatomically characterize neurotransmitters as well as drug receptors. We used this technique to investigate the presence of [^{3}H]-PCP binding sites in the brain.

In the first series of experiments, we investigated the effects of preincubation, ions, and temperature on [^{3}H]-PCP binding to rat brain sections. Preincubation at 4°C for 15 min in presence of a low concentration of NaCl doubled specific [^{3}H]-PCP binding to brain sections. Preincubations at 25°C did not seem to be as efficient in increasing the signal-to-noise ratio. The increase in [^{3}H]-PCP binding after preincubation is probably caused by the dissociation of putative endogenous ligands from the binding sites (*see* section 3.8). Following the preincubation, the highest [^{3}H]-PCP specific binding is obtained by an incubation at 4°C for 45 min in 5.0 m*M* Tris-HCl, pH 7.4, plus 50 m*M* sucrose. Incubations with monovalent and/or divalent cations markedly decrease [^{3}H]-PCP binding (Quirion et al., 1981a). Similar results have also been reported by Vignon et al. (1982). Saturation experiments and Scatchard analy-

TABLE 7
Characteristics of [^{3}H]-PCP Binding Sites Present in Mammalian Brain

Group	K_d[a]	B_{max}[b]	Enzyme and heat denaturation	Filter	Stereo-selec-tivity	Selec-tivity pattern	Distri-bution
Geary et al. (1983)	ND[c]	ND	ND	ND	*	*	ND
Hampton et al. (1982)	60–70 300–400	ND	*[d]	0.05% PEA[e]	*	*	ND
Jasinski et al. (1981)	ND	ND	ND	ND	*	*	ND
Kozlowski and Lebel (1983)	279.5	3.1	ND	NA[f]	*	*	*
McQuinn et al. (1981)	ND	ND	ND	*tert*-Amyl alcohol	ND	*	ND
Mendelsohn et al. (1984)	151.3	2.91	*	ND	*	*	*

Murray and Leid (1984)	27.2	0.87	*	0.1% Poly-L-lysine	*	*	*
Quirion et al. (1981a,b)	46	10.5 fmol/ Section	*	NA	*	*	*
Vignon et al. (1981)	250	2.4	*	Subtraction	*	*	*
Zukin and Sircar (1984)	ND	ND	ND	NA	ND	ND	*
Zukin et al. (1983)	85	2.0	*	0.01% Polylysine	*	*	*

[a] K_d is expressed as n*M*.
[b] B_{max} is expressed as pmol/mg protein.
[c] ND, no data available.
[d] *, indicates that experiments have been performed.
[e] PEA, polyethyleneimine.
[f] NA, not applicable.

sis of [^{3}H]-PCP binding revealed an apparently saturable single class of sites with K_d = 46 nM and B_{max} = 10.5 fmol/section (Quirion et al., 1981a).

A number of groups have reported on the relative potency of various PCP analogs in [^{3}H]-PCP binding assays (Hampton et al., 1982; Murray and Leid, 1984; Quirion et al., 1981a,b; Vincent et al., 1979; Zukin and Zukin, 1979). As shown in Table 8, [^{3}H]-PCP binding is potently inhibited by analogs such as *N*-ethylphenylcyclohexylamine (PCE), but not by others like ketamine. Moreover, their relative potencies obtained in [^{3}H]-PCP binding assays correlate very well with those reported in behavioral assays (Shannon, 1981, 1982a). More recently, Altura et al. (1983) also reported that the relative potencies of PCP and congeners in inducing contractions of isolated dog cerebral arteries are correlated with their activities in the [^{3}H]-PCP binding assay. Finally, using enantiomers of PCP, we have demonstrated the stereoselective nature of the [^{3}H]-PCP binding sites (Quirion et al., 1981b). This strongly suggests that [^{3}H]-PCP binding sites are relevant to the pharmacological effects induced by PCP and related compounds.

Among other classes of drugs, sigma opiates (*see* section 3.5), amantadine and certain calcium channel antagonists appear to interact with [^{3}H]-PCP binding sites. We observed that in the presence of amantadine, an antiviral drug with antiparkinsonian activity (Grelak et al., 1970), the affinity of [^{3}H]-PCP for its receptors is significantly increased (Quirion and Pert, 1982b). The number of sites (B_{max}) does not appear to be changed by amantadine. Interestingly, rimantadine, an analog of amantadine completely devoid of any central nervous system effects, is also totally inactive on the [^{3}H]-PCP binding site. It is tempting to speculate that some CNS effects of amantadine could be related to an interaction of this drug with the PCP–receptor complex.

Another family of drugs that appears to interact with [^{3}H]-PCP binding sites is the group of calcium channel antagonists including verapamil and its methoxy derivative, D-600. However, other calcium channel antagonists, such as nifedipine and nitrendipine, are unable to displace [^{3}H]-PCP binding (Quirion and Pert, 1982a). Similar results have been reported by Eldefrawi et al. (1982a,b) in crayfish muscle. Moreover, West et al. (1983) recently reported that [^{3}H]-ethylketocyclazocine binds to a lower affinity site in NCB-20 Hybrid neurotumor cells. This site is sensitive to PCP, its derivatives, calcium antagonists like verapamil, and to an alpha-adrenergic antagonist, WB-4101. Thus, it seems that PCP and certain calcium channel antagonists could bind to similar sites in brain.

TABLE 8
Relative Potencies of PCP and Derivatives in [^{3}H]-PCP Binding Assays and in Behavioral Tests and Bioassays

Drug	Binding[a]	Binding[b]	Binding[c]	Binding section[d]	Behavior[e]	Bioassay[f]
PCP	100	100	100	100	100	100
PCE	164	—	—	600	579	480
TCP	144	—	—	170	131	120
PHP	115	—	—	140	97	—
THP	77	—	—	80	87	60
PCDEA	45	—	—	30	83	24
Ketamine	23	2.8	26	10	10	18
TCM	14	—	—	5	7	—
PCC	4	—	—	<0.01	0	0
(+)PCMP	—	—	—	23	—	—
(–)PCMP	—	—	—	4.1	—	—
(±)SKF 10,047	20	—	—	30	29	176
(+)SKF 10,047	14	9.8	18	26	50	—
(–)SKF 10,047	32	3.7	4	37	16	—

For abbreviations, *see* Quirion et al. (1981a)
[a]From Zukin and Zukin (1979) and Zukin et al. (1983; 1984).
[b]From Hampton et al. (1982).
[c]From Murray and Leid (1984).
[d]From Quirion et al. (1981a,b, and unpublished results).
[e]From Shannon (1981; 1982a,b).
[f]From Altura et al. (1983).

TABLE 9
Autoradiographic Distribution of [^{3}H]-PCP Binding Sites in Rat Brain[a]

Structure	Density of [^{3}H]-PCP binding[b]
Olfactory bulb	+++
Frontal cortex	+++
Cingulate cortex	+++
Hippocampus	+++
Dentate gyrus	+++
Visual cortex	+++
Entorhinal cortex	+++
Ventral horn of spinal cord	+++
Striatum	++
Nucleus accumbens	++
Amygdala	+
Central gray	+
Interpeduncular nucleus	+
Substantia nigra	+
Periaqueductal gray	+
Cerebellum	+
Dorsal horn of spinal cord	+
Thalamus	—
Hypothalamus	—
White matter (corpus callosum, optic tract)	0

[a] Modified from Quirion et al. (1983).
[b] +++, very high; ++, high; +, moderate; —, low; 0, absent.

As shown in Table 9, [^{3}H]-PCP binding sites are widely distributed in rat brain. The visualization of the autoradiographic distribution of [^{3}H]-PCP receptor binding sites (Quirion et al., 1981a) demonstrates that labeling was most striking in cortical areas. The hippocampus and the dentate gyrus contain high levels of specific [^{3}H]-PCP binding, with concentrations in layers superficial to the pyramidal and granule cell layers. In the neocortex, labeling is dense in the outer five layers. In the cerebellum, it predominates in the molecular layer. The striatum and the nucleus accumbens contain moderate levels of [^{3}H]-PCP receptors. In the brainstem, the central gray, substantia nigra, interpeduncular nucleus, and periaqueductal gray are richer in binding sites than adjacent structures. Most of the thalamic and hypothalamic areas contain low densities of [^{3}H]-PCP sites. In the spinal cord, the ventral horn is richer in [^{3}H]-PCP binding sites than the dorsal horn. Overall, [^{3}H]-PCP receptor distribution seems appropriate for a drug with "schizophrenomimetic" properties (Snyder, 1980). Finally, [^{3}H]-PCP receptor distribution correlates well with regional brain metabolism localization (using 2-deoxyglucose as a probe) observed during ketamine anesthesia (Hammer et al., 1982).

3.4. Structure–Activity Relationships and Correlation of Binding Data With Biological Assays

The most critical criteria used to ensure that binding sites are relevant to biological actions of the ligand is to obtain a statistically rigorous correlation between the ability of a series of analogs to inhibit the binding of a labeled probe and to elicit behavioral, physiological, or pharmacological responses (Pert and Quirion, 1983). As shown in Table 8, very good correlations exist between the relative potency of PCP analogs in different binding assays (membrane preparations and brain sections) with their activity in behavioral assays (discriminative stimulus) and bioassays (dog cerebral artery). Moreover, various other groups have reported similar results in other tests, such as rotarod behavioral assays (Balster, 1980; Marwaha et al., 1981; Vincent et al., 1979), discriminative stimuli (Browne, 1982; Holtzman, 1980, 1982; Brady and Balster, 1982; Brady et al., 1982a), pupil diameter assays (Shannon and DeGregorio, 1981; Cone et al., 1984), guinea pig ileum bioassay (Gintzler et al., 1982), electrophysiological assays (Marwaha et al., 1980; Stringer and Guyenet, 1982; Stringer et al., 1983, 1984), and brain protein synthesis (Deutsch et al., 1984). Thus, all these data obtained in very different biological assays demonstrate that [^{3}H]-PCP binding sites are highly relevant to the pharmacological effects of PCP and congeners.

3.5. PCP and Sigma Opioid Receptor Binding Sites

As shown in Table 8, benzomorphan opiates that possess psychotomimetic sigma properties (Martin et al., 1976), such as SKF 10,047, cyclazocine, and pentazocine, interact with PCP binding sites (Hampton et al., 1982; Mendelsohn et al., 1984; Murray and Leid, 1984; Quirion et al., 1981a, 1982c, 1983; Zukin and Zukin, 1979, 1981; Zukin et al., 1983, 1984). Other opiate drugs are only weak inhibitors of [^{3}H]-PCP binding (Quirion et al., 1981a). Moreover, PCP-related drugs and sigma opiates share common properties in a variety of pharmacological tests, such as the discriminative stimulus assay (Brady and Balster, 1982; Brady et al.,

1982a; Holtzman, 1980, 1982; Shannon, 1982a,b; Shearman and Herz, 1982; Zukin et al., 1984), long-term potentiation electrophysiological assay (Stringer et al., 1983, 1984), and in vitro bioassay (Altura et al., 1983). This suggests that PCP and sigma opiates bind to closely related brain structures to elicit their effects (Quirion et al., 1981a, 1983). However, it remains to be determined whether both classes of drugs act precisely on the same receptor or on different subunits of a protein complex. Moreover, the existence of specific sigma opioid binding sites distinct from the PCP sites has recently been reported (Su, 1981; Tam, 1983). Thus, it seems likely that sigma opiates act on multiple classes of sites, one of which could be the PCP receptor.

3.6. Down-Regulation of PCP Binding Sites

Another approach used to study the possible relevance of binding sites to pharmacological actions of a drug is to characterize the effects of this drug-chronic treatment on its binding sites. Chronic PCP treatment caused a significant decrease in the number of [^{3}H]-PCP and [^{3}H]-spiperone (dopamine receptors) binding sites in rat brain sections. For [^{3}H]-PCP, we observed a 33% decrease in the number of sites and for [^{3}H]-spiperone, the B_{max} was 31% lower in treated than in controlled animals. No changes in the affinity (K_d) of the receptors for [^{3}H]-PCP and [^{3}H]-spiperone were observed. An acute PCP treatment did not modify K_d and B_{max} of any of the binding sites under study. It indicates that changes observed after chronic treatment are probably not related to the presence of PCP or metabolites on receptors when performing binding assays.

Chronic PCP treatment induced a decrease in the number of [^{3}H]-PCP binding sites in rat brain. Much evidence suggests that there is an inverse relationship between the numbers of receptors and the degree of receptor occupancy by agonist. PCP receptors also appear to be similarly regulated. This down-regulation of the number of PCP sites induced by chronic PCP treatment may be related to the development of tolerance (Castellani and Adams, 1981b; Chait and Balster, 1978; Murray, 1978; Pinchasi et al., 1977, 1978; Sturgeon et al., 1981) and dependence (Woolverton et al., 1980) to this drug observed in various species. However, changes in drug distribution, uptake, and degradation are also likely to be involved in the development of tolerance and dependence to PCP.

The diminution in the number of [^{3}H]-spiperone binding sites, an antagonist of the D-2 dopamine receptor, after chronic PCP treatment is of interest, especially in regard to the "schizophrenic-like" effects induced by PCP. Recently, Robertson (1982, 1983) obtained similar results in rat brain membrane preparations. Since a direct interaction of PCP on dopamine binding sites is unlikely, the decrease in D-2 dopamine binding sites in rat striatum might be related to the various effects of PCP on the dopaminergic system (*see* section 2.2). It is possible that some PCP binding sites might be located on dopaminergic terminals (French and Vantini, 1984; French et al., 1984). Activation of these binding sites stimulates the release of dopamine, which in a chronic situation would induce a down-regulation of the number of D-2 dopamine binding sites. Another explanation is that PCP can block the uptake of dopamine, thus inducing a decrease in the number of D-2 postsynaptic binding sites because of higher than normal concentrations of dopamine in the synaptic cleft.

It is also possible that chronic PCP treatment might affect other classes of receptors. Ward and Trevor (1981) have shown that such treatment decreased muscarinic receptors in rat brain. On the other hand, Schwartz et al. (1982) were not able to find any significant alterations in muscarinic, opiate, adrenergic, or serotonergic receptors following a chronic PCP treatment. In summary, we have shown that a chronic PCP treatment decreases the number of [^{3}H]-PCP and [^{3}H]-spiperone binding sites in rat brain. These results demonstrate the plasticity of the PCP binding site and suggest that this site is relevant to the pharmacological effects of PCP.

3.7. PCP-Related Radioligands

The relatively low affinity of [^{3}H]-PCP for the PCP receptor (K_d between 50 and 200 n*M*) is certainly associated with some of the problems related to PCP binding assays. However, two new PCP-related radioligands have recently become available and should greatly facilitate future studies on the precise molecular and biochemical characterization of the PCP receptor.

Vignon et al. (1983) have demonstrated that [^{3}H]-TCP [*N*-(1-[2-thienyl]-cyclohexyl)3,4-[^{3}H]-piperidine], a PCP derivative, is a much better ligand than PCP to characterize PCP receptors. First, [^{3}H]-TCP possesses a greater affinity for the PCP receptor (K_d of 7.4 instead of 64 n*M* for [^{3}H]-PCP). Thus, the dissociation rate of the [^{3}H]-TCP–receptor complex is much slower than with [^{3}H]-PCP ($t_{1/2}$ of 20 min instead of 1.5 min for PCP), allowing longer washing periods during a rapid filtration binding assay. Moreover, non-specific binding of [^{3}H]-TCP is almost negligible at concentrations near the K_d value and represents only 22% of total binding at 50 n*M* [^{3}H]-TCP.

Recently, we have also used [^{3}H]-TCP in our laboratories (Pilapil and Quirion, 1984). Under our incubation conditions, [^{3}H]-TCP binds to a single class of sites in rat and guinea pig brain membranes (K_d between 12 and 41 nM, depending on the brain region). Specifically bound [^{3}H]-TCP represents 75–80% of total binding at a [^{3}H]-TCP concentration between 5 and 50 nM. Binding to filters is negligible when filters are presoaked in 0.1% polyethyleneimine. As shown in Table 10, the order of potency of various PCP-related analogs in displacing either [^{3}H]-PCP or [^{3}H]-TCP is very similar, suggesting that both ligands label the same population of sites. Moreover, the distribution of [^{3}H]-TCP binding sites is almost identical to that of [^{3}H]-PCP in rat brain (Pilapil and Quirion, 1984). Thus, our results confirm those of Vignon et al. (1983), and indicate that [^{3}H]-TCP is a better ligand than [^{3}H]-PCP to characterize PCP receptors.

[^{3}H]-Dexoxadrol is another putative PCP-related ligand that recently became available. Dexoxadrol shares many pharmacological properties with PCP and ketamine (Brady et al., 1982b), and it has been demonstrated that dexoxadrol is able to displace [^{3}H]-PCP binding in membrane preparations (Hampton et al., 1982). We have recently used [^{3}H]-dexoxadrol in our laboratories and reported preliminary data on this new ligand (Pilapil and Quirion, 1984; Pilapil et al., 1985.)

Under our incubation conditions, [^{3}H]-dexoxadrol binds to an apparently saturable and single class of sites in rat and guinea pig brain membranes (K_d between 8 and 43 nM). Interestingly, nonspecifically bound [^{3}H]-dexoxadrol represents less than 25% of the total binding at concentrations up to 50 nM. The regional distribution of [^{3}H]-dexoxadrol binding sites in rat and guinea pig reveals important differences from that of [^{3}H]-TCP and [^{3}H]-PCP. As shown in Fig. 1, [^{3}H]-dexoxadrol binding is concentrated in the hippocampus, cortex, cerebellum, and hypothalamus in rat brain. In fact, the density of [^{3}H]-dexoxadrol binding sites appears to be high in the hypothalamus (Fig. 2), hippocampus (Figs. 2 and 3), and cortex (Figs. 2 and 3). Similar data have been obtained in membrane binding assays (Pilapil and Quirion, 1984). [^{3}H]-PCP binding sites are also concentrated in the hippocampus and the cortex, but not in the hypothalamus, which contains only limited densities of [^{3}H]-PCP binding sites (Quirion et al., 1981a; Zukin and Zukin, 1979).

Other brain regions containing moderate to high densities of [^{3}H]-dexoxadrol binding sites are the substantia nigra (Fig. 3), central gray (Fig. 3), thalamus (Fig. 2), amygdala (Fig. 2), and the pineal gland (Fig. 3), a CNS-related structure. Preliminary data

TABLE 10
Order of Potency of Various PCP-Related Compounds in [^{3}H]-PCP, [^{3}H]-TCP, and [^{3}H]-Dexoxadrol Binding Assays

Ligand	Order of potency
[^{3}H]-PCP[a]	PCE > TCP > PHP > PCP > THP > PCDEA > TCM > PCC
[^{3}H]-TCP	TCP > PCE > PCP > PHP > THP > PCDEA > TCM > PCC
[^{3}H]-Dexoxadrol	TCP > PHP > PCP > THP > PCE > PCDEA > TCM > PCC

[a]From Quirion et al. (1981a). For abbreviations, *see* Quirion et al. (1981a).

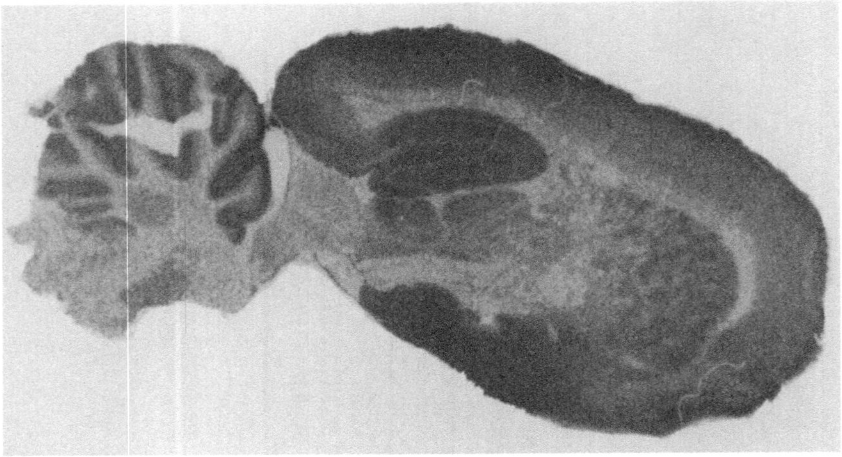

Fig. 1. Photomicrograph of [^{3}H]-dexoxadrol binding in rat brain saggittal section. Slide-mounted brain sections were incubated as described (Pilapil and Quirion, 1984). High densities of sites are present in the cortex, hippocampus, caudate-putamen, hypothalamus, and cerebellum.

show that [^{3}H]-dexoxadrol binding sites are also present in human forebrain, where they are concentrated in the cortex, caudate, and putamen (Fig. 4).

Another difference between [^{3}H]-dexoxadrol and [^{3}H]-PCP binding is related to the order of potency of various PCP derivatives in displacing the two radioligands. As shown in Table 10, PCE is much weaker against [^{3}H]-dexoxadrol than [^{3}H]-PCP. These data suggest that [^{3}H]-dexoxadrol labels PCP-related sites in certain brain regions like the hippocampus, but probably not in others like the hypothalamus.

3.8. Possible Existence of an Endogenous PCP-Like Ligand

The presence of highly specific and selective receptors for PCP, a totally synthetic drug, strongly suggests the existence of endogenous ligands for these receptors. We have already presented preliminary evidence suggesting that such a compound is present in the brain (O'Donohue et al., 1984; Quirion et al., 1983, 1984). Our data demonstrate that porcine brain contains an active factor that displaces [^{3}H]-PCP binding to rat brain membranes. On reverse-phase high-pressure liquid chromatography, the active

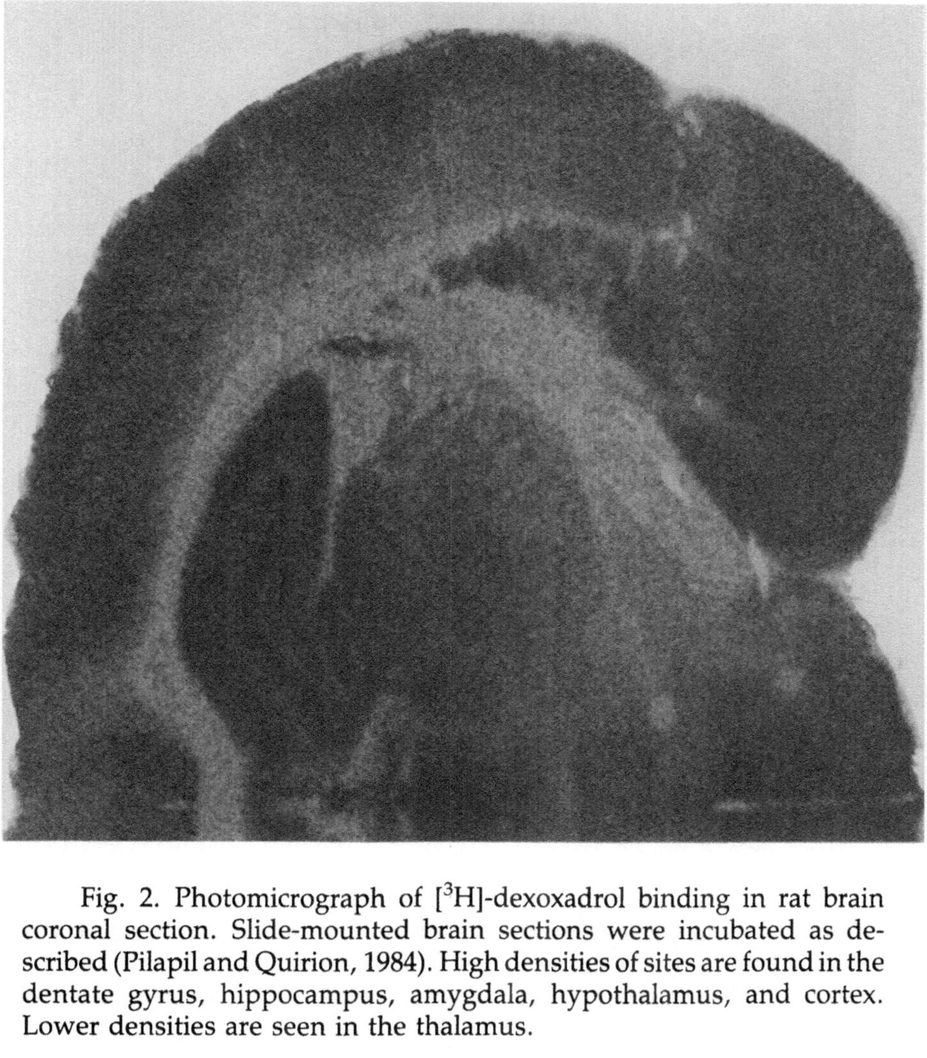

Fig. 2. Photomicrograph of [^{3}H]-dexoxadrol binding in rat brain coronal section. Slide-mounted brain sections were incubated as described (Pilapil and Quirion, 1984). High densities of sites are found in the dentate gyrus, hippocampus, amygdala, hypothalamus, and cortex. Lower densities are seen in the thalamus.

material elutes between 38 and 42% acetonitrile. Gel filtration chromatography of the factor predicts a molecular weight of approximately 3000 daltons. The endogenous substance is selective for PCP receptors since it does not interact with either benzodiazepine, neurotensin, or opioid receptors. The active material is heterogenously distributed in brain, with highest concentrations found in the hippocampus and cortex. It is likely to be a small peptide since various proteases markedly reduce the activity of this endogenous factor. Moreover, the material also possesses

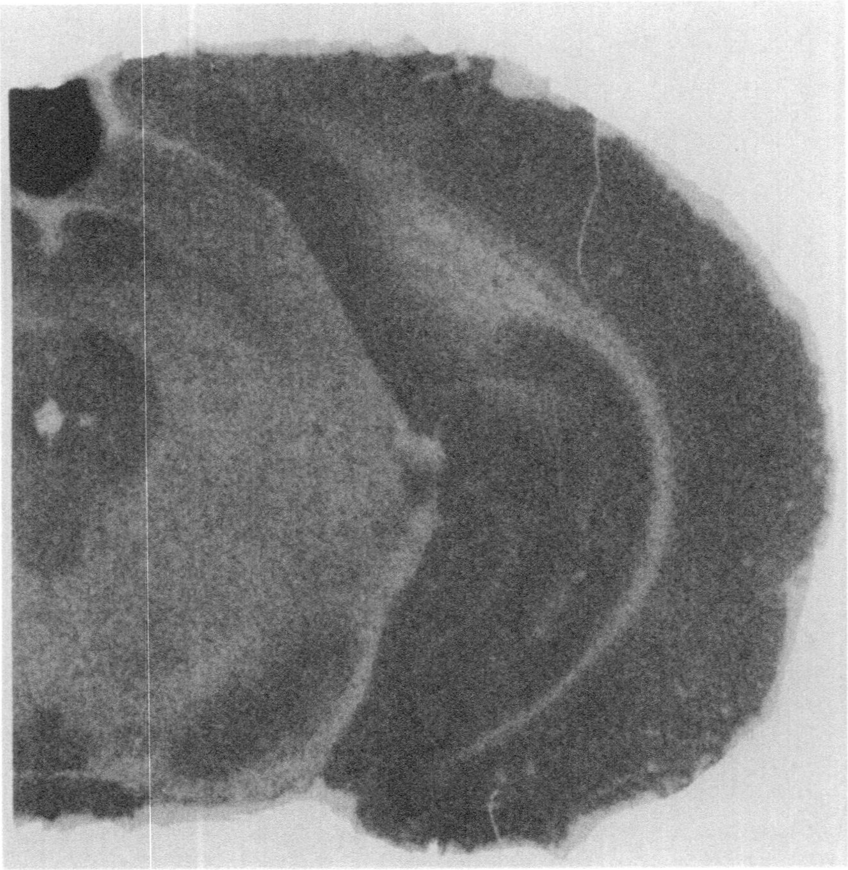

Fig. 3. Photomicrograph of [^{3}H]-dexoxadrol binding in rat brain coronal section. Slide-mounted sections were incubated as described (Pilapil and Quirion, 1984). Moderate to high densities of sites are located in the central gray, substantia nigra, interpeduncular nucleus, hippocampus, and cortex. Very high densities of sites are found in the pineal gland.

PCP-like activity in two bioassays (Quirion et al., 1984). Further characterization and the complete purification of this factor are currently underway in our laboratories.

3.9. Summary of Evidence for the Existence of PCP Receptors

In summary, much evidence currently supports the existence of specific PCP receptors in the CNS. These sites demonstrate high affinity for PCP derivatives, are heterogenously distributed in brain, are sensitive to enzymes and heat denaturation; and are down-regulated following chronic PCP treatment. The relative

Fig. 4. Photomicrograph of [^{3}H]-dexoxadrol binding in human hemisphere brain sections. Slide-mounted sections were incubated as described (Pilapil and Quirion, 1984). High densities of sites are present in the cortex, caudate, and putamen, and lower densities are found in the claustrum.

potency of PCP analogs in binding assays correlate very closely with their respective activity in various bioassays. This suggests that PCP receptors are likely to be the major site of action by which PCP induces its pharmacological effects.

4. Amphetamine Receptor Binding Sites

4.1. Neurochemical Effects of Amphetamine

Amphetamine (β-phenylisopropylamine) has powerful CNS

piratory center and induces tremor, restlessness, agitation, and sleeplessness, and increases motor activity. In the CNS, the (+)-isomer (dextroamphetamine) is three to four times more potent than the (–)-isomer.

In man, amphetamine induces wakefulness, alertness, elevation of mood, self-confidence, and euphoria, and decreases the sense of fatigue (Weiss and Latres, 1962). The improvement of mental and physical performances induced by amphetamine is certainly related to the abuse of this drug (Biel and Bopp, 1978). Moreover, amphetamine and similar drugs have been widely used (and abused) for their anorectic properties (Biel and Bopp, 1978; Weiner, 1980).

Amphetamine appears to exert most of its CNS effects by releasing biogenic amines from their storage sites in nerve terminals. The alerting and anorectic effects of amphetamine are probably mediated by the release of noradrenaline from central noradrenergic neurons. Some aspects of locomotor activity and the stereotyped behavior induced by amphetamine are probably caused by the release of dopamine. At much higher doses, disturbances of perception and psychotic behavior occur. These effects may be related to the release of serotonin from tryptaminergic neurons and of dopamine in the mesolimbic system (Weiner, 1980). In addition, amphetamine may exert direct agonistic effects on central serotonin receptors (Weiner, 1972).

4.2. Specific Amphetamine Receptor Binding Site

Although biochemical studies have demonstrated potent effects of amphetamine on the neuronal release, reuptake, and metabolism of biogenic amines, it is unclear whether any of these effects are responsible for the multiple pharmacological actions of this drug.

Recent data have suggested the existence of specific [^{3}H]-(+)-amphetamine binding sites in brain (Paul et al., 1982; Hauger et al., 1984). As shown in Table 11, [^{3}H]-(+)-amphetamine binding is saturable, stereospecific, and present in high densities in the hypothalamus and brainstem (Paul et al., 1982). Specific [^{3}H]-(+)-amphetamine binding in the hypothalamus is sensitive to enzymes and heat denaturation. Scatchard analysis reveals two distinct sites with K_d values of 93 and 300 n*M*, respectively. The effects of various psychotropic drugs and putative neurotransmitters in displacing [^{3}H]-(+)-amphetamine binding demonstrate that these sites are specific for amphetamine and related phenylethylamine derivatives. The relative affinities of a series of phenylethylamines

TABLE 11
Evidences for the Existence of Specific and Selective [^{3}H]-(+)-Amphetamine Receptors

Criteria	Reference
Saturability, reversibility	Paul et al. (1982)
Stereoselectivity	Paul et al. (1982)
Enzyme and heat denaturation	Paul et al. (1982)
Sensitivity to ions	Hauger et al. (1984)
Ligand selectivity pattern correlated to behavior	Paul et al. (1982)
Regional brain distribution	Paul et al. (1982)
Absence in peripheral tissues	Hauger et al. (1984)

for [^{3}H]-(+)-amphetamine binding sites in hypothalamic membranes is correlated to their potencies as anorectic agents. This suggests the presence of specific receptor sites in the hypothalamus that mediate the anorectic properties of amphetamines. In a subsequent study, Hauger et al. (1984) have extended their characterization of [^{3}H]-(+)-amphetamine binding sites. They have demonstrated that [^{3}H]-(+)-amphetamine binding is rapidly reversible, temperature sensitive, labile to preincubation in Tris buffer, inhibited by sodium ions, and unevenly distributed in various brain regions. Peripheral tissues such as liver, kidney, and heart have very low levels of specific [^{3}H]-(+)-amphetamine binding sites (Hauger et al., 1984).

However, a very recent report questioned the above findings on the existence of [^{3}H]-(+)-amphetamine receptors (Lesage et al., 1984). Lesage et al. (1984) demonstrated that dissociation of [^{3}H]-(+)-amphetamine from its binding sites occurs very rapidly (less than 5 s at 0°C). Thus, they suggested that the use of the rapid filtration method is inappropriate and should be replaced by a centrifugation assay. Moreover, they show that [^{3}H]-(+)-amphetamine binds to a complex mixture of sites, and suggest that monoamine oxidase A could be one of them, since harmaline possesses very high affinity for their [^{3}H]-(+)-amphetamine binding sites. Wood et al. (1984) have also suggested that amphetamine binding may involve tryptamine receptors since amphetamines stereospecifically displace [^{3}H]-tryptamine binding. Thus, the existence of specific and selective [^{3}H]-(+)-amphetamine receptors still remains to be more extensively demonstrated using biochemical, pharmacological, and behavioral assays. The development of radioactive ligands with higher affinities for the putative amphetamine receptor would certainly facilitate such studies.

5. Conclusions

In summary, the existence of specific PCP receptors in the brain is now very well documented. These sites are saturable, stereoselective, down-regulated following chronic PCP treatment, regionally distributed in the brain, and sensitive to proteolytic enzymes, heat denaturation, and ionic strength. The recent development of radioligands with better affinity for the PCP receptor should facilitate further characterization of these sites, as well as the complete purification of putative endogenous ligands. Finally, preliminary evidence suggests the presence of specific amphetamine binding sites in mammalian brain. However, more extensive studies are needed before any definitive conclusions can be reached.

Acknowledgments

Doctor Remi Quirion is a "Chercheur-Boursier" du Fonds de la Recherche en Sante du Quebec. The expert technical and secretarial assistance of Ms C. Pilapil and Mrs. J. Currie are gratefully acknowledged. Part of this research has been supported by the Douglas Hospital Research Centre.

References

Adams P. M. (1980) Interaction of phencyclidine with drugs affecting cholinergic neurotransmission. *Neuropharmacology* **19,** 151–153.

Aguayo L. G., Weinstein H., Maayani S., Glick S. D., Warnick J. E., and Albuquerque E. X. (1984) Discriminant effects of behaviorally active and inactive analogs of phencyclidine on membrane electrical excilability. *J. Pharmacol. Exp. Ther.* **228,** 80–87.

Albuquerque E. X., Tsai M. C., Aronstam R. S., Witkop B., and Eldefrawi M. E. (1980) Phencyclidine interactions with the ionic channel of the acetylcholine receptor and electrogenic membrane. *Proc. Natl. Acad. Sci. USA* **77,** 1224–1228.

Albuquerque E. X., Aquayo L. G., Warnick J. E., Weinstein H., Glick S. D., Maayani S., Ickowicz R. K., and Blaustein M. P. (1981) The behavioral effects of phencyclidine may be due to their blockade of potassium channels. *Proc. Natl. Acad. Sci. USA* **78,** 7792–7796.

Altura B. T. and Altura B. M. (1983) Cerebrovasospasms induced by phencyclidine are prevented by calcium antagonists and magnesium ions. *Magnesium* (in press).

Altura B. T., Quirion R., Pert C. B., and Altura B. M. (1983) Phencyclidine (angel dust) analogs and "sigma" opiate benzomorphans cause cerebral arterial spasm. *Proc. Natl. Acad. Sci. USA* **80,** 865–869.

Aronstam R. S., Eldefrawi M. E., Eldefrawi A. T., Albuquerque E. X., Jim K. F., and Triggle D. J. (1980) Sites of action of phencyclidine. III. Interaction with muscarinic receptors. *Mol. Pharmacol.* **18,** 179–184.

Arora R. C. and Meltzer H. Y. (1980) In vitro effects of phencyclidine and other psychomotor stimulants on serotonin uptake in human platelets. *Life Sci.* **27,** 1607–1613.

Ary T. E. and Kowinskey H. (1980) Basis of phencyclidine's ability to decrease the synaptosomal accumulation of [^{3}H]-catecholamines. *Eur. J. Pharmacol.* **61,** 401–405.

Bagchi S. P. (1981) Effects of phencyclidine on synaptosomal dopamine continuously appearing from phenylalanine: Sensitivity to reserpine. *Neuropharmacology* **20,** 845–851.

Balster R. L. (1980) The effects of phencyclidine and three analogs on motor performance in mice. *Pharmacology* **20,** 46–51.

Balster R. L. and Chait L. D. (1978) The effects of phencyclidine on amphetamine stereotypy in rats. *Eur. J. Pharmacol* **48,** 445–450.

Bayorh M. A. and Kopin I. J. (1982) Effect of phencyclidine (PCP) on adrenergic responses to spinal cord stimulation in pithed rats. *Eur. J. Pharmacol.* **85,** 15–21.

Berry S. C., Burton N. A., Anis N. A., and Lodge D. (1983) Stereoselective effects of two phencyclidine derivatives on *N*-methylaspartate excitation of spinal neurones in the cat and rat. *Eur. J. Pharmacol.* **96,** 261–267.

Bickford P. C., Palmer M. R., Rice K. C., Hoffer B. J., and Freedman R. (1981) Electrophysiological effects of phencyclidine on rat hippocampal pyramidal neurons. *Neuropharmacology* **20,** 733–742.

Biel J. H. and Bopp B. A. (1978) Amphetamines: Structure-Activity Relationships, in *Handbook of Psychopharmacology* (Iversen L. L., Iversen S. D., and Snyder S. H., eds.), pp. 1–39, Plenum, New York.

Blaustein M. P. and Ickowicz R. K. (1983) Phencyclidine, in nanomolar concentrations, binds to synaptosomes and blocks potassium channels. *Proc. Natl. Acad. Sci. USA* **80,** 3855–3859.

Boggan W. O., Evans M. G., and Wallis C. J. (1982) Effect of phencyclidine on [^{3}H]-QNB binding. *Life Sci.* **30,** 1193–1200.

Brady K. T. and Balster R. L. (1982) Discriminative stimulus properties of stereoisomers of cyclazocine in phencyclidine-trained squirrel monkeys. *Life Sci.* **31,** 541–549.

Brady K. T., Balster R. L., and May E. L. (1982a) Stereoisomers of *N*-allylnormetazocine: Phencyclidine-like behavioral effects in squirrel monkeys and rats. *Science* **215,** 178–180.

Brady K. T., Woolverton W. L., and Balster R. L. (1982b) Discriminative stimulus and reinforcing properties of etoxadrol and dexoxadrol in monkeys. *J. Pharmacol. Exp. Ther.* **220,** 56–62.

Browne R. G. (1982) Discriminative Stimulus Properties of Phencyclidine, in *Drug Discrimination: Application in CNS Pharmacology* (Colpoert F. C. and Slangen J. L., eds.), pp. 109–122, Elsevier Biomedical, Amsterdam.

Browne R. G. and Welch W. M. (1982) Stereo-selective antagonism of phencyclidine's discriminative properties by adenosine receptor agonists. *Science* **217,** 1157–1159.

Browne R. G., Welch W. M., Kozlowski M. R., and Duther G. (1983) Antagonism of PCP Discrimination by Adenosine Analogs, in *Phencyclidine and Related Arylcyclohexylamines* (Kamanka J. M., Domino E. F., and Geneste D., eds.), pp. 639–665, NPP Books, Ann Arbor, Michigan.

Calixto J. B., Aucelio J. G., and Duarte D. F. (1983) Inhibitory effects of ketamine on the isolated uteri of the rat: Evidence for the mechanism of action. *Can. J. Physiol. Pharmacol.* **61,** 1305–1311.

Castellani S. and Adams P. M. (1981a) Acute and chronic phencyclidine effects on locomotor activity stereotypy and ataxia in rats. *Eur. J. Pharmacol.* **73,** 143–154.

Castellani S. and Adams P. M. (1981b) Effects of dopaminergic drugs on phencyclidine-induced behavior in the rat. *Neuropharmacology* **20,** 371–374.

Chait L. P. and Balster R. L. (1978) The effect of acute and chronic phencyclidine on schedule-controlled behavior in the squirrel monkey. *J. Pharmacol. Exp. Ther.* **204,** 77–87.

Cone E. J., McQuinn R. L., and Shannon H. E. (1984) Structure-activity relationships studies of phencyclidine derivatives in rats. *J. Pharmacol. Exp. Ther.* **228,** 147–153.

Contreras C. M., Guzman-Flores C., Dorantes M. E., Ervin F. R., and Palmour R. (1981) Naloxone and phencyclidine: Interacting effects on the limbic system and behavior. *Physiol. Behav.* **27,** 1019–1026.

Deutsch D. G., Koul O., and Kersten R. S. (1984) Phencyclidine and analogues: Effects on brain protein synthesis. *J. Neurochem.* **42,** 407–411.

Doherty J. D., Simonovic M., So R., and Meltzer H. Y. (1980) The effect of phencyclidine on dopamine synthesis and metabolism in rat striatum. *Eur. J. Pharmacol.* **65,** 139–149.

Domino E. F., ed. (1981) *Phencyclidine: Historical and Current Perspectives.* NPP Books, Ann Arbor, Michigan.

Domino E. F. and Wilson A. E. (1972) Psychotropic drug influences on brain acetylcholine utilization. *Psychopharmacologia* (Berl.) **25,** 291–298.

Eldefrawi A. T., Miller E. R., Murphy D. L., and Eldefrawi M. E. (1982a) [^{3}H]-Phencyclidine interactions with the nicotinic acetylcholine receptor channel and its inhibition by psychotropic, antipsychotic, opiate, antidepressant, antibiotic, antiviral and antiarrhythmic drugs. *Mol. Pharmacol.* **22,** 72–81.

Eldefrawi M. E., El-Fakahany E. E., Murphy D. L., Eldefrawi A. T., and Triggle D. J. (1982b) High affinity binding of phencyclidine (PCP) to crayfish muscle. Displacement by calcium antagonists. *Biochem. Pharmacol.* **31,** 2549–2552.

El-Fakahany E. E., Triggle D. J., Eldefrawi A. T., and Eldefrawi M. E. (1984a) Distinction between high-affinity [^{3}H]-phencyclidine binding sites and muscarinic receptors in guinea-pig ileum muscle. *J. Pharmacol. Exp. Ther.* **229,** 447–454.

El-Fakahany E. E., Eldefrawi A. T., Murphy D. L., Aguayo L. G., Triggle D. J., Albuquerque E. X., and Eldefrawi M. E. (1984b) Interactions of phencyclidine with crayfish muscle membranes. *Mol. Pharmacol.* **25,** 369–378.

Fosset M., Renaud J. F., Lenoir M. C., Kamenka J. M., Geneste P., and Lazdunski M. (1979) Interaction of molecules of the phencyclidine series with cardiac cells. Association with the muscarinic receptor. *FEBS Lett.* **103,** 133–137.

Fratta W., Zivcovic B., and Gessa G. L. (1983) Phencyclidine decreases methionine-enkephalin content in rat striatum. *Neurosci. Lett.* **39,** 161–163.

Freed W. J., Bing L. A., and Wyatt R. J. (1984) Effects of neuroleptics on phencyclidine (PCP)-induced locomotor stimulation in mice. *Neuropharmacology* **23,** 175–181.

French E. D. and Vantini G. (1984) Phencyclidine-induced locomotor activity in the rat is blocked by 6-hydroxydopamine lesion of the nucleus accumbens: Comparisons to other psychomotor stimulants. *Psychopharmacology* **82,** 83–88.

French E., Vantini G., Contreras P., Pilapil C., and Quirion R. (1984) Phencyclidine (PCP) binding in the nucleus accumbens (NaC) and PCP-induced hyperactivity are decreased following 6-hydroxydopamine (6-OHDA) lesions of the mesolimbic dopamine system. *Soc. Neurosci. Abst.* **10,** 1205.

Gabrielovitz A., Kloog Y., Kalir A., Balderman D., and Sokolovsky M. (1980) Interaction of phencyclidine and its new adamantyl derivatives with muscarinic receptors. *Life Sci.* **26,** 89–95.

Geary L. E., Zimmerman D. M., Leander J. D., and Kabra V. (1983) Structure–activity studies of ligands interacting with rat cortex phencyclidine receptor. *Soc. Neurosci. Abst.* **9,** 819.

Gintzler A. R., Zukin R. S., and Zukin S. R. (1982) Effects of phencyclidine

(PCP) and its derivatives on enteric neurons. *Brit. J. Pharmacol.* **75,** 261–267.

Glick S. D. and Guido R. A. (1982) Naloxone antagonism of the thermoregulatory effects of phencyclidine. *Science* **217,** 1272–1273.

Grelak R. P., Clark R., Stump J. M., and Vernier V. G. (1970) Amantadine dopamine interactions: Possible mode of action in parkinsonism. *Science* **169,** 203–204.

Haggerty G. C., Richter J. A., and Forney R. B. (1983) The effects of an in vivo administration of phencyclidine on sodium-dependent high-affinity choline uptake in rat hippocampus and striatum in vitro. *Neuropharmacology* **22,** 1389–1395.

Hammer R. P., Jr., Herkenham M., Pert C. B., and Quirion R. (1982) Correlation of regional brain metabolism with receptor localization during ketamine anesthesia: Combined autoradiographic 2-[^{3}H]-deoxy-*p*-glucose receptor binding technique. *Proc. Natl. Acad. Sci. USA* **79,** 3067–3070.

Hampton R. Y., Medzihradsky F., Woods J. H., and Dahlstrom P. J. (1982) Stereospecific binding of [^{3}H]-phencyclidine in brain membranes. *Life Sci.* **30,** 2147–2154.

Hauger R. L., Hulihan-Giblin B., Skolnick P., and Paul S. M. (1984) Characteristics of [^{3}H]-(+)amphetamine binding sites in the rat central nervous system. *Life Sci.* **34,** 771–782.

Holtzman S. G. (1980) Phencyclidine-like discriminative effects of opioids in the rat. *J. Pharmacol. Exp. Ther.* **214,** 614–619.

Holtzman S. G. (1982) Phencyclidine-like discriminative stimulus properties of opioids in the squirrel monkey. *Psychopharmacology* **77,** 295–300.

Hsu L. L., Smith R. C., Rolsten C., and Leelavathi D. E. (1980) Effects of acute and chronic phencyclidine on neurotransmitter enzymes in rat brain. *Biochem. Pharmacol.* **29,** 2542–2546.

Jasinski D. J., Cone E. J., Gorodetzky C. W., Risner M. D., Shannon H. E., Su T. P., and Vaupel D. B. (1979) Progress Report From the NIDA Addiction Research Center, in *Proceedings of the Committee of Problems of Drug Dependence* (National Institute of Drug Abuse Research Monograph) (R. Paterson, ed.) vol. 29, pp. 61–69, Rockville City, Maryland.

Johnson K. M. and Oeffinger K. C. (1981) The effect of phencyclidine on dopamine metabolism in the mouse brain. *Life Sci.* **28,** 361–369.

Johnson S. W., Haroldsen P. E., Hoffer B. J., and Freedman R. (1984) Presynaptic dopaminergic activity of phencyclidine in rat caudate. *J. Pharmacol. Exp. Ther.* **229,** 322–332.

Kamenka J. M., Chiche B., Goudal R., Geneste P., Vignon J., Vincent J. P., and Lazdunski M. (1982) Chemical synthesis and molecular pharmacology of hydroxylated 1-(1-phenylyclohexyl.)piperidine derivatives. *J. Med. Chem.* **25,** 431–435.

Kozlowski M. R. and Lebel L. A. (1983) Characterization of [^{3}H]phencyclidine (PCP) binding to rat hippocampal sections using autoradiography. *Soc. Neurosci. Abst.* **9,** 816.

Lazdunski M., Bidard J. N., Romey G., Tourneur Y., Vignon J., and Vincent J. P. (1983) The Different Sites of Actions of Phencyclidine and its Analogues in Nervous Tissues, in *Phencyclidine and Related Arylcyclohexylamines* (Kamenka J. M., Domino E. F., and Geneste P., eds.), pp. 85–105, NPP Books, Ann Arbor, Michigan.

Leonard B. E. and Tonge S. R. (1969) The effects of some hallucinogenic drugs upon the metabolism of noradrenaline. *Life Sci.* **8,** 815–825.

Lesage A., Benedetti M. S., and Rumigny J. F. (1984) High affinity binding site for (+)-amphetamine in rat hypothalamus: Fact or artefact? *Neurochem. Int.* **6,** 283–286.

Leventer S. M. and Johnson K. M. (1983) Effects of phencyclidine on the release of radioactivity from striatal slices labelled with [^{3}H]-choline. *J. Pharmacol. Exp. Ther.* **225,** 332–337.

Lodge D. and Anis N. A. (1982) Effects of phencyclidine on excitatory amino acid activation of spinal interneurones in the cat. *Eur. J. Pharmacol.* **77,** 203–204.

Maayani S. and Weinstein H. (1980) "Specific binding" of [^{3}H]-phencyclidine: Artifacts of the rapid filtration method. *Life Sci.* **26,** 2011–2022.

Maayani S., Weinstein H., Ben-Zvi N., Cohen S., and Sokolovsky M. (1974) Psychotomimetics as anticholinergic agents. I. Cyclohexylpiperidine derivatives: Anticholinesterase activity and antagonistic activity to acetylcholine. *Biochem. Pharmacol.* **23,** 1263–1281.

Martin L. L. and Smith D. J. (1982) Ketamine inhibits serotonin synthesis and metabolism in vivo. *Neuropharmacology* **21,** 119–125.

Martin L. L., Bouchal R. L., and Smith D. J. (1982) Ketamine inhibits serotonin uptake in vivo. *Neuropharmacology* **21,** 113–118.

Martin J. R., Berman M. H., Krewsun I., and Small S. F. (1979) Phencyclidine-induced stereotyped behavior and serotonergic syndrome in rat. *Life Sci.* **24,** 1699–1704.

Martin W. R., Eades C. G., Thompson J. A., Huppler R. E., and Gilbert P. E. (1976) The effects of morphine and nalorphine-like drugs in the nondependent and morphine-dependent chronic spinal dogs. *J. Pharmacol. Exp. Ther.* **197,** 517–532.

Marwaha J. (1982) Candidate mechanisms underlying phencyclidine-induced psychosis: An electrophysiological, behavioral and biochemical study. *Biol. Psychiatry* **17,** 155–198.

Marwaha J., Palmer M. R., Woodward D. J., Hoffer B. J., and Freedman R. (1980) Electrophysiological evidence for presynaptic actions of phencyclidine on noradrenergic transmission in rat cerebellum. *J. Pharmacol. Exp. Ther.* **215,** 606–613.

McQuinn R. L., Cone E. J., Shannon H. E., and Su T. P. (1981) Structure–activity relationships of the cycloalkyl ring of phencyclidine. *J. Med. Chem.* **24,** 1429–1432.

Meibach R. C., Glick S. D., Cox R., and Maayani S. (1979) Localization of phencyclidine-induced changes in brain energy metabolism. *Nature* (Lond.), **282,** 625–626.

Meltzer H. Y., Sturgeon R. D., Simonovic M., and Fessler R. G. (1981) Phencyclidine as an Indirect Dopamine Agonist, in *PCP (Phencyclidine): Historical and Current Perspectives* (Domino E. F., ed.), pp. 207–242, NPP Books, Ann Arbor, Michigan.

Mendelsohn L. G., Kalra V., and Johnson B. G. (1984) The sigma opioid receptor: Characterization and co-identity with the phencyclidine receptor. *Soc. Neurosci. Abst.* **10,** 958.

Murray T. F. (1978) The effects of phencyclidine on operant behavior in the rat: Biphase effect and tolerance development. *Life Sci.* **22,** 175–202.

Murray T. F. (1983) A Comparison of Phencyclidine With Other Psychoactive Drugs on Cholinergic Dynamics in the Rat Brain, in *Phencyclidine and Related Arylcyclohexylamines* (Kamenka J. M., Domino E. F., and Geneste P., eds.), pp. 547–561, NPP Books, Ann Arbor, Michigan.

Murray T. F. and Cheney D. L. (1981) The effect of phencyclidine on the turnover rate of acetylcholine in various regions of rat brain. *J. Pharmacol. Exp. Ther.* **217,** 733–737.

Murray T. F. and Horita A. (1979) Phencyclidine-induced stereotyped behavior in rats: Dose–response effects and antagonism by neuroleptics. *Life Sci.* **24,** 2217–2226.

Murray T. F. and Leid M. E. (1984) Interaction of dextrorotary opioids with phencyclidine recognition sites in rat brain membranes. *Life Sci.* **34,** 1899–1911.

Nabeshima T., Yamada K., Yamaguchi K., Hiromatsu M., Furukawa H., and Kameyama T. (1983a) Effects of lesions in the striatum, nucleus accumbens and medial raphe on phencyclidine-induced stereotyped behaviors and hyperactivity in rats. *Eur. J. Pharmacol.* **91,** 455–462.

Nabeshima T., Yamaguchi K., Yamada K., Hiramatsu M., Furakawa H., and Kameyama T. (1983b) Phencyclidine-induced stereotyped behaviors in rats following specific neurotoxin lesions of the striatum. *Eur. J. Pharmacol.* **93,** 229–234.

Nabeshima T., Hiramatsu M., Amano M., Furukawa A., and Kameyama T. (1983c) Decrease of methionine-enkephalin levels in mouse brain induced by phencyclidine. *Eur. J. Pharmacol.* **86,** 271–273.

Nabeshima T., Hiramatsu M., Amano M., Furukawa H., and Kameyama T. (1983d) Chronic phencyclidine increases methionine-enkephalin level in mouse striatum. *Neurosci. Lett.* **37,** 69–74.

O'Donohue T. L., Pert C. B., French E. D., Pert A., Dimaggio D. A.,

Everist H., and Quirion R. (1984) Evidence for a Central Nervous System Ligand to the Phencyclidine Receptor, in *Peptides: Proceedings of the Eighth American Peptide Symposium* (Hruby V. and Rich D., eds.), pp. 433–436, Pierce Chemical, Rockford, Illinois.

Palmer M. R., Bickford P. C., Hoffer B. J., and Freedman R. (1983) Electrophysiological Evidence for Presynaptic Actions of Phencyclidine on Noradrenergic Transmission in Rat Cerebellum and Hippocampus, in *Phencyclidine and Related Arylcyclohexylamines* (Kamenka J. M., Domino E. E., and Geneste P., eds.), pp. 443–469, NPP Books, Ann Arbor, Michigan.

Paul S. M., Hulihan-Giblin B., and Skolnick P. (1982) (+)-Amphetamine binding to rat hypothalamus: Relation to anorexic potency of phenylethylamines. *Science* **218,** 487–489.

Peat M. A. and Gibb J. W. (1983) The effects of phencyclidine on glutamic acid decarboxylase activity in several regions of the rat brain. *Neurosci. Lett.* **35,** 301–306.

Pert C. B. and Snyder S. H. (1973) Opiate receptor: Demonstration in nervous tissue. *Science* **179,** 1011–1014.

Pert C. B. and Quirion R. (1983) The phencyclidine receptor. *Trends Pharmacol. Sci.* **4,** 12–13.

Petersen R. C. and Stillman R. C. (1978) Phencyclidine: An Overview, in *Phencyclidine Abuse: An Appraisal* (Petersen R. and Stillman R. C., eds.), pp. 1–7, NIDA Research Monograph 21, Rockville, Maryland.

Pilapil C. and Quirion R. (1984) Comparative receptor binding properties of [^{3}H]-TCP and [^{3}H]-dexoxadrol, two phencyclidine (PCP)-related ligands. *Soc. Neurosci. Abst.* **10,** 959.

Pilapil C., Contreras P., O'Donohue T. L. and Quirion R. (1985) Autoradiographic distribution of [^{3}H]-dexoxadrol, a phencyclidine-related ligand, binding sites in rat and human brain. *Neurosci. Lett.* (in press).

Pinchasi I., Maayani S., and Sokolovsky M. (1977) Tolerance to phencyclidine derivatives: In vivo and in vitro studies. *Biochem. Pharmacol.* **26,** 1671–1679.

Pinchasi J., Maayani S., Egozi Y., and Sokolovsky M. (1978) On the interaction of drugs with the cholinergic nervous system. II. Cross tolerance between phencyclidine derivatives and cholinergic drugs. *Psychopharmacology* **56,** 37–40.

Quirion R. and Pert C. B. (1982a) Certain calcium antagonists are potent displacers of [^{3}H]-phencyclidine binding in rat brain. *Eur. J. Pharmacol.* **83,** 155–156.

Quirion R. and Pert C. B. (1982b) Amantadine modulates phencyclidine binding sites sensitivity in rat brain. *Experientia* **38,** 955–956.

Quirion R. and Pert C. B. (1983a) Properties of the phencyclidine (PCP) receptors. *Prog. NeuroPsychopharmacol. Biol. Psychiatry* **7,** 451–456.

Quirion R. and Pert C. B. (1983b) Modulation of Phencyclidine Receptor Sensitivity, in *Proceedings of the Committee on Problem of Drug Dependence* (National Institute of Drug Abuse Research Monograph) (Harris L. S., ed.), pp. 217–223, Washington, D.C.

Quirion R., Hammer R. P., Herkenham M., and Pert C. B. (1981a) The phencyclidine ("angel dust") and σ "opiate" receptor: Its visualization by tritium sensitive film. *Proc. Natl. Acad. Sci. USA* **78,** 5881–5885.

Quirion R., Rice K., Skolnick P., Paul S., and Pert C. B. (1981b) Stereospecific displacement of [^{3}H]-phencyclidine (PCP) receptor binding by an enantiomeric pair of PCP analogs. *Eur. J. Pharmacol.* **74,** 107–108.

Quirion R., Bayhor M. A., Zerbe R. L., and Pert C. B. (1982a) Chronic phencyclidine treatment decreases phencyclidine and dopamine receptors in rat brain. *Pharmacol. Biochem. Behav.* **17,** 699–702.

Quirion R., Hammer R. P., Jr., Herkenham M., and Pert C. B. (1982b) Autoradiographic Localization of the Phencyclidine/Sigma Opiate Receptor in Rat Brain, in *Proceedings of the Committee on Problems of Drug Dependence* (National Institute of Drug Abuse Research Monograph) (C. S. Harris, ed.), vol. 41, pp. 178–183, Rockville, Maryland.

Quirion R., O'Donohue T. L., Everist H., Pert A., and Pert C. B. (1983) Phencyclidine Receptors and Possible Existence of an Endogenous Ligand, in *Phencyclidine and Related Arylcyclohexylamines* (Kamenka J. M., Domino E. F., and Geneste P., eds.), pp. 667–684, NPP Books, Ann Arbor, Michigan.

Quirion R., Dimaggio D. A., French E. D., Contreras P. C., Shiloach J., Pert C. B., Everist H., Pert A., and O'Donohue T. L. (1984) Evidence for an endogenous peptide ligand for the phencyclidine receptor. *Peptides* **5,** 967–973.

Raja S. N. and Guyenet P. G. (1980) Effects of phencyclidine on the spontaneous activity of monoaminergic neurons. *Eur. J. Pharmacol.* **62,** 229–233.

Raja S. M. and Guyenet P. G. (1982) Actions of phencyclidine on synaptic transmission in the hippocampus. *Brain Res.* **236,** 289–304.

Robertson H. A. (1982) Chronic phencyclidine, like amphetamine, produces a decrease in [^{3}H]-spiroperidol binding in rat striatum. *Eur. J. Pharmacol.* **78,** 363–365.

Robertson H. A. (1983) Chronic D-amphetamine and phencyclidine: Effects on dopamine agonist and antagonist binding sites in the extrapyramidal and mesolimbic systems. *Brain Res.* **267,** 179–182.

Schwartz R. D., Moerschbaecher J. M., Thompson D. M., and Keller K. J. (1982) Effects of chronic phencyclidine in fixed-ratio responding: No relation to neurotransmitter receptor binding in rat cerebral cortex. *Pharmacol. Biochem. Behav.* **16,** 647–652.

Shannon H. E. (1981) Evaluations of phencyclidine analogs on the basis of their discriminative stimulus properties in the rat. *J. Pharmacol. Exp. Ther.* **216,** 543–551.

Shannon H. E. (1982a) Pharmacological analysis of the phencyclidine-like discriminative stimulus properties of narcotic derivatives in rats. *J. Pharmacol. Exp. Ther.* **222,** 146–151.

Shannon H. E. (1982b) Phencyclidine-like discriminative stimuli of (+)- and (–)-*N*-allylnormetazocine in rats. *Eur. J. Pharmacol.* **84,** 225–228.

Shannon H. E. and DeGregorio C. M. (1981) Effects of *N*-substituted analogs of phencyclidine on avoidance behavior in the rat. *J. Pharmacol. Exp. Ther.* **218,** 55–62.

Shearman G. T. and Herz A. (1982) Non-opioid psychotomimetic-like discriminative stimulus properties of *N*-Allylnormetazocine (SKF 10,047) in the rat. *Eur. J. Pharmacol.* **82,** 167–172.

Shore P. A. (1976) Action of amfonelic acid and other non-amphetamine stimulants on the dopamine neuron. *J. Pharm. Pharmacol.* **28,** 855–857.

Smith R. C., Meltzer H. Y., Arora R. C., and Davis J. M. (1977) Effects of phencyclidine on [^{3}H]-catecholamine and [^{3}H]-serotonin uptake in synaptosomal preparations from rat brain. *Biochem. Pharmacol.* **26,** 1435–1439.

Smith R. C., Doddamane C. L., Hsu L., Ho B. T., Tansey W., Taylor D., and Biggs C. (1981) Acute Versus Chronic Administration of Phencyclidine Effects on Behavior and Brain Biochemistry, in *PCP (Phencyclidine): Historical and Current Perspectives,* (Domino E. F., ed.), pp. 243–292, NPP Books, Ann Arbor, Michigan.

Snyder S. H. (1980) Phencyclidine. *Nature* (Lond.) **285,** 355–356.

Stringer J. L. and Guyenet P. G. (1982) Effects of phencyclidines on hippocampal pyramidal cells. *Brain Res.* **252,** 343–352.

Stringer J. L., Hackett J. T., and Guyenet P. G. (1984) Long-term potentiation blocked by phencyclidine and cyclazocine in vitro. *Eur. J. Pharmacol.* **98,** 381–398.

Stringer J. L., Greenfield J., Hackett J. T. and Guyenet P. G. (1983) Blockade of long-term potentiation by phencyclidine or σ opiates in the hippocampus in vivo and in vitro. *Brain Res.* **280,** 127–138.

Sturgeon R. D., Fessler R. G., London S. F., and Meltzer H. Y. (1981) A comparison of the effects of neuroleptics on phencyclidine-induced behaviors in the rat. *Eur. J. Pharmacol.* **76,** 37–53.

Su T. P. (1981) Psychotomimetic opioid binding: Specific binding of [^{3}H]-SKF 10,0047 to etorphine-inaccessible sites in guinea pig brain. *Eur. J. Pharmacol.* **75,** 81–82.

Tam S. W. (1983) Naloxone-inaccessible σ receptor in rat central nervous system. *Proc. Natl. Acad. Sci. USA* **80,** 6703–6707.

Taube H. D., Montel H., Hau G., and Starke K. (1975) Phencyclidine and ketamine: Comparison with the effect of cocaine on the noradrenergic neurons of the rat brain cortex. *Naunyn-Schmiedebergs Arch. Pharmacol.* **291,** 47–54.

Temma K., Ahera T., Brody T. M., and Rech R. H. (1983) Negative chronotropic and positive inotropic actions of phencyclidine on isolated atrial muscle in guinea pigs and rats. *J. Pharmacol. Exp. Ther.* **226,** 885–892.

Tonge S. R. and Leonard B. E. (1971) Variation in hydroxytryptamine metabolism in the rat: Effects on the neurochemical response to phencyclidine. *J. Pharm. Pharmacol.* **23,** 711–712.

Tourneur Y., Romey G., and Lazdunski M. (1982) Phencyclidine blockade of sodium and potassium channels in neuroblastoma cells. *Brain Res.* **245,** 154–158.

Vaupel D. B. (1983) Naltrexone fails to antagonize the σ effects of PCP and SKF 10,047 in the dog. *Eur. J. Pharmacol.* **92,** 269–274.

Verebey K., Kogan M. J., and Mule S. J. (1981) Phencyclidine-induced stereotypy in rats: Effects of methadone, apomorphine, and naloxone. *Psychopharmacology* **75,** 44–47.

Vickroy T. W. and Johnson K. M. (1981) Stimulation of tyrosine hydroxylation by phencyclidine in vitro. *Eur. J. Pharmacol.* **71,** 463–473.

Vignon J. and Lazdunski M. (1984) Structure–function relationships in the inhibition of synaptosomal dopamine uptake by phencyclidine and analogues: Potential correlation with binding site identified with [^{3}H]-phencyclidine. *Biochem. Pharmacol.* **33,** 700–702.

Vignon J., Chicheportiche R., Chicheportiche M., Kamenka J. M., Geneste P., and Lazdunski M. (1983) [^{3}H]-TCP: A new tool with high affinity for the PCP receptor in rat brain. *Brain Res.* **280,** 194–197.

Vignon J., Vincent J. P., Bidard J. N., Kamenka J. M., Geneste P., Monier S., and Lazdunski M. (1982) Biochemical properties of the brain phencyclidine receptor. *Eur. J. Pharmacol.* **81,** 531–542.

Vincent J. P., Vignon J., Kartalovski B., and Lazdunski M. (1980a) Binding of phencyclidine to rat brain membranes: Technical aspect. *Eur. J. Pharmacol.* **68,** 73–77.

Vincent J. P., Vignon J., Kartalovski B., and Lazdunski M. (1980b) Compared properties of central and peripheral binding sites for phencyclidine. *Eur. J. Pharmacol.* **68,** 79–82.

Vincent J. P., Vignon J., Kartalovski B., and Lazdunski M. (1981) Receptor Sites for Phencyclidine in Mammalian Brain and Peripheral Organs, in *PCP (Phencyclidine): Historical and Current Perspectives* (Domino E. F., ed.), pp. 83–103, NPP Books, Ann Arbor, Michigan.

Vincent J. P., Cavey D., Kamenka J. M., Geneste P., and Lazdunski M. (1978) Interaction of phencyclidine with muscarinic and opiate receptors in the central nervous system. *Brain Res.* **152,** 176–182.

Vincent J. P., Kartalovski B., Geneste P., Kamenka J. M., and Lazdunski M. (1979) Interaction of phencyclidine ("angel dust") with a specific receptor in rat brain membranes. *Proc. Natl. Acad. Sci. USA* **76,** 4678–4682.

Vincent J. P., Bidard J. N., Lazdunski M., Romey G., Tournear Y., and

Vignon J. (1983) Identification and properties of phencyclidine-binding sites in nervous tissues. *Fed. Proc.* **42,** 2570–2573.

vonVoigtlander P. F., Tang A. A., Ochoa M. C., Franklin S. R., and Triezenberg H. J. (1983) A Search for Phencyclidine Antagonists, in *Phencyclidine and Related Arylcyclohexylamines* (Kamanka J. M., Domino E. F., and Geneste P., eds.), pp. 657–638, NPP Books, Ann Arbor, Michigan.

Wagner G. C., Franko C. M., and Tomie A. (1984) Interactions of naloxone and haloperidol with phencyclidine: Effects on milk intake. *Pharmacol. Biochem. Behav.* **20,** 379–382.

Walters J. R. and Roth R. H. (1976) Dopaminergic neurons: An in vivo system for measuring drug interactions with presynaptic receptors. *Naunyn-Schmiedebergs Arch. Pharmacol.* **296,** 5–14.

Ward D. and Trevor A. (1981) Phencyclidine-induced alteration in rat muscarinic cholinergic receptor regulation. *Eur. J. Pharmacol.* **74,** 189–193.

Weiner N. (1972) Pharmacology of Central Nervous System Stimulants, in *Drug Abuse: Proceedings of the International Conference* (Zarafonetes C. J. D., ed.), pp. 243–251, Lea and Febiger, Philadelphia, Pennsylvania.

Weiner N. (1980) Norepinephrine, Epinephrine and the Sympathomimetic Amines, in *The Pharmacological Basis of Therapeutics* (Gilman A. G., Goodman L. S., and Gilman A., eds.), pp. 138–175, MacMillan, New York.

Weiss B. and Latres V. G. (1962) Enhancement of human performance by caffeine and the amphetamines. *Pharmacol. Rev.* **14,** 1–36.

West R. E., McLawhon R. W., Dawson G., and Miller R. J. (1983) [^{3}H]-Ethylketocyclazocine binding to NCB-20 hybrid neurotumor cells. *Mol. Pharmacol.* **23,** 486–492.

Wood P. L., Pilapil C., Lafaille F., Nair N. P. V., and Glennon R. A. (1984) Unique [^{3}H]-tryptamine binding sites in rat brain: Distribution and pharmacology. *Arch. Int. Pharmacodyn. Ther.* **268,** 194–201.

Woolverton W. L. and Balster R. L. (1979) Tolerance to the behavioral effects of phencyclidine: The importance of behavioral and pharmacological variables. *Psychopharmacology* **64,** 19–24.

Woolverton W. L., Martin B. R., and Balster R. L. (1980) Modification of the behavioral effects of phencyclidine by repeated drug exposure and body weight changes. *Pharmacol. Biochem. Behav.* **12,** 761–766.

Zukin S. R. and Sircar R. (1984) Quantitative localization of sigma opiate/phencyclidine receptors in rat brain by light microscopy autoradiography. *Soc. Neurosci. Abst.* **10,** 584

Zukin R. S. and Zukin S. R. (1981) Demonstration of [^{3}H]-cyclazocine binding to multiple opiate receptor sites. *Mol. Pharmacol.* **20,** 246–254.

Zukin S. R. and Zukin R. S. (1979) Specific [^{3}H]-phencyclidine binding in rat central nervous system. *Proc. Natl. Acad. Sci. USA* **76,** 5372–5376.

Zukin S. R., Fitz-Syage M. L., Nichtenhauser R., and Zukin R. S. (1983) Specific binding of [^{3}H]-phencyclidine in rat central nervous tissue: Further characterization and technical considerations. *Brain Res.* **258,** 277–284.

Zukin S. R., Brady K. T., Slifer B. L., and Balster R. L. (1984) Behavioral and biochemical stereoselectivity of sigma opiate/PCP receptors. *Brain Res.* **294,** 174–177.

Chapter 15
Receptor Binding in Drug Discovery and Development

Michael Williams and Paul L. Wood

1. Introduction

The ultimate goal of any drug development program is the application of basic research, both chemical and biological, to the discovery of therapeutic modalities that can prevent, alleviate, or cure the malfunctions in the mammalian homeostatic mechanism that lead to the disease state.

The translation of such basic research to the clinic has, to a large extent, been dependent on serendipity, and although more recent developments involving a strictly mechanistic approach, such as the H_2 blockers, cimetidine and ranitidine, have been successful therapeutically, the very complexity of the mammalian organism can infinitely complicate an apparently straightforward enzyme or receptor-based approach to drug development. Indeed, as noted by Gross (1983), drug research has become increasingly more complex and considerably less productive in recent years, with breakthroughs in terms of Ehrlich's "magic bullet" being few and far between.

Drug discovery and the subsequent development process is dependent on two major elements; novel, creative chemistry, and adequate and relevant biological evaluation. Balanced against these two entities is the necessity for compounds in a given area and breakthroughs at the cutting edge of science that provide new approaches to particular therapeutic goals. Although the phrase "screening" may be considered antithetical to the intellectual aesthetic, and a general schism is considered to exist between basic (academic) and applied (industrial) research in pharmacology (Maxwell, 1984), it is difficult to envisage a rational drug development program in which this element is not of crucial importance. The generation of the structure–activity relationship (SAR) is the cornerstone in the development of an understanding of the recognition site(s) to which a potential therapeutic agent is directed. Much of what constitutes modern pharmacology, both classical and molecular, is simply SAR generation. As Maxwell (1984) has further emphasized, without drugs, basic research in pharmacology is limited, and without basic research the development of new compounds is hindered. A healthy synergism between the two disciplines is therefore crucial.

The medicinal chemist has several avenues with which strategies to synthesize new compounds can be developed. Compounds known to be effective in a given situation (and more likely than not to have been discovered fortuitously) can be improved (Cohen, 1983) either by the use of advanced molecular modeling techniques (Loftus, 1986), or by empirical modification of a molecule in light of recent developments in the basic research area. An example of the latter would be the efforts to make a classical tricyclic antidepressant, with alpha-2 antagonist activity, this rationale being based on the observation (Kendall et al., 1982a) that the combination of a classical antidepressant and an alpha-2 blocker, such as yohimbine, can accelerate the down-regulation of cortical beta-adrenoceptors, a tentative biochemical correlate of clinical antidepressant activity (Charney et al., 1981).

Another approach would be to base synthetic efforts on the proposed neuromodulatory agent involved in the particular disease state. This 'mechanistic' approach is exemplified by the H_2 blocker, cimetidine (Black et al., 1972), and the angiotensin-converting enzyme (ACE) inhibitors (Cushman and Ondetti, 1980). However, it may be noted that neither the structure of the histamine receptor nor of the active site of ACE is known at the present time (Goodford, 1984). A third approach to drug development, which may be termed chemical pharmacognosy, is to take newly synthesized, patentable structures and screen them for potential biological activity in a random manner. The serendipity involved in such an approach is paramount, but has led to the discovery of such drugs as chlorpromazine (Delay and Deniker, 1952) and the anxioselective anxiolytics buspirone (Taylor et al., 1984) and TVX Q 7821 (Traber et al., 1984).

Once compounds such as these have been shown to be clinically effective, the emphasis shifts to the elucidation of possible mechanisms of action.

Minor modifications to existing heterocyclic compounds can dramatically affect their intrinsic biological properties. Chemical modification of the benzodiazepine nucleus has resulted in an analgesic, tifluadom (Romer et al., 1982), and a putative antidepressant, AHR 9377 (Kinnier et al., 1984). Similarly, the tricylic, cyproheptadine, has several distinct properties (Williams and Martin, 1982) that can be modified by substituents and the stereochemistry of the molecule (Clineschmidt et al., 1979; Randall et al., 1979).

In searching for novel therapeutic agents, the biological evaluation of newly synthesized entities is of paramount importance. This can take several formats; the in vivo pharmacometrics screen, in which compounds are evaluated in test paradigms in which known active compounds work (and that will present problems with novel compounds that do not have similar bioavailability or pharmacokinetic properties), and are tailored for a given research area, i.e., cardiovascular or CNS. Other less empirical approaches involving more selective screens, such as the Cook-Davidson anticonflict model, or the Porsolt swim test for antidepressants, are also useful in defining compound activity, but are time consuming. In the end, the use of such approaches for the detection of pharmacological activity will, in general, result in compounds with the same limitations of those currently available.

An inevitable issue in the strategy of drug development is whether those elusive entities that may represent novel therapeutic agents already exist, and, because of the human factors (cost, tunnel vision, the ephemeral nature of 'state-of-the-art,' and so on), which precluded their being fully evaluated, are gathering dust in the sample collections of various drug houses.

In view of the pitfalls and limitations of the preclinical drug evaluation process, it is clearly desirable that any biological evaluation procedure should maximize the potential for the identification of novel and relevant activity in new chemical entities that can in time be translated into a therapeutic modality. In the last 10 yr or so, it has become possible to characterize the recognition sites, both receptors and enzymes, through which drugs act, using radioligand binding assays (Snyder, 1983; Williams and U'Prichard, 1984). These assays, as documented in detail in the present volume, are rapid, cost effective, and require minimal amounts of compound (10–20 mg) to provide a broad range of in vitro biological data. In addition, the SARs generated using radioligand binding reflect directly on the ligand–receptor interaction and not on events distal to the recognition process or to biovailability or pharmacodynamic issues. Binding assays have also revolutionized the concept of drug action inasmuch as drug receptors, such as the benzodiazepine receptor, have been identified solely on the basis of ligand binding. This in turn has led to the search for endogenous factors related to anxiety (Squires, 1984).

2. Validation of Radioligand Binding Assays

Despite the obvious advantages of radioligand binding as a screening technique, the methodology has not been as widely accepted in the pharmaceutical industry as would have been imagined. This has, to a certain extent, been because of a natural reliance on proven animal and tissue test procedures and the inevitable question that arises in relation to test tube screening: Does it have any relevance to what occurs in the animal? At the Merck Institute, two compounds, MK 212 (Clineschmidt, 1979) and MK 801 (Clineschmidt et al., 1982; Martin and Papp, 1984), were effective as serotonin agonist and anticonvulsant, respectively, in appropriate animal test models, but were devoid of significant activity in relevant binding assays (Williams and U'Prichard, 1984). Radioligand binding assays, like other biological paradigms, have limitations. A major problem is that they measure potency rather than efficacy; thus, without appropriate manipulations (which are often empirical) involving the use of temperature or pH shifts, guanine nucleotides, or cation effects, it is not always easy to distinguish agonists from antagonists, although as a generalization agonists are usually more active in binding assays in which they are used as ligands, whereas the converse is true of antagonists. In addition, binding assays do not measure bioavailability or duration of drug action. These assays also are incapable of contributing significantly to the understanding of spare receptors, such issues being in the realm of in vivo studies.

The use of batteries of receptor binding assays as in the case of the benzodiazepine receptor complex (Williams, 1984), do, however, permit the distinction between partial agonists, agonists, agonist/antagonists, antagonists, and inverse agonists for this receptor (Wood et al., 1984a).

The physiological and pharmacological criteria originally proposed by Cuatrecasas and Hollenberg (1976) have been reviewed by Hrdina (1985) in this volume and have also been stressed by Laduron (1984) who has carefully delineated the difference between a receptor and an "acceptor," the latter being a binding site uninvolved in signal transmission and hence lacking physiological relevance. For instance, although both cimetidine and ranitidine are highly effective H_2 receptor blockers in the clinic, in vitro they bind respectively to imidazole and furane recognition sites that appear unrelated to H_2 receptors (Laduron, 1984).

The overinterpretation of binding data based solely on multiphasic displacement or saturation curves leading to the description of "new" receptor subtypes in the total absence of any physiological data has contributed significantly to the skepticism of the more classically trained pharmacologist to the technique. For example, the type III anticonvulsant benzodiazepine recognition site, described by Bowling and De Lorenzo (1982), has eluded detection by at least 11 other research groups, whereas the five or more dopamine recognition sites "detected" by ligand binding (Seeman, 1980) are more realistically confined to two tangible entities as defined by their interactions with adenylate cyclase (Stoof and Kebabian, 1984).

The issue of receptor subtypes, as opposed to differing association/affinity states of a coupling complex (Olsen, 1982; Williams, 1983; Martin I. L. et al., 1983), or that of drug receptors such as those cited for ranitidine and cimetidine above, or for the trazodone analog, tetrahydrotrazodone (Kendall et al., 1983), which has a pharmacology suggestive of a serotonin-1 receptor, but is not entirely resolved in this context (Laduron, 1984), underline the importance of interpreting binding data in light of data obtained by other pharmacological techniques. The recent description of muscarinic receptor subtypes (Hammer and Giachetti, 1980) and various calcium antagonist "receptors" (Murphy et al., 1983) underlines the rapid progress that is occurring with this methodology and the need to exercise caution in the interpretation of data in the absence of that derived by other biochemical, tissue, and in vivo techniques.

Despite such considerations, of which any competent "binder" should be aware, binding assays are an invaluable tool for drug screening. Such assays have been used to identify novel dopamine agonists, such as EOE and PHNO (Martin G. E. et al., 1983, 1984), the cyproheptadine antiemetic, L-646,462 (Williams et al., 1984), and the nonbenzodiazepine anxioselective, anxiolytic, CGS 9896 (Yokoyama et al., 1982). The ability of binding assays to predict potential drug side effects, such as the EPS liability of neuroleptics that may be associated with their anticholinergic activity (Snyder et al., 1976), have not been validated in the clinic. As with all biological evaluation procedures, there is a constant need to reevaluate the meaning and the criteria involved in binding studies in light of the translation of certain intrinsic properties of new entities to clinical effectiveness.

3. Binding as a Tool in Drug Screening

3.1. General Considerations

The criteria applied to the use of binding assays for screening tools differs somewhat from that used in basic research. For instance, in the initial evaluation the question is whether or not the compound is active, usually at a screening dose between 10^{-6} and $10^{-4}M$, depending on the affinity of known active compounds for the receptor in question. Compounds active in the nanomolar range are required for monoamine receptors. However, in the area of excitatory amino acids, micromolar activity may be of interest. Active compounds may then be tested for potency and selectivity. The issue of a new compound having a unique bi- or multi-phasic displacement curve over a 6 log unit range is one that is addressed after a compound has been shown to have in vivo activity consistent with its binding profile. Although one is never certain of the situation, in vivo binding sites that differ from one another on the order of a few nanomoles may have therapeutic relevance. Binding studies (Sokoloff et al., 1980) have shown that pre- and postsynaptic dopamine receptors differ in their affinities for agonist radioligands on the order of 2 n*M*. In schizophrenic patients, low doses of apomorphine can ameliorate their symptoms, whereas higher doses exacerbate the condition (Tamminga et al., 1978). Such data would therefore indicate that receptor subpopulations may be titrated in vivo.

One criterion for establishing activity is that a compound at $10^{-6}M$ should produce a >70% displacement of specific ligand binding. Following from this, an IC_{50} should be generated, and if the compound is of interest, specificity can be examined at the in vitro level. Specificity is, at best, a working term, and it would be difficult to define just how selective a compound should be using a mechanistic approach to drug development. There is still no way of ensuring that a dopamine agonist devoid of alpha-adrenoceptor activity is less attractive than a compound that has such activity. In addition, the relative "straightforwardness" of the ligand binding methodology often reveals a lack of selectivity that has no apparent consequence in vivo. Ketanserin selectively labels serotonin-2 receptors in rat frontal cortex. However, in vivo, the compound has alpha-adrenergic blocking activity that is physiologically significant (Fozard, 1982). In vitro, radioligand spiperone can bind to both serotonin-2 receptors in rat frontal cortex and neuroleptic/ dopamine-2 receptors in striatum. In vivo, both receptor types must be considered as possible target sites for the overall actions of this class of neuroleptic.

3.2. Practical Aspects

Ligand receptor complex formation is dependent on two primary factors:

(1) the time of incubation required to achieve equilibrium, a process subject to the Law Of Mass Action, and
(2) the conditions under which the equilibrated complex is isolated.

3.2.1. Equilibrium Binding

In contrast to the rapidity of ligand interactions with neurotransmitter recognition sites under in vivo conditions, equilibrium association times for similar ligands in binding assays are generally incompatible with any proposed physiological event. Binding of the nicotinic cholinergic antagonist, dihydroerythroidine, reaches equilibrium in 5 min (Williams and Robinson, 1984), whereas binding of the peptide TRH (thyrotropin releasing hormone) requires periods of up to 6 h (Simasko and Horita, 1982). Quite clearly, the in vitro system is, at best, an approximation of in vivo events, and it cannot be sufficiently reemphasized that binding assays must be validated using appropriate physiological correlates. Despite such caveats, the technique of radioligand binding is no more inappropriate to the task of drug discovery than the comparable use of in vitro tissue bath preparations.

3.2.2. Receptor-Ligand Isolation

Receptor–radioligand complexes can be isolated by either vacuum filtration, using 25–30 lb of negative pressure, or by centrifugation, using either a high-speed (21,000 rpm) refrigerated centrifuge (Beckman J-2-21-M) or microfuge (Beckman 12 or Eppendorf).

3.2.2.1. FILTRATION. Filtration isolation of receptor–ligand complexes involves the use of glass fiber filter paper (Whatman GF/B or other grade) held in one of three types of apparatus; a Millipore Filtration Assembly (Model 1225); a Sandbeck filter box (Bennett, 1978), or a 24-channel Cell Harvester (Brandel M-24).

The Sandbeck box, which holds 45 filter disks, has associated with it problems because of uneven vacuum application, a result of the drying of the filter disks. This can produce very tangible reproducibility problems (Hung et al., 1982). The Millipore apparatus is, to some extent, free of such problems, given that filtration times are short (~7–10 s/tube), but is inconvenient in regard to setting up because it uses individual filter disks that have to be individually washed. The Brandel Cell Harvester, on the other hand, is to radioligand binding what the automated scintillation counter was to radioactivity determination some 30 yr ago. It permits a rapidity and ease of sample throughput that is not readily achieved by other filtration methods, but should be evaluated for use in any particular binding assay. It is inappropriate for use, for instance, in both peripheral and central benzodiazepine receptor binding assays for reasons as yet undetermined. In this context, it may be noted that although theoretical criteria have been established for ligand complex stability given a certain K_d value (Bennett, 1978), these do not always hold true in practice. Thus, although flunitrazepam binding is very robust, and its subnanomolar K_d is particularly suited to filtration isolation, for some reason, a good deal of variability is seen when using the Brandel Cell Harvester (Williams and Risley, unpublished; Wood et al., unpublished). Conversely, the use of filtration procedures for the isolation of micromolar affinity glutamate receptor–ligand complexes that theoretically should be isolated by centrifugation can still produce viable data (Baudry and Lynch, 1981).

3.2.2.2. CENTRIFUGATION. Because of the cost of equipment and the amount of actual sample handling involved, centrifugation is not a popular technique. However, when assay conditions do not permit the use of filtration, centrifugation must be used. Typically, as noted above either a refrigerated centrifuge or a microfuge may be used.

The use of the larger machine provides more solid tissue pellets from the assay incubation, but involves centrifugation periods of up to 15 min, lengthy wash and transfer produres for tissue solubilization, and scintillation fluid addition. Whereas part of the labor involved with this procedure has been circumvented by the manufacture of scintillation vials that can be centrifuged, and in which the actual assays can be run, it is still very difficult, given an assay in the hundreds of samples, to obtain adequate centrifugation facilities. Microfugation can circumvent this problem to some extent. Newer microfuges, such as the Beckman 12, can produce 12,000*g* centrifugation forces in minutes, and with their ability to hold 72 samples at a time, they improve the throughput, especially since centrifugation times are usually in the order of 2 min. Problems associated with the microfuge, however, are; that incubation volumes are restricted to 2 mL or less, depending on the number of samples that are required to be run in a given experiment; that unless the machine is used in a cold room, spins to isolate receptor–ligand complexes usually involve some degree of heat generation that may be deleterious to the complex association; and that pellets are less robust than those obtained by centrifugation at higher gravity forces and are more subject to loss during washing off of residually associated, but not bound, radioactivity. An additional problem is the transfer of the pellet to scintillation vials. To avoid the use of quenching tissue solubilizers, such as Protosol (NEN), it is customary to cut the end of the microfuge tube off and transfer it and the tissue pellet to a scintillation vial. More recently, a common practice has been to carry out the experiment in scintillation vials to remove this step, and the vials can indeed stand up to centrifugation.

3.2.3. Other Considerations

In using radioligand binding assays on a routine, daily basis it is essential that the signal-to-noise ratio (% specific binding) must be sufficiently high and robust that the assay can be used to determine binding for large numbers of samples. The inevitable "drifts" and day-to-day variability between tissue triplicates can cause very tangible problems with data interpretation, especially when counts on the order of fractions of a hundred decide on the slopes of displacement curves. In addition, unstable tissue preparations that tend to cause upward or downward "drifts" tend to cause problems in routine practice, as do unstable ligands. In the latter instance, experience will usually indicate the useful laboratory life of a given ligand, and it is then appropriate to keep available stocks within the framework of anticipated usage and replace the ligand on a routine basis. For instance, monoamine ligands, such as serotonin and apomorphine, have a useful life on the order of 3 mo, whereas other ligands, such as cyclohexyladenosine and diazepam, are very robust. Although it may seem wasteful, it is usually less expensive to adopt this approach to ligand stability than attempt to purify existing stocks in which additional impurities and the increased possibility of radiolysis may negate the attempts at purification.

Another aspect of the routine use of ligand binding assays is that brain tissue, rather than heart or renal membrane preparations, are used for primary evaluation of new compounds, this being a reflection of the greater number of binding sites and the ease of tissue preparation. Secondary evaluation of active compounds in their appropriate target tissues can be done more rigorously with known active compounds when the inherent problems associated with such tissue preparations (low-specific binding, extended tissue-preparation times) do not deter from the actual identification of active compounds.

3.2.4. Sample Preparation

The preparation of compound samples for evaluation in ligand binding assays is a major component in terms of time for any assay and very often requires more time than the actual execution of the assay. Unlike compounds whose solubility and stability are well documented, newer compounds usually require a good deal of evaluation before they can be put into solution in solvents that are compatible with biological test systems. Although alcohol and other, more exotic solvents such as DMSO may be used, it is necessary to evaluate a given binding assay for compatibility with such a solvent, and run appropriate controls to compensate for the effects of these solvents on baseline parameters of the binding assay. It should be borne in mind that not all binding assays are "created equal," and the lack of effect of DMSO at 0.5% in one assay should not be extrapolated to another assay. One final point in regard to sample preparation is the actual number of samples that can be run in a given experiment. Claims in the hundreds of samples per assay have been made (Creese et al., 1976) that have made routine assays involving 10–30 compounds appear very unambitious, if not nihilistic. It may be noted that to weigh out unknown compounds and get them into solution can take a considerable period of time, often on the order of days. Once the data have been generated, it also requires a work-up to determine percent displacement and IC_{50} or K_i values, where appropriate. Thus an assay involving a run of 100 or so given compounds that takes a day to actually perform can involve a day of preparation and a day of data analysis. The tip of the iceberg, as always, can conceal a large mass. An additional important point is that the reliability of each and every assay should be assessed using appropriate reference standards. Progressive drifts in the activity of reference compounds are the best indication of either ligand or tissue instability.

4. Data Analysis

In performing the binding assay, one runs triplicate determinations of each observation involving; total counts, nonspecific counts, and "drug-dependent" counts. It is usual procedure to run total and nonspecific counts in every group of 36 tubes, to check for intraexperimental variation in baseline parameters. For screening purposes, a new compound would be run at a single concentration in the micromolar range, and if producing a displacement of greater than 50% at the screening concentration, an IC_{50} value would be determined by running 5–7 concentrations of the compound. In obtaining the IC_{50}, it is also necessary to determine the Hill number (n_H), an indication of the interaction of the compound with a single, or more than one, type of binding site(s) in the membrane preparation. If the Hill slope or Hill coefficient differs significantly from unity then the K_i value cannot be determined.

The data obtained from the binding experiment, not unexpectedly, is in the form of a scintillation counter printout, and data analysis involves the editing and transformation of such data to obtain the desired inhibition percentages and IC_{50} values for the compounds evaluated. Ideally, this should involve extensive computerization. Thus, the triplicates should have their mean and standard deviation calculated, the nonspecific counts for the experiment subtracted, and the final specific binding value expressed as a percent of the control values. Such "number crunching" can greatly facilitate throughput, and where spurious values are generated, is open to editing in an objective manner by the experimenter. Unfortunately, communication between scintillation counter manufacturers and those who make computers compatible with this task is less than optimal, and the on-line or batch evaluation of scintillation counter data by a computer is the exception rather than the rule. The direct analysis of data that does not involve the reentering of the hard data can significantly reduce data analysis time, although removing the possibility of error in the transfer of data. (M. Vogel, personal communication).

4.1. Computer Analysis of Displacement Curves

Once data have been stored in a computer, they can be mathematically analyzed by the reduction of sigmoid dose–response curves. To this end, the inhibition data of a displacement curve can be subjected to linear transformation so that straight lines are generated for linear regression analysis. These transforms are:

(1) Probits: The % bound data can be transformed to a probit, or probability unit, which is related to the percent area under the normalized probability curve. These conversions can be performed with standard tables of probit values (i.e., Documenta Geigy.)

(2) Logits: The logit transformation (Rodbard and Lewald, 1970; Rodbard and Frazier, 1975) is defined as:

$$\text{logit} = \ln[\%\ \text{bound}/(100 - \%\ \text{bound})]$$

This transformation has the advantage that no statistical tables are required for analyzing the data. Furthermore, there is no

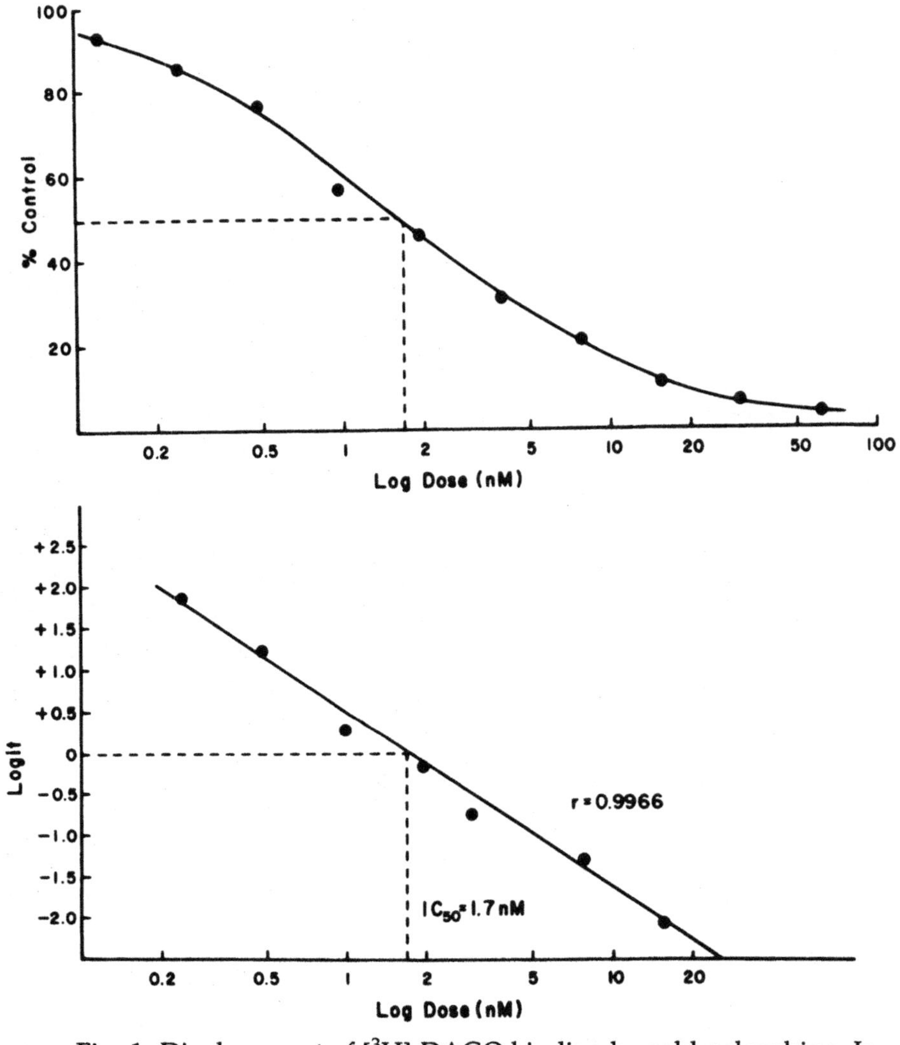

Fig. 1. Displacement of [^{3}H]-DAGO binding by cold nalorphine. In the top figure, the data is presented as log dose (concentration) vs percent inhibition of specific binding. In the lower figure, the same data has undergone logit transformation for the determination of an IC_{50} value.

mathematically based relationship between the cumulative normal distribution (probit) and receptor models describing sigmoid dose–response curves. This does not appear to be the case for the logit transform (Hafner et al., 1977). An example of the logit transform is presented in Fig. 1, in which [^{3}H]-DAGO binding in the presence of increasing concentrations of cold nalorphine is analyzed.

Since the logit–log dose model is sensitive to errors at the extremes of the curve (Rodbard and Frazier, 1975), points below

10% and above 90% of control binding can be excluded (*see* Fig. 1, Program 3 in Chapter 11 of this volume). Such logit–log dose plots also involve the determination of the Hill coefficient. For displacement curves:

$$Y = \log[\% \text{ bound}/(100 - \% \text{ bound})]$$

Thus a plot of Y-log dose will yield a slope that is the Hill coefficient (n_H).

In the case of competitive inhibition, IC_{50} values can be converted to K_i values according to the relationship described by Cheng and Prusoff (1973):

$$K_i = IC_{50} \; / \; [1 = (\text{ligand})/K_d]$$

A preliminary test to assure competitive inhibition is to determine if 0.1 × IC_{50} represents 10% displacement and 10 × IC_{50} represents 90% displacement. However, a more reliable method to determine the type of inhibition is to express the data as a Dixon plot [(1/B) vs (I)], together with plots of [(L)/B vs (I)] (Cornish-Bowden, 1974), or as Hill plots [log((100-B)/B) vs log(I)], along with plots of IC_{50} vs [L] (Peterson, 1975); where B= percent of control specific binding; I= inhibitor; L= ligand. The profiles of these plots are presented in these publications and enable the distinction between competitive, uncompetitive, mixed, and noncompetitive inhibition. An alternative approach is to determine the K_d and B_{max} of the radioligand in the presence of several fixed concentrations of inhibitor. In this case, an inspection of the effects of the inhibitor on these parameters yields an unambigous distinction between the different forms of inhibition (Table 1).

A more recent concern is in the area of peptide receptors. The displacement of bound peptide radioligands involves heterologous binding analysis, in which the rate of peptide-receptor association is largely controlled by diffusion (Burgen, 1966; Pliska, 1983). Indeed, the association rate constants for most peptide ligands are about three orders of magnitude below those of classical drug–receptor interactions (Pliska, 1983). The long incubation times used for such assays are associated with many technical and theoretical problems: (1) stability of the radioligand; (2) stability of the receptor/membrane preparation; (3) receptor desensitization; (4) deformations in the displacement curves as a result of the large differences in the K_d between the tracer and cold displacer; and (5) pseudoirreversible binding of some iodinated peptide ligands. The popular, but unsatisfactory, solution to this problem is to use

TABLE 1
Types of Inhibition of Radioligand Binding In Vitro

Type	K_d	B_{max}	Example
Competitive	Increased	Unchanged	Diazepam/[^{3}H]-FNZ[a]
Noncompetitive	Unchanged	Decreased	Mianserin/[^{3}H]-5-HT[b]
Uncompetitive	Decreased =	Decreased	?
Mixed	Decreased ≠	Decreased	CGS 9896/[^{3}H]-FNZ[a,c]

[a]Petrack et al., 1983.
[b]Helmeste and Tang, 1984.
[c]FNZ, flunitrazepam.

shorter incubation times that achieve 50–75% of equilibrium binding. The problem in this case is that all of the mathematical models used in displacement analyses are based upon the assumption of equilibrium between radioligand and receptor.

5. In Vivo Binding

In vivo radioligand binding studies allow for the measurement of the relative bioavailability of the compound under study. Bioavailability is a collective term that reflects both accessibility to the bloodstream and passage of the blood–brain barrier. An additional aspect of in vivo binding is that the microenvironment of the receptor that is disrupted for in vitro studies is maintained, particularly with regard to a physiological ionic and guanyl nucleotide milieu. One negative aspect is that undefined concentrations of endogenous receptor ligands will be present also. Another is the possibility that the drug under examination may in fact alter the bioavailability of the ligand, preventing its access to the brain rather than acting as a true displacer inside the blood–brain barrier. Alternatively, a compound can restrict egress and cause an apparent stimulation in binding. Two examples of the in vivo approach are, however, given below:

(1) [^{3}*H*]-Spiperone binding. The binding of [^{3}H]-spiperone in vivo has proven to be an extremely sensitive means of measuring the accessability of dopamine antagonists to CNS receptors in the whole animal (Laduron et al., 1978). Using this approach, substituted benzamides have been shown to pos-

sess a selective affinity for dopamine receptors in the septum and hippocampus (Kohler et al., 1979, 1981; Bischoff et al., 1982). The atypical neuroleptic, clozapine, also expresses a small degree of selectivity for dopamine receptors in the septum and olfactory tubercle (Kohler et al., 1981; LeFur et al., 1980). Binding of [^{3}H]-spiperone in the frontal cortex appears to involve mainly serotonergic (S-2) receptors, and not dopaminergic receptors (LeFur et al., 1980). This is, therefore, a useful in vivo assay system to evaluate the bioavailability of S-2 receptor antagonists.

(2) [^{3}*H*]-Flunitrazepam. The binding of [^{3}H]-flunitrazepam in vivo has been used to demonstrate bioavailability of BZ agonists, antagonists, and partial agonists (Chang and Snyder, 1978; Peterson et al., 1984). This assay has also been used to evaluate drugs that enhance [^{3}H]-flunitrazepam binding, rather than displace it. These drugs can be divided into two classes: (1) drugs that enhance binding only in vivo: buspirone (Oakley and Jones, 1983; Wood et al., 1984b); clozapine (Wood et al., 1984b); and mianserin (Jeevanjee et al., 1984); and (2) drugs that enhance binding both in vitro and in vivo: tofizopam (Saano, 1982; Mennini et al., 1982); EMD 28422 (Skolnick et al., 1980); and GABAergics including muscimol, THIP, isoguavacine, and progabide (Koe, 1983; Wood et al., 1984b). The issue of the enhancement of radioligand binding is discussed further below.

6. Receptors and Drug Treatment

An additional approach to the in vivo receptor situation is that termed *ex vivo*, in which receptor characteristics are manipulated in vivo, usually by a chronic drug treatment. Under such conditions, certain classes of drugs, especially those that are active in the brain, have been found to selectively alter receptor binding.

6.1. Down-Regulation

The phenomenon of a decrease in receptor number is referred to as down-regulation, and may be considered to be functionally analogous to receptor desensitization. Thus, treatment with agonist ligands would be expected to cause a decrease in receptor number. This has been documented for several receptor agonists (Creese and Sibley, 1981). More interesting are the effects of compounds such as antidepressants that have several distinct mechanisms of action that are reflected physiologically in their therapeutic mode of action. Chronic treatment with antidepressants such as iprindole, desmethylimipramine, and amitryptyline causes a down-regulation of cortical beta adrenoceptors (Bannerjee et al., 1977; Sugrue, 1981). This biochemical measurement is accepted as an indication of antidepressant potential, despite a large body of evidence that atypical antidepressants such as trazodone (Taylor et al., 1981), and drug treatments unrelated to antidepressants (Kendall et al., 1982b), have no effect and decrease binding, respectively. Serotonin-2 receptors have also been found to undergo down-regulation in response to chronic antidepressant drug treatment (Peroutka and Snyder, 1980), although the significance of this phenomenon has been questioned (Blackshear and Sanders-Bush, 1983).

6.2. Up-Regulation

Increases in receptor number following drug treatment are referred to as upregulation. As may be expected, ligand antagonists, by blocking the effects of endogenous agonist ligands, may be expected to cause subsensitivity and thus result in a compensatory increase in receptor number. Antagonists do increase receptor number (Creese and Sibley, 1981) and more recently the adenosine antagonists, caffeine and theophylline, have been found to increase adenosine A-1 receptor number (Murray, 1982; Boulenger et al., 1983). Recent work (Pilc and Lloyd, 1984) has also revealed that antidepressants can selectively upregulate $GABA_B$ receptors in rodent brain.

7. Receptor Autoradiography

The light microscopic–autoradiographic approach to receptor binding is unique in that it can anatomically define the location of a receptor population of interest (Kuhar, 1982). This allows analysis of displacement curves from many brain regions within a single rat brain, thereby allowing comparative analysis of regional differences in drug–receptor interactions with great precision. These relative differences in affinities have successfully correlated clinical actions and side effects with specific anatomical sites (Kuhar,

1982). Another impressive use of autoradiography is concerned with the study of peptide receptors that tend to have much more restricted localizations than classical neurotransmitters. For example, the binding of [^{3}H]-vasopressin is discretely localized to the paraventricular and supraoptic nuclei (Yamamura et al., 1983). Clearly, binding studies with whole brain homogenate or hypothalamic homogenate can mask such small-density receptor populations, because of large nonspecific backgrounds.

Examination of adenosine A-1 and A-2 receptor distribution in rat brain using [^{3}H]-cyclohexyladenosine and [^{3}H]-5'-*N*-ethylcarboxamidoadenosine (NECA), respectively, in conjunction with autoradiographic visualization has provided additional evidence for the striking enrichment of A-2 receptors in the striatum (Snowhill and Williams, 1985). As the technology for computerized imaging of autoradiographs improves, and no doubt becomes more accessible financially, it seems likely that the regional evaluation of receptor distribution using homogenate binding will become increasingly redundant. The technique still requires improvement, however. Quenching of tritium radiation caused by the constituents of the tissue slice is one drawback; another is the lack of automation in analyzing the final autoradiograph. One experiment can generate sufficient data for 15 h or more of one-on-one computer analysis time. Quite clearly, the autoradiographic equivalent of the scintillation counter requires development. Although the technical feasibility appears daunting at the present time, many felt the same way about the domestic applicability of the compact digital audio disk.

8. Methods to Reveal Receptor Subtypes

With the majority of radioligands, some crossreactivity with other receptor types is inevitable in a membrane preparation. This can be an advantage in some cases, because it can be used to uncover a new receptor or isoreceptor population. However, in most situations in which the investigator is attempting to perform SAR studies of a defined receptor, these crossreactivities can cause significant errors in the evaluation of new compounds. To overcome these problems, several approaches are available:

8.1. Specific Radioligands

In some research areas such as the monoamines, a number of very specific radioligands—rauwoulscine (Perry and U'Prichard,

1981) and SCH 23390 (Billard et al., 1984)—are available, and should be utilized whenever possible.

However, inherent to new research areas is the lack of such radioligands, and more novel approaches have to be evaluated.

8.2. Receptor Masking

This approach is used in cases in which specific cold ligands can be used to suppress the binding of a radioligand to associated recognition sites, while maintaining binding to a site for which no specific radioligand is available. A good example of this approach is the opioid receptor area where specific ligands are available for the mu and delta opioid sites, but not for the kappa site. In this case, a radioligand like [^{3}H]-ethylketazocine that binds to all three receptor populations can be used to label kappa receptors by selective suppression of binding to mu and delta sites by coincubation with selective cold ligands for these sites (James and Goldstein, 1984).

8.3. Computer Analysis of Multiphasic Displacement Curves

Nonlinear computer analysis of a series of multiphasic competitive displacement curves can reveal differences in radioligand displacement curves that can be related to the possible existence of receptor subtypes (Munson and Rodbard, 1980). Some caution is necessary in regard to this approach, however, because receptors by definition are functional entities subserving physiologically relevant roles in mammalian tissue, not mere "blips" in test tube biochemical studies.

8.4. Selective Incubation Conditions

The choice of incubation buffer and its ionic composition and pH, the temperature of the incubation, the inclusion of guanyl nucleotides, and pretreatment of membranes with various detergents are just a few of the parameters that can be altered to specifically suppress ligand binding, to enhance binding to a given site, to distinguish isoreceptor populations, or to unmask receptor sites. The selective suppression of mu and delta opioid receptor binding by a combination of low incubation temperatures, and inclusion of sodium in the incubation buffer, is an example of how incubation conditions can be manipulated to maximize labeling of kappa opioid receptors (Wood, 1985). The importance of the procedures used to prepare and pretreat the membrane preparation is also illustrated by the case of [^{3}H]-GABA binding, in which rat brain crude synaptic membranes must be extensively washed, subjected to a freeze–thaw cycle, and treated with the detergent Triton X-100 to ensure the removal of the endogenous GABA that would compete for binding to $GABA_A$ sites (Enna and Snyder, 1975).

8.5. Receptor Inactivation

Under controlled conditions of alkylation, selective inactivation of receptor populations can be achieved. This is the case with phenoxybenzamine pretreatment of brain striatal membranes that can selectively eliminate the high-affinity binding of [^{3}H]-spiperone to CNS dopamine receptors, although leaving [^{3}H]-apomorphine binding sites unaltered (Creese et al., 1982). Similarly, mu and delta opioid receptor binding sites are more rapidly inactivated by *N*-ethylmaleimide than are kappa sites (Wood and Charleson, 1982). A variant of this approach is to selectively block, with specific cold ligands, the inactivation of a defined receptor population, while other sites are inactivated. After this pretreatment, the membranes are extensively washed to remove the alkylating agent and the cold ligand, leaving a preparation with a specific population of receptors for displacement studies. Such elegant studies have been performed with the alkylating agent, beta-FNA, to obtain membrane preparations with a relatively pure kappa opioid receptor population (James et al., 1982).

9. Conclusions

From the above considerations, it is apparent that receptor binding has much to offer in terms of evaluating new compounds for drug development. As noted in section 2, it is necessary to validate these assays using appropriate classical biochemical, tissue, or in vivo assays that would constitute secondary screens. In some instances, however, ligand binding assays have revealed properties in compounds that are relatively new from a mechanistic standpoint. One example is that of the putative anxiolytic, etazolate, that can increase [^{3}H]-diazepam binding in vitro, presumably by "unmasking" benzodiazepine recognition sites. Although various theories may be propounded to explain the mechanistic rationale for such changes (Williams, 1984), it is uncertain how such compounds may produce any pharmacological effects in a clinical situation, especially when endogenous ligands have to be contended with.

For the future it is apparent that the more information that can be obtained for a known compound in regard to its role as a potential therapeutic agent, the better the chance will be to design more selective agents for dealing with various disease states and tissue malfunctions. As already noted, it is not really known to what extent ancillary properties in a given molecule aid in its presumed primary action or contribute to its unwanted "side effects." Fine tuning of receptor molecules appears feasible, and it is to be anticipated that the development of more sophisticated ligand binding approaches such as those outlined above and their extension from CNS tissues to those in the periphery, will aid in the development of novel, potent, therapeutic entities.

On a final note, it should be cautioned that radioligand binding is an excellent tool in aiding understanding of tissue function at the molecular level and in defining possible drug properties of new compounds. It is not an end unto itself. The generation of new receptor subtypes based on extensive tissue manipulations at the in vitro level, without any concern for physiological aspects, can only serve to limit and ultimately discredit the technique. Both nature and computers are wonderful entities. The mystery of life is, however, elusive, and hopefully will not be reduced to a multiphasic displacement curve in a radioligand binding assay.

References

Bannerjee S. P., Kung L. S., Riggi S. J., and Chanda S. K. (1977) Development of β-adrenergic receptor subsensitivity by antidepressants. *Nature* (Lond.) **268,** 455–458.

Baudry M. and Lynch G. (1981) Characterization of two [^{3}H]-glutamate binding sites in rat hippocampal membranes. *J. Neurochem.* **36,** 811–820.

Bennett J. P. (1978) Methods in Binding Studies, in *Neurotransmitter-Receptor Binding* (Yamamura H. I., Enna S. J., and Kuhar M. J., eds.), pp. 57–90, Raven, New York.

Billard W., Ruperto V., Crosby G., Iorio L. C., and Barnett A. (1984) Characterization of the binding of ^{3}H-SCH 23390, a selective D-1 receptor antagonist ligand, in rat striatum. *Life Sci.* **35,** 1885–1891.

Bischoff S., Bittiger H., Delini-Stula A., and Ortmann R. (1982) Septo-hippocampal system: Target for substituted benzamides? *Eur. J. Pharmacol.* **79,** 225–232.

Black J. W., Duncan W. A. M., Durant C. J., Ganellin C. R., and Parsons E. M. (1972) Definition and antagonism of histamine H_2-receptors. *Nature* (Lond.) **236,** 385–390.

Blackshear M. A. and Sanders-Bush E. (1982) Serotonin receptors after acute and chronic treatment with mianserin. *J. Pharmacol. Exp. Ther.* **221,** 303–308.

Boulenger J-P., Patel J., Post R. M., Parma A. M., and Marangos P. J. (1983) Chronic caffeine consumption increases the number of brain adenosine receptors. *Life Sci.* **32,** 1135–1142.

Bowling A. C. and De Lorenzo R. J. (1982) Micromolar affinity benzodiazepine receptors: Identification and characterization in central nervous system. *Science* **216,** 1247–1250.

Burgen A. S. V. (1966) The drug–receptor complex. *J. Pharm. Pharmacol.* **18,** 137–148.

Chang R. S. L. and Snyder S. H. (1978) Benzodiazepine receptors: Labelling in intact animals with [^{3}H]flunitrazepam. *Eur. J. Pharmacol.* **48,** 213–218.

Charney D. S., Menkes D. B., and Heninger G. R. (1981) Receptor sensitivity and the mechanism of action of antidepressant treatment. *Arch. Gen. Psychiatry.* **38,** 1160–1180.

Cheng Y. C. and Prusoff W. H. (1973) Relationship between the inhibition constant (K_i) and the concentration of the inhibitor which causes 50 percent inhibition (IC_{50}) of an enzymic reaction. *Biochem. Pharmacol.* **22,** 3099–3108.

Clineschmidt B. V. (1979) MK 212: A novel serotonomimetic. *Gen. Pharmacol.* **10,** 287–291.

Clineschmidt B. V., McKendry M. A., Papp N. L., Pflueger A. B., Stone C. A., Totaro J. A., and Williams M. (1979) Stereospecific antidopaminergic and anticholinergic actions of the enantiomers of (±) 1-cyclopropylmethyl-4-(3-triflurometheyl-5H-dibenzo[a,b]cyclohepten-5-ylidene)piperidine(CTC), a derivative of cyproheptadine. *J. Pharmacol. Exp. Ther.* **208,** 460–467.

Clineschmidt B. V., Williams M., Witoslawski J. J., Bunting P. R., Risley E. A., and Totaro J. A. (1982) Restoration of shock suppressed behavior by treatment with (+)-5-methyl-10,11-dihydro-5H-dibenzo[a,b]-cyclohepten-5,10-amine (MK 801), a substance with potent anticonvulsant, central sympathomimetic and apparent anxiolytic properties. *Drug Develop. Res.* **2,** 147–163.

Cohen N. C. (1983) Towards the rational design of new leads in drug research. *Trends Pharmacol. Sci.* **4,** 503–506.

Cornish-Bowden A. (1974) A simple graphical method for determining the inhibition constants of mixed, uncompetitive and non-competitive inhibitors. *Biochem. J.* **137,** 143–144.

Creese I. and Sibley D. R. (1981) Receptor adaptation to centrally acting drugs. *Ann. Rev. Pharmacol. Toxicol.* **21,** 357–391.

Creese I., Burt D. R., and Snyder S. H. (1976) Dopamine receptors and average clinical doses. *Science* **194,** 546.

Creese I., Hamblin M. W., Leff S. E., and Sibley D. R. (1982) CNS dopamine receptors, in *Handbook of Psychopharmacology,* vol. 17 (Iversen L. L, Iversen S. D., and Snyder S. H., eds.), Plenum, New York.

Cuatrecasas P. and Hollenberg M. D. (1976) Membrane receptor and hormone action. *Adv. Prot. Chem.* **30,** 251–451.

Cushman D. W. and Ondetti M. A. (1980) Inhibitors of angiotensin-converting enzyme for treatment of hypertension. *Biochem. Pharmacol.* **29,** 1871–1877.

Delay J. and Deniker P. (1952) Trente-huit cas de psychoses traitees par la cure prolongee et continue de 4560RP. *Le Congres des Al et Neurol. de Langue Fr. In., Compte rendu de Congres,* Masson et Cie, Paris.

Enna S. J. and Snyder S. H. (1975) Properties of gamma-aminobutyric (GABA) receptor binding in rat brain synaptic membrane fractions. *Brain Res.* **100,** 81–97.

Fozard J. R. (1982) Mechanism of the hypotensive effect of ketanserin. *J. Cardiovasc. Pharmacol.* **4,** 829–838.

Goodford P. J. (1984) Drug design by the method of receptor fit. *J. Med. Chem.* **27,** 557–564.

Gross F. (1983) Introduction: The Present Situation of the Search for New Drugs, in *Decision Making In Drug Research* (Gross F., ed.) pp. 1–3, Raven, New York.

Hafner D., Heinen E., and Noack E. (1977) Mathematical analysis of concentration–response relationships. Method for the evaluation of the ED50 and the number of binding sites per receptor molecule using the logit transformation. *Arzn. Forsch.* **27,** 1871–1874.

Hammer R. and Giachetti A. (1980) Selective muscarinic receptor antagonists. *Trends Pharmacol. Sci.* **5,** 18–20.

Helmeste D. M. and Tang S. W. (1984) Temperature-sensitive high-affinity [^{3}H]serotonin binding: Characterization and effects of antidepressant treatment. *Life Sci.* **35,** 759–767.

Hrdina P. General Principles of Receptor Binding, in *Neuromethods,* vol. 4 (Boulton A. A., Baker G. B., Hrdina P., eds.), Humana, Clifton, New Jersey (in press).

Hung C-R., Hong J-S., and Bondy S. C. (1982) The prevention of an artifact in receptor binding assay by an improved technique. *Life Sci.* **30,** 1713–1720.

James I. F. and Goldstein A. (1984) Site-directed alkylating of multiple opioid receptors. I. Binding selectivity. *Mol. Pharmacol.* **25,** 337–342.

James I. F., Chavkin C., and Goldstein A. (1982) Preparation of brain membranes containing a single type of opioid receptor highly selective for dynorphin. *Proc. Natl. Acad. Sci. USA* **79,** 7570–7574.

Jeevanjee F., Johnson A. M., Loudon J. M., and Nicholass J. M. (1984) Enhancement of [^{3}H]-flunitrazepam binding by mianserin in vivo. *Neurosci. Lett.* **46,** 305–309.

Kendall D. A., Duman R., Slopis J., and Enna S. J. (1982a) The influence of adrenocorticotropin and yohimbine on antidepressant-induced declines in rat brain neurotransmitter receptor binding and function. *J. Pharmacol. Exp. Ther.* **222,** 566–571.

Kendall D. A., Taylor D. P., and Enna S. J. (1982b) ^{3}H-Tetrahydrotrazodone binding: Association with serotonin-1 receptors. *Mol. Pharmacol.* **23,** 594–599.

Kinnier W. J., Tabor R. D., and Norrell L. Y. (1984) Neurochemical properties of AHR 9377: A novel inhibitor of norepinephrine reuptake. *Biochem. Pharmacol.* **33,** 3001–3005.

Koe K. (1983) Enhancement of benzodiazepine binding by progabide (SL 76002) and SL 75102. *Drug Dev. Res.* **3,** 421–432.

Kohler C., Ogren S. V., Haglund L., and Angeby T. (1979) Regional displacement by sulpiride of [^{3}H]spiperone binding in vivo. Biochemical and behavioral evidence for a preferential action on limbic and nigral dopamine receptors. *Neurosci. Lett.* **13,** 51–56.

Kohler C., Haglund L., Ogren S. O., and Angeby T. (1981) Regional blockade by neuroleptic drugs of in vivo ^{3}H-spiperone binding in the rat brain. Relation to blockade of apomorphine induced hyperactivity and stereotypes. *J. Neural Transm.* **52,** 163–173.

Kuhar M. J. (1982) Localization of drug and neurotransmitter receptors in brain by light microscopic autoradiography. *Handbook Psychopharmacol.* **15,** 299–320.

Laduron P. M. (1984) Criteria for receptor sites in binding studies. *Biochem. Pharmacol.* **33,** 833–839.

Laduron P. M., Janssen P. F. M., and Leysen J. E. (1978) Spiperone: A ligand of choice for neuroleptic receptors 2. Regional distribution and in vivo displacement of neuroleptic drugs. *Biochem. Pharmacol.* **27,** 317–321.

LeFur G., Guilloux F., and Uzan A. (1980) In vivo blockade of dopaminergic receptors from different rat brain regions by classical and atypical neuroleptics. *Biochem. Pharmacol.* **29,** 267–270.

Loftus P. (1986) Computer Based Approaches to Drug Design, in *Drug Discovery and Development* (Williams M. and Malick J. B., eds.), Humana, Clifton, New Jersey, in press.

Martin G. E. and Papp N. L. (1984) Blockade of MK-801 induced ipsiversive turning in 6-OHDA-lesioned rats by alpha l-autoreceptor antagonists. *Pharmacol. Biochem. Behav.* **20,** 893–897.

Martin G. E., Williams M., Clineschmidt B. V., Yarbrough G. G., Jones J. H., and Haubrich D. R. (1983) Potent central dopamine-like activity of a novel ergoline, EOE. *Life Sci.* **30,** 1847–1856.

Martin G. E., Williams M., Pettibone D. J., Yarbrough G. G., Clineschmidt B. V., and Jones J. H. (1984) Pharmacologic profile of a novel, potent direct acting dopamine agonist, (+)-4-propyl-9-

hydroxynapthoxazine [(+)-PHNO]. *J. Pharmacol. Exp. Ther.* **230,** 569–576.

Martin I. L., Brown C. L., and Doble A. (1983) Multiple benzodiazepine receptors—structures in the brain or structures in the mind? *Life Sci.* **32,** 1925–1933.

Maxwell R. A. (1984) The state of the art of the science of drug discovery—an opinion. *Drug Develp. Res.* **4,** 375–389.

Mennini T., Abbiati A., Caccia S., Cotecchia S., Gomez A., and Garattini S. (1982) Brain levels of tofizopam in the rat and relationship with benzodiazepine receptors. *Naunyn Schmiedebergs Arch. Pharmacol.* **321,** 112–115.

Munson P. J. and Rodbard D. (1980) Ligand: A versatile approach for characterization of ligand binding systems. *Anal. Biochem.* **107,** 220–239.

Murphy K. M. M., Gould R. J., Largent B. L., and Snyder S. H. (1983) A unitary mechanism of calcium antagonist drug action. *Proc. Natl. Acad. Sci. USA* **80,** 860–864.

Murray T. F. (1982) Upregulation of rat cortical adenosine receptors following chronic administration of theophylline. *Eur. J. Pharmacol.* **82,** 113–114.

Oakley N. R. and Jones B. J. (1983) Buspirone enhances [^{3}H]flunitazepam binding in vivo. *Eur. J. Pharmacol.* **87,** 499–500.

Olsen R. W. (1982) Drug interactions at the GABA receptor–ionophore complex. *Ann. Rev. Pharmacol. Toxicol.* **22,** 245–277.

Peroutka S. J. and Snyder S. H. (1980) Regulation of serotonin 2 (5HT2) receptors labelled with ^{3}H-spiroperidol by chronic treatment with the antidepressant, amitriptyline. *J. Pharmacol. Exp. Ther.* **215,** 582–587.

Perry B. D. and U'Prichard D. C. (1981) [^{3}H]-rauwoulscine (α-yohimbine): A specific antagonist radioligand for brain α-2 adrenergic receptors. *Eur. J. Pharmacol.* **76,** 461–464.

Peterson N. A. (1975) Discrimination between various types of enzyme inhibition by means of Hill plots. *Physiol. Chem. Physics* **7,** 277–282.

Peterson E. N., Jensen L. H., Honore T., Braestrup C., Kehr W., Stephens D. N., Wachtel H., Seidelman D., and Schmiechen K. L. (1984) ZK91296, a partial agonist at benzodiazepine receptors. *Psychopharmacology* **83,** 240–248.

Petrack B., Czernik A. J., Cassidy J. P., Bernard P., and Yokoyama N. (1983) Benzodiazepine receptor ligands with opposing pharmacologic actions. *Adv. Biochem. Psychopharmacol.* **38,** 129–137.

Pilc A. and Lloyd K. G. (1984) Chronic antidepressants and GABA "B" receptors: A GABA hypothesis of antidepressant drug action. *Life Sci.* **35,** 2149–2154.

Pliska V. (1983) Displacement reactions employing heterologous tracer ligands in peptide receptor studies: A review. *J. Receptor Res.* **3,** 227–238.

Randall W. C., Anderson P. S., Cresson E. L., Hunt C. A., Lyon T. F., Rittle K. E., Remy D. C., Springer J. A., Hirschfield J. M., Hoogsteen K., Williams M., Risley E. A., and Totaro J. A. (1979) Synthesis, assignment of absolute configuration and receptor binding studies relevant to the neuroleptic activities of a series of chiral 3-substituted cyproheptadine atropisomers. *J. Med. Chem.* **22,** 1221–1230.

Rodbard D. and Lewald J. E. (1970) Computer analysis of radioligand assay and radioimmunoassay data. *Acta Endocrinol.* **147,** 79–103.

Rodbard D. and Frazier G. R. (1975) Statistical analysis of radioligand assay data. *Methods Enzymol.* **37,** 3–22.

Romer D., Busher H. H., Hill R. C., Maurer R., Petcher T. J., Zuegner H., Benson W., Finnier E., Milowski W., and Theis P. W. (1982) Tifluadom, an opioid benzodiazepine. *Nature* (Lond.) **298,** 759–760.

Saano V. (1982) Tofizopam selectively increases the affinity of benzodiazepine binding sites for [^{3}H]flunitrazepam but not for beta[^{3}H]carboline-3-carboxylic acid ethyl ester. *Pharmacol. Res. Comm.* **14,** 971–981.

Seeman P (1980) Brain dopamine receptors. *Pharmacol. Rev.* **32,** 229–313.

Simasko S. M. and Horita A. (1982) Characterization and distribution of ^{3}H-(3MeHis^2)thyrotropin releasing hormone receptors in rat brain. *Life Sci.* **30,** 1793–1799.

Skolnick P., Lock K. L., Paugh B., Marangos P., Windsor R., and Paul S. (1980) Pharmacologic and behavioral effects of EMD 28422: A novel purine which enhances [^{3}H]diazepam binding to brain benzodiazepine receptors. *Pharmacol. Biochem. Behav.* **12,** 685–689.

Snowhill E. W. and Williams M. (1985) Autoradiographic evaluation of adenosine A-1 and A-2 receptors in rat brain. *J Neurochem* **44,** S88A.

Snyder S. H. (1983) Neurotransmitter receptor binding and drug discovery. *J. Med. Chem.* **26,** 1667–1672.

Snyder S. H., Greenberg D. A., and Yamamura H. I. (1976) Antischizophrenic drugs and brain cholinergic receptors. *Am. J. Psychiatry* **31,** 58–61.

Sokoloff P., Matres M. P., and Schwartz J. C. (1980) Three classes of dopamine receptors (D-2, D-3, D-4) identified by binding studies with [^{3}H]-apomorphine and [^{3}H]-domperidone. *Naunyn Schmiedeberg's Arch. Pharmacol.* **315,** 89–95.

Squires R. F. (1984) Benzodiazepine Receptors, in *Handbook of Neurochemistry,* vol. 6 (Lajtha A., ed.), Plenum, New York.

Stoof J. C., and Kebabian J. W. (1984) Minireview: Two dopamine receptors: Biochemistry, physiology and pharmacology. *Life Sci.* **35,** 2281–2296.

Sugrue M. F. (1981) Chronic Antidepressant Administration and Adaptive Changes in Central Monoaminergic Systems, in *Antidepressants: Neurochemical, Behavioral and Clinical Perspectives* (Enna S. J., Malick J. B., and Richelson E. eds.), pp. 13–30, Raven, New York.

Tamminga C. A., Schaeffer M. H., Smith R. C., and Davis J. M. (1978) Schizophrenic symptoms improve with apomorphine. *Science* **200,** 567–570.

Taylor D. P., Allen L. E., Ashworth E. M., Becker J. A., Hyslop D. A., and Riblet L. A. (1981) Treatment with trazodone plus phenoxybenzamine accelerates development of decreased type 2 serotonin binding in rat cortex. *Neuropharmacology* **20,** 513–516.

Taylor D. P., Allen L. E., Becker J. A., Crane M., Hyslop D. K., and Riblet L. A. (1984) Changing concepts of the biochemical action of the anxioselective drug, buspirone. *Drug Develop. Res.* **4,** 95–107.

Traber J., Davies M. A., Dompert W. U., Glaser T., Schuurman T., and Seidel P. R. (1984) Brain serotonin receptors as a target for the putative anxiolytic TVX Q 7821. *Brain Res. Bull.* **12,** 741–744.

Williams M. (1983) Anxioselective anxioytics. *J. Med. Chem.* **26,** 619–628.

Williams M. (1984) Molecular aspects of the action of benzodiazepine and non-benzodiazepine anxiolytics. *Prog. Neuropyschopharmacol. Biol. Psychiatry* **8,** 209–247.

Williams M. and Martin G. E. (1982) Selectivity of cyproheptadine as assessed by radioligand binding. *J. Pharm. Pharmacol.* **34,** 58–59.

Williams M. and Robinson J. L. (1984) Binding of the nicotinic cholinergic antagonist, dihydro-beta-erythroidine (DBE) to rat brain tissue. *J. Neurosci.* **4,** 2906–2911.

Williams M. and U'Prichard D. C. (1984) Drug discovery at the molecular level: A decade of radioligand binding in retrospect. *Ann. Rep. Med. Chem.* **19,** 283–292.

Williams M., Martin G. E., Remy D. C., Hichens M., Mangel A. W., Taylor D. A., Yarbrough G. G., Bendesky R. J., King S. W., Robinson J. L., Totaro J. A., and Clineschmidt B. V. (1984) L-646,462, a cyproheptadine related antagonist of dopamine and serotonin with selectivity for peripheral systems. *J. Pharmacol. Exp. Ther.* **229,** 775–781.

Wood P. L. (1985) Multiple Opioid Receptors in the Central Nervous System, in *Neuromethods,* vol. 4 (Boulton A. A., Baker G. B., Hrdina P., eds.), Humana, New Jersey (in press).

Wood P. L. and Charleson S. (1982) High affinity [^{3}H]ethylketazocine binding: Evidence for specific kappa receptors. *Neuropharmacology* **21,** 215–219.

Wood P. L., Loo P., Braunwalder A., Yokoyama N., and Cheney D. L. (1984a) In vitro characterization of benzodiazepine receptor agonists, antagonists, inverse agonists and agonist/antagonists. *J. Pharmacol. Exp. Ther.* **231,** 572–576.

Wood P. L., McPherson S. E., and Braunwalder A. (1984b) In vivo flunitrazepam binding: Actions of buspirone and clozapine. *Soc. Neurosci. Abstr.* **7,** 117–12.

Yamamura H. I., Gee K. W., Brinton R. E., Davis T. P., Hadley M., and Wamsley J. K. (1983) Light microscopic autoradiographic visualization of [^{3}H]arginine vasopressin binding sites in rat brain. *Life Sci.* **32,** 1919–1924.

Yokoyama N., Ritter B., and Neubert A. (1982) 2-Arylpyrazolo[4,3c] quinolin-3-ones: Novel agonist, partial agonist and antagonist of benzodiazepines. *J. Med. Chem.* **25,** 337–341.

Index

GPSR Compliance

The European Union's (EU) General Product Safety Regulation (GPSR) is a set of rules that requires consumer products to be safe and our obligations to ensure this.

If you have any concerns about our products, you can contact us on ProductSafety@springernature.com

In case Publisher is established outside the EU, the EU authorized representative is:

Springer Nature Customer Service Center GmbH
Europaplatz 3
69115 Heidelberg, Germany

Batch number: 10761503

Printed by Printforce, the Netherlands